Emerging Techniques in Imaging, Modelling and Visualization for Cardiovascular Diagnosis and Therapy

Emerging Techniques in Imaging, Modelling and Visualization for Cardiovascular Diagnosis and Therapy

Editors

Mihaela Pop
Cristian A. Linte

MDPI • Basel • Beijing • Wuhan • Barcelona • Belgrade • Manchester • Tokyo • Cluj • Tianjin

Editors
Mihaela Pop
Sunnybrook Research Institute
Toronto, ON, Canada

Cristian A. Linte
Rochester Institute of Technology
New York, NY, USA

Editorial Office
MDPI
St. Alban-Anlage 66
4052 Basel, Switzerland

This is a reprint of articles from the Special Issue published online in the open access journal *Applied Sciences* (ISSN 2076-3417) (available at: https://www.mdpi.com/journal/applsci/special_issues/Emerging_Imaging_Cardiovascular_Therapy).

For citation purposes, cite each article independently as indicated on the article page online and as indicated below:

LastName, A.A.; LastName, B.B.; LastName, C.C. Article Title. *Journal Name* **Year**, *Volume Number*, Page Range.

ISBN 978-3-0365-7100-3 (Hbk)
ISBN 978-3-0365-7101-0 (PDF)

Contents

About the Editors vii

Cristian A. Linte and Mihaela Pop
Applied Sciences—Special Issue on Emerging Techniques in Imaging, Modelling and Visualization for Cardiovascular Diagnosis and Therapy
Reprinted from: *Appl. Sci.* **2023**, *13*, 984, doi:10.3390/app13020984 1

Francesco Galati, Sébastien Ourselin and Maria A. Zuluaga
From Accuracy to Reliability and Robustness in Cardiac Magnetic Resonance Image Segmentation: A Review
Reprinted from: *Appl. Sci.* **2022**, *12*, 3936, doi:10.3390/app12083936 5

S. M. Kamrul Hasan and Cristian A. Linte
Learning Deep Representations of Cardiac Structures for 4D Cine MRI Image Segmentation through Semi-Supervised Learning
Reprinted from: *Appl. Sci.* **2022**, *12*, 12163, doi:10.3390/app122312163 25

Fumin Guo, Matthew Ng, Idan Roifman, and Graham Wright
Cardiac Magnetic Resonance Left Ventricle Segmentation and FunctionEvaluation Using a Trained Deep-Learning Model
Reprinted from: *Appl. Sci.* **2022**, *12*, 2627, doi:10.3390/app12052627 53

Mathilde Merle, Florent Collot, Julien Castelneau, Pauline Migerditichan, Mehdi Juhoor, Buntheng Ly, et al.
MUSIC: Cardiac Imaging, Modelling and Visualisation Software for Diagnosis and Therapy
Reprinted from: *Appl. Sci.* **2022**, *12*, 6145, doi:10.3390/app12126145 69

Shu Wang, Carlo Saija, Justin Choo, Zhanchong Ou, Maria Birsoan, Sarah Germanos, et al.
Cardiac Radiofrequency Ablation Simulation Using a 3D-Printed Bi-Atrial Thermochromic Model
Reprinted from: *Appl. Sci.* **2022**, *12*, 6553, doi:10.3390/app12136553 87

Carlos Albors, Èric Lluch, Juan Francisco Gomez, Nicolas Cedilnik, Konstantinos A Mountris, Tommaso Mansi, et al.
Meshless Electrophysiological Modeling of Cardiac Resynchronization Therapy—Benchmark Analysis with Finite-Element Methods in Experimental Data
Reprinted from: *Appl. Sci.* **2022**, *12*, 6438, doi:10.3390/app12136438 101

Jermiah J. Joseph, Clara Sun, Ting-Yim Lee, Daniel Goldman, Sanjay R. Kharche and Christopher W. McIntyre
Structure (Epicardial Stenosis) and Function (Microvascular Dysfunction) That Influence Coronary Fractional Flow Reserve Estimation †
Reprinted from: *Appl. Sci.* **2022**, *12*, 4281, doi:10.3390/app12094281 129

Timothy J. Hunter, Jermiah J. Joseph, Udunna Anazodo, Sanjay R. Kharche, Christopher W. McIntyre and Daniel Goldman
Atrial Fibrillation and Anterior Cerebral Artery Absence Reduce Cerebral Perfusion: A De Novo Hemodynamic Model †
Reprinted from: *Appl. Sci.* **2022**, *12*, 1750, doi:10.3390/app12031750 145

Johane H. Bracamonte, Sarah K. Saunders, John S. Wilson, Uyen T. Truong and Joao S. Soares
Patient-Specific Inverse Modeling of In Vivo Cardiovascular Mechanics with Medical Image-Derived Kinematics as Input Data: Concepts, Methods, and Applications
Reprinted from: *Appl. Sci.* **2022**, *12*, 3954, doi:10.3390/app12083954 . . . . **159**

Tanjib Rahman, Kévin Moulin and Luigi E. Perotti
Cardiac Diffusion Tensor Biomarkers of Chronic Infarction Based on In Vivo Data
Reprinted from: *Appl. Sci.* **2022**, *12*, 3512, doi:10.3390/app12073512 . . . . **231**

Monica-Simina Mihuta, Corina Paul, Adrian Ciulpan, Farah Dacca, Iulian Puiu Velea, Ioana Mozos and Dana Stoian
Subclinical Atherosclerosis Progression in Obese Children with Relevant Cardiometabolic Risk Factors Can Be Assessed through Carotid Intima Media Thickness
Reprinted from: *Appl. Sci.* **2021**, *11*, 10721, doi:10.3390/app112210721 . . . . **247**

Peter Lin, Terenz Escartin, Melissa Larsen, Matthew Ng, Mengyuan Li, Jennifer Barry, et al.
MR Imaging and Electrophysiological Features of Doxorubicin-Induced Fibrosis: Protocol Development in a Small Preclinical Pig Study with Histological Validation †
Reprinted from: *Appl. Sci.* **2022**, *12*, 11620, doi:10.3390/app122211620 . . . . **273**

Patrick Carnahan, John Moore, Daniel Bainbridge, Elvis C.S. Chen and Terry M. Peters
Multi-View 3D Transesophageal Echocardiography Registration and Volume Compounding for Mitral Valve Procedure Planning
Reprinted from: *Appl. Sci.* **2022**, *12*, 4562, doi:10.3390/app12094562 . . . . **289**

About the Editors

Mihaela Pop

Dr Pop received a PhD in medical biophysics from the University of Toronto (2010). Currently, she is a Visiting Scientist at Sunnybrook Research Institute Toronto (Canada) and Inria - Sophia Antipolis (France). Dr. Pop's research interests are in the medical biophysics field, combining preclinical experimentation with image-based biophysical simulations, for which she received funding from the CIHR project grant. Her projects have been focused on implementing advanced cardiac imaging methods (e.g. high resolution MRI and optical fluorescence via voltage sensitive dyes to map action potential propagation) and image analysis tools for tissue characterization, along with developing preclinical models of pathology for image-guided electrophysiology interventions. Furthermore, her expertise includes data integration into personalized 3D predictive heart models for a better risk stratification of arrhythmia and the improvement of cardiac therapy outcome. Her research has disseminated more than 140 publications (abstracts, peer-reviewed journal and conference papers, and co-edited proceedings). She was co-founder and co-chair of the international STACOM workshop (2010–2021), and is in the Board of Directors of the Functional Imaging and Modelling of the Heart conference since 2017.

Cristian A. Linte

Cristian A. Linte is an Associate Professor in Biomedical Engineering and the Center for Imaging Science at Rochester Institute of Technology. Dr. Linte's research focuses on the development, implementation, evaluation, and pre-clinical translation of medical image computing, visualization, and navigation cyber-infrastructure to provide intelligent solutions in support of computer-assisted diagnosis and therapy. Dr. Linte's research group at RIT has been home for more than 25 graduate and undergraduate students in imaging science, engineering, mathematical modeling, and computer science. His research has been disseminated in more than 120 peer reviewed journal articles and conference proceedings, and has been recognized at several international conferences, with more than USD 4 million in funding support from funding agencies in the United States (NIH and NSF) and Canada (Natural Sciences and Engineering research Council, Canadian Institutes of Health research, and Heart and Stroke Foundation of Canada). In addition to his active involvement in IEEE world, Dr. Linte has served as chair, lead organizer, and editor of several international workshops, conference, journals and journal Special Issues, and conference proceedings.

applied sciences

MDPI

Editorial

Applied Sciences—Special Issue on Emerging Techniques in Imaging, Modelling and Visualization for Cardiovascular Diagnosis and Therapy

Cristian A. Linte [1,*] and Mihaela Pop [2,*]

1 Biomedical Engineering and Imaging Science, Rochester Institute of Technology, Rochester, NY 14623, USA
2 Schulich Heart Research Program, Sunnybrook Research Institute, Toronto, ON 4MN 3M5, Canada
* Correspondence: calbme@rit.edu (C.A.L.); mihaela.pop@utoronto.ca (M.P.)

Citation: Linte, C.A.; Pop, M. Applied Sciences—Special Issue on Emerging Techniques in Imaging, Modelling and Visualization for Cardiovascular Diagnosis and Therapy. *Appl. Sci.* **2023**, *13*, 984. https://doi.org/10.3390/app13020984

Received: 9 January 2023
Accepted: 9 January 2023
Published: 11 January 2023

Editors' Foreword

Ongoing developments in computing and data acquisition, along with continuous advances in medical imaging technology, computational modelling, robotics and visualization have revolutionized many medical specialties and, in particular, diagnostic and interventional cardiology. As a result, the diagnosis and treatment of many cardiac conditions that previously relied on invasive tests or procedures have been reshaped by breakthroughs in medical imaging and visualization. A concrete example is cardiac surgery, which, for many decades, was only conducted via a highly invasive sternotomy (i.e., open chest), with the patient connected to a heart–lung machine. By slowly complementing and eventually substituting the need for a direct view of the surgical field with medical imaging, open chest access was later replaced by a smaller incision (i.e., mini-thoracotomy) between the ribs for some procedures, or, for others, by access ports through which laparoscopic or robotic instruments were introduced, enabling the surgeon to operate on the heart under real-time visualization provided by a laparoscopic video camera. Ultimately, the navigation of catheters via percutaneous access through the peripheral vasculature and into the heart became the least invasive means to deliver cardiac therapy, yet it relies solely on medical imaging for guidance, as clinicians have no direct visual access to the sites or tissues they manipulate during therapy.

Hence, effective minimally invasive approaches to diagnose, plan therapy or treat cardiac conditions rely heavily or almost entirely on medical imaging and, therefore, require the development of reliable, accurate and robust tools and techniques at the interface of medical image computing, modelling and visualization. These research contributions are often the result of multi-disciplinary collaborations among scientists and professionals, spanning basic and translational research, clinical practice, medical (bio)physics, engineering, mathematics and computer science.

Several examples include, but are not limited to, the development of the following: advanced techniques in cardiovascular imaging to investigate structure-function interaction and identify pathology; image analysis algorithms and artificial intelligence (AI)-based classification methods to better characterize tissue and physiological signals; computational modelling platforms that enable the characterization and visualization of normal or pathologic anatomy, geometry, morphology and mechanical properties of the heart and coronary vessels, including applications relevant to 3D printing; the personalized, non-invasive in silico modelling-based assessment of cardiovascular function and simulation-based planning and optimization of treatments; novel pre-clinical experimental models and clinical approaches employed in electro-anatomical mapping for image-aided cardiac ablation, electroporation or resynchronization therapy; and, last but not least, innovative image-guided interventional procedures for cardiovascular applications.

The goal of this Special Issue was to disseminate emerging techniques and innovative solutions that comprehensively address unmet needs in cardiovascular disease and have the potential to be translated into the clinical arena to help improve the timeliness and accuracy of disease diagnosis, as well as the precision and efficacy of therapy delivery, toward achieving optimal patient outcomes. This Issue consists of thirteen scientific contributions: ten research articles, two review articles and one technical note, spanning several topics, from novel algorithms and platforms for medical image computing to biomechanical and hemodynamic modelling and to functional assessment in response to drug therapy, as briefly highlighted below.

Magnetic resonance imaging (MRI) provides unrivaled images of cardiac anatomy and function, thanks to its exquisite capability of capturing soft tissue contrast. As a result, a vast body of work has focused on the development of optimal image segmentation algorithms designed to extract various cardiac structures and key features of interest for better diagnosis or superior therapy planning. The review article by Galati et al. [1] provides a comprehensive overview of deep learning-based image segmentation algorithms designed to operate on cardiac MRI images and focuses specifically on the critical importance of the accuracy, reliability and robustness of cardiac image segmentation tools prior to their deployment in clinical practice. In the cardiac image segmentation arena, Guo et al. [2] also propose a deep learning model for left ventricle myocardium and blood pool segmentation from cine cardiac MR images and the subsequent functional evaluation based on clinical indices, such as stroke volume and ejection fraction. Moreover, Hasan et al. [3] describe a novel semi-supervised approach for learning the deep representations of cardiac structures that enables the highly accurate segmentation of 4D cine cardiac MRI images using as little as 1% annotated data.

To facilitate the integration of multi-modality cardiac imaging data with measured electrical activity toward enabling more accurate predictive modelling, Merle et al. [4] introduce a novel and robust software platform, created as part of a newly established consortium with international hospitals, which is dedicated to cardiovascular diagnosis and therapy guidance and features a plethora of AI/deep learning-based methods for cardiac image computing, modelling and visualization. This integrative platform has great utility in routine clinical procedures, especially in the catheter electrophysiology lab.

The interesting studies by Wang et al. [5] and Albors et al. [6] provide a deeper dive into the cardiac modelling and simulation fields for two different applications—cardiac ablation and resynchronization therapy. The former study [5] describes a cardiac ablation simulator that consists of an X-ray compatible 3D-printed, bi-atrial model contained in a custom-made enclosure for RFA simulation using a new soft tissue-mimicking polymer. The group used this phantom to perform a full simulation of a radiofrequency ablation procedure in the cardiac catheterization laboratory and demonstrated the effective delivery and visualization of radiofrequency ablation lesions. The latter study [6] describes a first attempt at the development of a meshless in silico model of cardiac electrophysiology designed to predict patient response to cardiac resynchronization therapy (CRT), as a means to optimize electrode placement during CRT procedures. Such virtual simulation-based approaches will continue to receive considerable attention, as they provide non-invasive methods to improve therapy outcome.

The contributions by Joseph et al. [7] and Hunter et al. [8] focus on different aspects of cardiovascular modelling. Specifically, using mathematical models, these authors explore vascular hemodynamics, whose understanding is critical when making decisions with respect to the spectrum of therapies. For instance, the treatment of coronary stenosis is decided based on the fractional flow reserve diagnostic index, whose estimation requires high-risk surgery. As such, the work by Joseph et al. [7] proposes an extensive mathematical description of the coronary vasculature that provides non-invasive estimates of coronary fractional flow reserve, which could be used to predict a patient eligibility for subsequent therapy. The study by Hunter et al. [8], on the other hand, is founded on the premise that cardiac arrhythmia may reduce cerebral blood perfusion and describes a novel cardio-cerebral lumped parameter hemodynamic model to investigate the role of the circle of Willis variants on cerebral blood flow dynamics under atrial fibrillation conditions.

Another venture into the vast field of cardiac biomechanical modelling is the exhaustive contribution by Bracamonte et al. [9], which provides a comprehensive review of the field of patient-specific inverse modelling of cardiovascular mechanics based on image-derived kinematic data.

Furthermore, the use of imaging as a biomarker for quantifying cardiac disease has become a popular topic in computer-integrated diagnosis. Driven by the goal to non-invasively characterize cardiac tissue that may have undergone chronic myocardial infarction, the pre-clinical work by Rahman et al. [10] describes the utility of cardiac diffusion tensor MR imaging to identify the microstructural-based biomarkers of myocardial infarction by evaluating the diffusion tensor invariants, eigenvalues and radial diffusivity in different myocardial areas (i.e., scar, border zone and healthy myocardium) of several porcine subjects. In their clinical study, Mihuta et al. [11] employed the ultrasound imaging of the carotid artery to quantify the carotid intima-media thickness as a potential biomarker indicative of atherosclerotic progression in children and young adults, with the overall goal to provide a more complete evaluation of their cardiometabolic risk. Lastly, the work by Lin et al. [12] also illustrates the development of a novel assessment protocol tested on several porcine subjects which integrates MR imaging and electrophysiology measurements to assess the effect of chemotherapy on cardiac function by quantifying several imaging-based biomarkers and assessing the presence of drug-induced tissue fibrosis or electrical remodelling.

Finally, to further attest to the popularity gained by ultrasound imaging for cardiovascular applications and, specifically, for therapy planning and guidance, the work by Carnahan et al. [13] describes the development of a novel method to register multi-view 3D transesophageal echocardiography images to enable volume compounding as a means to generate extended field-of-view images that can be used to plan mitral valve procedures.

In sum, while we acknowledge that the contributions disseminated in this Special Issue barely scratch the surface and only briefly address a very few niches of the vast field of cardiac image computing, modelling and visualization, we hope our readers find these pieces sufficiently intriguing to foster their curiosity and to dig deeper and seek additional literature on the topics of their interest.

Author Contributions: C.A.L. and M.P. both contributed to the preparation, writing and editing of this editorial. All authors have read and agreed to the published version of the manuscript.

Funding: This research received no external funding.

Acknowledgments: This Issue would not be possible without the contributions of various talented authors, committed reviewers and the dedicated editorial team at Applied Sciences. We thank all authors for publishing their high-quality research results in this Special Issue. We would also like to express our sincere thanks to all reviewers and editors for their feedback, comments and suggestions that helped the authors improve their manuscripts. Finally, we would like to take the opportunity to acknowledge the outstanding support provided by the editorial team at Applied Sciences.

Conflicts of Interest: The authors declare no conflict of interest.

References

1. Galati, F.; Ourselin, S.; Zuluaga, M.A. From Accuracy to Reliability and Robustness in Cardiac Magnetic Resonance Image Segmentation: A Review. *Appl. Sci.* **2022**, *12*, 3936. [CrossRef]
2. Hasan, S.M.K.; Linte, C.A. Learning Deep Representations of Cardiac Structures for 4D Cine MRI Image Segmentation through Semi-Supervised Learning. *Appl. Sci.* **2022**, *12*, 12163. [CrossRef]
3. Guo, F.; Ng, M.; Roifman, I.; Wright, G. Cardiac Magnetic Resonance Left Ventricle Segmentation and Function Evaluation Using a Trained Deep-Learning Model. *Appl. Sci.* **2022**, *12*, 2627. [CrossRef]
4. Merle, M.; Collot, F.; Castleneau, J.; Migerditichan, P.; Juhoor, M.; Ly, B.; Ozenne, V.; Quesson, B.; Zemsemi, N.; Coudiere, Y.; et al. MUSIC: Cardiac Imaging, Modelling and Visualisation Software for Diagnosis and Therapy. *Appl. Sci.* **2022**, *12*, 6145. [CrossRef]
5. Wang, S.; Saija, C.; Choo, J.; Ou, Z.; Birsoan, M.; Germanos, S.; Rothwell, J.; Vakili, B.; Kotadia, I.; Xu, Z.; et al. Cardiac Radiofrequency Ablation Simulation Using a 3D-Printed Bi-Atrial Thermochromic Model. *Appl. Sci.* **2022**, *12*, 6553. [CrossRef]

6. Albors, C.; Lluch, E.; Gomes, J.F.; Cedilnik, N.; Mountris, K.A.; Mansi, T.; Khamzin, S.; Dokuchaev, A.; Solovyova, O.; Pueyo, E.; et al. Meshless Electrophysiological Modeling of Cardiac Resynchronization Therapy—Benchmark Analysis with Finite-Element Methods in Experimental Data. *Appl. Sci.* **2022**, *12*, 6438. [CrossRef]
7. Joseph, J.; Sun, C.; Lee, T.Y.; Goldman, D.; Kharche, S.R.; McIntyre, C.W. Structure (Epicardial Stenosis) and Function (Microvascular Dysfunction) That Influence Coronary Fractional Flow Reserve Estimation. *Appl. Sci.* **2022**, *12*, 4281. [CrossRef]
8. Hunter, T.J.; Joseph, J.J.; Anazodo, U.; Kharche, S.R.; McIntyre, C.W.; Goldman, D. Atrial Fibrillation and Anterior Cerebral Artery Absence Reduce Cerebral Perfusion: A De Novo Hemodynamic Model. *Appl. Sci.* **2022**, *13*, 1750. [CrossRef]
9. Bracamonte, J.H.; Saunders, S.K.; Wilson, J.S.; Truong, U.T.; Soares, J.S. Patient-Specific Inverse Modeling of In Vivo Cardiovascular Mechanics with Medical Image-Derived Kinematics as Input Data: Concepts, Methods, and Applications. *Appl. Sci.* **2022**, *12*, 3954. [CrossRef]
10. Rahman, T.; Moulin, K.; Perotti, L.E. Cardiac Diffusion Tensor Biomarkers of Chronic Infarction Based on In Vivo Data. *Appl. Sci.* **2022**, *12*, 3512. [CrossRef] [PubMed]
11. Mihuta, M.S.; Paul, C.; Ciulpan, A.; Dacca, F.; Velea, I.P.; Mozos, I.; Stoian, D. Sub-clinical Atherosclerosis Progression in Obese Children with Relevant Cardiometabolic Risk Factors Can Be Assessed through Carotid Intima Media Thickness. *Appl. Sci.* **2022**, *11*, 10721. [CrossRef]
12. Lin, P.; Escartin, T.; Larsen, M.; Ng, M.; Li, M.; Barry, J.; Roifman, I.; Pop, M. MR Imaging and Electrophysiological Features of Doxorubicin-Induced Fibrosis: Protocol Development in a Small Preclinical Pig Study with Histological Validation. *Appl. Sci.* **2022**, *12*, 11620. [CrossRef]
13. Carnahan, P.; Moore, J.; Bainbridge, D.; Chen, E.C.S.; Peters, T.M. Multi-View 3D Transesophageal Echocardiography Registration and Volume Compounding for Mitral Valve Procedure Planning. *Appl. Sci.* **2022**, *12*, 4562. [CrossRef]

applied sciences

MDPI

Review

From Accuracy to Reliability and Robustness in Cardiac Magnetic Resonance Image Segmentation: A Review

Francesco Galati [1,*], Sébastien Ourselin [2] and Maria A. Zuluaga [1,2,*]

1 Data Science Department, EURECOM, 06410 Biot, France
2 School of Biomedical Engineering and Imaging Sciences, King's College London, London WC2R 2LS, UK; sebastien.ourselin@kcl.ac.uk
* Correspondence: francesco.galati@eurecom.fr (F.G.); zuluaga@eurecom.fr (M.A.Z.); Tel.: +33-(0)4-9300-8125 (M.A.Z.)

Abstract: Since the rise of deep learning (DL) in the mid-2010s, cardiac magnetic resonance (CMR) image segmentation has achieved state-of-the-art performance. Despite achieving inter-observer variability in terms of different accuracy performance measures, visual inspections reveal errors in most segmentation results, indicating a lack of reliability and robustness of DL segmentation models, which can be critical if a model was to be deployed into clinical practice. In this work, we aim to bring attention to reliability and robustness, two unmet needs of cardiac image segmentation methods, which are hampering their translation into practice. To this end, we first study the performance accuracy evolution of CMR segmentation, illustrate the improvements brought by DL algorithms and highlight the symptoms of performance stagnation. Afterwards, we provide formal definitions of reliability and robustness. Based on the two definitions, we identify the factors that limit the reliability and robustness of state-of-the-art deep learning CMR segmentation techniques. Finally, we give an overview of the current set of works that focus on improving the reliability and robustness of CMR segmentation, and we categorize them into two families of methods: quality control methods and model improvement techniques. The first category corresponds to simpler strategies that only aim to flag situations where a model may be incurring poor reliability or robustness. The second one, instead, directly tackles the problem by bringing improvements into different aspects of the CMR segmentation model development process. We aim to bring the attention of more researchers towards these emerging trends regarding the development of reliable and robust CMR segmentation frameworks, which can guarantee the safe use of DL in clinical routines and studies.

Keywords: cardiac image segmentation; reliability and robustness; deep learning; cardiac magnetic resonance imaging

Citation: Galati, F.; Ourselin, S.; Zuluaga, M.A. From Accuracy to Reliability and Robustness in Cardiac Magnetic Resonance Image Segmentation: A Review. *Appl. Sci.* **2022**, *12*, 3936. https://doi.org/10.3390/app12083936

Academic Editors: Mihaela Pop and Cristian A. Linte

Received: 15 March 2022
Accepted: 8 April 2022
Published: 13 April 2022

1. Introduction

Cardiovascular diseases (CVDs) are the leading cause of death globally and a major contributor to disability [1]. In 2019, an estimate of 17.9 million people died from CVDs, representing 32% of all global deaths and 38% of premature deaths (under the age of 70) due to non-communicable diseases [2]. It is projected that, by 2035, the number of people with CVD will increase by 30%, reaching over 130 million people and a prevalence rate of 45.1% [3]. As a consequence, there are important efforts in place to improve prevention, early diagnosis and management of CVDs [4].

In this context, cardiovascular magnetic resonance (CMR) imaging has been positioned as a reference for quantitative cardiac analysis, due to its non-invasive nature and its superior spatiotemporal resolution that allows imaging the cardiac chambers and great vessels with a great level of detail [5]. Quantitative cardiac analysis from CMR requires an accurate segmentation of the heart. Manual delineation of the cardiac anatomical structures can take a trained expert around 20 min per subject, which is lengthy, monotonous, and prone to subjective errors [6]. Therefore, alongside the advances in CMR imaging, there has been a substantial part of research devoted to the development of techniques for automatic CMR segmentation [7–9].

Before the emergence of deep learning (DL), traditional techniques, such as thresholding, edge-based and region-based approaches, model-based (e.g., active shape and appearance models) and atlas-based segmentation methods, represented the state-of-the-art performance in CMR segmentation [7]. The main drawback of traditional techniques is that they require significant user expertise, in the form of feature engineering, encoded prior knowledge or posterior user intervention, to reach good accuracy.

Over the last ten years, benefiting from advanced computer hardware and greater availability of public datasets, DL-based techniques emerged as the reference method for CMR segmentation [9], outperforming previous approaches and demonstrating the capacity to reproduce the analysis of experts [10]. In fact, DL currently represents a real chance of developing CMR segmentation frameworks to assist, automate and accelerate routine clinical procedures and large-scale population studies. Nevertheless, despite their success and high reported accuracy, they still lack the necessary reliability and robustness to be safely translated into practice. As highlighted by recent studies [11], unlike experts, even the top-performing DL methods sometimes generate anatomically impossible segmentation results. If a model were to be deployed in clinical practice, such segmentation errors would represent a risk. With DL algorithms unable to provide guarantees on the quality of their results, the task of inspecting, detecting errors, correcting them and validating the segmentation results is left to the responsibility of an expert. The development of additional mechanisms to enable their use in subsequent quantitative cardiac analyses is highly desirable.

The goal of this paper is threefold. Firstly, we motivate the need to shift research from targeting high accuracy to new performance goals by showing that the accuracy objective has currently been met. Second, we provide formal definitions of robustness and reliability and summarize the major challenges that DL-based CMR segmentation methods face when trying to meet these two criteria. Finally, we present a review of the current and ongoing research for reliable and robust CMR segmentation.

The remainder of the paper is organized as follows: Section 2 motivates this work by illustrating the improvements brought by DL-based algorithms in CMR segmentation over the last decade. Section 3 formalizes the concepts of reliability and robustness and presents the challenges faced by DL-based methods that hinder the reliability and robustness of the CMR segmentations. Section 4 reviews current methods addressing reliability and robustness and categorizes the proposed solutions into two families, Quality Control (QC) and Model Improvement (MI) techniques. Although sharing the same objective, QC techniques are typically external tools that do not require any modification in model architecture or training procedure, allowing an effortless integration into state-of-the-art segmentation pipelines. MI techniques, instead, are harder to integrate into existing pipelines, as their functioning is related to an inner modification of the models. Finally, discussion and conclusions are presented in Section 6.

2. Evolution of CMR Segmentation Performance (2009–2021)

We motivate the need to shift from a focus on accuracy, as the main performance criterion, towards other criteria, i.e., reliability and robustness, by studying the evolution of CMR segmentation methods' accuracy over approximately a decade. To this end, we focus on fully-automated cardiac segmentation methods from short-axis (SA) CMR acquisitions. SA CMR segmentation has been widely studied, thanks to the large number of labelled SA CMR datasets available through multiple segmentation challenges and within the UK Biobank [12], a large-scale biomedical database containing in-depth genetic and health information from half a million participants.

We analyze the performance of 50 CMR segmentation methods, published since 2009, the year where the Sunnybrook Cardiac MR Left Ventricle Segmentation Challenge (https://www.cardiacatlas.org/studies/sunnybrook-cardiac-data/, accessed on

7 April 2022). took place. This challenge is the first ever reported CMR segmentation challenge. A large number of the here-reported works were developed in the context of this and four other CMR segmentation challenges. In chronological order, these are: the LV Segmentation Challenge (http://www.cardiacatlas.org/challenges/lv-segmentation-challenge, accessed on 7 April 2022) in 2011 [13], the Right Ventricle (RV) Segmentation Challenge (https://rvsc.projets.litislab.fr, accessed on 7 April 2022) in 2012 [14], the Automated Cardiac Diagnosis Challenge (https://www.creatis.insa-lyon.fr/Challenge/acdc, accessed on 7 April 2022) in 2017 [11] (ACDC), and the Multi-Centre, Multi-Vendor & Multi-Disease Cardiac Image Segmentation Challenge (https://www.ub.edu/mnms, accessed on 7 April 2022) in 2020 [15] (M&Ms).

Table 1 presents the SA CMR segmentation methods considered in our study and specifies the cardiac structures each method extracts, i.e., the left ventricle (LV), the right ventricle (RV) and left ventricular myocardium (MYO). Figure 1 presents SA CMR segmentation methods' progress in performance measured with the Dice Score Coefficient (DSC). The methods are discriminated per segmented cardiac structure (LV, RV and MYO). Furthermore, we differentiate between DL-based (blue) and non-DL methods (orange).

Table 1. Fully automated SA CMR segmentation methods published between 2009 and 2021 with the segmented structure of interest (LV, RV or MYO). ALL denotes that a method segments the three cardiac sub-structures.

No.	Ref.	Challenge	No.	Ref.	Challenge
1	Jolly et al. [16]	LV	25	Baumgartner et al. [17]	ALL
2	Huang et al. [18]	LV	26	Grinias and Tziritas [19]	ALL
3	Schaerer et al. [20]	LV	27	Khened et al. [21]	MYO
4	Ou et al. [22]	RV	28	Jang et al. [23]	ALL
5	Margeta et al. [24]	MYO	29	Isensee et al. [25]	ALL
6	Jolly et al. [26]	MYO	30	Yang et al. [27]	ALL
7	Liu et al. [28]	LV	31	Attar et al. [29]	ALL
8	Wang et al. [30]	RV	32	Calisto and Lai-Yuen [31]	ALL
9	Constantinidès et al. [32]	LV	33	Scannell et al. [33]	ALL
10	Hu et al. [34]	LV	34	Liu et al. [35]	ALL
11	Zuluaga et al. [36]	RV	35	Li et al. [37]	ALL
12	Ngo and Carneiro [38]	LV	36	Huang et al. [39]	ALL
13	Queirós et al. [40]	LV	37	Li et al. [41]	ALL
14	Tufvesson et al. [42]	LV	38	Simantiris and Tziritas [43]	ALL
15	Avendi et al. [44]	LV	39	Full et al. [45]	ALL
16	Tran Phi Vu [46]	ALL	40	Ma [47]	ALL
17	Tan et al. [48]	LV	43	Zhang et al. [49]	ALL
18	Patravali et al. [50]	ALL	42	Carscadden et al. [51]	ALL
19	Tan et al. [52]	MYO	43	Khader et al. [53]	ALL
20	Wolterink et al. [54]	ALL	44	Saber et al. [55]	ALL
21	Rohé et al. [56]	ALL	45	Kong and Shadden [57]	ALL
22	Zotti et al. [58]	ALL	46	Acero et al. [59]	ALL
23	Khened et al. [60]	ALL	47	Parreño et al. [61]	ALL
24	Bai et al. [6]	ALL	48	Zhou et al. [62]	ALL

We observe that, up to 2015, methods were exclusively not DL-based, mostly focused on LV segmentation, and with an important performance gap between the LV and the RV and MYO. The latter may be explained by the LV's relatively lower variability in shape than the other cardiac structures. In 2015, in the context of the Kaggle Second Annual Data Science Bowl (https://www.kaggle.com/c/second-annual-data-science-bowl, accessed on 7 April 2022), the top-performing methods relied on deep learning technologies (https://github.com/woshialex/diagnose-heart, accessed on 7 April 2022). After this milestone, the scientific community shifted quickly towards DL. After 2016, only one non-DL CMR segmentation method [19] has been reported.

An immediate consequence of this change of techniques is the jump in performance for all cardiac structures. This is more evident for MYO and RV, which had the lowest DSCs, improving from average DSCs of 0.71 and 0.64, respectively before 2015, to both achieving 0.85 after 2015. LV segmentation reports an improvement from 0.88 average DSC to 0.91. Since then, the number of methods has exploded. However, performance improvements have stalled and, in some cases, deteriorated. This is the case of the general performance in the M&Ms Challenge [15], which assessed how well methods could cope with changes in the properties of the input images (e.g., different origins, scanner vendors and protocols). The result was a drop in the performance, as observed from the RV trend line or the very low performing methods (e.g., point 34) in Figure 1.

Finally, while most DL-based methods in Figure 1 report a very high accuracy, close to the inter-observer variability, Bernard et al. [11] demonstrated that DL-based methods, even the best performing ones [25], produced CMR segmentations with implausible anatomical configurations. The authors go then to suggest the adoption of new performance evaluation metrics that are more resilient to abnormalities. In the following, we show that the problems here identified, i.e., performance drops or implausible segmentations, can be addressed by accounting for reliability and robustness.

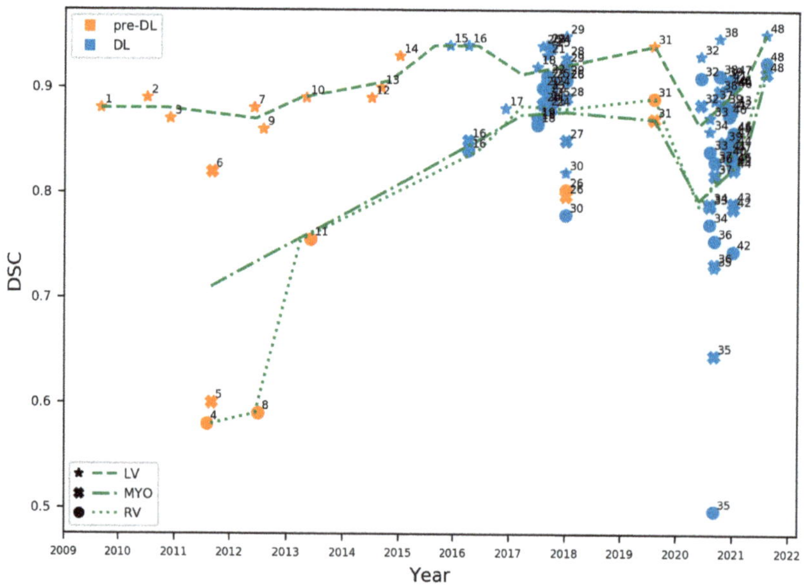

Figure 1. Dice Score Coefficients (DSCs) obtained between 2009 and 2021 for LV, RV, and MYO. Methods that do not use deep learning appear in orange, DL-based methods in blue. Green lines indicate the performance trend over the years, estimated as an average of DSCs within a window of 290 days. Interpretation of numbered labels in Table 1.

3. Robustness and Reliability: New Challenges in CMR Segmentation

In this section, we first provide formal definitions of reliability and robustness. Based on these definitions, we then identify the main factors that can hinder the reliability and robustness of DL-based CMR segmentation methods.

3.1. Definitions

The literature offers several definitions for reliability and robustness, as they can have slightly different interpretations associated with the domain where they are used, or they are often interchangeably used with related terms, such as stability [63] or safety [64]. In this work, we consider a CMR segmentation method as a computer system, thus we adhere to the following definitions from the IEEE Standard Glossary of Software Engineering Terminology [65].

3.1.1. Reliability

The ability of a system to perform its required functions under some stated conditions for a specified period of time.

3.1.2. Robustness

The degree to which a system can function correctly in the presence of invalid inputs.

3.2. Challenges to Reliable Segmentation

Following the definitions in Section 3.1, we identify two factors that can hinder the reliability of a DL-based segmentation method: overfitting and loss formulation.

3.2.1. Overfitting

The first and most basic condition that a reliable segmentation model should meet is that its performance is consistent from training to testing. Failing to do so is commonly referred to as *overfitting* or poor generalization. Two main factors are linked to overfitting: model complexity and data collection. Model complexity is related to the number of parameters in a model (e.g., the number of weights in a network), whereas data collection refers to the task of collecting and pre-processing data to train a model. In this study, we assume that the best architectures for fulfilling segmentation in the presence of an adequate number of training samples have already been identified. Therefore, we consider that overfitting can only be caused by poor data collection. In other words, the CMR segmentation methods presented in Section 2 should have a consistent training vs. testing performance as long as good data collection is guaranteed.

The data collection process that can guarantee the reliability of the model during testing needs to meet two conditions. First, it requires collecting a large number of samples. Being CMR segmentation typically fulfilled in a supervised manner, this also implies that the collected samples require annotations. Second, the collected data should be representative of the phenomenon under study. Failing to do so is commonly known as *data bias*.

3.2.2. Loss Formulation

State-of-the-art CMR segmentation is performed through supervised learning techniques. During supervised training, the loss functions measure the dissimilarity between the ground truth and the predicted segmentation. There is a vast offer of loss functions for medical image segmentation (e.g., the cross-entropy loss, the soft-Dice loss) [66], which can be used independently or combining multiple losses together. An inherent disadvantage of most of these loss functions is that they are typically pixel-wise objective functions, which measure dissimilarity in terms of correctly classified pixels over the total. This formulation does not optimize the model towards the final problem task since it does not reward segmentation results that better reflect the anatomy, i.e., the shape of the heart. Instead, it favors similarity among pixel intensities and, eventually, it leads to incomplete and unrealistic segmentation results both at training and at inference. In particular, predictions may contain holes inside the structures, abnormal concavities, or duplicated regions, typically located in the most basal and apical slices [67]. Being caused by intrinsic limitations of DL-based algorithms, anatomical failures can occur at inference without any possibility of inferring the quality of the model outcome. Therefore, the model becomes unpredictable, intractable for model verification, and ultimately unreliable.

3.3. *Challenges to Robust Segmentation*

Robustness is associated with performance in face of invalid inputs. We identify two sources that can lead to invalid inputs, thus affecting the robustness of a DL-based segmentation method: domain shift and data acquisition.

3.3.1. Domain Shift

Domain shift, or distribution shift, refers to a change in the data distribution between the one observed at training dataset, and the one the model encounters at inference, i.e., when deployed. Domain shift represents a critical risk for supervised deployed models because it has been shown that the inference error increases proportionally to the difference between samples from the two distributions [68]. In a strict sense, domain shifted data do not constitute an invalid input because it is still representative of the phenomenon under study. In this work, we follow a computer system approach where we consider domain shifted data as deviated from the "specifications" in which the model is developed or trained. As such, it does not affect reliability. However, the model is expected to perform well even in the presence of the domain shifted data, i.e., they should be robust. In CMR segmentation, this drift can be caused by numerous factors, such as changes in demographics, modalities, acquisition protocols and scanner vendors or simply anatomical variability or, even, an adversarial attack that may alter the statistical properties of the input [69]. The M&Ms challenge [15] was designed to assess the capacity of existing methods to cope with CMR domain shift. The result was an overall drop in performance showing a lack of robustness in existing methods.

3.3.2. Data Acquisition

Data acquisition may deteriorate the quality of an image and its visual appearance, but differently from domain shift, it does not alter the image's statistical properties. Several factors affect the quality of a CMR image during its acquisition. Some of them are under the control of the clinician (e.g., the number of acquired slices), some depend on the subject being scanned (e.g., bulk or respiratory motion), and some are out of control (e.g., arrhythmias, blood flow or magnetic field inhomogeneities) [70]. When the quality is compromised, CMR images may contain artifacts like ghosting, blurring and smearing. During manual labelling, these images can be discarded for training. At inference, low-quality input images may not be possible to discard. Potentially, they could be the only information available for a patient. However, these low-quality inputs images may lead to poor segmentation results, if the segmentation model is not capable of handling invalid inputs.

4. Methods for Improved Reliability and Robustness

Two different approaches have arisen aiming to improve the reliability and the robustness of state-of-the-art DL-based segmentation methods. We distinguish between techniques limited to identify failures of the segmentation model, which hinder its reliability or robustness, and techniques that adopt countermeasures to improve the segmentation performance. In the former case, which we denote quality control (QC), the developed tools raise a flag when the system (i.e., the segmentation model) under analysis incurs into a lack of reliability or robustness, without necessarily explaining the cause or source of failure. In the latter case, models are improved in their architecture, acting on the sources of failures to eradicate them, and as a result to increase reliability and robustness. We denote this category as model improvement (MI) techniques.

4.1. *Quality Control Techniques*

QC techniques grade the quality of either input CMR images or segmentation outputs, allowing for recognizing anomalous scenarios, but without performing any action to correct the identified problem. Therefore, they improve reliability and/or robustness by signalling the identified anomalies to the users for them to act upon the problem. Most of these frameworks are not conceived to depend on a specific segmentation architecture, but they can adapt to the different segmentation pipelines available in the literature.

We identify two types of QC techniques, depending on when they are used. We denote as *pre-analysis QC* [71–77] those methods that act exclusively on the inputs of a DL-based model, i.e., before the model is executed, thus aiming specifically to improve robustness. *Post-analysis QC* [76–88] refers to those methods that act on the outputs of the model to detect a malfunction, thus addressing reliability. Pre- and post-analysis mechanisms are not mutually exclusive. They can be combined in an end-to-end framework. Moreover, pre-analysis QC tools can be combined with further processing steps that mitigate the erroneous detected inputs.

4.1.1. Pre-Analysis QC Tools

Pre-analysis QC tools aim to identify erroneous inputs, addressing robustness by discarding them from the segmentation pipeline. The first barrier to overcome by this type of methods is to define quality itself. Some methods aim to detect predefined types of artifacts using learning-based approaches [73], heuristic techniques [71] or a combination of both [72,75]. Other works, instead, follow a more qualitative definition that is based on a cardiologist's input [74,76,77]. In this category, machine learning classifiers provided with a set of qualitative labels (e.g., good/bad, discard/keep) are trained to emulate experts criteria, aiming to flag low quality. At inference, these models automatically retrieve the binary feedback, which replaces experts' decisions in high-throughput pipelines.

In one of the first QC works, Miao et al. [71] assess a perceptual difference model that quantitatively evaluates image quality of large volumes of magnetic resonance images to rate different image reconstruction algorithms. Lorch et al. [72] use box-, line-, histogram-, and texture-based features to train a random decision forest algorithm to distinguish between motion-corrupted and artifact-free images. Zhang et al. [73] aim to identify missing apical and/or basal LV slices in CMR images by using generative adversarial networks (GANs). This is achieved in two stages. First, adversarial examples are generated and exploited to extract high-level features from the CMR images. The features are then used to detect missing basal and apical slices. Such process improves not only robustness to adversarial examples, but also generalization performance for original examples. Oksuz et al. [74] exploit different levels of k-space synthetic corruption to detect CMR images with low perceptual quality, defined as the mean of the individual ratings assigned by human observers. The authors use a data augmentation technique to handle the severe class imbalance between good-quality and motion-corrupted images, training two deep learning architectures to increase their robustness in the classification task. In [70,75], Tarroni et al. present a quality control pipeline for CMR images in the UK Biobank dataset, capable of detecting three problematic scenarios to warn a human operator. The scenarios are low heart coverage, high inter-slice motion and low cardiac image contrast.

Finally, some recent works have succeeded at integrating QC tools within a more complex cardiac analysis pipeline. Machado et al. [76] use a ResNet [89] to classify CMR images as analyzable or non-analyzable. The network is trained with a dataset of 225 images labelled by an expert cardiologist. Those considered as analyzable move in forward in a cardiac analysis pipeline (see Section 4.1.2). Ruijsink et al. [77] present a DL-based pipeline for automated analysis of cardiac function. Inside the pipeline, two convolutional neural networks (CNNs) are trained to perform pre-analysis QC: a two-dimensional CNN with a recurrent long short-term memory layer for motion artifacts detection, and a two-dimensional CNN for detecting erroneous planning of the 4-chamber view. Flagged images are discarded from the subsequent segmentation step that serves as input to the cardiac function analysis.

4.1.2. Post-Analysis QC Tools

Post-analysis QC tools focus on the assessment of the segmentation outputs of a model. In this sense, we consider these tools as targeting reliability, as the quality of the segmentation output is the final indicator of the model's performance.

Methods under this category follow two main approaches to performance assessment. They act either as binary classifiers, assigning correct/incorrect labels to a segmentation, or as regressors, which attempt to infer well-known validation metrics, such as the Dice Score or the Hausdorff Distance (HD), or uncertainty estimates.

Among regressors, Kohlberger et al. [82] train an SVM regressor from DSCs measured against ground truth to build confidence measures and rank candidate segmentation models against each other. Valindria et al. [83] propose the Reverse Classification Accuracy (RCA), a registration-based method relying on the spatial overlap between predicted segmentations and reference atlases as a pseudo-measure of the performance of a segmentation model on new data. The technique has been extensively validated in the UK Biobank [84], despite being computationally expensive at inference time or prone to failure at the registration stage [90].

Robinson et al. [85] rely on a CNN to predict the DSC of unseen segmented data. The authors are the first to observe that it is difficult to obtain a balanced set of labelled data reflecting the complete feasible distribution of DSCs. Hann et al. [86] use an ensemble of neural networks to segment the LV from T1 magnetic resonance, while providing an estimate of the DSC of the predicted segmentation using multiple linear regression. Fournel et al. [87] question the usefulness of 3D DSCs as the sole measure of segmentation quality, as it excludes specific information related to the single slices, which is actually fundamental when analysing the base and the apex. The authors overcome this limitation by performing simultaneously quality control at 2D-level and 3D-level using a CNN capable of predicting both 3D and 2D DSCs. Galati and Zuluaga [88] use a convolutional autoencoder that reconstructs input segmentation masks into pseudo ground truth masks. Pseudo DSC and HD are then measured between the segmentations and their reconstructions that act as surrogate measures of the quality of the segmentation results.

Among the classifiers, Albà et al. [78] use statistical, pattern and fractal descriptors in a random forest classifier, which detect segmentation failures to be corrected or removed from subsequent analyses. Puyol-Antón et al. [79] use the uncertainty information captured in the evidence lower bound (ELBO) produced by a Bayesian CNN to identify incorrect segmentations, which can be rejected or flagged for revision by an expert. In [80], segmentation uncertainty is first assessed at the voxel level by using the multi-class entropy and Monte Carlo dropout. After deriving uncertainty maps, a CNN is trained to detect image regions containing local segmentation failures that potentially need correction by an expert. The authors differentiate tolerated errors, which lay within the range of inter-observer variability, and the segmentation failures, which are flagged to be corrected by an expert. Gonzalez et al. [81] propose combining self-supervision loss terms and post hoc uncertainty estimations into a reliable and lightweight novelty score that allows anomalous samples' identification.

The RCA [83], a regressor approach, has been embedded into the method proposed in [76], where the authors build a cardiac analysis pipeline that integrates both pre- (see Section 4.1.1) and post-analysis QC. For the latter, they estimate several quality metrics between pairs of segmentations, before and after being processed by RCA. Based on these values, an SVM binary classifier is trained to discriminate between poor and good quality segmentations. As [76], Ruijsink et al. [77] integrate pre- and post-analysis QC in a unified end-to-end pipeline. When dealing with post-analysis, they attempt to determine inconsistencies by making comparisons between long and short-axis views, LV and RV volumes, end-diastole and end-systole phases. They implement two support vector machine (SVM) classification algorithms to detect abnormalities in the obtained volume and strain curves.

Table 2 summarizes the main characteristics of the reported post-analysis QC tools. In addition to the distinction among classifiers and regressors (*Regression*), we highlight whether a proposed method formulates the problem in a traditional supervised manner, thus requiring QC labels (*no QC labels*). Given the cost of data labelling, it can be disadvantageous to require QC labels on top of the labels required to train the segmentation algorithm. Classification methods typically exploit qualitative (e.g., correct/incorrect) labels, whereas regressors require quantitative labels (e.g., DSC), which can be difficult to obtain [85]. To avoid these, a final set of methods avoid the use of QC labels by considering alternative self-supervised techniques or registration-based approaches as the RCA. Finally, Table 2 also highlights whether a given method allows the identification of the specific areas of segmentation failure, or it just gives an estimation of the general quality (*detection*).

Table 2. Post-analysis QC methods and their three main characteristics: performing regression or classification(regression), the need of quality control labels (no QC labels) and if they detect the element causing the error within the image (detection).

Method	Regression	No QC Labels	Detection
Albà et al. [78]	✗	✗	✗
Puyol-Antón et al. [79]	✗	✗	✗
Sander et al. [80]	✗	✗	✓
Gonzales et al. [81]	✗	✓	✗
Kohlberger et al. [82]	✓	✗	✗
Valindria et al. [83]	✓	✓	✗
Machado et al. [76]	✗	✗	✗
Ruijsink et al. [77]	✗	✗	✗
Robinson et al. [85]	✓	✗	✗
Hann et al. [86]	✓	✗	✗
Fournel et al. [87]	✓	✗	✓
Galati and Zuluaga [88]	✓	✓	✓

4.2. Model Improvement Techniques

We denote model improvement (MI) techniques as those methods that directly address the limitations of DL-based approaches leading to poor reliability or robustness. Differently from QC techniques, where an external algorithmic tool flags problematic situations, MI techniques solve the lack of reliability or robustness by explicitly correcting the model. Another key difference w.r.t. QC tools, which can be plugged in most of the segmentation models as an external module, is that MI techniques imply modifications to the models or the overall analysis pipelines. In the following, we first present MI techniques for improved reliability and robustness classifying them based on the specific problem they tackle (Section 3). The section concludes with an ablation analysis of the presented MI techniques to illustrate their contributions to the performance of CMR segmentation methods.

4.2.1. Overfitting

As discussed in Section 3.2.1, the necessary complexity of DL-based models to guarantee a high-performance accuracy has been established. Therefore, MI techniques to reduce overfitting firstly consist of strategies to enlarge the available datasets, when further data collection is not possible. Chen et al. [91] apply geometrical operations to the source training data in order to simulate various possible data distributions across different domains. This data augmentation strategy was also adopted by Full et al. [45] in the context of the M&Ms Challenge.

Other MI techniques assume it is not possible to sufficiently increase (artificially or through further data collection) the size of the training set that it avoids overfitting and propose to control the complexity of the highly complex models through regularization. Among them, Khened et al. [21] present a DenseNet-based FCN architecture with long skip and short-cut connections to increase parameter efficiency. Guo et al. [92] integrate continuous kernel cut and bound optimization into a CNN, building a unified max-flow framework with improved generalization capabilities.

4.2.2. Loss Formulation

MI techniques mitigating the lack of reliability induced by typical loss functions aim at re-formulating the training procedure through the definition of additional objective losses that take into account anatomical constraints. Many of these works rely on *shape priors*, embedding prior expertise knowledge into the segmentation model. A second set of works takes inspiration from control theory, proposing *automatic correction* schemes that make use of high-level feedback systems.

Shape Priors

Zotti et al. [93] extend the well-established U-net architecture [94] through the formulation of a probabilistic framework, which allows the embedding of a cardiac shape prior, in the form of a 3D volume encoding the probability of a voxel to belong to a certain "cardiac class" (LV, RV, or MYO), and the definition of a loss function tailored to the cardiac anatomy. Clough et al. [95] propose a loss function that measures the topological correspondence between predicted segmentations and prior shape knowledge. This is done by using the differentiable properties of persistent homology, which compares topologies in terms of their Betti numbers. Wyburg et al. [96] enforce topology preservation by combining a segmentation network with spatial transformers and diffeomorphic displacement fields. In this way, the network learns to warp a binary prior, completing the segmentation task with the desired topological characteristics.

Automatic Correction

Girum et al. [67] formulate the segmentation problem as a two systems task: the first is a U-Net inspired encoder–decoder CNN predicting segmentations from the input images, the second is a fully convolutional network (FCN) working as a context feedback system. Once fed with segmentations, the FCN outputs encoded features which are integrated back into the decoder of the CNN. This context feedback loop helps the model extract high-level image features and fix uncertainties over time.

Ruijsink et al. [97] build from their previously proposed QC technique [77] to embed anatomical awareness into CMR segmentation models. The authors assume that the QC information provided by the QC tool encapsulates expertise biophysical knowledge that can be used to provide feedback to the network. As such, predictions flagged as high quality by the QC tool are fed back into the network model to reinforce its anatomical awareness. Painchaud et al. [98] present a segmentation framework that guarantees anatomical criteria by warping the predictions of a given model towards the closest anatomically valid cardiac shape with the use of a constrained Variational Autoencoder (cVAE). This warping step acts as the correction procedure, effectively leading to a reduced number of anatomical errors in the segmentation results. Finally, Galati and Zuluaga [99] use the information from an autoencoder-based post-analysis QC tool as a proxy of a model's performance in unseen cardiac images [88]. The QC tool allows the automatic identification of Out-of-Distribution (OoD) data, which cause failures of the segmentation model. The information is then used as feedback to refine the training of the segmentation model, thus adapting to the OoD data.

4.2.3. Data Acquisition

Methods trying to mitigate data acquisition problems to improve the robustness of CMR segmentation models have mostly focused on improving the image quality at the image reconstruction phase. Among these, Schlemper et al. [100] propose two different methods to segment the heart directly from the k-space of dynamic MRI data, bypassing middle reconstruction stages. The first method relies on an end-to-end synthesis network that exploits the spatiotemporal redundancy of the input to generate the segmentations directly from the input k-space. The second method is conceived for heavily undersampled and aliased images, where there may be a loss of geometrical information and the first approach fails. It uses an autoencoder and a predictor network. The autoencoder is trained to encode and decode segmentations. The predictor learns to map undersampled images to latent encodings. The predicted encodings are used by the autoencoder to decode the corresponding segmentation maps. Huang et al. [101] propose a method that takes as input the undersampled k-space data from CMR scans to solve the reconstruction and segmentation problems simultaneously. The reconstruction is derived from the fast iterative shrinkage-thresholding algorithm (FISTA), while the segmentation is based on a U-Net architecture. Combining the two modules into a joint single-step, the reconstructed image becomes a set of differentiable parameters for the segmentation module itself, allowing the two to mutually benefit from each other through backpropagation. Finally, Oksuz et al. [102] propose to detect, correct and segment CMR images with motion artifacts, integrating reconstruction and segmentation in a unique framework, which combines a spatiotemporal 2D+time CNN for artifact detection, a convolutional recurrent neural network for reconstruction and a classical U-net for segmentation. The full framework is trained by incorporating terms from all three subnetworks into an overall loss function.

4.2.4. Domain Shift

Domain adaptation is the umbrella term used to refer to the techniques addressing the domain shift problem [103,104]. Within our work, we consider domain adaptation as an MI technique that aims at improving robustness to domain-shifted inputs. It consists of combining labelled source domain data, i.e., data from the original training distribution, with target domain one, i.e., the domain shifted data, typically in an unsupervised manner that avoids labelling the target domain, where in principle no annotated data are available.

Different alternatives have been explored to improve the generalization capacity of CMR segmentation models to an unseen domain, where the unseen domain can be a different image modality, such as computed tomography [105–107], a different magnetic resonance sequence, such as late gadolinium enhancement [108], or the same modality with varying statistical properties (e.g., different vendors and/or centers) [99]. Chen et al. [105,106] present an unsupervised domain adaptation framework, named SIFA. This framework adapts a segmentation network to an unlabeled domain by aligning source and target domains from both image and feature perspectives. Adversarial learning is enforced at multiple levels in the pipeline, guiding the two adaptive perspectives through a shared feature encoder to exploit their mutual benefits. Ouyang et al. [107] introduce an unsupervised domain adaptation method specifically designed to compensate for the drawback of domain adversarial training when only a small number of target samples is available. This result is achieved by introducing prior regularization on a shared domain-invariant latent space of the source and target domain images, which is exploited during segmentation. Chen et al. [108] tackle the problem of domain adaptation by using a common feature generator to fuse the feature spaces of source and target data into a combined feature domain. This new space is kept domain-invariant via indirect double-sided adversarial learning.

4.2.5. Ablation Analysis of MI Techniques

We analyzed the reported performance accuracy of the different MI techniques and their ablated versions. By ablated version, we refer to the backbone architecture of each method without MI. Figure 2 summarizes the reported DSC and HD of the different methods. We observe a clear trend of improvement when using MI: there is an DSC increase, whereas the HD is reduced. Although the reported methods use different backbone architectures, configurations and datasets, which limit a direct comparison, there is a clear trend that suggests that MI techniques addressing robustness and reliability do have a positive impact in the performance of CMR segmentation methods.

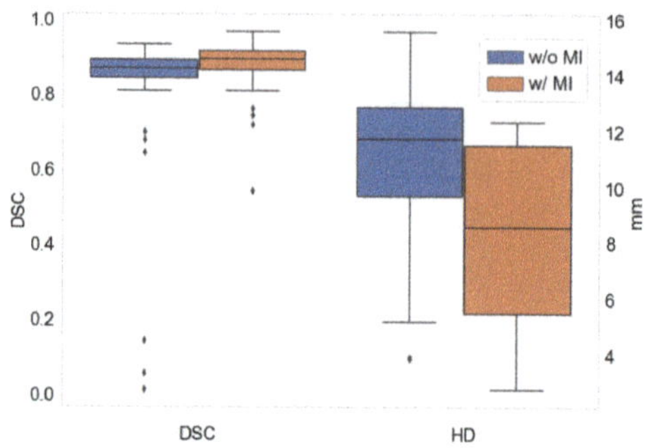

Figure 2. Average DSC (**left**) and HD (**right**) with (w/) the use of MI techniques and without (w/o) them.

5. Discussion

After tracing DL history for CMR segmentation (Section 2), we have highlighted the shortcomings that currently prevent this technology from meeting some of the requirements to be safely deployed and used in clinical routine and cardiac analysis pipelines [109]. In this work, we focus on two main factors: a lack of reliability and robustness of many state-of-the-art methods. After providing formal definitions for the two terms, we have identified and discuss the elements that lead to poor reliability and/or robustness and we presented a wide range of works that have recently been published tackling both problems in CMR segmentation.

In this survey, we proposed to categorize the existing literature into two families: quality control and model improvement techniques. Quality control techniques can be seen as simpler strategies that only aim at flagging situations where a model may be incurring poor reliability or robustness, without aiming to fix the problem. Their main advantage is that these methods are typically external modules that can be promptly attached to an existing segmentation pipeline. However, they leave the problem to the expert, who needs to decide how to address the identified situation. Therefore, QC tools contribute to reducing the analysis time for the expert and providing some safety guarantees, through the generation of alerts, but do not contribute to improving CMR segmentation performance.

Model improvement techniques, instead, bring specific improvements in several aspects of the segmentation model development process, with the final goal of addressing the limitations of DL models that lead to poor reliability or robustness. As such, these type of methods are not only capable of identifying a potential problem, as QC tools do, but they can also act on it and aim to fix it. This being a more complex problem to tackle, it may explain why the number of existing QC methods is larger than MI techniques. A second possible explanation to this may be that the development of QC techniques has been strongly driven by the need to fully automate the processing pipelines of large databases, such as the UK Biobank.

A current limiting factor to further research on new QC and MI techniques addressing robustness and reliability is the lack of a common and well-established framework for their evaluation. QC techniques use different types of outputs, such as quantitative scores or a wide range of qualitative labels, with no clear mapping among them. MI techniques, as discussed in Section 4.2.5, rely on different backbone architectures and configurations that cannot be directly compared. The heterogeneity of existing solutions for both categories of methods challenges an objective and consistent evaluation. Moreover, as demonstrated by Bernard et al. [11], current performance measures, such as the DSC or HD, are not well-suited to identify errors which are associated with poor reliability and robustness. Progress in the field should therefore be accompanied with the investigation of better evaluation strategies.

6. Conclusions

In this paper, we present an overview of the state-of-the-art methods in CMR segmentation deep learning techniques, focusing on the changes of performances preceding and succeeding their rise. As we show, DL models have reached their maturity, achieving performance comparable to experts. Therefore, efforts to develop new models that optimize performance accuracy seem unnecessary. Instead, we observe that works specifically tackling reliability and robustness are rather limited and the field is quite young. We hope that our review can increase the awareness of these two important challenges of CMR segmentation and more research work will focus on developing methods that can efficiently solve them, thus enabling the translation of accurate, reliable, and robust CMR segmentation pipelines into the clinic.

Author Contributions: Conceptualization and methodology, F.G. and M.A.Z.; investigation, F.G.; resources, M.A.Z.; data curation, F.G.; writing—original draft preparation, review and editing, F.G., S.O. and M.A.Z.; visualization, F.G.; supervision, project administration, and funding acquisition, M.A.Z. All authors have read and agreed to the published version of the manuscript.

Funding: This work is supported by the French government, through the 3IA Côte d'Azur Investments in the future project managed by the National Research Agency (ANR) (ANR-19-P3IA-0002).

Institutional Review Board Statement: Not applicable.

Informed Consent Statement: Not applicable.

Data Availability Statement: We have uploaded at https://zenodo.org/record/6451469 (accessed on 7 April 2022) the data and code needed to build Figures 1 and 2.

Conflicts of Interest: The authors declare no conflict of interest.

References

1. Roth, G.A.; Mensah, G.A.; Johnson, C.O.; Addolorato, G.; Ammirati, E.; Baddour, L.M.; Barengo, N.C.; Beaton, A.Z.; Benjamin, E.J.; Benziger, C.P.; et al. Global burden of cardiovascular diseases and risk factors, 1990–2019: Update from the GBD 2019 study. *J. Am. Coll. Cardiol.* **2020**, *76*, 2982–3021. [CrossRef] [PubMed]
2. World Health Organization. Cardiovascular Diseases (CVDs) Fact Sheet. 2021. Available online: https://www.who.int/news-room/fact-sheets/detail/cardiovascular-diseases-(cvds) (accessed on 1 February 2022).
3. Nelson, S.; Whitsel, L.; Khavjou, O.; Phelps, D.; Leib, A. Projections of Cardiovascular Disease Prevalence and Costs. Available online: https://www.heart.org/-/media/Files/Get-Involved/Advocacy/CVD-Predictions-Through-2035.pdf (accessed on 7 April 2022) .
4. World Health Organization. *Global Action Plan for the Prevention and Control of NCDs 2013–2020*; WHO: Geneva, Switzerland, 2013.
5. van der Geest, R.J.; Reiber, J.H. Quantification in cardiac MRI. *J. Magn. Reson. Imaging* **1999**, *10*, 602–608. [CrossRef]
6. Bai, W.; Sinclair, M.; Tarroni, G.; Oktay, O.; Rajchl, M.; Vaillant, G.; Lee, A.M.; Aung, N.; Lukaschuk, E.; Sanghvi, M.M.; et al. Automated cardiovascular magnetic resonance image analysis with fully convolutional networks. *J. Cardiovasc. Magn. Reson.* **2018**, *20*, 65. [CrossRef] [PubMed]
7. Petitjean, C.; Dacher, J.N. A review of segmentation methods in short axis cardiac MR images. *Med. Image Anal.* **2011**, *15*, 169–184. [CrossRef]
8. Zhuang, X. Challenges and methodologies of fully automatic whole heart segmentation: A review. *J. Healthc. Eng.* **2013**, *4*, 371–407. [CrossRef]
9. Chen, C.; Qin, C.; Qiu, H.; Tarroni, G.; Duan, J.; Bai, W.; Rueckert, D. Deep learning for cardiac image segmentation: A review. *Front. Cardiovasc. Med.* **2020**, *7*, 25. [CrossRef]
10. Litjens, G.; Kooi, T.; Bejnordi, B.E.; Setio, A.A.A.; Ciompi, F.; Ghafoorian, M.; Van Der Laak, J.A.; Van Ginneken, B.; Sánchez, C.I. A survey on deep learning in Medical Image Analysis. *Med. Image Anal.* **2017**, *42*, 60–88. [CrossRef]
11. Bernard, O.; Lalande, A.; Zotti, C.; Cervenansky, F.; Yang, X.; Heng, P.A.; Cetin, I.; Lekadir, K.; Camara, O.; Ballester, M.A.G.; et al. Deep Learning Techniques for Automatic MRI Cardiac Multi-Structures Segmentation and Diagnosis: Is the Problem Solved? *IEEE Trans. Med. Imaging* **2018**, *37*, 2514–2525. [CrossRef]
12. Sudlow, C.; Gallacher, J.; Allen, N.; Beral, V.; Burton, P.; Danesh, J.; Downey, P.; Elliott, P.; Green, J.; Landray, M.; et al. UK Biobank: An Open Access Resource for Identifying the Causes of a Wide Range of Complex Diseases of Middle and Old Age. *PLoS Med.* **2015**, *12*, e1001779. [CrossRef]
13. Suinesiaputra, A.; Cowan, B.R.; Al-Agamy, A.O.; Elattar, M.A.; Ayache, N.; Fahmy, A.S.; Khalifa, A.M.; Medrano-Gracia, P.; Jolly, M.; Kadish, A.H.; et al. A collaborative resource to build consensus for automated left ventricular segmentation of cardiac MR images. *Med. Image Anal.* **2014**, *18*, 50–62. [CrossRef]

14. Petitjean, C.; Zuluaga, M.A.; Bai, W.; Dacher, J.; Grosgeorge, D.; Caudron, J.; Ruan, S.; Ayed, I.B.; Cardoso, M.J.; Chen, H.; et al. Right ventricle segmentation from cardiac MRI: A collation study. *Med. Image Anal.* **2015**, *19*, 187–202. [CrossRef] [PubMed]
15. Campello, V.M.; Gkontra, P.; Izquierdo, C.; Martín-Isla, C.; Sojoudi, A.; Full, P.M.; Maier-Hein, K.; Zhang, Y.; He, Z.; Ma, J.; et al. Multi-Centre, Multi-Vendor and Multi-Disease Cardiac Segmentation: The M&Ms Challenge. *IEEE Trans. Med. Imaging* **2021**, *40*, 3543–3554. [PubMed]
16. Jolly, M.; Xue, H.; Grady, L.J.; Guehring, J. Combining Registration and Minimum Surfaces for the Segmentation of the Left Ventricle in Cardiac Cine MR Images. In *Medical Image Computing and Computer-Assisted Intervention-MICCAI 2009, Proceedings of the International Conference on Medical Image Computing and Computer-Assisted Intervention, 12th International Conference, London, UK, 20–24 September 2009*; Lecture Notes in Computer Science Series; Springer: Berlin/Heidelberg, Germany, 2009; Volume 5762, pp. 910–918.
17. Baumgartner, C.F.; Koch, L.M.; Pollefeys, M.; Konukoglu, E. An Exploration of 2D and 3D Deep Learning Techniques for Cardiac MR Image Segmentation. In *Statistical Atlases and Computational Models of the Heart, Proceedings of the International Workshop on Statistical Atlases and Computational Models of the Heart, ACDC and MMWHS Challenges-8th International Workshop, STACOM 2017, Held in Conjunction with MICCAI 2017, Quebec City, QC, Canada, 10–14 September 2017*; Lecture Notes in Computer Science Series; Springer: Cham, Switzerland, 2017; Volume 10663, pp. 111–119.
18. Huang, S.; Liu, J.; Lee, L.C.; Venkatesh, S.K.; Teo, L.L.S.; Au, C.; Nowinski, W.L. An Image-Based Comprehensive Approach for Automatic Segmentation of Left Ventricle from Cardiac Short Axis Cine MR Images. *J. Digit. Imaging* **2011**, *24*, 598–608. [CrossRef] [PubMed]
19. Grinias, E.; Tziritas, G. Fast Fully-Automatic Cardiac Segmentation in MRI Using MRF Model Optimization, Substructures Tracking and B-Spline Smoothing. In *Statistical Atlases and Computational Models of the Heart, Proceedings of the International Workshop on Statistical Atlases and Computational Models of the Heart, ACDC and MMWHS Challenges-8th International Workshop, STACOM 2017, Held in Conjunction with MICCAI 2017, Quebec City, QC, Canada, 10–14 September 2017*; Lecture Notes in Computer Science; Springer: Cham, Switzerland, 2017; Volume 10663, pp. 91–100.
20. Schaerer, J.; Casta, C.; Pousin, J.; Clarysse, P. A dynamic elastic model for segmentation and tracking of the heart in MR image sequences. *Med. Image Anal.* **2010**, *14*, 738–749. [CrossRef]
21. Khened, M.; Kollerathu, V.A.; Krishnamurthi, G. Fully convolutional multi-scale residual DenseNets for cardiac segmentation and automated cardiac diagnosis using ensemble of classifiers. *Med. Image Anal.* **2019**, *51*, 21–45. [CrossRef]
22. Ou, Y.; Sotiras, A.; Paragios, N.; Davatzikos, C. DRAMMS: Deformable registration via attribute matching and mutual-saliency weighting. *Med. Image Anal.* **2011**, *15*, 622–639. [CrossRef]
23. Jang, Y.; Hong, Y.; Ha, S.; Kim, S.; Chang, H. Automatic Segmentation of LV and RV in Cardiac MRI. In *Statistical Atlases and Computational Models of the Heart, Proceedings of the International Workshop on Statistical Atlases and Computational Models of the Heart, ACDC and MMWHS Challenges-8th International Workshop, STACOM 2017, Held in Conjunction with MICCAI 2017, Quebec City, QC, Canada, 10–14 September 2017*; Lecture Notes in Computer Science; Springer: Cham, Switzerland, 2017; Volume 10663, pp. 161–169.
24. Margeta, J.; Geremia, E.; Criminisi, A.; Ayache, N. Layered Spatio-temporal Forests for Left Ventricle Segmentation from 4D Cardiac MRI Data. In *Statistical Atlases and Computational Models of the Heart, Imaging and Modelling Challenges, Proceedings of the International Workshop on Statistical Atlases and Computational Models of the Heart, Toronto, ON, Canada, 22 September 2011*; Lecture Notes in Computer Science; Springer: Berlin/Heidelberg, Germany, 2011; Volume 7085, pp. 109–119.
25. Isensee, F.; Jaeger, P.F.; Full, P.M.; Wolf, I.; Engelhardt, S.; Maier-Hein, K.H. Automatic Cardiac Disease Assessment on cine-MRI via Time-Series Segmentation and Domain Specific Features. In *Statistical Atlases and Computational Models of the Heart, Proceedings of the International Workshop on Statistical Atlases and Computational Models of the Heart, ACDC and MMWHS Challenges-8th International Workshop, STACOM 2017, Held in Conjunction with MICCAI 2017, Quebec City, QC, Canada, 10–14 September 2017*; Lecture Notes in Computer Science; Springer: Cham, Switzerland, 2017; Volume 10663, pp. 120–129.
26. Jolly, M.; Guetter, C.; Lu, X.; Xue, H.; Guehring, J. Automatic Segmentation of the Myocardium in Cine MR Images Using Deformable Registration. In *Statistical Atlases and Computational Models of the Heart, Imaging and Modelling Challenges, Proceedings of the International Workshop on Statistical Atlases and Computational Models of the Heart, Toronto, ON, Canada, 22 September 2011*; Lecture Notes in Computer Science; Springer: Berlin/Heidelberg, Germany, 2011; Volume 7085, pp. 98–108.
27. Yang, X.; Bian, C.; Yu, L.; Ni, D.; Heng, P. Class-Balanced Deep Neural Network for Automatic Ventricular Structure Segmentation. In *Statistical Atlases and Computational Models of the Heart, Proceedings of the International Workshop on Statistical Atlases and Computational Models of the Heart, ACDC and MMWHS Challenges-8th International Workshop, STACOM 2017, Held in Conjunction with MICCAI 2017, Quebec City, QC, Canada, 10–14 September 2017*; Lecture Notes in Computer Science; Springer: Cham, Switzerland, 2017; Volume 10663, pp. 152–160.
28. Liu, H.; Hu, H.; Xu, X.; Song, E. Automatic Left Ventricle Segmentation in Cardiac MRI Using Topological Stable-State Thresholding and Region Restricted Dynamic Programming. *Acad. Radiol.* **2012**, *19*, 723–731. [CrossRef]
29. Attar, R.; Pereañez, M.; Gooya, A.; Zhang, X.A.L.; de Vila, M.H.; Lee, A.M.; Aung, N.; Lukaschuk, E.; Sanghvi, M.M.; Fung, K.; et al. Quantitative CMR population imaging on 20,000 subjects of the UK Biobank imaging study: LV/RV quantification pipeline and its evaluation. *Med. Image Anal.* **2019**, *56*, 26–42. [CrossRef]
30. Wang, C.W.; Peng, C.W.; Chen, H.C. A simple and fully automatic right ventricle segmentation method for 4-dimensional cardiac MR images. In Proceedings of the MICCAI RV Segmentation Challenge, Nice, France, 1–5 October 2012.

31. Calisto, M.G.B.; Lai-Yuen, S.K. AdaEn-Net: An ensemble of adaptive 2D-3D Fully Convolutional Networks for medical image segmentation. *Neural Netw.* **2020**, *126*, 76–94. [CrossRef]
32. Constantinides, C.; Roullot, E.; Lefort, M.; Frouin, F. Fully automated segmentation of the left ventricle applied to cine MR images: Description and results on a database of 45 Subjects. In Proceedings of the Annual International Conference of the IEEE Engineering in Medicine and Biology Society, EMBC 2012, San Diego, CA, USA, 28 August–1 September 2012; IEEE: Piscataway, NJ, USA, 2012; pp. 3207–3210.
33. Scannell, C.M.; Chiribiri, A.; Veta, M. Domain-Adversarial Learning for Multi-Centre, Multi-Vendor, and Multi-Disease Cardiac MR Image Segmentation. In *Statistical Atlases and Computational Models of the Heart, M&Ms and EMIDEC Challenges, Proceedings of the International Workshop on Statistical Atlases and Computational Models of the Heart, 11th International Workshop, STACOM 2020, Held in Conjunction with MICCAI 2020, Lima, Peru, 4 October 2020*; Lecture Notes in Computer Science; Springer: Berlin/Heidelberg, Germany, 2020; Volume 12592, pp. 228–237.
34. Hu, H.; Liu, H.; Gao, Z.; Huang, L. Hybrid segmentation of left ventricle in cardiac MRI using gaussian-mixture model and region restricted dynamic programming. *Magn. Reson. Imaging* **2013**, *31*, 575–584. [CrossRef]
35. Liu, X.; Thermos, S.; Chartsias, A.; O'Neil, A.; Tsaftaris, S.A. Disentangled Representations for Domain-Generalized Cardiac Segmentation. In *Statistical Atlases and Computational Models of the Heart, M&Ms and EMIDEC Challenges, Proceedings of the International Workshop on Statistical Atlases and Computational Models of the Heart, 11th International Workshop, STACOM 2020, Held in Conjunction with MICCAI 2020, Lima, Peru, 4 October 2020*; Lecture Notes in Computer Science; Springer: Berlin/Heidelberg, Germany, 2020; Volume 12592, pp. 187–195.

36. Zuluaga, M.A.; Cardoso, M.J.; Modat, M.; Ourselin, S. Multi-atlas Propagation Whole Heart Segmentation from MRI and CTA Using a Local Normalised Correlation Coefficient Criterion. In *Statistical Atlases and Computational Models of the Heart, M&Ms and EMIDEC Challenges, Proceedings of the International Workshop on Statistical Atlases and Computational Models of the Heart, 11th International Workshop, STACOM 2020, Held in Conjunction with MICCAI 2020, Lima, Peru, 4 October 2020*; Lecture Notes in Computer Science; Springer: Berlin/Heidelberg, Germany, 2020; Volume 7945, pp. 174–181.
37. Li, L.; Zimmer, V.A.; Ding, W.; Wu, F.; Huang, L.; Schnabel, J.A.; Zhuang, X. Random Style Transfer Based Domain Generalization Networks Integrating Shape and Spatial Information. In *Statistical Atlases and Computational Models of the Heart, M&Ms and EMIDEC Challenges, Proceedings of the International Workshop on Statistical Atlases and Computational Models of the Heart, 11th International Workshop, STACOM 2020, Held in Conjunction with MICCAI 2020, Lima, Peru, 4 October 2020*; Lecture Notes in Computer Science; Springer: Berlin/Heidelberg, Germany, 2020; Volume 12592, pp. 208–218.
38. Ngo, T.A.; Carneiro, G. Fully Automated Non-rigid Segmentation with Distance Regularized Level Set Evolution Initialized and Constrained by Deep-Structured Inference. In Proceedings of the 2014 IEEE Conference on Computer Vision and Pattern Recognition, Columbus, OH, USA, 23–28 June 2014; pp. 3118–3125.
39. Huang, X.; Chen, Z.; Yang, X.; Liu, Z.; Zou, Y.; Luo, M.; Xue, W.; Ni, D. Style-Invariant Cardiac Image Segmentation with Test-Time Augmentation. In *Statistical Atlases and Computational Models of the Heart, M&Ms and EMIDEC Challenges, Proceedings of the International Workshop on Statistical Atlases and Computational Models of the Heart, 11th International Workshop, STACOM 2020, Held in Conjunction with MICCAI 2020, Lima, Peru, 4 October 2020*; Lecture Notes in Computer Science; Springer: Berlin/Heidelberg, Germany, 2020; Volume 12592, pp. 305–315.
40. Queiros, S.F.; Barbosa, D.; Heyde, B.; Morais, P.; Vilaça, J.L.; Friboulet, D.; Bernard, O.; D'hooge, J. Fast automatic myocardial segmentation in 4D cine CMR datasets. *Med. Image Anal.* **2014**, *18*, 1115–1131. [CrossRef] [PubMed]
41. Li, H.; Zhang, J.; Menze, B.H. Generalisable Cardiac Structure Segmentation via Attentional and Stacked Image Adaptation. In *Statistical Atlases and Computational Models of the Heart, M&Ms and EMIDEC Challenges, Proceedings of the International Workshop on Statistical Atlases and Computational Models of the Heart, 11th International Workshop, STACOM 2020, Held in Conjunction with MICCAI 2020, Lima, Peru, 4 October 2020*; Lecture Notes in Computer Science; Springer: Berlin/Heidelberg, Germany, 2020; Volume 12592, pp. 297–304.
42. Tufvesson, J.; Hedström, E.; Steding-Ehrenborg, K.; Carlsson, M.; Arheden, H.; Heiberg, E. Validation and development of a new automatic algorithm for time resolved segmentation of the left ventricle in magnetic resonance imaging. *J. Cardiovasc. Magn. Reson.* **2015**, *17*, 1–3. [CrossRef]
43. Simantiris, G.; Tziritas, G. Cardiac MRI Segmentation With a Dilated CNN Incorporating Domain-Specific Constraints. *IEEE J. Sel. Top. Signal Process.* **2020**, *14*, 1235–1243. [CrossRef]
44. Avendi, M.R.; Kheradvar, A.; Jafarkhani, H. A combined deep-learning and deformable-model approach to fully automatic segmentation of the left ventricle in cardiac MRI. *Med. Image Anal.* **2016**, *30*, 108–119. [CrossRef] [PubMed]
45. Full, P.M.; Isensee, F.; Jäger, P.F.; Maier-Hein, K. Studying Robustness of Semantic Segmentation Under Domain Shift in Cardiac MRI. In *Statistical Atlases and Computational Models of the Heart, M&Ms and EMIDEC Challenges, Proceedings of the International Workshop on Statistical Atlases and Computational Models of the Heart, 11th International Workshop, STACOM 2020, Held in Conjunction with MICCAI 2020, Lima, Peru, 4 October 2020*; Lecture Notes in Computer Science; Springer: Berlin/Heidelberg, Germany, 2020; Volume 12592, pp. 238–249.
46. Tran, P.V. A Fully Convolutional Neural Network for Cardiac Segmentation in Short-Axis MRI. *arXiv* **2016**, arXiv:1604.00494.
47. Ma, J. Histogram Matching Augmentation for Domain Adaptation with Application to Multi-centre, Multi-vendor and Multi-disease Cardiac Image Segmentation. In *Statistical Atlases and Computational Models of the Heart, M&Ms and EMIDEC Challenges, Proceedings of the International Workshop on Statistical Atlases and Computational Models of the Heart, 11th International Workshop, STACOM 2020, Held in Conjunction with MICCAI 2020, Lima, Peru, 4 October 2020*; Lecture Notes in Computer Science; Springer: Berlin/Heidelberg, Germany, 2020; Volume 12592, pp. 177–186.
48. Tan, L.K.; Liew, Y.M.; Lim, E.; McLaughlin, R.A. Cardiac left ventricle segmentation using convolutional neural network regression. In Proceedings of the 2016 IEEE EMBS Conference on Biomedical Engineering and Sciences (IECBES), Kuala Lumpur, Malaysia, 4–8 December 2016; pp. 490–493.
49. Zhang, Y.; Yang, J.; Hou, F.; Liu, Y.; Wang, Y.; Tian, J.; Zhong, C.; Zhang, Y.; He, Z. Semi-supervised Cardiac Image Segmentation via Label Propagation and Style Transfer. In *Statistical Atlases and Computational Models of the Heart, M&Ms and EMIDEC Challenges, Proceedings of the International Workshop on Statistical Atlases and Computational Models of the Heart, 11th International Workshop, STACOM 2020, Held in Conjunction with MICCAI 2020, Lima, Peru, 4 October 2020*; Lecture Notes in Computer Science; Springer: Berlin/Heidelberg, Germany, 2020; Volume 12592, pp. 219–227.
50. Patravali, J.; Jain, S.; Chilamkurthy, S. 2D-3D Fully Convolutional Neural Networks for Cardiac MR Segmentation. In *Statistical Atlases and Computational Models of the Heart, Proceedings of the International Workshop on Statistical Atlases and Computational Models of the Heart, ACDC and MMWHS Challenges-8th International Workshop, STACOM 2017, Held in Conjunction with MICCAI 2017, Quebec City, QC, Canada, 10–14 September 2017*; Lecture Notes in Computer Science; Springer: Cham, Switzerland, 2017; Volume 10663, pp. 130–139.

51. Carscadden, A.; Noga, M.; Punithakumar, K. A Deep Convolutional Neural Network Approach for the Segmentation of Cardiac Structures from MRI Sequences. In *Statistical Atlases and Computational Models of the Heart, M&Ms and EMIDEC Challenges, Proceedings of the International Workshop on Statistical Atlases and Computational Models of the Heart, 11th International Workshop, STACOM 2020, Held in Conjunction with MICCAI 2020, Lima, Peru, 4 October 2020*; Lecture Notes in Computer Science; Springer: Berlin/Heidelberg, Germany, 2020; Volume 12592, pp. 250–258.
52. Tan, L.K.; Liew, Y.M.; Lim, E.; McLaughlin, R.A. Convolutional neural network regression for short-axis left ventricle segmentation in cardiac cine MR sequences. *Med. Image Anal.* **2017**, *39*, 78–86. [CrossRef]
53. Khader, F.; Schock, J.; Truhn, D.; Morsbach, F.; Haarburger, C. Adaptive Preprocessing for Generalization in Cardiac MR Image Segmentation. In *Statistical Atlases and Computational Models of the Heart, M&Ms and EMIDEC Challenges, Proceedings of the International Workshop on Statistical Atlases and Computational Models of the Heart, 11th International Workshop, STACOM 2020, Held in Conjunction with MICCAI 2020, Lima, Peru, 4 October 2020*; Lecture Notes in Computer Science; Springer: Berlin/Heidelberg, Germany, 2020; Volume 12592, pp. 269–276.
54. Wolterink, J.M.; Leiner, T.; Viergever, M.A.; Isgum, I. Automatic Segmentation and Disease Classification Using Cardiac Cine MR Images. In *Statistical Atlases and Computational Models of the Heart, Proceedings of the International Workshop on Statistical Atlases and Computational Models of the Heart, ACDC and MMWHS Challenges-8th International Workshop, STACOM 2017, Held in Conjunction with MICCAI 2017, Quebec City, QC, Canada, 10–14 September 2017*; Lecture Notes in Computer Science; Springer: Cham, Switzerland, 2017; Volume 10663, pp. 101–110.
55. Saber, M.; Abdelrauof, D.; Elattar, M. Multi-center, Multi-vendor, and Multi-disease Cardiac Image Segmentation Using Scale-Independent Multi-gate UNET. In *Statistical Atlases and Computational Models of the Heart, M&Ms and EMIDEC Challenges, Proceedings of the International Workshop on Statistical Atlases and Computational Models of the Heart, 11th International Workshop, STACOM 2020, Held in Conjunction with MICCAI 2020, Lima, Peru, 4 October 2020*; Lecture Notes in Computer Science; Springer: Berlin/Heidelberg, Germany, 2020; Volume 12592, pp. 259–268.
56. Rohé, M.; Sermesant, M.; Pennec, X. Automatic Multi-Atlas Segmentation of Myocardium with SVF-Net. In *Statistical Atlases and Computational Models of the Heart, Proceedings of the International Workshop on Statistical Atlases and Computational Models of the Heart, ACDC and MMWHS Challenges-8th International Workshop, STACOM 2017, Held in Conjunction with MICCAI 2017, Quebec City, QC, Canada, 10–14 September 2017*; Lecture Notes in Computer Science; Springer: Cham, Switzerland, 2017; Volume 10663, pp. 170–177.
57. Kong, F.; Shadden, S.C. A Generalizable Deep-Learning Approach for Cardiac Magnetic Resonance Image Segmentation Using Image Augmentation and Attention U-Net. In *Statistical Atlases and Computational Models of the Heart, M&Ms and EMIDEC Challenges, Proceedings of the International Workshop on Statistical Atlases and Computational Models of the Heart, 11th International Workshop, STACOM 2020, Held in Conjunction with MICCAI 2020, Lima, Peru, 4 October 2020*; Lecture Notes in Computer Science; Springer: Berlin/Heidelberg, Germany, 2020; Volume 12592, pp. 287–296.
58. Zotti, C.; Luo, Z.; Humbert, O.; Lalande, A.; Jodoin, P.M. GridNet with Automatic Shape Prior Registration for Automatic MRI Cardiac Segmentation. In *Statistical Atlases and Computational Models of the Heart, Proceedings of the International Workshop on Statistical Atlases and Computational Models of the Heart, ACDC and MMWHS Challenges-8th International Workshop, STACOM 2017, Held in Conjunction with MICCAI 2017, Quebec City, QC, Canada, 10–14 September 2017*; Lecture Notes in Computer Science; Springer: Cham, Switzerland, 2017; Volume 10663, pp. 73–81.
59. Acero, J.C.; Sundaresan, V.; Dinsdale, N.K.; Grau, V.; Jenkinson, M. A 2-Step Deep Learning Method with Domain Adaptation for Multi-Centre, Multi-Vendor and Multi-Disease Cardiac Magnetic Resonance Segmentation. In *Statistical Atlases and Computational Models of the Heart, M&Ms and EMIDEC Challenges, Proceedings of the International Workshop on Statistical Atlases and Computational Models of the Heart, 11th International Workshop, STACOM 2020, Held in Conjunction with MICCAI 2020, Lima, Peru, 4 October 2020*; Lecture Notes in Computer Science; Springer: Berlin/Heidelberg, Germany, 2020; Volume 12592, pp. 196–207.
60. Khened, M.; Varghese, A.; Krishnamurthi, G. Densely Connected Fully Convolutional Network for Short-Axis Cardiac Cine MR Image Segmentation and Heart Diagnosis Using Random Forest. In *Statistical Atlases and Computational Models of the Heart, Proceedings of the International Workshop on Statistical Atlases and Computational Models of the Heart, ACDC and MMWHS Challenges-8th International Workshop, STACOM 2017, Held in Conjunction with MICCAI 2017, Quebec City, QC, Canada, 10–14 September 2017*; Lecture Notes in Computer Science; Springer: Cham, Switzerland, 2017; Volume 10663, pp. 140–151.
61. Parreño, M.; Paredes, R.; Albiol, A. Deidentifying MRI Data Domain by Iterative Backpropagation. In *Statistical Atlases and Computational Models of the Heart, M&Ms and EMIDEC Challenges, Proceedings of the International Workshop on Statistical Atlases and Computational Models of the Heart, 11th International Workshop, STACOM 2020, Held in Conjunction with MICCAI 2020, Lima, Peru, 4 October 2020*; Lecture Notes in Computer Science; Springer: Berlin/Heidelberg, Germany, 2020; Volume 12592, pp. 277–286.
62. Zhou, R.; Guo, F.; Azarpazhooh, M.R.; Hashemi, S.; Cheng, X.; Spence, J.D.; Ding, M.; Fenster, A. Deep Learning-Based Measurement of Total Plaque Area in B-Mode Ultrasound Images. *IEEE J. Biomed. Health Inform.* **2021**, *25*, 2967–2977. [CrossRef] [PubMed]
63. Bousquet, O.; Elisseeff, A. Stability and generalization. *J. Mach. Learn. Res.* **2002**, *2*, 499–526.
64. Bahr, N.J. *System Safety Engineering and Risk Assessment: A Practical Approach*; CRC Press: Boca Raton, FL, USA, 2014.
65. IEEE Standards Coordinating Committee. IEEE Standard Glossary of Software Engineering Terminology. *CA IEEE Comput. Soc.* **1990**, *169*, 132.
66. Ma, J.; Chen, J.; Ng, M.; Huang, R.; Li, Y.; Li, C.; Yang, X.; Martel, A.L. Loss odyssey in medical image segmentation. *Med. Image Anal.* **2021**, *71*, 102035. [CrossRef]

67. Girum, K.B.; Créhange, G.; Lalande, A. Learning With Context Feedback Loop for Robust Medical Image Segmentation. *IEEE Trans. Med. Imaging* **2021**, *40*, 1542–1554. [CrossRef] [PubMed]
68. Sun, B.; Feng, J.; Saenko, K. Return of frustratingly easy domain adaptation. In Proceedings of the AAAI Conference on Artificial Intelligence, Phoenix, AZ, USA, 12–17 February 2016; Volume 30.
69. Gao, R.; Liu, F.; Zhang, J.; Han, B.; Liu, T.; Niu, G.; Sugiyama, M. Maximum Mean Discrepancy Test is Aware of Adversarial Attacks. In Proceedings of the 38th International Conference on Machine Learning, ICML 2021, Virtual Event, 18–24 July 2021; Volume 139, pp. 3564–3575.
70. Tarroni, G.; Oktay, O.; Bai, W.; Schuh, A.; Suzuki, H.; Passerat-Palmbach, J.; de Marvao, A.; O'Regan, D.P.; Cook, S.; Glocker, B.; et al. Large-scale Quality Control of Cardiac Imaging in Population Studies: Application to UK Biobank. *Sci. Rep.* **2020**, *10*, 2408. [CrossRef]
71. Miao, J.; Huo, D.; Wilson, D.L. Quantitative image quality evaluation of MR images using perceptual difference models. *Med. Phys.* **2008**, *35*, 2541–2553. [CrossRef] [PubMed]
72. Lorch, B.; Vaillant, G.; Baumgartner, C.; Bai, W.; Rueckert, D.; Maier, A. Automated detection of motion artefacts in MR imaging using decision forests. *J. Med. Eng.* **2017**, *2017*. [CrossRef] [PubMed]
73. Zhang, L.; Gooya, A.; Frangi, A.F. Semi-supervised Assessment of Incomplete LV Coverage in Cardiac MRI Using Generative Adversarial Nets. In *Simulation and Synthesis in Medical Imaging*; Springer: Berlin/Heidelberg, Germany, 2017; Volume 10557, pp. 61–68.
74. Öksüz, I.; Ruijsink, B.; Puyol-Antón, E.; Clough, J.R.; Cruz, G.; Bustin, A.; Prieto, C.; Botnar, R.M.; Rueckert, D.; Schnabel, J.A.; et al. Automatic CNN-based detection of cardiac MR motion artefacts using k-space data augmentation and curriculum learning. *Med. Image Anal.* **2019**, *55*, 136–147. [CrossRef] [PubMed]
75. Tarroni, G.; Oktay, O.; Bai, W.; Schuh, A.; Suzuki, H.; Passerat-Palmbach, J.; de Marvao, A.; O'Regan, D.P.; Cook, S.; Glocker, B.; et al. Learning-Based Quality Control for Cardiac MR Images. *IEEE Trans. Med. Imaging* **2019**, *38*, 1127–1138. [CrossRef] [PubMed]
76. Machado, I.; Puyol-Antón, E.; Hammernik, K.; Cruz, G.; Ugurlu, D.; Ruijsink, B.; Castelo-Branco, M.; Young, A.; Prieto, C.; Schnabel, J.A.; et al. Quality-Aware Cine Cardiac MRI Reconstruction and Analysis from Undersampled K-Space Data. In *Statistical Atlases and Computational Models of the Heart. Multi-Disease, Multi-View, and Multi-Center Right Ventricular Segmentation in Cardiac MRI Challenge, Proceedings of the International Workshop on Statistical Atlases and Computational Models of the Heart, 12th International Workshop, STACOM 2021, Held in Conjunction with MICC, 12th International Workshop, Strasbourg, France, 27 September 2021*; Springer: Berlin/Heidelberg, Germany, 2022; pp. 12–20.
77. Ruijsink, B.; Puyol-Antón, E.; Oksuz, I.; Sinclair, M.; Bai, W.; Schnabel, J.A.; Razavi, R.; King, A.P. Fully Automated, Quality-Controlled Cardiac Analysis From CMR: Validation and Large-Scale Application to Characterize Cardiac Function. *JACC Cardiovasc. Imaging* **2020**, *13*, 684–695. [CrossRef] [PubMed]
78. Albà, X.; Lekadir, K.; Pereañez, M.; Medrano-Gracia, P.; Young, A.A.; Frangi, A.F. Automatic initialization and quality control of large-scale cardiac MRI segmentations. *Med. Image Anal.* **2018**, *43*, 129–141. [CrossRef]
79. Puyol-Antón, E.; Ruijsink, B.; Baumgartner, C.F.; Masci, P.G.; Sinclair, M.; Konukoglu, E.; Razavi, R.; King, A.P. Automated quantification of myocardial tissue characteristics from native T1 mapping using neural networks with uncertainty-based quality-control. *J. Cardiovasc. Magn. Reson.* **2020**, *22*, 60. [CrossRef]
80. Sander, J.; de Vos, B.D.; Isgum, I. Automatic segmentation with detection of local segmentation failures in cardiac MRI. *Sci. Rep.* **2020**, *10*, 21769. [CrossRef]
81. González, C.; Mukhopadhyay, A. Self-supervised Out-of-distribution Detection for Cardiac CMR Segmentation. In Proceedings of the Medical Imaging with Deep Learning, Lübeck, Germany, 7–9 July 2021; Volume 143, pp. 205–218.
82. Kohlberger, T.; Singh, V.K.; Alvino, C.V.; Bahlmann, C.; Grady, L.J. Evaluating Segmentation Error without Ground Truth. In *Medical Image Computing and Computer-Assisted Intervention-MICCAI 2012, Proceedings of the International Conference on Medical Image Computing and Computer-Assisted Intervention, 15th International Conference, Nice, France, 1–5 October 2012*; Lecture Notes in Computer Science; Springer: Berlin/Heidelberg, Germany, 2012; Volume 7510, pp. 528–536.
83. Valindria, V.V.; Lavdas, I.; Bai, W.; Kamnitsas, K.; Aboagye, E.O.; Rockall, A.G.; Rueckert, D.; Glocker, B. Reverse Classification Accuracy: Predicting Segmentation Performance in the Absence of Ground Truth. *IEEE Trans. Med. Imaging* **2017**, *36*, 1597–1606. [CrossRef]
84. Robinson, R.; Valindria, V.V.; Bai, W.; Oktay, O.; Kainz, B.; Suzuki, H.; Sanghvi, M.M.; Aung, N.; Paiva, J.M.; Zemrak, F.; et al. Automated quality control in image segmentation: Application to the UK Biobank cardiovascular magnetic resonance imaging study. *J. Cardiovasc. Magn. Reson.* **2019**, *21*, 18. [CrossRef] [PubMed]
85. Robinson, R.; Oktay, O.; Bai, W.; Valindria, V.V.; Sanghvi, M.M.; Aung, N.; Paiva, J.M.; Zemrak, F.; Fung, K.; Lukaschuk, E.; et al. Real-Time Prediction of Segmentation Quality. In *Medical Image Computing and Computer Assisted Intervention-MICCAI 2018, Proceedings of the International Conference on Medical Image Computing and Computer-Assisted Intervention, 21st International Conference, Granada, Spain, 16–20 September 2018*; Lecture Notes in Computer Science; Springer: Berlin/Heidelberg, Germany, 2018; Volume 11070, pp. 578–585.
86. Hann, E.; Popescu, I.A.; Zhang, Q.; Gonzales, R.A.; Barutçu, A.; Neubauer, S.; Ferreira, V.M.; Piechnik, S.K. Deep neural network ensemble for on-the-fly quality control-driven segmentation of cardiac MRI T1 mapping. *Med. Image Anal.* **2021**, *71*, 102029. [CrossRef] [PubMed]

87. Fournel, J.; Bartoli, A.; Bendahan, D.; Guye, M.; Bernard, M.; Rauseo, E.; Khanji, M.Y.; Petersen, S.E.; Jacquier, A.; Ghattas, B. Medical image segmentation automatic quality control: A multi-dimensional approach. *Med. Image Anal.* **2021**, *74*, 102213. [CrossRef] [PubMed]
88. Galati, F.; Zuluaga, M.A. Efficient Model Monitoring for Quality Control in Cardiac Image Segmentation. In *Functional Imaging and Modeling of the Heart, Proceedings of the International Conference on Functional Imaging and Modeling of the Heart, 11th International Conference, FIMH 2021, Stanford, CA, USA, 21–25 June 2021*; Lecture Notes in Computer Science; Springer: Berlin/Heidelberg, Germany, 2021; pp. 101–111.
89. He, K.; Zhang, X.; Ren, S.; Sun, J. Deep Residual Learning for Image Recognition. In Proceedings of the IEEE Conference on Computer Vision and Pattern Recognition (CVPR), Las Vegas, NV, USA, 27–30 June 2016.
90. Zuluaga, M.A.; Burgos, N.; Mendelson, A.F.; Taylor, A.M.; Ourselin, S. Voxelwise atlas rating for computer assisted diagnosis: Application to congenital heart diseases of the great arteries. *Med. Image Anal.* **2015**, *26*, 185–194. [CrossRef] [PubMed]
91. Chen, C.; Bai, W.; Davies, R.H.; Bhuva, A.N.; Manisty, C.H.; Augusto, J.B.; Moon, J.C.; Aung, N.; Lee, A.M.; Sanghvi, M.M.; et al. Improving the generalizability of convolutional neural network-based segmentation on CMR images. *Front. Cardiovasc. Med.* **2020**, *7*, 105. [CrossRef]
92. Guo, F.; Ng, M.; Goubran, M.; Petersen, S.E.; Piechnik, S.K.; Neubauer, S.; Wright, G. Improving cardiac MRI convolutional neural network segmentation on small training datasets and dataset shift: A continuous kernel cut approach. *Med. Image Anal.* **2020**, *61*, 101636. [CrossRef]
93. Zotti, C.; Luo, Z.; Lalande, A.; Jodoin, P.M. Convolutional Neural Network With Shape Prior Applied to Cardiac MRI Segmentation. *IEEE J. Biomed. Health Inform.* **2019**, *23*, 1119–1128. [CrossRef]
94. Ronneberger, O.; Fischer, P.; Brox, T. U-Net: Convolutional Networks for Biomedical Image Segmentation. In *Medical Image Computing and Computer-Assisted Intervention-MICCAI 2015, Proceedings of the International Conference on Medical Image Computing and Computer-Assisted Intervention 18th International Conference, Munich, Germany, 5–9 October 2015*; Springer: Berlin/Heidelberg, Germany, 2015; pp. 234–241.
95. Clough, J.; Byrne, N.; Oksuz, I.; Zimmer, V.A.; Schnabel, J.A.; King, A. A Topological Loss Function for Deep-Learning based Image Segmentation using Persistent Homology. *arXiv* **2020**, arXiv:1910.01877.
96. Wyburd, M.K.; Dinsdale, N.K.; Namburete, A.I.L.; Jenkinson, M. TEDS-Net: Enforcing Diffeomorphisms in Spatial Transformers to Guarantee Topology Preservation in Segmentations. In *Medical Image Computing and Computer Assisted Intervention-MICCAI 2021, Proceedings of the International Conference on Medical Image Computing and Computer-Assisted Intervention, 24th International Conference, Strasbourg, France, 27 September–1 October 2021*; Springer: Berlin/Heidelberg, Germany, 2021; pp. 250–260.
97. Ruijsink, B.; Puyol-Antón, E.; Li, Y.; Bai, W.; Kerfoot, E.; Razavi, R.; King, A.P. Quality-Aware Semi-supervised Learning for CMR Segmentation. In *Statistical Atlases and Computational Models of the Heart, M&Ms and EMIDEC Challenges, Proceedings of the International Workshop on Statistical Atlases and Computational Models of the Heart, 11th International Workshop, STACOM 2020, Held in Conjunction with MICCAI 2020, Lima, Peru, 4 October 2020*; Lecture Notes in Computer Science; Springer: Berlin/Heidelberg, Germany, 2020; Volume 12592, pp. 97–107.
98. Painchaud, N.; Skandarani, Y.; Judge, T.; Bernard, O.; Lalande, A.; Jodoin, P.M. Cardiac Segmentation With Strong Anatomical Guarantees. *IEEE Trans. Med. Imaging* **2020**, *39*, 3703–3713. [CrossRef]
99. Galati, F.; Zuluaga, M.A. Using Out-of-Distribution Detection for Model Refinement in Cardiac Image Segmentation. In *Statistical Atlases and Computational Models of the Heart. Multi-Disease, Multi-View, and Multi-Center Right Ventricular Segmentation in Cardiac MRI Challenge, Proceedings of the International Workshop on Statistical Atlases and Computational Models of the Heart, 12th International Workshop, STACOM 2021, Held in Conjunction with MICC, 12th International Workshop, Strasbourg, France, 27 September 2021*; Springer: Berlin/Heidelberg, Germany, 2022; pp. 374–382.
100. Schlemper, J.; Oktay, O.; Bai, W.; de Castro, D.C.; Duan, J.; Qin, C.; Hajnal, J.V.; Rueckert, D. Cardiac MR Segmentation from Undersampled k-space Using Deep Latent Representation Learning. In *Medical Image Computing and Computer Assisted Intervention-MICCAI 2018, Proceedings of the International Conference on Medical Image Computing and Computer-Assisted Intervention, 21st International Conference, Granada, Spain, 16–20 September 2018*; Lecture Notes in Computer Science; Springer: Berlin/Heidelberg, Germany, 2018; Volume 11070, pp. 259–267.
101. Huang, Q.; Yang, D.; Yi, J.; Axel, L.; Metaxas, D.N. FR-Net: Joint Reconstruction and Segmentation in Compressed Sensing Cardiac MRI. In *Functional Imaging and Modeling of the Heart, Proceedings of the International Conference on Functional Imaging and Modeling of the Heart, 10th International Conference, FIMH 2019, Bordeaux, France, 6–8 June 2019*; Lecture Notes in Computer Science; Springer: Berlin/Heidelberg, Germany, 2018; Volume 11504, pp. 352–360.
102. Oksuz, I.; Clough, J.R.; Ruijsink, B.; Anton, E.P.; Bustin, A.; Cruz, G.; Prieto, C.; King, A.P.; Schnabel, J.A. Deep Learning-Based Detection and Correction of Cardiac MR Motion Artefacts During Reconstruction for High-Quality Segmentation. *IEEE Trans. Med. Imaging* **2020**, *39*, 4001–4010. [CrossRef] [PubMed]
103. Torralba, A.; Efros, A.A. Unbiased look at dataset bias. In Proceedings of the 2011 IEEE Conference on Computer Vision and Pattern Recognition, Washington, DC, USA, 20–25 June 2011; pp. 1521–1528.
104. Saito, K.; Kim, D.; Sclaroff, S.; Darrell, T.; Saenko, K. Semi-supervised domain adaptation via minimax entropy. In Proceedings of the IEEE/CVF International Conference on Computer Vision, Seoul, Korea, 27 October–2 November 2019; pp. 8050–8058.
105. Chen, C.; Dou, Q.; Chen, H.; Qin, J.; Heng, P.A. Synergistic Image and Feature Adaptation: Towards Cross-Modality Domain Adaptation for Medical Image Segmentation. *Proc. AAAI Conf. Artif. Intell.* **2019**, *33*, 865–872. [CrossRef]

106. Chen, C.; Dou, Q.; Chen, H.; Qin, J.; Heng, P.A. Unsupervised Bidirectional Cross-Modality Adaptation via Deeply Synergistic Image and Feature Alignment for Medical Image Segmentation. *IEEE Trans. Med. Imaging* **2020**, *39*, 2494–2505. [CrossRef] [PubMed]
107. Ouyang, C.; Kamnitsas, K.; Biffi, C.; Duan, J.; Rueckert, D. Data efficient unsupervised domain adaptation for cross-modality image segmentation. In *Medical Image Computing and Computer Assisted Intervention-MICCAI 2019, Proceedings of the International Conference on Medical Image Computing and Computer-Assisted Intervention, 22nd International Conference, Shenzhen, China, 13–17 October 2019*; Lecture Notes in Computer Science; Springer: Berlin/Heidelberg, Germany, 2019; pp. 669–677.
108. Chen, J.; Zhang, H.; Zhang, Y.; Zhao, S.; Mohiaddin, R.; Wong, T.; Firmin, D.; Yang, G.; Keegan, J. Discriminative consistent domain generation for semi-supervised learning. In *Medical Image Computing and Computer Assisted Intervention-MICCAI 2019, Proceedings of the International Conference on Medical Image Computing and Computer-Assisted Intervention, 22nd International Conference, Shenzhen, China, 13–17 October 2019*; Lecture Notes in Computer Science; Springer: Berlin/Heidelberg, Germany, 2019; pp. 595–604.
109. Floridi, L. Establishing the rules for building trustworthy AI. *Nat. Mach. Intell.* **2019**, *1*, 261–262. [CrossRef]

applied sciences

MDPI

Article

Learning Deep Representations of Cardiac Structures for 4D Cine MRI Image Segmentation through Semi-Supervised Learning

S. M. Kamrul Hasan [1,*] and Cristian A. Linte [1,2]

1 Center for Imaging Science, Rochester Institute of Technology, Rochester, NY 14623, USA
2 Department of Biomedical Engineering, Rochester Institute of Technology, Rochester, NY 14623, USA
* Correspondence: sh3190@rit.edu

Abstract: Learning good data representations for medical imaging tasks ensures the preservation of relevant information and the removal of irrelevant information from the data to improve the interpretability of the learned features. In this paper, we propose a semi-supervised model—namely, combine-all in semi-supervised learning (CqSL)—to demonstrate the power of a simple combination of a disentanglement block, variational autoencoder (VAE), generative adversarial network (GAN), and a conditioning layer-based reconstructor for performing two important tasks in medical imaging: segmentation and reconstruction. Our work is motivated by the recent progress in image segmentation using semi-supervised learning (SSL), which has shown good results with limited labeled data and large amounts of unlabeled data. A disentanglement block decomposes an input image into a domain-invariant spatial factor and a domain-specific non-spatial factor. We assume that medical images acquired using multiple scanners (different domain information) share a common spatial space but differ in non-spatial space (intensities, contrast, etc.). Hence, we utilize our spatial information to generate segmentation masks from unlabeled datasets using a generative adversarial network (GAN). Finally, to reconstruct the original image, our conditioning layer-based reconstruction block recombines spatial information with random non-spatial information sampled from the generative models. Our ablation study demonstrates the benefits of disentanglement in holding domain-invariant (spatial) as well as domain-specific (non-spatial) information with high accuracy. We further apply a structured L_2 similarity (SL_2SIM) loss along with a mutual information minimizer (MIM) to improve the adversarially trained generative models for better reconstruction. Experimental results achieved on the STACOM 2017 ACDC cine cardiac magnetic resonance (MR) dataset suggest that our proposed (CqSL) model outperforms fully supervised and semi-supervised models, achieving an 83.2% performance accuracy even when using only 1% labeled data. We hypothesize that our proposed model has the potential to become an efficient semantic segmentation tool that may be used for domain adaptation in data-limited medical imaging scenarios, where annotations are expensive. Code, and experimental configurations will be made available publicly.

Keywords: augmentation; cardiac segmentation; domain invariant features; disentangled representation; generative adversarial network; image quality; mutual information; reconstruction; variational autoencoder

Citation: Hasan, S.M.K.; Linte, C.A. Learning Deep Representations of Cardiac Structures for 4D Cine MRI Image Segmentation through Semi-Supervised Learning. *Appl. Sci.* **2022**, *12*, 12163. https://doi.org/10.3390/app122312163

Academic Editor: Yu-Dong Zhang and Qi-Huang Zheng

Received: 7 October 2022
Accepted: 22 November 2022
Published: 28 November 2022

1. Introduction

1.1. Background and Problem Statement

The emerging success of deep convolutional neural networks (CNNs) has rendered them the de facto model in solving high-level computer vision tasks [1–3]. However, such approaches mostly rely on large amounts of annotated data for training, the acquisition of which is expensive and laborious, especially for medical imaging/diagnostic radiology data. To address the need for high performance, there has been a growing trend in using a limited amount of annotated data along with an abundance of unlabeled data in a semi-supervised learning (SSL) setting.

The recent dominant body of research that has proposed SSL methods in deep learning features various approaches, including an auxiliary loss term defined on un-annotated data (consistency regularization) [4,5], adversarial networks [6], generating pseudo-labels [7,8] based on model predictions on weakly augmented un-annotated data, self-training [9,10], adversarial learning [11] and domain adaptation [12]. Here we acknowledge their latest accomplishments in the field of domain adaptation, semi-supervised learning and interpretable representation learning by disentanglement and briefly discuss some of their yet outstanding limitations.

1.2. Ongoing Efforts and Related Work

Semi-Supervised Learning: Semi-supervised learning (SSL) [13,14] has experienced much research attention thanks to the increasing availability of large-scale unlabeled data. Semi-supervised learning aims to revamp the model performance by learning from a small portion of labeled data along with optimizing an additional unsupervised loss on a larger portion of unlabeled data, assumed to be sampled from similar distributions, depending on the type of information that needs to be captured from the unlabeled data. Commonly, the rationale of SSL is based on generative models and adversarial networks. The integration of consistency regularization in SSL has shed light on standard baselines recently. By optimizing this loss term, the model imposes several assumptions/constraints on the decision boundary to avoid high-density regions of unannotated data.

Generative adversarial networks: Moreover, generative adversarial learning can be adapted to semi-supervised learning for semantic segmentation [15–17] as well as by generating pseudo pixel-level predictions [18,19]. Adversarial networks use a critic to predict the pixel-level distribution of the data, which acts as an adversarial loss term with the goal to provide the generator with learnable useful visual features from the unlabeled data for medical image synthesis [20]. Nonetheless, learning high-dimensional data can be difficult. Autoencoders struggle with multi-modal data distributions, and generative models rely on computationally demanding models, which are especially difficult to train.

Mutual information estimation: Recent work on representation learning has focused on mutual information estimation [21]. As mutual information maximization has been shown to be effective at capturing the salient attributes of data, being able to disentangle these attributes is another desirable property. For example, it may be beneficial to remove data attributes that are irrelevant to a given task, such as illumination conditions in object recognition.

Disentanglement learning: Some newly introduced techniques have dedicated considerable attention to disentangle representation with generative modeling [22,23]. In disentangled representation, information is represented as a collection of (independent) factors [24], each of which corresponds to a meaningful aspect of the data [25,26]. A current line of research has argued that disentangled representations are beneficial for a variety of tasks, including (semi-)supervised learning of downstream tasks, few-shot learning [27], and exploratory medical data analysis. Additionally, these representations also make it easier for later processes to only use the relevant parts of the data as input.

Unpaired image-to-image translation: Image-to-image translation was first proposed by Isola et al. in [28] in their conditional GAN paper. Furthermore, CycleGAN [29] tackles the problem of the above paired image translation approach by introducing a cycle-consistency loss to retrieve the original images by exploiting a cycle of translation. Later work [30] improved CycleGAN from one-to-one mapping to multimodal image generation. Nevertheless, in medical applications, image synthesis without explicit anatomy design constrain may lead to volatile anatomical structures and artifacts. Moreover, these methods are not aimed at medical image segmentation.

Domain Adaptation: Domain adaptation, a form of transfer learning, encodes the distribution knowledge from a certain source domain to a different but related target domain, and thus, alleviates the domain shift discrepancy in real world applications [31]. Various methods have been proposed, including style and content-disentanglement [32], and adversary based approaches [33,34]. As described later, in this work, we disentangle the most interpretable segmentation-aware spatial (skeleton) information.

Normalization layers: Inspired by instance normalization (IN) [35], conditional batch-normalization [36] and adaptive IN (AdaIN) [37] bring significant improvement in image generation. Later on, feature-wise linear modulation (FiLM) [38] and spatially adaptive denormalization (SPADE) [39] shed additional light over other normalization layers in image synthesis. In our proposed work, we also show how we can adapt both SPADE as well as FiLM normalization as part of a residual and common decoder, respectively.

Variational autoencoder-based models: There have been several recent works involving disentangled learning with variational autoencoder (VAE) [24,40,41]. In contrast to these previous works, we will attempt to demonstrate the use of a VAE as a disentangled representation by sampling the sentiency code to separate the domain-specific information from the domain-invariant latent code.

1.3. Overview of the Proposed Method

To further address some of the shortcomings associated with existing methods, our efforts focus on learning meaningful spatial features utilizing a disentangler with a mutual information minimizer (MIM) to improve the adversarially trained generative models for improving semi-supervised segmentation and reconstruction results.

Our proposed method builds on several recent and key research findings in the fields of generative models, semi-supervised learning, and representation learning via disentanglement. We believe that the proposed framework's reliance on as little as 1% labeled data for training, in concert with the high segmentation accuracy achieved, comparable to the fully or semi-supervised models, renders the proposed work an attractive solution for medical image segmentation, where access to vast expert-annotated data is expensive and often difficult to gain access to.

We approach this problem using a method that is based on disentangled representations and utilizes data from multiple scanners with varying intensities and contrast (Figure 1). Our method is intended to address multi-scanner unlabeled-data issues, such as intensity differences, and a lack of sufficient annotated data. Learning good data representations for medical imaging tasks ensures the preservation of relevant information and the removal of irrelevant information from the data to improve the interpretability of the learned features. Our model disentangles the input image into spatial and non-spatial space. These spatial features are represented as categorical feature maps, with each category corresponding to input pixels that are spatially similar and are from the same organ part. This semantic similarity aids in learning to be generalized the anatomical representation to any modality from different scanners. Furthermore, the non-spatial features capture the image's global intensity information, which aids the renderer in painting the anatomy in the reconstructed image. Finally, because annotating data is time-consuming and expensive, the ability to learn this decomposition through disentanglement using a small number of labels is critical in medical image analysis.

In light of these needs, here we propose a semi-supervised (C*q*SL) model for learning disentangled representations that combines recent developments in semi-supervised learning–generative models and adversarial learning. We aim to factorize the representation of an image pair into two parts: a shared representation that captures the common information between images and an exclusive representation that contains the specific information of each image. Furthermore, in order to achieve representation disentanglement, we propose to minimize mutual information between shared and exclusive representations. Moreover, we use feature-wise linear modulation (FiLM) [38] to distinguish the domain-invariant information from the domain-specific information, as well as a spatially adaptive normalization (SPADE) [39]-based decoder to guide the synthesis of more texture information to restrain the posterior collapse of the VAE and spatial information.

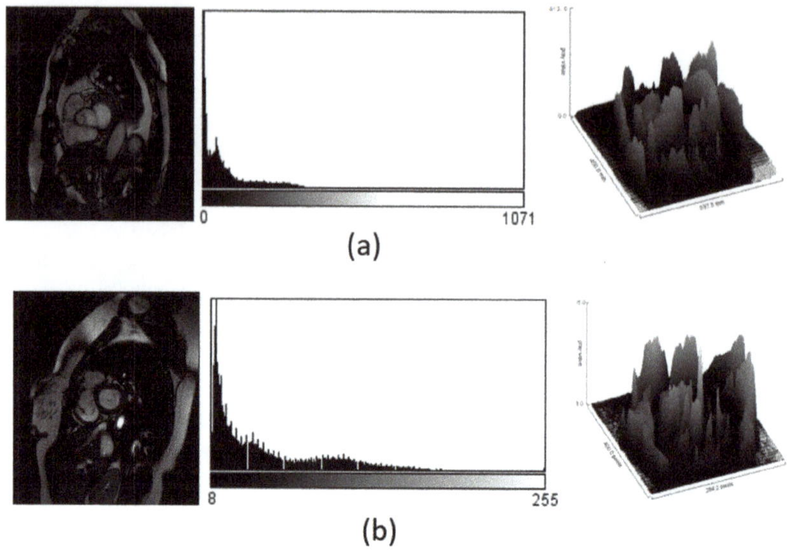

Figure 1. Images, histograms and surface plots of two 3D cardiac images featuring all slices of two random patients from the ACDC dataset are illustrated in (**a**,**b**). From left to right: cardiac MR image in 4 dimensions, histogram plot, and surface plot.

To illustrate its adequacy, our model is applied to two of the foremost critical tasks in medical imaging—segmentation of cardiac structures and reconstruction of the original image—and both assignments are handled by the same model. Our model leverages a large amount of unannotated data from the ACDC (https://www.creatis.insa-lyon.fr/Challenge/acdc/databases.html, accessed on 2 October 2021) dataset to learn the interpretable representations through judicious choices of common factors that serve as strong prior knowledge for more complicated problems—the segmentation of cardiac structures. Figure 2 shows a simplified data view of our proposed model.

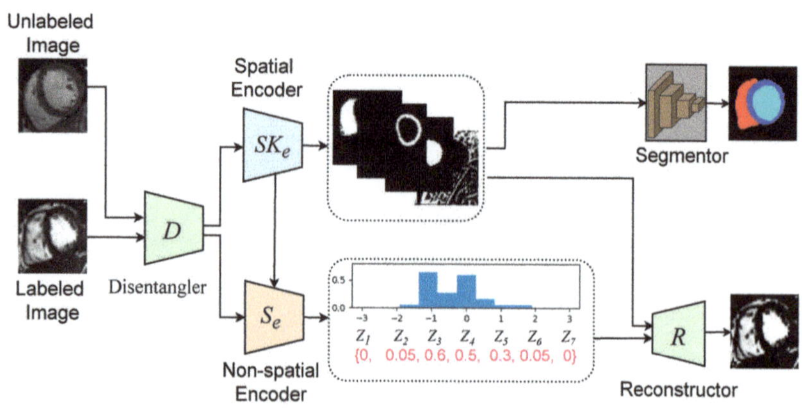

Figure 2. A simplified schematic overview of the proposed model.

1.4. Contributions

Our proposed work makes several contributions summarized as follows:

1. We combine recent developments in disentangled representation learning with strong prior knowledge about medical imaging data that features a decomposition into "skeleton (spatial)" and "sentiency (non-spatial)", to ensure that the spatial information is not mixed up with the non-spatial information.
2. We alter the usual cross-entropy loss to down-weigh the loss applied to well-classified samples in order to overcome the foreground–background class imbalance problem. Specifically, we exploit a novel supervised loss—the weighted-soft-background-focal (WSBF) loss, which focuses the training on a set of hard examples to ensure that this loss can differentiate between easy/hard examples.
3. We employ both qualitative and quantitative tests to evaluate the usefulness of our framework, which show that our model outperformed fully supervised methods, even when using only 1% labeled data for training.

The paper is organized as follows: Section 1 establishes the general background and motivation of the work, reviews the related literature on latest developments in the field of domain adaptation, semi-supervised learning and representation learning, and provides an overview of the proposed work; Section 2 describes our proposed methodology; Section 3 presents our quantitative and qualitative results achieved using our proposed method for both image segmentation and reconstruction, along with the associated ablation studies; Section 4 concludes the paper with a summary of our contributions and promising future research directions.

2. Methods

2.1. CqSL Model Overview

We propose a model that combines the concept of variational generative and adversarial learning, and disentangled interpretation learning in a semi-supervised learning scheme, which is suited for domain-adapted segmentation as well as reconstruction.

We define the learning task as follows: given an (unknown) data distribution $p(x,y)$ over images and segmentation masks, we define a source domain having a training set, $\mathcal{D}_{\mathcal{L}} = \{(x_i^l, y_i^l)\}_{i=1}^{n_l}$ with n_l labeled examples, and another domain having a training set, $\mathcal{D}_{\mathcal{UL}} = \{(x_j^{ul})\}_{j=1}^{n_{ul}}$ with n_{ul} unlabeled examples, which are sampled as independent, identically distributed variables from $p(x,y)$ and $p(x)$ distribution. Empirically, we want to minimize the target risk $\in_t (\phi,\theta) = min_{\phi,\theta} \quad \mathcal{L}_{\mathcal{L}}(\mathcal{D}_{\mathcal{L}}, (\phi,\theta)) + \gamma\mathcal{L}_{\mathcal{UL}}(\mathcal{D}_{\mathcal{UL}}, (\phi,\theta))$, where $\mathcal{L}_{\mathcal{L}}$ is the supervised loss for segmentation, $\mathcal{L}_{\mathcal{UL}}$ is the unsupervised loss defined on unlabeled images and ϕ, θ denotes the learnable parameters of the overall network.

We propose to solve the task by learning domain-specific and domain-invariant features that are discriminative of the semgentor and reconstructor. Figure 3 shows the proposed model comprised of five components—(1) disentanglement component, (2) a disentangled variational autoencoder (DVAE), (3) a mask segmentor identifier (SI), (4) a mask discriminator identifier (DI), and (5) a reconstructor R.

The disentangler D (Figure 3a) is designed to factorize the representation of an image pair into two parts: a shared spatial representation (skeleton, SK_e) that captures the common information between images and an exclusive non-spatial representation (sentiency, S_e) that contains the specific information of each image. The skeleton block SK_e is a modified U-Net++ [42] type architecture (EPU-Net++) (Figure 4 and Section 2.1.1) and is responsible for capturing the domain-invariant features (f_{SK}). The sentiency block S_e is a DVAE (Figure 3b) type architecture, which takes both the input image and the domain-invariant features (f_{SK}) as the input to map domain-specific features (f_{SE}) using the reparameterized trick [43].

The reconstruction block consists of two decoders: the SPADE-based decoder takes the (f_{SE}) feature from the sentiency block and proceeds directly to the reconstructor R (Figure 3d), while the FiLM-based decoder works as another disentangler, which untangles a segmentor identifier (SI) (Figure 3c), used for segmentation and extracted features, which then proceed directly to the reconstructor R. The reconstructor R aims to recover the original image from both (f_{SK}, f_{SE}). A mutual information minimizer (Figure 3a block) is applied between (SK_e and S_e) to enhance the disentanglement. A supervised trainer is trained on the labeled data to predict the segmentation mask distribution optimizing a supervised loss. An unsupervised trainer is trained on the unlabeled data, optimizing unsupervised losses (Algorithm 1 specifies the overall training procedure). Both the unsupervised and supervised trainers share the same block, as mentioned above.

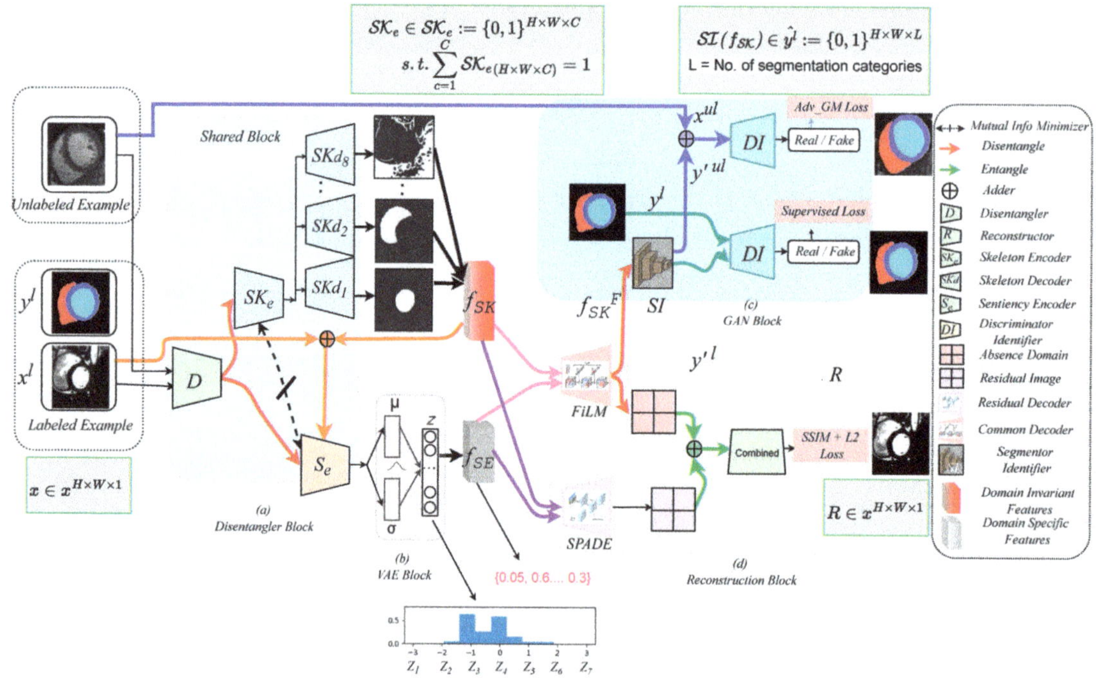

Figure 3. Illustration of ***CqSL* framework:** Our model makes use of both labeled as well as unlabeled images. The first block **(a)** crops the input images to a specific dimension. Then, we disentangle the latent features of the images via a disentangled block. An input image is first encoded to a multi-channel spatial representation, $SKd_{n=1,2...8}$. Then, SKd_n can be fed into a segmentation network SI to generate a multi-class segmentation mask. **(c)** We train a generative network, which predicts semantic labels for both labeled and unlabeled data. **(b)** A sentiency encoder S_e uses the factor SKd_n and the input image to generate a latent vector z representing the imaging modality using a variational autoencoding block. **(d)** The decoder networks combine the two representations SKd_n and z to reconstruct the input image.

Algorithm 1 C*q*SL mini-batch training.

Input:
Training set of labeled data $x^l, y^l, c^l \in \mathcal{D}_{\mathcal{L}}$
Training set of unlabeled data x^{ul}, size m, $\in \mathcal{D}_{\mathcal{UL}}$
Disentanglement Learned parameters: (ϕ, θ), generator G; segmentor S; disentangler D; discriminator identifier DI, mutual information estimator M, and reconstructor R.
Require:
Shared disentangler D, Shared encoder SK_d^k, S_e and decoder
for *each epoch* **do**
  for *each step* **do**
    Sample mini-batch from $x_i^l; x_1^l, \ldots, x_{n_l}^l$; through $\mathcal{D}_{\mathcal{L}}(x)$
    Sample mini-batch from $x_j^{ul}; x_1^{ul}, \ldots, x_{n_{ul}}^{ul}$; through $\mathcal{D}_{\mathcal{UL}}(x)$
    Compute model outputs for the labeled inputs
    $\hat{y}^l \leftarrow \mathcal{W}_{\phi,\theta}(\mathcal{I}_{\mathcal{L}})$
    Compute model outputs for the unlabeled inputs
    $\hat{y}^{ul} \leftarrow \mathcal{W}_{\phi,\theta}(\mathcal{I}_{\mathcal{UL}})$
    Calculate *mutual information* between the disentangled feature pair (f_{sk}, f_{se}) with M_i:
    Update the mask discriminator identifier DI along its gradient:

$$\nabla_{\phi DI} \frac{1}{|\mathcal{I}_{\mathcal{L}}|} \sum_{i \in \mathcal{I}_{\mathcal{L}}} \left[L_{DI}(x_i^l, y_i^l, \hat{y}_i^l) \right] +$$

$$\gamma \frac{1}{|\mathcal{I}_{\mathcal{UL}}|} \sum_{i \in \mathcal{I}_{\mathcal{UL}}} \left[L_{DI}(x_j^{ul}, \hat{y}_j^{ul}) \right]$$

    Update the segmentation mask generator SI and VAE encoder along its gradient:

$$\nabla_{\theta SI} \frac{1}{|\mathcal{I}_{\mathcal{L}}|} \sum_{i \in \mathcal{I}_{\mathcal{L}}} \left[L_{SI}(x_i^l, y_i^l, \hat{y}_i^l) \right] +$$

$$\nabla_{\theta SE} \frac{1}{|\mathcal{I}_{\mathcal{L}}|} \sum_{i \in \mathcal{I}_{\mathcal{L}}} \left[L_{S_e}(x_i^l, \mathcal{F}(x_i^l), \sim z_{dim}^l) \right] +$$

$$\gamma \frac{1}{|\mathcal{I}_{\mathcal{UL}}|} \sum_{j \in \mathcal{I}_{\mathcal{UL}}} \left[L_G(x_j^{ul}, \hat{y}_j^{ul}) \right] +$$

$$\nabla_{\theta SE} \frac{1}{|\mathcal{I}_{\mathcal{UL}}|} \sum_{i \in \mathcal{I}_{\mathcal{UL}}} \left[L_{S_e}(x_j^{ul}, \mathcal{F}(x_j^{ul}), \sim z_{dim}^{ul}) \right]$$

  end for
end for

2.1.1. Disentanglement

Referring to Figure 3a, the disentangler block factorizes the image features into spatial (skeleton/physique) features, as well as non-spatial (sentiency) features that carry residual information. The skeleton block is a modified U-Net type architecture—EvoNorm-Projection-UNet++ (EPU-Net++) as shown in Figure 4. We attach eight different decoders at the common bottleneck layer of EPU-Net++. Each decoder captures bottleneck features from 2D cropped images and transforms them into different feature maps consisting of a number of binary channels which are then combined together to form eight most effective channels: $x_{ST} \xrightarrow{(0,1)_{(h \times w \times c)}} \{\sum_{i=1}^{i=8} f_{SK_i}\}$. These feature maps are responsible for capturing the domain-invariant features and contain cardiac structures (myocardium, the left and the right ventricle), effective for segmentation and some surrounding structures, effective for reconstruction (Figure 5).

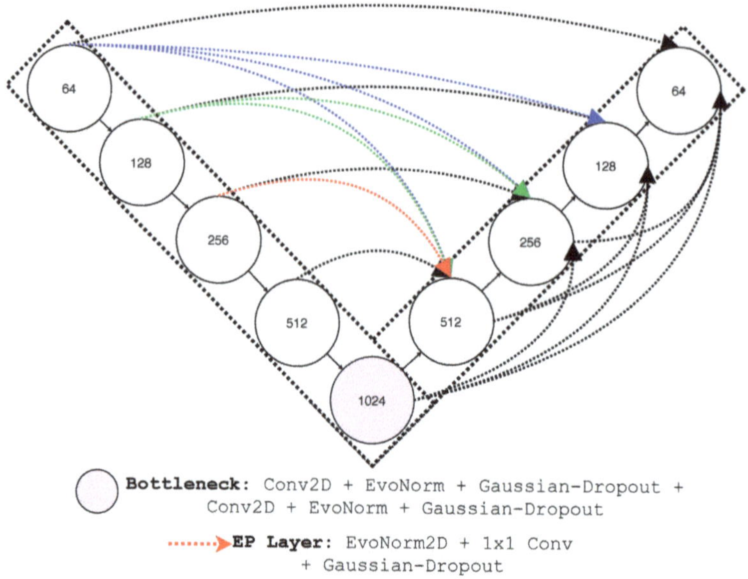

Figure 4. Illustration of EPU-Net++ Block: skip connections are replaced with a long projection block.

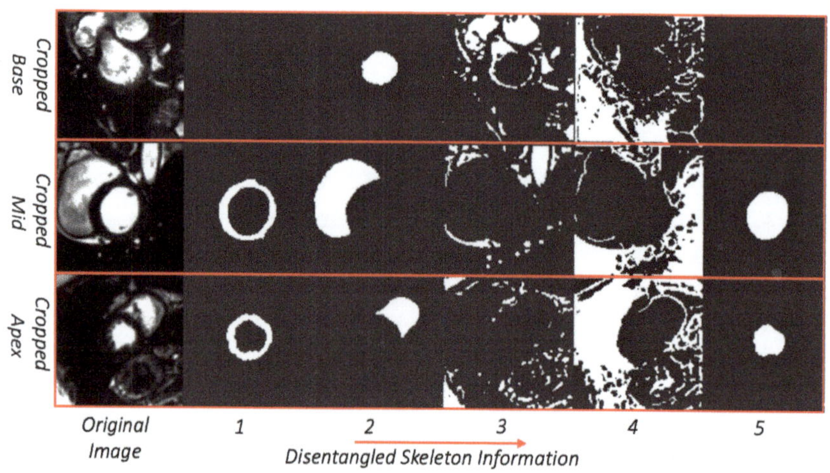

Figure 5. Representative examples showing the 5 (out of 8) most semantic disentangled multi-channel binary maps of the spatial information generated from the skeleton decoder from the base to apex (top to bottom rows). Some channels indicate anatomical portions that are well-defined, such as the myocardium, left ventricle or the right ventricle, while others represent the remaining anatomy needed to characterize the input image.

We use a separate neural network for capturing the sentiency information i.e., domain-specific information. We combine the crop image and the domain-invariant features to penalize the deviation of latent features from the prior distribution employing *Kullback–Leibler divergence* by applying a VAE architecture (Figure 3b) with the following objective function:

$$L_{vae} = \sum \left| \left(p(z_i) \log \frac{p(z_i)}{p(z_i|x_i^{ul}, f_{SK_i})} \right) \right| \tag{1}$$

A VAE learns a low dimensional latent space such that the acquired latent representations fit a prior distribution that is predetermined to be an isotropic multivariate Gaussian $p(z) = \mathcal{N}(0,1)$. An encoder and a decoder make up a VAE. Given an input, the encoder guesses the Gaussian distribution's parameters. In order to enable learning through back propagation, this distribution is then sampled using the reparameterization technique, and the resulting sample is sent through the decoder to reconstruct the input.

We use disentangled features as the prior distribution in a VAE (Equation (1)) to remove class-irrelevant features (e.g., background pixels) and ensure that domain-invariant features are well-disentangled from class-specific features, because the image-only a priori aligns the latent features to a normal distribution.

2.1.2. Mutual Information Minimizer

To better exploit the disentanglement, we add a regularization term based on mutual information (MI), denoted as MIM, which measures the "amount of information" learned from knowledge of random variable Y about the other random variable X [44]. In this paper, we adopt the *mutual information neural estimator (MINE)* [45], $MI(f_{SK}, f_{SE})$:

$$\frac{1}{N}\sum_{i=1}^{N} M(\alpha,\beta,\theta) - \log\left(\frac{1}{N}\sum_{i=1}^{N} \exp^{M(\alpha,\beta',\theta)}\right) \tag{2}$$

where (α,β) are sampled from the joint distribution of (f_{SK}, f_{SE}) and β' is sampled from the marginal distribution.

The mutual information can be expressed as the difference of two entropy terms $MIM(X;Y) = H(X) - H(X|Y)$; we seek to minimize the MI between domain-invariant and domain-specific features (f_{SK}, f_{SE}), whereas we make an assumption that the information content does not vary much between intra-domains (Figure 3a).

2.1.3. Segmentation

The mask segmentor identifier (SI) (Figure 3c) takes the output from the FiLM decoder f^F_{SK} as input and generates predicted segmentation mask $SI(f_{SK}) = \hat{y}^l \in \{0,1\}^{(H\times W\times L)}$, where L is the number of categories (RV, LV, LV-Myo, and background) in the training dataset. We exploit a novel supervised loss, weighted soft background focal (WSBF) loss, $L^{\mathcal{L}}_{SI(seg)} = \mathcal{L}_{WSFL} + \mathcal{L}_{BFD}$ for the base model, which is a combination of background focal dice loss (BFD) and weighted soft focal loss (WSFL):

$$\begin{aligned} L^{\mathcal{L}}_{SI(seg)} = {} & \Big[\alpha_0 + y(\alpha_1 - \alpha_0)\Big]\,|y-\hat{y}|^{\gamma}.w_{map}.CE(y,\hat{y}) + \\ & \sum_c \left[2 - \frac{2\sum y\hat{y} + \epsilon}{\sum(y+\hat{y}) + \epsilon} - \frac{2\sum \overline{y}\overline{\hat{y}} + \epsilon}{\sum(\overline{y}+\overline{\hat{y}}) + \epsilon}\right]^{\frac{1}{\gamma}} \end{aligned} \tag{3}$$

where α_0 and α_1 are designed to account for class imbalance and are treated as hyper-parameters, the term $|y-\hat{y}|^{\gamma}$ is used to down-weigh examples with backgrounds, where γ varies in the range $[1,3]$. The term $CE(y,\hat{y}) = -y\log\hat{y} - (1-y)\log(1-\hat{y})$ denotes the cross-entropy loss.

On the other hand, the data with no corresponding segmentation masks are trained by minimizing the unsupervised loss via a KL divergence based on least-squares GAN [46]. However, since the least-squares loss is not sufficiently robust, we introduce a new divergence loss function by incorporating it into a Geman–McClure model [47] fashion called *adversarial-Geman–McClure (adv-GM)* loss between the ground truth of real mask y^l and prediction on unlabeled data y^{ul}:

$$L^{\mathcal{U}}_{SI(adv\text{-}GM)} = \frac{DI(SI(f_{SK}(x^{ul})))^2 + (DI(\hat{y}^{ul}) - 1)^2}{2\beta + DI(SI(f_{SK}(x^{ul})))^2 + (DI(\hat{y}^{ul}) - 1)^2} \tag{4}$$

where β is the scale factor which varies in the range of $[0,1]$ and we set $\beta = 0.5$ in our experiment.

2.1.4. Image Reconstruction

To better capture the anatomical shape and the intensity information in the synthetic image, we propose a two-branched reconstruction architecture featuring two separate decoders: one is conditioned with FiLM [38], and the other with SPADE [39] (Figure 6a) and both are then concatenated to produce a realistic image. The FiLM decoder consists of multiple FiLM layers, a gamma-beta predictor, and convolutional layers with 3×3 kernel and (8, 8, 8, 8, 1) channels in the stride of 1. Each convolution layer is followed by batch normalization layer along with a Leaky-ReLU layer.

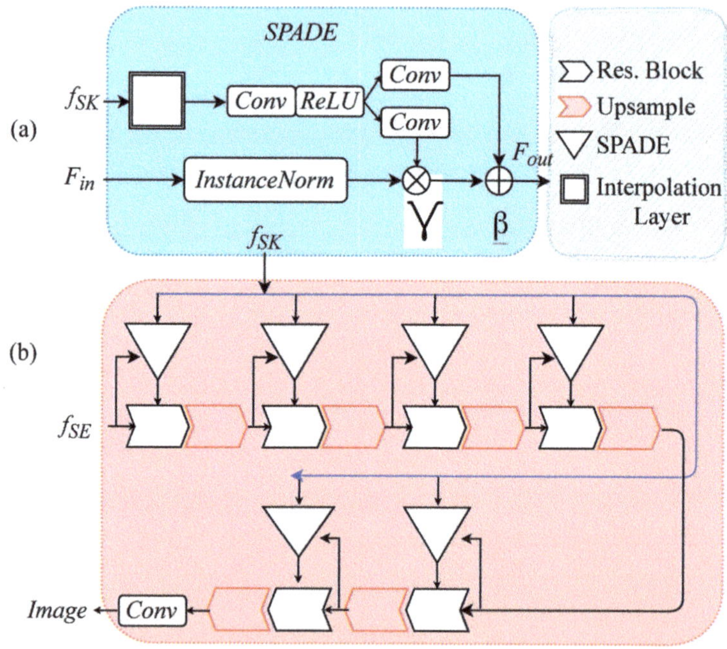

Figure 6. Detailed architecture of SPADE block: (**a**) shape-aware normalization block where the spatial tensors, γ and β are multiplied and added to the input features; (**b**) decoder block f_{SES} with shape-aware normalization.

To better retain the non-spatial information in the MR image, we integrate the shape knowledge into the idea of SPADE [39] and form a shape-aware normalization layer (see Figure 6). SPADE first normalizes the input feature F_{in} with a scale α and a shift μ learned from sampled z using an instance-normalization (InstanceNorm) layer, inspired by [38] and then denormalizes it based on a spatial representation f_{SK} through learnable parameters γ and β. f_{SK} is then interpolated to match the texture dimension of the sampled z from the sentiency encoder and used as a semantic mask for SPADE:

$$F_{out} = \frac{F_{in} - \mu}{\alpha} \times \gamma(f_{SK}) + \beta(f_{SK}) \tag{5}$$

where F_{in} and F_{out} denote the output feature maps. γ and β are learned from f_{SK} by three Conv layers. Thus, the learned shape information precludes washing away the anatomical information, which encourages the image synthesis to be more accurate. The first convolution layer inside the SPADE block (Figure 6) encodes the interpolated f_{SK}, and the other two convolution layers learn the spatial tensors γ and β. Simultaneously, an instance normalization layer is applied to the intermediate feature map, which is then modulated by the scale and shift parameters γ and β learned from sampled z to produce the output. Finally, the output of the two decoders is re-entangled in order to reconstruct an image.

2.2. Objective Functions

The training objective function consists of multiple losses for labeled and unlabeled data, each weighted by some scalar term λ:

$$\begin{aligned} L_{total} = \lambda_{seg}\, L^{\mathcal{L}}_{SI(seg)} + \lambda_{adv-GM}\, \{L^{\mathcal{L}}_{SI,DI(adv-GM)} \\ + L^{\mathcal{U}}_{SI,DI^{u}(adv-GM)}\} + \lambda_{vae} L_{vae} \\ + \lambda_{SL_2SIM}\, \{\, L^{\mathcal{L}}_{SL_2SIM} + L^{\mathcal{U}}_{SL_2SIM}\} \\ + \lambda_{MIM}\, MIM(f_{SK}, f_{SE}) \end{aligned} \tag{6}$$

where λ_t is the weight for the loss of type t. In this paper, we empirically set the weights as $\lambda_{vae} = 0.01$, $\lambda_{seg} = 10$, $\lambda_{adv-GM} = 10$, $\lambda_{SL_2SIM} = 0.01$, $\lambda_{MIM} = 1$.

2.2.1. Segmentation Loss

Since the model is trained on both labeled and unlabeled data, the segmentation loss L_{seg} includes both supervised and unsupervised losses:

$$L_{seg} = L_{sup} + L_{usup} \tag{7}$$

Supervised Loss. Our supervised cost is based on the combination of the two following functions: (1) the weighted soft focal loss, and (2) the background focal dice loss mentioned in Equation (3) ($L_{sup} = L^{\mathcal{L}}_{SI(seg)}$).

Unsupervised Loss. The discriminator identifier is adversarially trained for the labeled and unlabeled data and updated along with adversarial-Geman–McClure (adv-GM) loss $L_{usup} = L^{\mathcal{L}}_{SI,DI(adv-GM)} + L^{\mathcal{U}}_{SI,DI^{u}(adv-GM)}$. For labeled data, the adversarial loss is

$$\begin{aligned} L^{\mathcal{L}}_{SI,DI(adv-GM)} = \\ \frac{\mathbb{E}_{x\sim x_i^l}[DI(SI(f_{SK_i}(x_i^l)))^2] + \mathbb{E}_{y\sim y_i^l}[(DI(y_i^l)-1)^2]}{2\beta + \mathbb{E}_{x\sim x_i^l}[DI(SI(f_{SK_i}(x_i^l)))^2] + \mathbb{E}_{y\sim y_i^l}[(DI(y_i^l)-1)^2]} \end{aligned} \tag{8}$$

Similarly, for the unlabeled data, the adversarial loss is

$$\begin{aligned} L^{\mathcal{U}}_{SI,DI^{u}(adv-GM)} = \frac{\mathbb{E}_{x\sim x_i^{ul}}[DI^{u}(SI(f_{SK_i}(x_i^{ul})))^2]}{2\beta + \mathbb{E}_{x\sim x_i^{ul}}[DI^{u}(SI(f_{SK_i}(x_i^{ul})))^2]} \\ \frac{+\,\mathbb{E}_{y\sim \hat{y}_i^{ul}}[(DI^{u}(y_i^{ul})-1)^2]}{+\,\mathbb{E}_{y\sim \hat{y}_i^{ul}}[(DI^{u}(y_i^{ul})-1)^2]} \end{aligned} \tag{9}$$

VAE Loss. For the smooth texture detail of the input data, the VAE learns factorized representations to optimize a KL-divergence loss, given an image x_i^{ul}, and its decomposed skeleton feature f_{SK} (Equation (1)).

2.2.2. Reconstruction Loss

We adopt a novel reconstruction loss as a combination of structural similarity (SSIM) and L_2 loss–SL_2SIM in order to enforce the similarity between recovered image and original image for better learning the distribution of images.

SL_2SIM Loss. Since the image intensities vary across imaging scanners, as a result, there are high chances that the generative model will tend to *mode collapse*. This structural

L_2 similarity (SL_2SIM) loss provides a similarity measure between the input image and the reconstructed image based on high light-dark variance, contrast, and structural similarity. The concatenated FiLM and SPADE decoder learn the parameters to reconstruct the input image using a novel combination of structured similarity loss and L_2 loss. For labeled data, the reconstruction loss is

$$L^{\mathcal{L}}_{SL_2SIM} = \mathbb{E}_{x_i \sim x_i^l}\Big[1 - SL_2SIM\Big\{x_i^l,\ (\mathcal{F}(f_{SK_i}, f_{SE_i}) \oplus\ \mathcal{S}(f_{SK_i}, f_{SE_i}))\Big\} + \alpha \sum_{i=1}^{n_l} \Big|\Big|\Big\{x_i^l -\ (\mathcal{F}(f_{SK_i}, f_{SE_i}) \oplus\ \mathcal{S}(f_{SK_i}, f_{SE_i}))\Big\}\Big|\Big|_2^2\Big] \tag{10}$$

Similarly, for unlabeled data, the reconstruction loss is

$$L^{\mathcal{U}}_{SL_2SIM} = \mathbb{E}_{x_i \sim x_i^{ul}}\Big[1 - SL_2SIM\Big\{x_i^{ul},\ (\mathcal{F}(f_{SK_i}, f_{SE_i}) \oplus\ \mathcal{S}(f_{SK_i}, f_{SE_i}))\Big\} + \alpha \sum_{i=1}^{n_{ul}} \Big|\Big|\Big\{x_i^{ul} -\ (\mathcal{F}(f_{SK_i}, f_{SE_i}) \oplus\ \mathcal{S}(f_{SK_i}, f_{SE_i}))\Big\}\Big|\Big|_2^2\Big] \tag{11}$$

where SL_2SIM is the structure similarity index term and α is a regularized term.

2.3. Experiments

2.3.1. Datasets

We validate the effectiveness of C*q*SL on a widely adopted cardiac image segmentation challenge dataset by conducting several comparisons to other baseline models. We use the STACOM 2017 *Automated Cardiac Diagnosis Challenge (ACDC)* dataset (https://www.creatis.insa-lyon.fr/Challenge/acdc/databases.html, accessed on 2 October 2021), consisting of short-axis cardiac cine-MR images acquired for 100 patients (1920 labeled and 23,530 unlabeled images) divided into 5 subgroups: normal (NOR), myocardial infarction (MINF), dilated cardiomyopathy (DCM), hypertrophic cardiomyopathy (HCM), and abnormal right ventricle (ARV), available through the 2017 MICCAI-ACDC STACOM challenge [48]. The images were acquired over a 6 year period using two MRI scanners of different magnetic strengths (1.5 T and 3.0 T). The images were acquired using the SSFP sequence with spatial resolution 1.37 to 1.68 mm^2/pixel and 28 to 40 frames per cardiac cycle. We split the dataset into three sets—training (70), validation (15), and test (15).

2.3.2. Implementation Details

Input: All the cine cardiac images employed slice-wise normalization in the range $[0, 1]$ by subtracting the mean slice intensity from each pixel intensity, then dividing it by the difference between the maximum and minimum slice intensity. All images were resampled to 1.37 mm^2/pixel. Images were cropped to $192 \times 192 \times 1$ pixels before feeding to the models. We applied data augmentation on-the-fly during training as shown in Figure 7, which includes random rotations up to 90 degrees, random zooms up to 20%, random horizontal shifts up to 20%, random horizontal and/or vertical flips, and noise addition (Figure 7).

Baselines Architecture: As the disentangled encoder in the skeletal block, we use a modified U-Net-like architecture, EPU-Net++, and as a sentiency encoder, we use VAE. As the reconstruction block, we use FiLM- and SPADE-based decoder as used in [49].

Generator–Discriminator Network: Our segmentation generator network consists of 3 convolution layers with 3×3 kernel and {64, 64, 1} channels in the stride of 1. Each convolution layer is followed by a batch normalization [50] layer along with a Leaky-

ReLU [51] except the last layer. We use the structure similar to DCGAN [52] for the discriminator network.

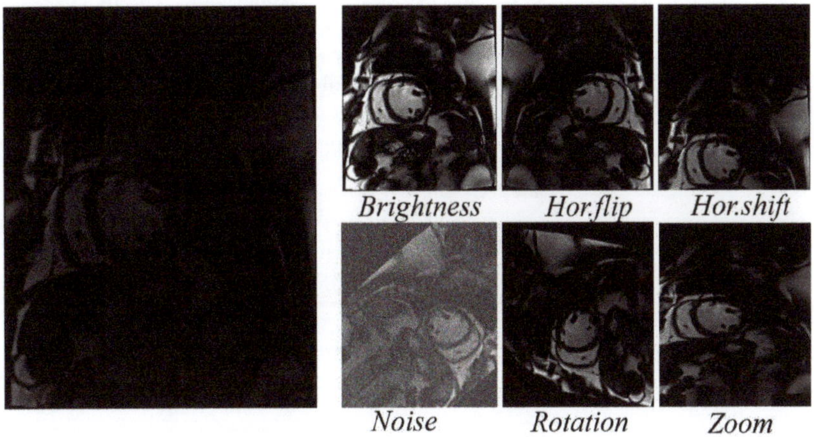

Figure 7. Example images of applying data augmentation via affine transformations.

EvoNorm-Projection skip connections: In our skeleton encoder, we replace the standard skip connection with a normalized-projection operation using $EvoNorm2D + 1 \times 1 - Conv + Gaussian - dropout$, as in Figure 4. This new normalization layer adds together two types of statistical moments–batch variance, and instance variance, both of which capture both the global and local information across images without having any explicit activation function [53]. The proposed projection operation helps in reducing the learnable weights and also allows intricate learnability of cross-channel information.

Additional Factors: The performance of semi-supervised models trained for image segmentation can be significantly impacted by the proper selection of regularizer, optimizer, and hyper-parameters. The model implemented in Keras was initialized with the He normal initializer and trained for 100 epochs with a batch size of 4. We trained all the components iteratively with the Adam optimizer with a 0.0001 learning rate to minimize the objective function. All experiments were conducted on a machine equipped with two NVIDIA RTX 2080 Ti GPU (each 11GBs memory). The detailed training procedure is presented in Algorithm 1.

Training: In our semi-supervised setup, we trained the network on varying proportions of labeled data: 1%, 10%, 20%, 30%, 50%, and 90% as a labeled set and used the rest of the data as the training unlabeled set to hold $|\mathcal{D}_{\mathcal{L}}| \leq |\mathcal{D}_{\mathcal{UL}}|$. In Section 3, we include an ablation study to investigate the importance of adding different loss components in our model C*q*SL which is comprised of all the three loss functions: WSBF , MIM, Adv-GM. (Definitions are provided in Sections 2.1.2 and 2.1.3.)

We experimented an ablation study containing four of the variants of our proposed model C*q*SL. The variants are described as follows: ^{1}C*q*SL, without weighted-soft focal loss (WSFL); ^{2}C*q*SL, without adversarial-Geman–McClure loss (Adv-GM); ^{3}C*q*SL, dice and cross-entropy loss only; and ^{4}C*q*SL, without mutual information minimizer loss (MIM). Here, we utilize the same backbones as the baselines with the only exceptions being different loss functions. To clarify our point, in ^{1}C*q*SL, we removed the weighted soft focal loss (WSFL) from the weighted soft background focal loss (WSBF), while keeping the background focal dice loss (BFD), mutual information minimizer loss (MIM) and adversarial-Geman–McClure (adv-GM) the same as before. In ^{2}C*q*SL, we removed our Geman–McClure version of adversarial loss, while keeping the regular adversarial loss, weighted soft background focal loss (WSBF), and mutual information minimizer loss (MIM) the same as before. Similarly, in ^{3}C*q*SL, we used $DICE + CE$ loss rather than using our novel weighted soft background focal loss (WSBF) while keeping the mutual information minimizer loss (MIM) and adversarial-Geman–McClure (adv-GM) the same as before. Finally, in ^{4}CqSL, we removed our mutual information minimizer loss (MIM) loss, while keeping the weighted soft background focal loss (WSBF), and adversarial Geman–McClure (adv-GM) the same as before. Additionally, the sentiency block, S_e and the skeleton block, SK_e were in place. We evaluated the performance of all four CqSL semi-supervised variants as summarized in Tables 1–3 in the Results section, and, as illustrated later, the ^{1}CqSL variant performed best, but for the sake of consistency, we asses and compare the performance of all four implemented variants.

Table 1. Quantitative evaluation of RV blood pool segmentation results achieved using four semi-supervised variants of the proposed CqSL model in terms of mean Dice score (%) with std. dev., Jaccard index, Hausdorff distance (mm), precision (%) and recall (%) rate evaluated for varying proportions of labeled data on the ACDC dataset compared across several frameworks.

	Right Ventricle (RV)				
	Dice	Jaccard	HD	Prec.	Rec.
U-Net-90%	80.50 ± 8.45	72.03 ± 9.77	8.89 ± 8.45	90.09	94.35
U-Net-50%	79.21 ± 8.49	70.26 ± 10.69	8.90 ± 6.12	85.32	90.11
U-Net-30%	72.32 ± 10.60	66.10 ± 14.75	10.19 ± 7.43	79.50	83.45
U-Net-20%	61.29 ± 16.59	55.65 ± 18.90	12.88 ± 7.32	67.19	74.50
U-Net-10%	54.90 ± 19.66	46.89 ± 20.05	14.58 ± 9.03	60.55	63.02
U-Net-1.0%	39.02 ± 21.22	32.10 ± 22.22	15.90 ± 9.12	43.02	44.15
GAN-90%	79.0 ± 8.15	70.59 ± 10.89	9.55 ± 6.35	85.09	90.12
GAN-50%	78.76 ± 8.98	70.16 ± 11.18	9.88 ± 6.44	84.32	89.43
GAN-30%	73.97 ± 10.87	67.01 ± 13.04	10.23 ± 6.98	79.93	84.97
GAN-20%	69.92 ± 11.45	63.65 ± 16.88	11.66 ± 7.14	79.12	84.12
GAN-10%	66.33 ± 13.21	60.18 ± 19.23	11.99 ± 7.88	74.12	78.34
GAN-1.0%	62.43 ± 13.23	56.43 ± 22.12	13.43 ± 8.11	69.12	73.33
GAN+REC-90%	78.78 ± 8.11	71.13 ± 9.77	9.12 ± 6.46	86.09	90.23
GAN+REC-50%	78.98 ± 8.88	70.13 ± 11.13	9.78 ± 6.66	85.12	90.54
GAN+REC-30%	74.83 ± 10.67	68.67 ± 14.06	10.01 ± 6.98	80.12	85.32
GAN+REC-20%	71.14 ± 11.18	66.65 ± 16.44	11.34 ± 7.05	80.23	84.23
GAN+REC-10%	69.24 ± 13.78	63.23 ± 17.71	11.80 ± 7.23	75.13	79.12
GAN+REC-1.0%	64.19 ± 12.22	59.33 ± 21.01	12.91 ± 7.54	70.34	74.67
CqSL-90%	83.0 ± 6.33	77.77 ± 11.66	8.1 ± 6.00	90.78	95.12
CqSL-50%	82.72 ± 8.29	76.15 ± 11.0	8.21 ± 6.04	88.44	94.26
CqSL-30%	81.59 ± 7.20	73.27 ± 12.14	8.28 ± 6.10	85.19	92.62
CqSL-20%	81.44 ± 6.12	75.33 ± 11.52	8.56 ± 6.11	83.14	93.79
CqSL-10%	79.21 ± 9.76	71.45 ± 12.91	9.82 ± 6.78	82.40	90.93
CqSL-1.0%	75.50 ± 10.87	70.55 ± 12.58	9.87 ± 6.72	80.55	83.68
^{1}CqSL-90%	81.88 ± 6.0	74.31 ± 11.65	8.5 ± 6.15	90.12	91.97
^{1}CqSL-50%	82.03 ± 6.45	75.22 ± 11.24	8.49 ± 6.10	88.11	93.44
^{1}CqSL-30%	79.25 ± 8.11	73.16 ± 8.14	8.77 ± 6.22	83.62	92.05
^{1}CqSL-20%	80.21 ± 7.54	73.19 ± 11.04	9.01 ± 6.34	83.69	91.05
^{1}CqSL-10%	78.58 ± 9.22	71.12 ± 11.25	9.48 ± 6.57	82.21	91.01
^{1}CqSL-1.0%	73.90 ± 11.88	68.58 ± 13.89	9.85 ± 6.71	79.54	84.54
^{2}CqSL-90%	81.03 ± 7.11	74.37 ± 11.48	8.74 ± 6.25	88.39	92.28
^{2}CqSL-50%	80.65 ± 7.26	73.36 ± 12.06	8.54 ± 6.23	86.78	93.05
^{2}CqSL-30%	78.02 ± 9.36	72.66 ± 10.55	9.35 ± 6.65	82.88	91.96
^{2}CqSL-20%	79.55 ± 8.10	73.0 ± 11.54	9.65 ± 6.63	83.02	89.15
^{2}CqSL-10%	78.33 ± 8.96	68.54 ± 12.89	9.77 ± 6.34	80.56	91.55
^{2}CqSL-1.0%	71.21 ± 11.76	63.45 ± 15.91	11.82 ± 7.12	76.40	81.93
^{3}CqSL-90%	81.13 ± 7.33	73.04 ± 12.11	8.93 ± 6.33	86.02	90.17
^{3}CqSL-50%	79.34 ± 8.56	71.23 ± 12.87	9.05 ± 6.66	84.34	91.24
^{3}CqSL-30%	76.77 ± 10.11	72.04 ± 11.26	9.66 ± 6.73	82.0	90.88
^{3}CqSL-20%	79.01 ± 8.58	71.89 ± 12.88	9.52 ± 6.46	81.66	87.56
^{3}CqSL-10%	76.55 ± 8.25	68.55 ± 13.23	10.12 ± 6.89	81.02	88.72
^{3}CqSL-1.0%	70.41 ± 11.86	64.77 ± 15.70	12.11 ± 7.23	74.44	80.21
^{4}CqSL-90%	79.83 ± 8.23	70.33 ± 12.66	9.25 ± 6.34	84.54	90.02
^{4}CqSL-50%	79.02 ± 8.88	72.68 ± 12.26	9.36 ± 6.23	85.20	90.22
^{4}CqSL-30%	75.38 ± 9.75	70.49 ± 12.0	9.52 ± 6.54	80.33	88.59
^{4}CqSL-20%	75.77 ± 9.05	69.88 ± 13.22	10.19 ± 6.77	81.02	88.78
^{4}CqSL-10%	72.24 ± 10.65	66.70 ± 13.56	10.55 ± 6.75	79.79	85.47
^{4}CqSL-1.0%	68.97 ± 13.90	63.19 ± 16.50	12.88 ± 7.43	72.13	77.59

Table 2. Quantitative evaluation of LV blood pool segmentation results achieved using four semi-supervised variants of the proposed C*q*SL model in terms of mean Dice score (%) with std. dev., Jaccard index, Hausdorff distance (mm), precision (%) and recall (%) rates evaluated for varying proportions of labeled data on the ACDC dataset compared across several frameworks.

	Left Ventricle (LV)				
	Dice	Jaccard	HD	Prec.	Rec.
U-Net-90%	88.03 ± 6.81	85.09 ± 6.98	5.16 ± 5.92	97.88	98.79
U-Net-50%	86.88 ± 6.09	84.67 ± 5.36	5.29 ± 6.20	97.01	98.19
U-Net-30%	82.98 ± 8.66	80.10 ± 8.19	6.89 ± 6.75	89.66	91.05
U-Net-20%	81.29 ± 8.91	79.78 ± 9.02	8.22 ± 8.23	87.50	89.77
U-Net-10%	79.49 ± 9.56	71.29 ± 11.26	9.56 ± 9.82	83.33	86.14
U-Net-1.0%	42.56 ± 19.76	37.02 ± 21.45	14.35 ± 10.12	45.53	46.17
GAN-90%	86.15 ± 6.45	81.23 ± 8.01	5.53 ± 5.08	90.57	92.87
GAN-50%	85.34 ± 7.03	81.26 ± 8.12	5.91 ± 6.03	88.34	89.43
GAN-30%	84.03 ± 8.16	80.22 ± 9.11	6.89 ± 7.03	87.23	88.87
GAN-20%	81.90 ± 8.59	79.12 ± 10.82	7.12 ± 7.33	86.19	88.12
GAN-10%	81.78 ± 8.16	76.67 ± 14.13	8.02 ± 7.54	83.15	87.43
GAN-1.0%	75.02 ± 12.32	70.22 ± 15.12	10.89 ± 9.12	80.22	83.12
GAN+REC-90%	88.06 ± 6.11	81.94 ± 8.12	5.73 ± 5.22	91.19	93.35
GAN+REC-50%	86.19 ± 6.89	81.02 ± 8.23	5.76 ± 5.43	90.54	91.65
GAN+REC-30%	85.53 ± 7.36	80.34 ± 9.12	6.78 ± 6.34	89.76	90.34
GAN+REC-20%	83.89 ± 8.19	79.34 ± 10.22	6.88 ± 7.05	87.19	89.53
GAN+REC-10%	83.29 ± 7.16	77.56 ± 13.05	7.58 ± 8.33	85.55	89.02
GAN+REC-1.0%	76.02 ± 11.22	71.32 ± 14.22	10.04 ± 9.12	80.12	84.43
C*q*SL-90%	92.77 ± 4.98	85.67 ± 7.31	4.53± 4.98	96.12	99.75
C*q*SL-50%	92.25 ± 5.12	83.98 ± 7.98	5.23 ± 5.03	95.91	97.95
C*q*SL-30%	90.10 ± 5.89	82.91 ± 8.12	5.93 ± 5.23	93.50	93.79
C*q*SL-20%	88.98 ± 6.33	81.26 ± 8.78	6.21 ± 5.04	90.14	92.90
C*q*SL-10%	88.33 ± 6.39	79.92 ± 9.21	6.17 ± 6.44	89.35	92.95
C*q*SL-1.0%	83.21 ± 7.12	77.94 ± 10.51	7.0 ± 5.98	86.96	91.36
^{1}C*q*SL-90%	92.21 ± 5.13	83.66 ± 7.45	4.88 ± 3.21	95.03	97.33
^{1}C*q*SL-50%	91.0 ± 5.55	81.61 ± 8.05	5.16 ± 4.09	94.12	96.13
^{1}C*q*SL-30%	89.56 ± 5.97	81.23 ± 7.89	5.89 ± 6.98	92.22	92.80
^{1}C*q*SL-20%	87.28 ± 6.91	80.32 ± 8.12	6.55 ± 5.23	89.89	91.0
^{1}C*q*SL-10%	87.89 ± 6.44	79.15 ± 9.30	6.05 ± 5.33	89.03	92.55
^{1}C*q*SL-1.0%	81.78 ± 7.22	75.36 ± 9.20	7.88 ± 5.44	84.55	89.17
^{2}C*q*SL-90%	91.45 ± 5.86	83.31 ± 7.23	4.90 ± 4.90	95.13	96.73
^{2}C*q*SL-50%	90.22 ± 5.12	80.78 ± 8.34	5.54 ± 4.55	93.02	96.04
^{2}C*q*SL-30%	89.11 ± 5.89	81.14 ± 8.10	5.88 ± 5.11	91.14	92.89
^{2}C*q*SL-20%	87.02 ± 6.98	81.12 ± 8.77	6.74 ± 5.28	89.11	90.58
^{2}C*q*SL-10%	87.15 ± 6.93	79.02 ± 8.87	6.44 ± 4.87	88.53	92.47
^{2}C*q*SL-1.0%	80.80 ± 8.12	75.06 ± 10.04	8.01 ± 6.12	85.54	90.20
^{3}C*q*SL-90%	91.03 ± 5.57	82.44 ± 7.87	5.32 ± 4.77	95.31	95.55
^{3}C*q*SL-50%	89.79 ± 5.02	79.15 ± 8.04	5.12 ± 5.12	93.44	95.18
^{3}C*q*SL-30%	89.24 ± 6.15	81.02 ± 7.95	5.71 ± 5.18	92.26	91.11
^{3}C*q*SL-20%	88.19 ± 5.53	80.52 ± 8.12	6.80 ± 5.05	88.78	89.10
^{3}C*q*SL-10%	86.56 ± 6.15	79.55 ± 8.45	6.56 ± 6.54	87.98	92.01
^{3}C*q*SL-1.0%	79.58 ± 9.25	73.20 ± 10.87	8.64 ± 7.01	85.77	91.05
^{4}C*q*SL-90%	90.55 ± 5.88	80.19 ± 8.25	6.55 ± 6.12	93.12	95.55
^{4}C*q*SL-50%	89.10 ± 6.15	79.01 ± 8.77	5.54 ± 5.88	92.11	93.22
^{4}C*q*SL-30%	88.01 ± 6.43	79.89 ± 8.00	5.86 ± 6.43	91.54	91.02
^{4}C*q*SL-20%	87.78 ± 5.53	80.13 ± 7.72	6.91 ± 5.16	88.17	90.56
^{4}C*q*SL-10%	86.0 ± 6.39	80.10 ± 8.90	6.92 ± 5.12	85.67	93.34
^{4}C*q*SL-1.0%	78.13 ± 8.66	74.19 ± 11.20	9.56 ± 8.05	84.66	89.10

Table 3. Quantitative evaluation of LV-Myocardium segmentation results achieved using four semi-supervised variants of the proposed C*q*SL model in terms of mean Dice score (%) with std. dev., Jaccard index, Hausdorff distance (mm), precision (%) and recall (%) evaluated for varying proportions of labeled data on the ACDC dataset compared to segmentation across several frameworks.

	LV-Myocardium (LV-Myo)				
	Dice	Jaccard	HD	Prec.	Rec.
U-Net-90%	86.93 ± 5.56	84.50 ± 5.20	4.97 ± 3.76	92.32	96.54
U-Net-50%	85.82 ± 6.32	82.25 ± 7.66	5.16 ± 5.77	90.19	95.66
U-Net-30%	77.29 ± 9.19	75.49 ± 7.90	6.56 ± 5.65	87.11	89.56
U-Net-20%	76.56 ± 9.16	71.78 ± 16.20	7.69 ± 5.45	83.57	88.34
U-Net-10%	66.23 ± 15.90	60.63 ± 19.87	10.10 ± 8.55	59.34	62.08
U-Net-1.0%	29.47 ± 20.29	25.39 ± 22.50	13.95 ± 9.12	32.25	34.54
GAN-90%	84.50 ± 6.14	79.03 ± 9.17	5.89 ± 4.23	88.12	89.14
GAN-50%	81.21 ± 7.49	74.12 ± 11.77	5.45 ± 5.14	85.55	88.01
GAN-30%	78.67 ± 9.61	75.88 ± 12.75	5.19 ± 6.15	84.33	86.10
GAN-20%	77.88 ± 9.89	72.45 ± 15.91	6.01 ± 7.65	83.32	85.12
GAN-10%	75.23 ± 11.19	70.33 ± 17.19	7.87 ± 8.55	76.44	81.33
GAN-1.0%	66.02 ± 20.10	62.55 ± 20.87	12.67 ± 9.72	71.43	76.23
GAN+REC-90%	85.34 ± 6.42	77.44 ± 12.13	5.34 ± 4.37	88.44	90.33
GAN+REC-50%	82.33 ± 7.49	75.16 ± 13.16	5.81 ± 4.73	87.32	89.10
GAN+REC-30%	79.77 ± 9.21	74.10 ± 14.77	5.91 ± 5.12	86.76	88.34
GAN+REC-20%	78.43 ± 9.11	73.32 ± 15.11	6.12 ± 6.14	84.12	87.43
GAN+REC-10%	76.18 ± 11.18	72.21 ± 15.80	7.23 ± 7.34	79.43	83.53
GAN+REC-1.0%	67.52 ± 18.12	64.22 ± 19.33	12.12 ± 9.34	72.43	78.44
C*q*SL-90%	89.33 ± 5.11	82.03 ± 7.33	5.20 ± 5.11	93.98	96.01
C*q*SL-50%	87.77 ± 6.19	79.12 ± 9.0	5.88 ± 5.43	93.33	93.17
C*q*SL-30%	85.89 ± 7.07	77.72 ± 11.92	6.23 ± 6.14	91.20	92.25
C*q*SL-20%	85.55 ± 7.22	76.95 ± 12.9	6.85 ± 7.04	90.01	91.09
C*q*SL-10%	84.14 ± 7.64	72.76 ± 13.01	7.07 ± 8.01	88.84	90.88
C*q*SL-1.0%	77.65 ± 9.26	74.20 ± 11.87	10.88 ± 8.45	83.22	88.10
^{1}C*q*SL-90%	88.98 ± 6.01	81.78 ± 7.63	6.11 ± 6.10	94.13	95.33
^{1}C*q*SL-50%	86.55 ± 6.22	78.31 ± 9.46	5.74 ± 5.34	93.41	94.11
^{1}C*q*SL-30%	86.23 ± 7.62	77.43 ± 11.89	6.43 ± 6.29	91.88	91.0
^{1}C*q*SL-20%	85.10 ± 6.98	76.09 ± 12.77	6.80 ± 6.25	88.87	91.09
^{1}C*q*SL-10%	84.56 ± 8.01	72.11 ± 13.54	8.13 ± 7.03	89.73	90.16
^{1}C*q*SL-1.0%	75.54 ± 9.89	73.01 ± 11.56	10.05 ± 8.43	80.89	85.44
^{2}C*q*SL-90%	88.44 ± 6.43	81.03 ± 7.89	6.65 ± 5.24	92.0	95.32
^{2}C*q*SL-50%	86.01 ± 6.69	79.28 ± 10.02	5.65 ± 5.27	93.19	92.66
^{2}C*q*SL-30%	84.93 ± 8.01	78.52 ± 11.61	6.88 ± 5.86	90.42	93.53
^{2}C*q*SL-20%	85.33 ± 5.73	77.11 ± 11.59	6.32 ± 7.32	89.82	92.38
^{2}C*q*SL-10%	83.02 ± 8.33	71.67 ± 14.04	8.71 ± 8.10	87.77	91.45
^{2}C*q*SL-1.0%	75.0 ± 10.10	72.55 ± 11.18	10.20 ± 8.88	81.01	86.56
^{3}C*q*SL-90%	87.33 ± 7.22	80.73 ± 8.10	6.43 ± 5.50	92.31	94.52
^{3}C*q*SL-50%	86.43 ± 6.32	78.56 ± 10.22	5.76 ± 5.40	91.34	92.11
^{3}C*q*SL-30%	83.10 ± 8.66	78.15 ± 10.78	5.92 ± 6.11	88.82	91.63
^{3}C*q*SL-20%	83.00 ± 6.02	75.44 ± 13.10	6.65 ± 7.63	90.31	92.11
^{3}C*q*SL-10%	82.88 ± 9.01	72.00 ± 14.66	7.98 ± 8.34	86.11	90.87
^{3}C*q*SL-1.0%	73.19 ± 11.56	70.04 ± 12.93	10.78 ± 8.54	77.50	83.39
^{4}C*q*SL-90%	87.44 ± 7.71	81.24 ± 7.45	6.12 ± 5.11	91.32	92.65
^{4}C*q*SL-50%	86.01 ± 6.81	76.12 ± 10.64	6.01 ± 6.12	89.32	91.88
^{4}C*q*SL-30%	81.98 ± 10.01	76.65 ± 11.44	5.32 ± 5.44	87.11	92.33
^{4}C*q*SL-20%	84.01 ± 7.44	75.15 ± 13.19	6.72 ± 6.41	88.43	91.66
^{4}C*q*SL-10%	81.97 ± 10.66	73.43 ± 13.78	6.69 ± 6.87	84.77	86.32
^{4}C*q*SL-1.0%	71.21 ± 11.76	69.25 ± 13.16	11.82 ± 9.23	75.40	82.56

2.4. Evaluation Metrics

To evaluate the performance of the semantic segmentation of cardiac structures, we use the standard metrics, including Dice score, Jaccard index, Hausdorff distance (HD), precision (Prec), and recall (Rec).

1. **Dice and Jaccard Coefficients:** The Dice score is used to measure the percentage of overlap between manually segmented boundaries and automatically segmented boundaries of the structures of interest. Given the set of all pixels in the image, set of foreground pixels by automated segmentation S_1^a, and the set of pixels for ground truth S_1^g, the Dice score can be compared with $[S_1^a, S_1^g] \subseteq \Omega$, when a vector of ground truth labels T_1 and a vector of predicted labels P_1 as

$$Dice(T_1, P_1) = \frac{2|T_1 \cap P_1|}{|T_1| + |P_1|} \tag{12}$$

The Dice score will measure the similarity between two sets, T_1 and P_1, and $|T_1|$ denotes the cardinality of the set T_1 with the range of D(T_1,P_1) ϵ [0, 1].
The Jaccard index or Jaccard similarity coefficient is another metric which aids in the evaluation of the overlap in two sets of data. This index is similar to the Dice coefficient but mathematically different and typically used for different applications. For the same set of pixels in the image, Jaccard index can be written by the following expression:

$$Jaccard(T_1, P_1) = \frac{|T_1 \cap P_1|}{|T_1 + P_1|} \tag{13}$$

2. **Precision and Recall**
Precision and recall are two other metrics used to measure the segmentation quality which are sensitive to under- and over-segmentation. High values of both precision and recall indicate that the boundaries in both segmentation agree in location and level of detail. Precision and recall can be written as

$$Precision = \frac{TP}{TP + FP} \tag{14}$$

$$Recall = \frac{TP}{TP + FN} \tag{15}$$

where TP denotes true positive rate when a prediction-target mask pair has a score which exceeds some predefined threshold value; FP denotes the false positive rate when a predicted mask has no associated ground truth mask; and FN denotes the false negative rate when a ground truth mask has no associated predicted mask.
3. **Hausdorff distance (HD)**: Hausdorff distance (HD) measures the maximum distance between the two surfaces. Let S_A and S_B be surfaces corresponding to two binary segmentation masks, A and B, respectively. The Hausdorff distance (HD) is defined as

$$HD = \max\left(\max_{p \epsilon S_A} d(p, S_B), \max_{q \epsilon S_B} d(q, S_A)\right) \tag{16}$$

where $d(p, S) = \min q \epsilon S d(p, q)$ is the minimum Euclidean distance of point p from the points $q \; \epsilon \; S$.
4. **Image Quality Metrics:**
PSNR: The peak signal-to-noise ratio (PSNR) is the most commonly used quality assessment technique for determining the quality of lossy image compression codec reconstruction. The signal is the original data, and the noise is the error caused by the distortion.
5. **Clinical Indices:** To assess the performance of the ventricles, different indices have been used in the literature [54], such as left ventricular volume (LVV), left ventricular myocardial mass (LVM), stroke volume (SV), and ejection fraction (EF). The left ventricular volume (LVV) is defined as the volume enclosed by the LV blood pool and the myocardial mass is equal to the volume of the myocardium, multiplied by the density of the myocardium:

$$\text{Myo-Mass} = \text{Myo-Volume (cm}^3) \times 1.06 \text{ (gram/cm}^3) \tag{17}$$

Stroke volume (SV) is defined as the volume ejected during systole and is equal to the difference between the end-diastolic volume (EDV) and the end-systolic volume (ESV):

$$SV = EDV - ESV \times 100\% \tag{18}$$

The ejection fraction (EF) is an important cardiac parameter quantifying the cardiac output and defined as the ratio of the SV to the EDV:

$$EF = \frac{SV}{EDV} \times 100\% \tag{19}$$

3. Results

3.1. Image Segmentation Assessment

We tested our C*q*SL model on varying proportions of labeled and unlabeled data available through the STACOM 2017 ACDC cine cardiac MRI dataset. Training and validation segmentation accuracies for three different classes (RV, LV, and LV-Myo) are shown in Figure 8 for 100 epochs. Note that the validation curves show similar trends as the training curves (Figure 8).

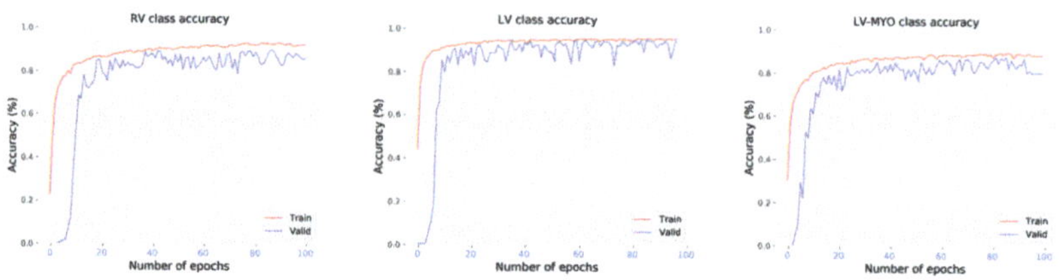

Figure 8. Representative accuracy curves showing the training and validation accuracy of three different classes (RV blood-pool, LV blood-pool, and LV-Myocardium).

The C*q*SL experimental results were compared against a fully supervised U-Net model trained from scratch, as reported in Tables 1–3. Furthermore, to explore the effectiveness of each component in our model, we propose three different semi-supervised ablations, i.e., model **I:** only a GAN architecture (Figure 3c); model **II: I** + reconstruction (Figure 3c,d); model **III: II** + disentangler block (Figure 3a–d), which are also reported in Tables 1–3. The detailed comparison of our model can be seen in Table 4. The segmentation performance is evaluated both qualitatively and quantitatively. As shown in Tables 1–3, our proposed model significantly improves the segmentation performance of right ventricle (RV), left ventricle blood-pool (LV), and LV-Myocardium, respectively on varying proportions of annotated data in terms of the Dice and Jaccard indices, Hausdorff distance, precision and recall rates. Our C*q*SL model achieves a high dice score (±std. dev.) of 75.50 ± 10.9% for the RV, 83.21 ± 7.1% for the LV blood-pool and 77.65 ± 9.3% for the LV-Myocardium even if we use only 1% labeled data.

Table 4. Our proposed C*q*SL model achieves 84.9% accuracy, significantly outperforming other baselines. We incrementally add each component, aiming to study their effectiveness on the final results; (model **I:** only a GAN architecture (Figure 3c); model **II:** GAN + reconstruction (Figure 3c,d); model **III:** GAN + reconstruction + disentangled block (Figure 3a–d). ↑ denotes higher the value better the result; ↓ denotes lower the value better the result.

	Average				
Models	Dice ↑	Jaccard ↑	HD ↓	Prec. ↑	Rec. ↑
Model I: GAN	76.56 ± 9.97	71.74 ± 14.54	8.26 ± 7.37	82.87 ± 7.66	85.78 ± 6.34
Model II: GAN + REC	77.82 ± 9.87	73.10 ± 13.92	8.11 ± 6.74	83.84 ± 7.12	87.06 ± 5.65
Model III: GAN + REC + DISENTANGLE (C*q*SL)	**84.92 ± 6.55**	**77.85 ± 11.06**	**7.20 ± 6.06**	**87.76 ± 5.45**	**89.56** ± 5.04

Figure 9 illustrates a qualitative segmentation output that compared C*q*SL and two others semi-supervised models, i.e., model **I:** only a GAN architecture (Figure 3c); model **II: I** + reconstruction (Figure 3c,d). For simplicity, this comparison is based on 20% unlabeled training data. As demonstrated, when only 20% of the training annotation is employed, U-Net fails completely to segment the cardiac structures from base to apex, particularly

RV segmentation. As shown in the figure, the segmentation results improve with each consecutive addition of a distinct block. The GAN-only architecture performs badly, particularly during RV segmentation, whereas the addition of a reconstruction block improves performance. Finally, adding a disentangled block to the GAN and reconstruction block yielded the greatest results. Even the least performing version of our proposed CqSL model (^{4}C*q*SL) achieves an overall accuracy superior to the U-Net, GAN-only, as well as GAN+REC model, confirming that the proposed model is able to effectively learn correct features that ensure correct segmentation.

Figure 10 illustrates a qualitative segmentation output that compared C*q*SL and U-Net results with increasing proportion of unlabeled training data. For simplicity, we have shown two of our best performing models. As shown, when only 1% training annotation is used, U-Net completely fails to segment the cardiac structures. Under similar conditions, our model is still able to yield a high segmentation accuracy of LV, RV, and LV-Myocardium. When the amount of labeled data increases from 1% to 10%, the U-Net model still performs poorly, especially for RV segmentation. On the other hand, although the performance of our model improves significantly when utilizing more than 30% annotated data, its performance with even 1% labeled data is still satisfactory, comparable to that of semi-supervised models, and superior to U-Net's performance under similar conditions.

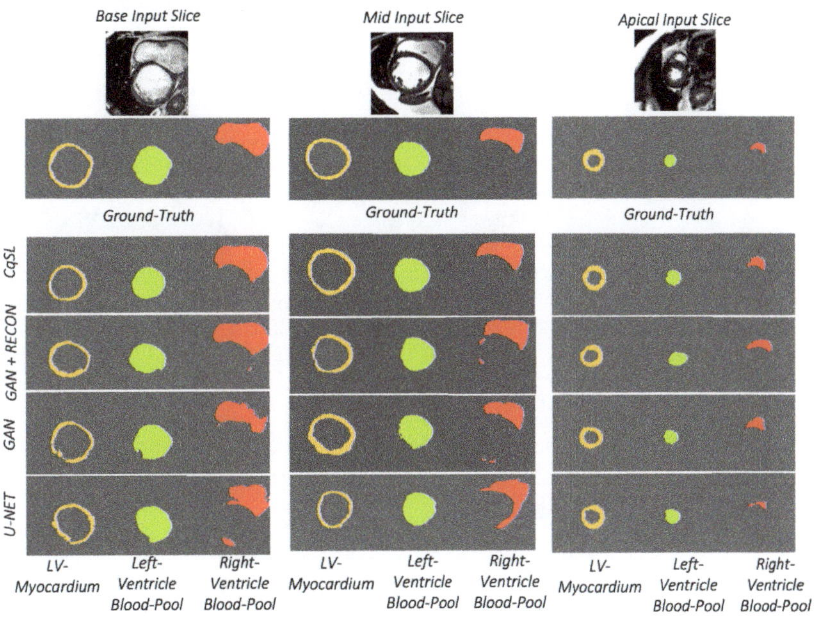

Figure 9. Representative results showing the comparison across several best performing networks, including C*q*SL for the semantic segmentation of full cardiac image dataset from the base to apex showing of RV blood-pool, LV blood-pool, and LV-Myocardium on 20% labeled data in red, green, and yellow respectively.

We assessed the performance of our proposed C*q*SL cardiac image segmentation method against the segmentation results yielded by the well-established, fully supervised U-Net architecture [55] in light of its effectiveness across various medical image segmentation applications, as well as its extensive use as a baseline method for comparison by the participants of the ACDC cardiac image segmentation challenge. Furthermore, to explore the effectiveness of each component in our model, we experiment on three different semi-supervised ablations, i.e., model **I:** only a GAN architecture; model **II:** GAN + reconstruction; and model **III:** GAN + reconstruction + disentangler block (C*q*SL).

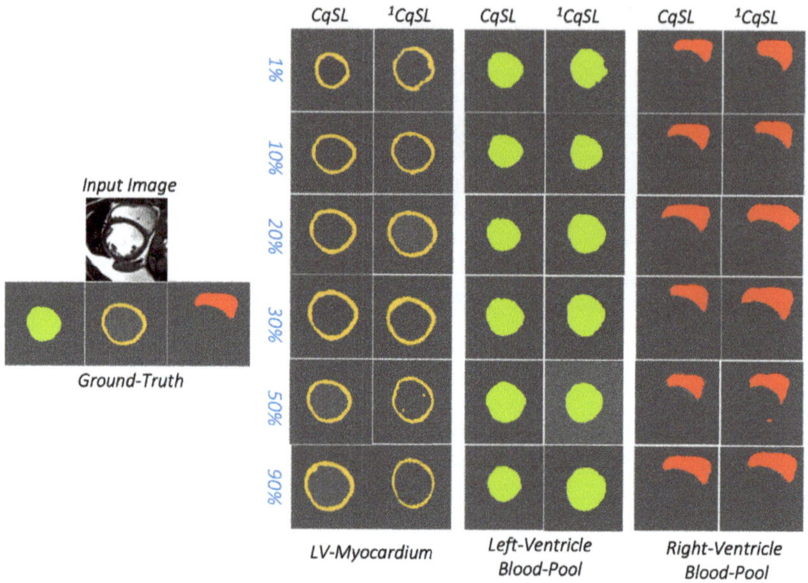

Figure 10. Representative results showing the semantic segmentation of RV, LV blood-pool, and LV-Myocardium on different proportion of labeled data in red, green, and yellow, respectively.

As shown in Figure 11, the accuracy of our C*q*SL models remains high when using as much as 50–90% unlabeled data, which essentially implies excellent performance with as little as as 10% annotated data. Nevertheless, both U-Net and C*q*SL models perform similar to each other when the amount of annotated data increases above 90%. We plot the mean accuracy for all the models in Figure 12 and confirm that under low amounts of annotated data conditions, even as low as 1%, our proposed C*q*SL model and all four of its semi-supervised variants (^{1}C*q*SL, ^{2}C*q*SL, ^{3}C*q*SL, and ^{4}C*q*SL) outperform GAN, GAN+REC, as well as U-Net models for LV, RV, and LV-Myocardium. The typical segmentation contours of complete cardiac image dataset for the mid and apical slices are shown in Figure 13.

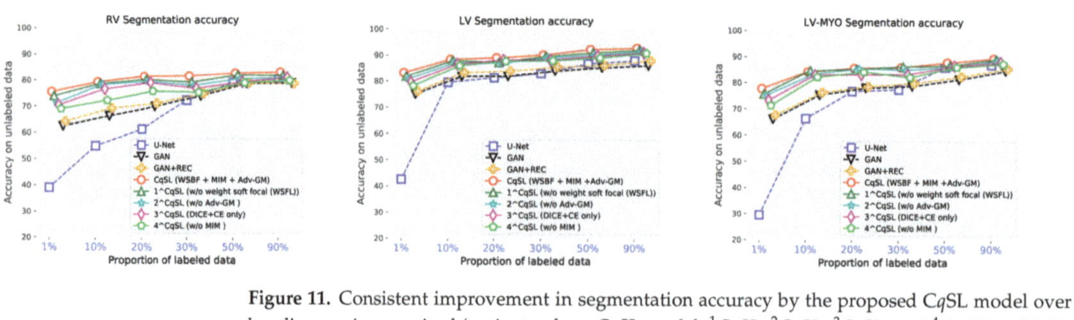

Figure 11. Consistent improvement in segmentation accuracy by the proposed C*q*SL model over baseline semi-supervised (variants of our C*q*SL model: ^{1}C*q*SL, ^{2}C*q*SL, ^{3}C*q*SL, and ^{4}C*q*SL) and fully supervised models in varying proportions of labeled training data.

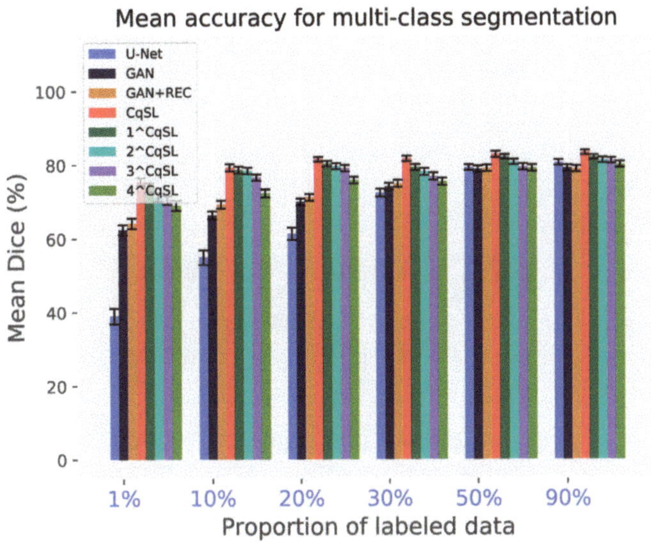

Figure 12. Evaluation on the robustness of CqSL in terms of mean accuracy over RV, LV, and LV-Myocardium segmentation tasks on varying amounts of labeled training samples. Note significant improvement in Dice score across all CqSL semi-supervised variants for as little as 1% unlabeled data.

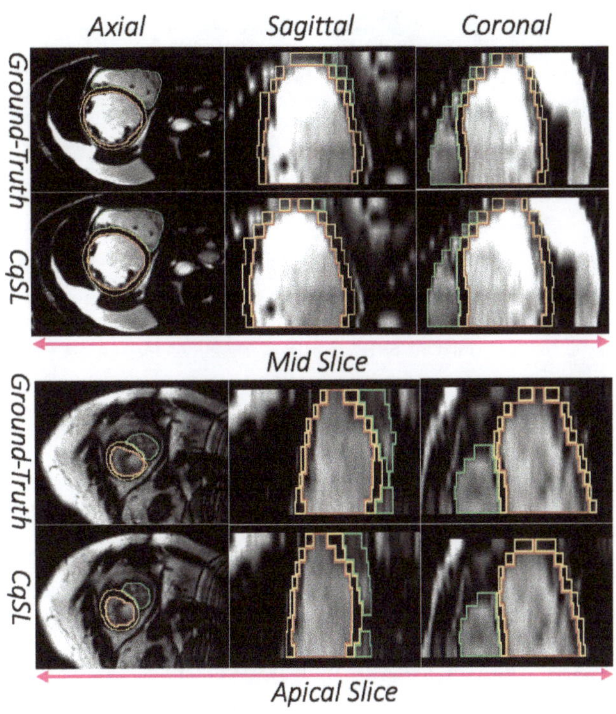

Figure 13. Representative segmentation contours of a complete cardiac cycle for the middle and apex slices showing RV and LV blood-pool, and LV-Myocardium in green, yellow, and brown, respectively, in three different view settings (axial, sagittal, and coronal).

3.2. Image Quality Assessment

Figure 14 illustrates a qualitative comparison between the original image slice and the reconstructed slices generated from our proposed approach on the ACDC dataset at the original 5 mm slice thickness. The comparison is augmented by the computed correlation coefficients (CC) and peak signal-to-noise ratio (PSNR) shown below each figure. As illustrated in Figure 14, our approach preserves the fine structural details and realistic textures while remaining visually comparable to the ground truth image. Aside from qualitative improvements, the proposed method's CC and PSNR values also prove that the synthesized image slices preserve the fine structural details.

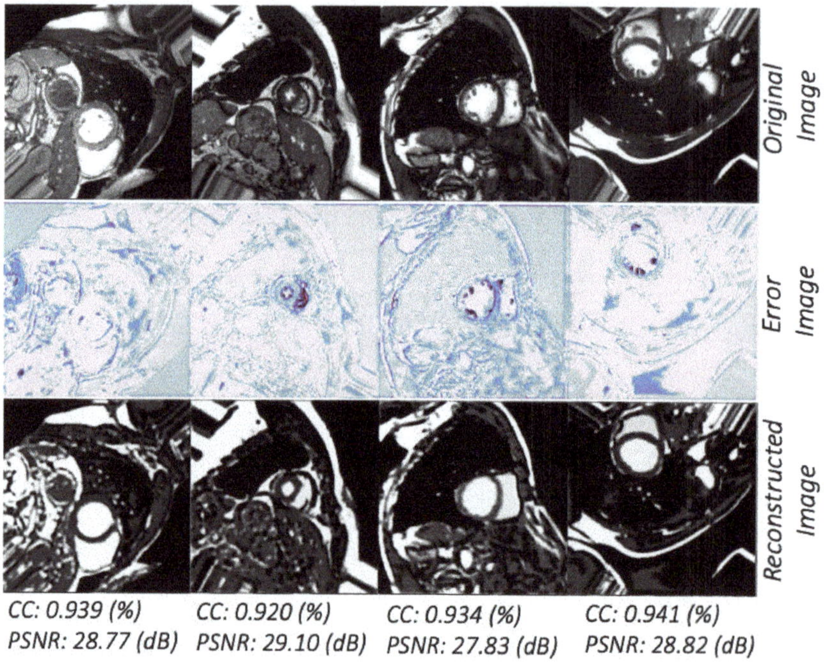

Figure 14. Qualitative comparison of the original and the reconstructed slices showing that the original images are well reconstructed by combining skeleton and sentiency information.The comparison is augmented by the computed correlation coefficients (CC) and peak signal-to-noise ratio (PSNR). The middle row illustrates the error images.

Table 5 shows the quantitative results of the objective quality metrics of reconstruction, indicating that the use of feature-wise linear modulation to remove domain-invariant information from the disentangled latent code guides the synthesis of more texture information. Starting with the spatial factor, we change the content of the spatial channels in Figure 15 to see how the decoder has learned a correlation between the position of each channel and different signal intensities of the skeleton parts. The sentiency factor remains constant in all of these experiments. The first two columns show the original input and the reconstruction. The third row is created by the RV spatial channels and disregarding (zeroing) the MYO and LV channel. In the fourth image, we swap the RV channels with those of LV. Finally, the fifth column is produced by considering all LV, MYO and RV channels.

Table 5. Image reconstruction assessment: correlation coefficient (CC) and PSNR comparison between reconstructed and input images based on 288 test sets.

	Reconstruction Quality	
	CC (%) n = 288	PSNR (dB) n = 288
Model II: GAN + REC	0.912	27.32
Model III: GAN + REC + DISENTANGLE (Proposed)	0.934	28.89

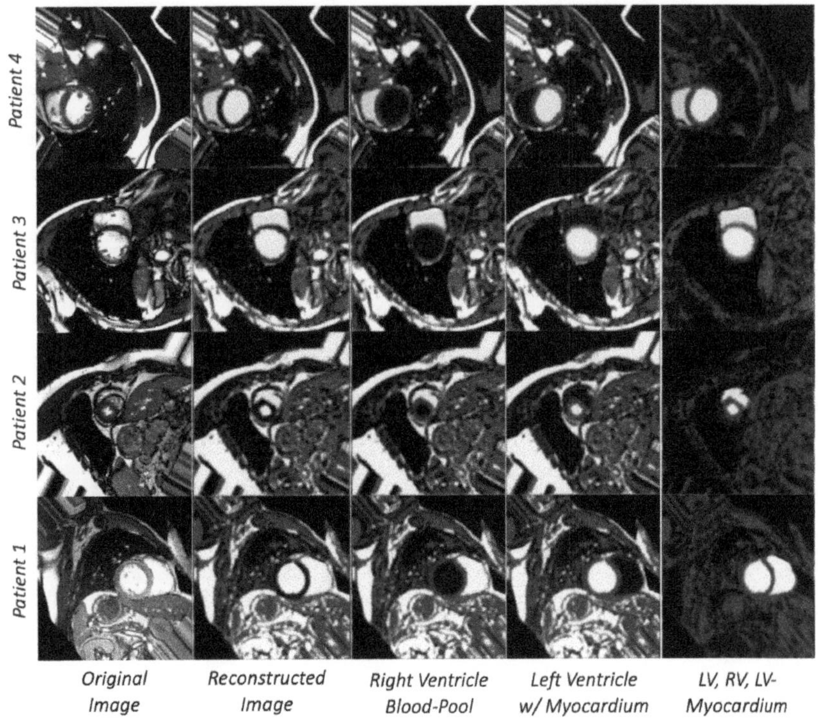

Figure 15. Reconstructions of a sample of input images when rearranging the spatial representation's channels. Rearranging the channels results in reconstructing only left ventricle blood-pool or only right ventricle blood-pool only or all the ventricular structures.

3.3. Clinical Parameter Estimation

The performance of our developed segmentation method was also reflected in the computed clinical indices. These clinical indices are computed using the Simpsons method and the agreement between the ground truth and the same parameters computed using the automated segmentation results is reported using correlation statistical analysis by mapping the predicted volumes of the testing set onto the ground truth volumes of the training set. As illustrated in Table 6 the agreement between our method's prediction and ground truth is high, characterized by a Pearson's correlation coefficient (rho) of 0.898 ($p < 0.01$) for LV-EF, 0.723 for RV-EF ($p < 0.1$) and 0.924 ($p < 0.01$) for Myo-mass. There was a slight over-estimation in the RV blood-pool segmentation also reflected in the clinical parameters estimation.

Table 6. The correlation between the CqSL-predicted and ground truth clinical indices is significantly higher than the correlation between the U-Net-predicted and same ground truth clinical indices ($\star\star$ ($p < 0.01$), $\star$ ($p < 0.1$)).

	Clinical Indices of Healthy Volunteers	
	UNet	CqSL
LV EF	0.487	0.898 $\star\star$
RV EF	0.371	0.723 $\star$
Myo mass	0.427	0.924 $\star\star$

Figure 16 shows a graphical comparison between the clinical parameters estimated from the cardiac features segmented via CqSL and the same homologous parameters estimated from the ground truth manual segmentations for both healthy volunteers and patients featuring various cardiac conditions. As shown, the clinical parameters estimated using our automatically segmented features show no statistically significant difference from those estimated based on the ground truth, manually segmented features.

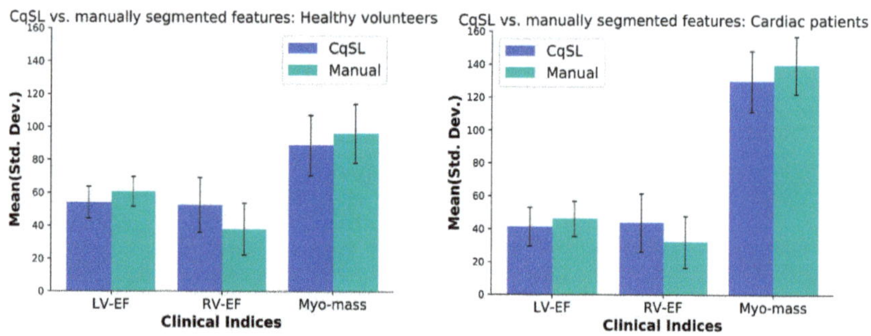

Figure 16. Graphical comparison showing no statistically significant differences between clinical parameters estimated using CqSL segmentation and same parameters estimated using the ground truth segmentation in terms of Mean (Std. Dev.) EF (mL/mL (%)) = ejection fraction, Myo-mass (in gm) = myocardial mass.

3.4. Ablation Studies

We perform an ablation study to investigate the effect of using different loss functions in our semi-supervised setting. We demonstrate the effect of different novel loss functions used in CqSL model: WSBF, MIM, and Adv-GM by assessing the model performance when each novel loss functions is removed. Figure 17 shows a graphical representation of the results achieved on the ACDC dataset. In Figure 10, we illustrate the qualitative results on the ACDC dataset to visualize the effect of using all of the loss components. We can observe that the best results are achieved when all of the loss components are used. Specifically, without MIM, the loss curve oscillates, while without WSBF, the output images deviate drastically from the ground truth. Both the quantitative and qualitative results show that the design of CqSL improves the preservation of the subject identity and enables more accurate segmentation of cardiac structures.

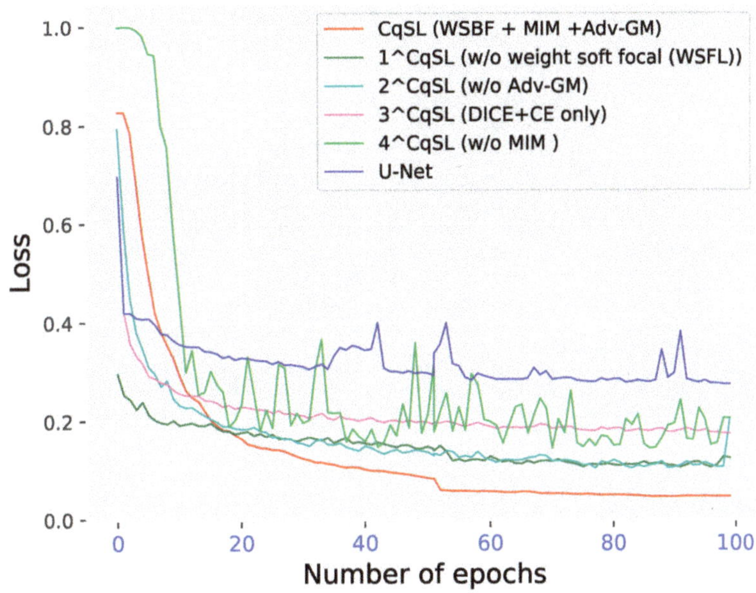

Figure 17. Empirical analysis showing the effect of different loss functions on the 2017 STACOM ACDC dataset. The significant reduction of total loss in C*q*SL (in red) suggests the best performing model with best learned features.

4. Conclusions and Future Work

In this paper, we propose a semi-supervised learning model (C*q*SL) that features multiple novel loss functions, including mutual information minimization (MIM), which minimizes the mutual information between the domain-invariant as well as domain-specific features. Empirically, we show that disentanglement with mutual information can improve the performance of the segmentation accuracy, while combined with an adversarial and a reconstruction block. Our novel use of total loss function enforces the network to capture both the spatial and intensity information. Our weighted soft focal loss can minimize the class imbalance problem by applying varying weights over different classes along with a modulating term. We apply the proposed model to cardiac image segmentation tasks with varying proportion of labeled data.

Our proposed C*q*SL model achieves 85% accuracy, significantly outperforming other baselines. We incrementally add each component, aiming to study their effectiveness on the final results: (model **I:** only a GAN architecture (Figure 3c); model **II:** GAN + reconstruction (Figure 3c,d); model **III:** GAN + reconstruction + disentangled block (Figure 3a–d).

In light of consistency, all four implemented C*q*SL variants are evaluated and compared to the baselines, but as shown in Tables 1–3, the first variant (^{1}C*q*S) performs best and hence it is deemed as the most suitable and recommended C*q*SL framework.

The experimental results reported in this manuscript show that the proposed *CqSL* framework outperforms semi-supervised learning with GANs [56] as well as fully supervised-type models when using as little as even 1% labeled data and display similar performance and comparable accuracy when employing more than 50% labeled data. Unlike these, we use adversarial-Geman–McClure (adv-GM) loss to force mask generation to be spatially aligned with the image. Furthermore, we discover that the semi-supervised segmentation approach of Hung et al. [18] obtains results slightly inferior to ours. Hung et al. reported that their adversarial model achieved a 80.63% accuracy when trained on 20% labeled data using the ACDC dataset, whereas our model achieved a 81.44% accuracy under similar training conditions.

Hence, the proposed method is the first to achieve significant performance for 4D cine cardiac MRI image segmentation with very minimal annotated data, specifically 1% of the training dataset. This is a key feature of the proposed work and hence a significant contribution to the medical (cardiac, in particular) image segmentation, as access to large amounts of expert-annotated ground truth imaging data is expensive in the medical field. Nevertheless, here we demonstrate that *CqSL* can still yield segmentation accuracy superior to other semi-supervised methods while requiring minimal annotated data for training.

Author Contributions: Conceptualization, S.M.K.H. and C.A.L.; methodology, S.M.K.H.; software, S.M.K.H.; validation, S.M.K.H.; formal analysis, S.M.K.H. and C.A.L.; resources, S.M.K.H. and C.A.L.; writing—original draft preparation, S.M.K.H.; writing—review and editing, S.M.K.H. and C.A.L.; visualization, S.M.K.H.; supervision, C.A.L.; funding acquisition, C.A.L. All authors have read and agreed to the published version of the manuscript.

Funding: Research reported in this publication was supported by the National Institute of General Medical Sciences Award No. R35GM128877 of the National Institutes of Health, and the Office of Advanced Cyber infrastructure Award No. 1808530 of the National Science Foundation.

Conflicts of Interest: The authors declare no conflict of interest.

References

1. Bhowmik, A.; Gumhold, S.; Rother, C.; Brachmann, E. Reinforced feature points: Optimizing feature detection and description for a high-level task. In Proceedings of the IEEE/CVF Conference on Computer Vision and Pattern Recognition, Seattle, WA, USA, 13–19 June 2020; pp. 4948–4957.
2. Li, S.; Wang, Z.; Liu, Z.; Tan, C.; Lin, H.; Wu, D.; Chen, Z.; Zheng, J.; Li, S.Z. Efficient Multi-order Gated Aggregation Network. *arXiv* **2022**, arXiv:2211.03295.
3. Ruan, J.; Xiang, S.; Xie, M.; Liu, T.; Fu, Y. MALUNet: A Multi-Attention and Light-weight UNet for Skin Lesion Segmentation. *arXiv* **2022**, arXiv:2211.01784.
4. Tack, J.; Yu, S.; Jeong, J.; Kim, M.; Hwang, S.J.; Shin, J. Consistency regularization for adversarial robustness. In Proceedings of the AAAI Conference on Artificial Intelligence, Virtually, 22 February–1 March 2022; Volume 36, pp. 8414–8422.
5. Sajjadi, M.; Javanmardi, M.; Tasdizen, T. Regularization with stochastic transformations and perturbations for deep semi-supervised learning. In Proceedings of the Advances in Neural Information Processing Systems, Barcelona, Spain, 5–10 December 2016; pp. 1163–1171.
6. Elakkiya, R.; Subramaniyaswamy, V.; Vijayakumar, V.; Mahanti, A. Cervical cancer diagnostics healthcare system using hybrid object detection adversarial networks. *IEEE J. Biomed. Health Inform.* **2021**, *26*, 1464–1471. [CrossRef] [PubMed]
7. Hasan, S.M.K.; Linte, C. STAMP: A Self-training Student-Teacher Augmentation-Driven Meta Pseudo-Labeling Framework for 3D Cardiac MRI Image Segmentation. In Proceedings of the Annual Conference on Medical Image Understanding and Analysis, Cambridge, UK, 27–29 July 2022; Springer: Berlin/Heidelberg, Germany, 2022; pp. 371–386.
8. Sohn, K.; Berthelot, D.; Li, C.L.; Zhang, Z.; Carlini, N.; Cubuk, E.D.; Kurakin, A.; Zhang, H.; Raffel, C. Fixmatch: Simplifying semi-supervised learning with consistency and confidence. *arXiv* **2020**, arXiv:2001.07685.
9. Xie, Q.; Luong, M.T.; Hovy, E.; Le, Q.V. Self-training with noisy student improves imagenet classification. In Proceedings of the IEEE/CVF Conference on Computer Vision and Pattern Recognition, Seattle, WA, USA, 13–19 June 2020; pp. 10687–10698.
10. Bai, W.; Oktay, O.; Sinclair, M.; Suzuki, H.; Rajchl, M.; Tarroni, G.; Glocker, B.; King, A.; Matthews, P.M.; Rueckert, D. Semi-supervised learning for network-based cardiac MR image segmentation. In Proceedings of the International Conference on Medical Image Computing and Computer-Assisted Intervention, Quebec City, QC, Canada, 10–14 September 2017; Springer: Berlin/Heidelberg, Germany, 2017; pp. 253–260.
11. Donahue, J.; Krähenbühl, P.; Darrell, T. Adversarial feature learning. *arXiv* **2016**, arXiv:1605.09782.
12. Saito, K.; Kim, D.; Sclaroff, S.; Darrell, T.; Saenko, K. Semi-supervised domain adaptation via minimax entropy. In Proceedings of the IEEE International Conference on Computer Vision, Seoul, Republic of Korea, 27 October–2 November 2019; pp. 8050–8058.
13. Gomes, H.M.; Grzenda, M.; Mello, R.; Read, J.; Le Nguyen, M.H.; Bifet, A. A survey on semi-supervised learning for delayed partially labelled data streams. *ACM Comput. Surv. (CSUR)* **2022**, *55*, 1–42. [CrossRef]
14. Hasan, S.M.K.; Linte, C.A. A Multi-Task Cross-Task Learning Architecture for Ad Hoc Uncertainty Estimation in 3D Cardiac MRI Image Segmentation. In Proceedings of the 2021 Computing in Cardiology (CinC), Brno, Czech Republic, 13–15 September 2021; Volume 48, pp. 1–4.
15. Chan, E.R.; Lin, C.Z.; Chan, M.A.; Nagano, K.; Pan, B.; De Mello, S.; Gallo, O.; Guibas, L.J.; Tremblay, J.; Khamis, S.; et al. Efficient geometry-aware 3D generative adversarial networks. In Proceedings of the IEEE/CVF Conference on Computer Vision and Pattern Recognition, New Orleans, LA, USA, 19–24 June 2022; pp. 16123–16133.
16. Souly, N.; Spampinato, C.; Shah, M. Semi supervised semantic segmentation using generative adversarial network. In Proceedings of the IEEE International Conference on Computer Vision, Venice, Italy, 22–29 October 2017; pp. 5688–5696.

17. Chen, C.; Dou, Q.; Chen, H.; Heng, P.A. Semantic-aware generative adversarial nets for unsupervised domain adaptation in chest X-ray segmentation. In Proceedings of the International Workshop on Machine Learning in Medical Imaging, Granada, Spain, 16 September 2018; Springer: Berlin/Heidelberg, Germany, 2018; pp. 143–151.
18. Hung, W.C.; Tsai, Y.H.; Liou, Y.T.; Lin, Y.Y.; Yang, M.H. Adversarial learning for semi-supervised semantic segmentation. *arXiv* **2018**, arXiv:1802.07934.
19. Zhang, Y.; Yang, L.; Chen, J.; Fredericksen, M.; Hughes, D.P.; Chen, D.Z. Deep adversarial networks for biomedical image segmentation utilizing unannotated images. In Proceedings of the International Conference on Medical Image Computing and Computer-Assisted Intervention, Quebec City, QC, Canada, 10–14 September 2017; Springer: Berlin/Heidelberg, Germany, 2017; pp. 408–416.
20. Chartsias, A.; Joyce, T.; Dharmakumar, R.; Tsaftaris, S.A. Adversarial image synthesis for unpaired multi-modal cardiac data. In Proceedings of the International Workshop on Simulation and Synthesis in Medical Imaging, Quebec City, QC, Canada, 10 September 2017; Springer: Berlin/Heidelberg, Germany, 2017; pp. 3–13.
21. Hjelm, R.D.; Fedorov, A.; Lavoie-Marchildon, S.; Grewal, K.; Bachman, P.; Trischler, A.; Bengio, Y. Learning deep representations by mutual information estimation and maximization. *arXiv* **2018**, arXiv:1808.06670.
22. Wang, Y.C.; Wang, C.Y.; Lai, S.H. Disentangled Representation with Dual-stage Feature Learning for Face Anti-spoofing. In Proceedings of the IEEE/CVF Winter Conference on Applications of Computer Vision, Waikoloa, HI, USA, 3–8 January 2022; pp. 1955–1964.
23. Siddharth, N.; Paige, B.; Van de Meent, J.W.; Desmaison, A.; Goodman, N.; Kohli, P.; Wood, F.; Torr, P. Learning disentangled representations with semi-supervised deep generative models. In Proceedings of the Advances in Neural Information Processing Systems, Long Beach, CA, USA, 4–9 December 2017; pp. 5925–5935.
24. Higgins, I.; Matthey, L.; Pal, A.; Burgess, C.; Glorot, X.; Botvinick, M.; Mohamed, S.; Lerchner, A. beta-vae: Learning Basic Visual Concepts with a Constrained Variational Framework. 2016. Available online: https://openreview.net/forum?id=Sy2fzU9gl (accessed on 2 October 2022).
25. Bengio, Y.; Courville, A.; Vincent, P. Representation learning: A review and new perspectives. *IEEE Trans. Pattern Anal. Mach. Intell.* **2013**, *35*, 1798–1828. [CrossRef]
26. Lipton, Z.C. The mythos of model interpretability. *Queue* **2018**, *16*, 31–57. [CrossRef]
27. Schölkopf, B.; Janzing, D.; Peters, J.; Sgouritsa, E.; Zhang, K.; Mooij, J. On causal and anticausal learning. *arXiv* **2012**, arXiv:1206.6471.
28. Isola, P.; Zhu, J.Y.; Zhou, T.; Efros, A.A. Image-to-image translation with conditional adversarial networks. In Proceedings of the IEEE Conference on Computer Vision and Pattern Recognition, Honolulu, HI, USA, 21–26 July 2017; pp. 1125–1134.
29. Zhu, J.Y.; Park, T.; Isola, P.; Efros, A.A. Unpaired image-to-image translation using cycle-consistent adversarial networks. In Proceedings of the IEEE International Conference on Computer Vision, Venice, Italy, 22–29 October 2017; pp. 2223–2232.
30. Huang, X.; Liu, M.Y.; Belongie, S.; Kautz, J. Multimodal unsupervised image-to-image translation. In Proceedings of the European Conference on Computer Vision (ECCV), Munich, Germany, 8–14 September 2018; pp. 172–189.
31. Shen, K.; Jones, R.M.; Kumar, A.; Xie, S.M.; HaoChen, J.Z.; Ma, T.; Liang, P. Connect, not collapse: Explaining contrastive learning for unsupervised domain adaptation. In Proceedings of the International Conference on Machine Learning, PMLR, Baltimore, MD, USA, 17–23 July 2022; pp. 19847–19878.
32. Gatys, L.A.; Ecker, A.S.; Bethge, M. Image style transfer using convolutional neural networks. In Proceedings of the IEEE Conference on Computer Vision and Pattern Recognition, Las Vegas, NV, USA, 37–30 June 2016; pp. 2414–2423.
33. Liu, A.H.; Liu, Y.C.; Yeh, Y.Y.; Wang, Y.C.F. A unified feature disentangler for multi-domain image translation and manipulation. In Proceedings of the Advances in Neural Information Processing Systems, Montreal, QC, Canada, 3–8 December 2018; pp. 2590–2599.
34. Tzeng, E.; Hoffman, J.; Saenko, K.; Darrell, T. Adversarial discriminative domain adaptation. In Proceedings of the IEEE Conference on Computer Vision and Pattern Recognition, Honolulu, HI, USA, 21–26 July 2017; pp. 7167–7176.
35. Ulyanov, D.; Vedaldi, A.; Lempitsky, V. Improved texture networks: Maximizing quality and diversity in feed-forward stylization and texture synthesis. In Proceedings of the IEEE Conference on Computer Vision and Pattern Recognition, Honolulu, HI, USA, 21–26 July 2017; pp. 6924–6932.
36. Dumoulin, V.; Shlens, J.; Kudlur, M. A learned representation for artistic style. *arXiv* **2016**, arXiv:1610.07629.
37. Huang, X.; Belongie, S. Arbitrary style transfer in real-time with adaptive instance normalization. In Proceedings of the IEEE International Conference on Computer Vision, Venice, Italy, 22–29 October 2017; pp. 1501–1510.
38. Perez, E.; Strub, F.; De Vries, H.; Dumoulin, V.; Courville, A. Film: Visual reasoning with a general conditioning layer. In Proceedings of the Thirty-Second AAAI Conference on Artificial Intelligence, New Orleans, LA, USA, 2–7 February 2018.
39. Park, T.; Liu, M.Y.; Wang, T.C.; Zhu, J.Y. Semantic image synthesis with spatially-adaptive normalization. In Proceedings of the IEEE Conference on Computer Vision and Pattern Recognition, Long Beach, CA, USA, 15–20 June 2019; pp. 2337–2346.
40. Marino, J. Predictive coding, variational autoencoders, and biological connections. *Neural Comput.* **2022**, *34*, 1–44. [CrossRef] [PubMed]
41. Kim, H.; Mnih, A. Disentangling by factorising. *arXiv* **2018**, arXiv:1802.05983.

42. Zhou, Z.; Rahman Siddiquee, M.M.; Tajbakhsh, N.; Liang, J. Unet++: A nested u-net architecture for medical image segmentation. In *Deep Learning in Medical Image Analysis and Multimodal Learning for Clinical Decision Support*; Springer: Berlin/Heidelberg, Germany, 2018; pp. 3–11.
43. Tian, R.; Mao, Y.; Zhang, R. Learning VAE-LDA models with rounded reparameterization trick. In Proceedings of the 2020 Conference on Empirical Methods in Natural Language Processing (EMNLP), Virtually, 16–20 November 2020; pp. 1315–1325.
44. Chen, X.; Duan, Y.; Houthooft, R.; Schulman, J.; Sutskever, I.; Abbeel, P. Infogan: Interpretable representation learning by information maximizing generative adversarial nets. In Proceedings of the Advances in Neural Information Processing Systems, Barcelona, Spain, 5–10 December 2016; pp. 2172–2180.
45. Peng, X.; Huang, Z.; Sun, X.; Saenko, K. Domain agnostic learning with disentangled representations. *arXiv* **2019**, arXiv:1904.12347.
46. Mao, X.; Li, Q.; Xie, H.; Lau, R.Y.; Wang, Z.; Paul Smolley, S. Least squares generative adversarial networks. In Proceedings of the IEEE International Conference on Computer Vision, Venice, Italy, 22–29 October 2017; pp. 2794–2802.
47. Ganan, S.; McClure, D. *Bayesian Image Analysis: An Application to Single Photon Emission Tomography*; American Statistical Association: Washington, DC, USA, 1985; pp. 12–18.
48. Bernard, O.; Lalande, A.; Zotti, C.; Cervenansky, F.; Yang, X.; Heng, P.A.; Cetin, I.; Lekadir, K.; Camara, O.; Ballester, M.A.G.; et al. Deep learning techniques for automatic MRI cardiac multi-structures segmentation and diagnosis: Is the problem solved? *IEEE Trans. Med. Imaging* **2018**, *37*, 2514–2525. [CrossRef] [PubMed]
49. Chartsias, A.; Joyce, T.; Papanastasiou, G.; Semple, S.; Williams, M.; Newby, D.E.; Dharmakumar, R.; Tsaftaris, S.A. Disentangled representation learning in cardiac image analysis. *Med. Image Anal.* **2019**, *58*, 101535. [CrossRef]
50. Ioffe, S.; Szegedy, C. Batch normalization: Accelerating deep network training by reducing internal covariate shift. In Proceedings of the International Conference on Machine Learning, PMLR, Lille, France, 7–9 July 2015; pp. 448–456.
51. Maas, A.L.; Hannun, A.Y.; Ng, A.Y.; et al. Rectifier nonlinearities improve neural network acoustic models. In Proceedings of the International Conference on Machine Learning, Atlanta, GA, USA, 16–21 June 2013; Volume 30, p. 3.
52. Radford, A.; Metz, L.; Chintala, S. Unsupervised representation learning with deep convolutional generative adversarial networks. *arXiv* **2015**, arXiv:1511.06434.
53. Liu, H.; Brock, A.; Simonyan, K.; Le, Q.V. Evolving Normalization-Activation Layers. *arXiv* **2020**, arXiv:2004.02967.
54. Frangi, A.F.; Niessen, W.J.; Viergever, M.A. Three-dimensional modeling for functional analysis of cardiac images, a review. *IEEE Trans. Med. Imaging* **2001**, *20*, 2–5. [CrossRef]
55. Ronneberger, O.; Fischer, P.; Brox, T. U-net: Convolutional networks for biomedical image segmentation. In Proceedings of the International Conference on Medical Image Computing and Computer-Assisted Intervention, Munich, Germany, 5–9 October 2015; Springer: Berlin/Heidelberg, Germany, 2015; pp. 234–241.
56. Luc, P.; Couprie, C.; Chintala, S.; Verbeek, J. Semantic segmentation using adversarial networks. *arXiv* **2016**, arXiv:1611.08408.

applied sciences

MDPI

Article

Cardiac Magnetic Resonance Left Ventricle Segmentation and Function Evaluation Using a Trained Deep-Learning Model

Fumin Guo [1,2,3,*], Matthew Ng [2,3], Idan Roifman[4] and Graham Wright [2,3]

1 Wuhan National Laboratory for Optoelectronics, Huazhong University of Science and Technology, Wuhan 430074, China
2 Sunnybrook Research Institute, University of Toronto, Toronto, ON M4N 3M5, Canada; matthewng.ng@mail.utoronto.ca (M.N.); gawright@sri.utoronto.ca (G.W.)
3 Department of Medical Biophysics, University of Toronto, Toronto, ON M4N 3M5, Canada
4 Sunnybrook Health Sciences Center, University of Toronto, Toronto, ON M4N 3M5, Canada; idan.roifman@sunnybrook.ca
* Correspondence: fumin.guo@utoronto.ca; Tel.: +1-(416)-480-6789

Abstract: Cardiac MRI is the gold standard for evaluating left ventricular myocardial mass (LVMM), end-systolic volume (LVESV), end-diastolic volume (LVEDV), stroke volume (LVSV), and ejection fraction (LVEF). Deep convolutional neural networks (CNNs) can provide automatic segmentation of LV myocardium (LVF) and blood cavity (LVC) and quantification of LV function; however, the performance is typically degraded when applied to new datasets. A 2D U-net with Monte-Carlo dropout was trained on 45 cine MR images and the model was used to segment 10 subjects from the ACDC dataset. The initial segmentations were post-processed using a continuous kernel-cut method. The refined segmentations were employed to update the trained model. This procedure was iterated several times and the final updated U-net model was used to segment the remaining 90 ACDC subjects. Algorithm and manual segmentations were compared using Dice coefficient (DSC) and average surface distance in a symmetric manner (ASSD). The relationships between algorithm and manual LV indices were evaluated using Pearson correlation coefficient (r), Bland-Altman analyses, and paired t-tests. Direct application of the pre-trained model yielded DSC of 0.74 ± 0.12 for LVM and 0.87 ± 0.12 for LVC. After fine-tuning, DSC was 0.81 ± 0.09 for LVM and 0.90 ± 0.09 for LVC. Algorithm LV function measurements were strongly correlated with manual analyses (r = 0.86–0.99, $p < 0.0001$) with minimal biases of −8.8 g for LVMM, −0.9 mL for LVEDV, −0.2 mL for LVESV, −0.7 mL for LVSV, and −0.6% for LVEF. The procedure required ∼12 min for fine-tuning and approximately 1 s to contour a new image on a Linux (Ubuntu 14.02) desktop (Inter(R) CPU i7-7770, 4.2 GHz, 16 GB RAM) with a GPU (GeForce, GTX TITAN X, 12 GB Memory). This approach provides a way to incorporate a trained CNN to segment and quantify previously unseen cardiac MR datasets without needing manual annotation of the unseen datasets.

Keywords: cardiac MRI; machine learning; left ventricle segmentation; cardiac function

Citation: Guo, F.; Ng, M.; Roifman, I.; Wright, G. Cardiac Magnetic Resonance Left Ventricle Segmentation and Function Evaluation Using a Trained Deep-Learning Model. *Appl. Sci.* **2022**, *12*, 2627. https://doi.org/10.3390/app12052627

Academic Editor: Mihaela Pop and Cristian A. Linte

Received: 13 January 2022
Accepted: 27 February 2022
Published: 3 March 2022

1. Introduction

Quantification of left ventricular (LV) function is crucial for risk stratification, diagnosis, and treatment of cardiac disease [1]. Cardiac magnetic resonance imaging (MRI) has been established as the gold standard for evaluating left ventricular function [2], including LV myocardial mass (LVMM), end-diastolic volume (LVEDV), end-systolic volume (LVESV), stroke volume (LVSV), and ejection fraction (LVEF). To generate these measurements, segmentation of the LV structures is required as a first step. Manual segmentation of cardiac MRI requires intensive efforts from users, depends on the experience of observers, introduces user variability, and is not compatible with efficient and high throughput cardiac imaging workflow [3].

Various methods have been developed for cardiac MR image analysis and demonstrated utility for use in research and clinical settings. Non-learning based methods [4,5] heavily rely on hand-crafted features with limited representation capability and generally provide suboptimal performance [6]. Recently, the development and use of deep convolutional neural networks (CNN) has achieved remarkable success for numerous cardiac imaging tasks [7]. With the availability of large annotated datasets and powerful computational platforms, these learning-based methods can automatically learn highly discriminative features through feature abstraction in a hierarchical manner. Recent studies [3,8,9] showed significant promise of using neural networks for heart segmentation in cardiac cine MRI, as recently reviewed [10]. Although promising, these studies mainly trained and tested CNNs on datasets acquired using the same scanner or at the same healthcare center, which represents a limited number of applications of deep learning in most research and clinical settings. Unfortunately, these [3,8] and other investigations [11] also demonstrated that deep-learning models trained on one domain (source) do not generalize well to a new domain (target) and direct application of a pre-trained model to a new dataset often yields degraded performance because of the well-known domain shift issue facing the community.

To facilitate translation of this important tool for widespread use in research and clinical care, it is urgently required to improve the generalizability of deep-learning methods to datasets collected using different imaging settings on different systems at various locations in patients with distinct diseases. Domain adaptation [12] aims to address this issue by fine-tuning a pre-trained model using a small amount of labeled data from a target domain, or by learning domain-invariant features or transforming data from the target domain to resemble the source domain. For example, previous studies [8,13] fine-tuned a pre-trained model using manually annotated datasets for new cardiac MRI segmentation tasks in a supervised manner. Other studies employed adversarial learning to transform data in a one domain (source) to resemble data in another domain (target) at the image level [14] or image-and-feature level [15] for unsupervised domain-adaptation-based cardiac image segmentation. Data augmentation [16] represents a very different approach to solving this problem by artificially enlarging the training datasets through extensive transformations to train a model that is robust to potential variations in new domains. Although commonly used classical data augmentation techniques (e.g., geometrical transformation, noise, contrast and blurring perturbation, histogram equalization and matching) have been widely used in various applications, other advanced and extensive augmentation techniques have also demonstrated effectiveness in addressing the domain shift issue [17,18]. In particular, recent studies using advanced data augmentation techniques demonstrated higher performance than several adversarial learning-based domain-adaptation methods for several medical image segmentation tasks [17,18].

Although the previous studies demonstrated some promise in tackling the domain shift issue for medical image segmentation, these algorithms have limitations. For example, domain adaptation using labeled data from a target domain requires a substantial amount of time and expertise for manual annotation and is not compatible with efficient research and clinical workflow. Adversarial learning that transforms a dataset from a source domain to resemble a target domain typically requires a large dataset from the target domain and a long time for re-training/fine-tuning. Data augmentation aims to learn non-domain specific features by performing extensive transformations to change the appearance of training datasets and often generates non-realistic datasets that do not resemble real world cases, which may or may not adversely affect the performance. Another potential issue associated with these techniques is the increased difficulty of algorithm interpretability because of the "black box" nature of deep-learning methods. Here, we proposed a different approach to tackling the domain shift issue. In particular, we employed a machine-learning method to automatically segment a subset of an unseen dataset without manual annotations to fine-tune a pre-trained deep-learning model to segment cardiac MRI datasets from a different domain. The proposed approach required 12 min to segment a relatively small dataset of 10 subjects for fine-tuning and to update the pre-trained model without affecting algorithm interpretability. Importantly, our approach yielded several commonly used and clinically relevant LV function measurements that are in strong agreement with expert manual analyses; this was not demonstrated in the previous studies. A preliminary version of this work has been published in conference proceedings [19] and there are substantial differences between the current work and the previous version [19]. In the current version, we reviewed some commonly used techniques (e.g., domain adaptation and data augmentation) that are developed to tackle the domain shift issue and discussed the advantages/limitations of these techniques. We also provided some details regarding the mathematical formulation and upper bound-based iterative optimization of the proposed continuous kernel-cut method. In addition, we implemented several state-of-the-art deep-learning segmentation models (DeepLabV3+ and an optimized style-intensity augmentation method) and performed comprehensive comparison between these methods and our approach. Furthermore, we discussed the study limitations and proposed some future work directions. These elements were not included in our previous work [19] and represents some of the major differences in the current work.

2. Methods

2.1. Cardiac MRI Datasets

We investigated two cine cardiac MR datasets from the Left Ventricle Segmentation Challenge (LVSC) held in 2009 [20] and the 2017 Automated Cardiac Diagnosis Challenge (ACDC) [9]. The LVSC dataset (https://www.cardiacatlas.org/\studies/sunnybrook-cardiac-data/, accessed 20 March 2021) consists of 45 subjects (mean age = 61 ± 15 years, age range = [23, 88] years; 32 male) enrolled in clinical studies at Sunnybrook Health Sciences Centre (Canada), including healthy volunteers (n = 9) and patients with hypertrophy (n = 12), or with heart failure with (n = 12) and without (n = 12) infarction. Two-dimensional short-axis cine images of the whole heart were obtained with as SSFP sequence (voxel size = 1.25–1.56 mm^2, slice thickness = 8–10 mm, inter-slice gap = 8 mm, 6–12 slices, 20 phases per cardiac cycle) on a 1.5T scanner (Signa, GE Healthcare, Milwaukee, WI, USA). For each subject, both the myocardium (LVM) and blood cavity (LVC) of the left ventricle in the cine images at the end-diastole were manually segmented by a cardiologist, and only the LV cavity was manually segmented at the end-systolic phase. Therefore, only the cine MRI datasets at the end-diastolic phase (n = 45 images) were used in this study.

The ACDC dataset (https://www.creatis.insa-lyon.fr/Challenge/acdc/, accessed 20 March 2021) comprises 100 participants (mean weight = 75 ± 17 kg; mean height = 171 ± 10 cm) acquired in clinical routine at the University Hospital of Dijon (France). The dataset covers five categories of well-defined pathologies (n = 20 subjects in each category): heart failure with myocardial infarction, dilated cardiomyopathy, hypertrophic cardiomyopathy, and abnormal right ventricle, as well as healthy subjects. Two-dimensional short-axis cine images covering the entire LV were acquired on 1.5T or 3.0T scanners (Siemens Aera and Siemens Trio, Siemens Medical Solutions, Germany) using an SSFP sequence (voxel size = 1.34–1.68 mm^2, slice thickness = 5–10 mm, inter-slice gap = 5 mm (sometimes), 6–18 slices, 28–40 phases per cardiac cycle). The dataset had substantial variability in image quality, including noise, motion and banding artefacts, MR low-frequency intensity fluctuation, and varying field-of-view. Manual segmentation of the LVM and LVC was performed on the cine images at both end-diastolic and end-systolic phases, which were double-checked by two independent experts to reach consensus.

We note that the manual segmentation of the LVSC dataset is not very consistent between subjects and there is substantial "noise" in manual annotations. In addition, the LVSC dataset contains cine images with LV cavity and myocardium segmentation only at the end-diastolic phase. We used the LVSC dataset for CNN pre-training, which provides additional opportunity to explore the tolerance to annotation noise and generalizability from end-diastolic phase to end-systolic phase for a deep-learning segmentation algorithm. The ACDC dataset was randomly divided into 10 and 90 subjects for CNN fine-tuning and testing, respectively. All data used in this study were anonymized and ethics approval for using these public datasets was exempted.

2.2. Algorithm Workflow

We used a 2D U-net [21] that comprised a symmetric contracting and expanding path with five levels. Each level consists of two blocks of 3 × 3 convolution and a rectified linear unit, followed by max-pooling in the contracting path or up-sampling in the expanding path; the number of feature maps was 16 in the top level and increased to 256 in the bottom level. The network was pre-trained on 45 images from the LVSC dataset for 200 epochs by minimizing the cross-entropy between model prediction and manual reference segmentation using an ADAM optimizer (learning rate = 10^{-4}). Spatial data augmentation, including translation (−50–50 pixels), random rotation (−50–50°), voxel size and intensity scaling (0.75–1.25 times), and elastic deformation, was performed in parallel. To further minimize overfitting and improve CNN segmentation generalizability, Monte-Carlo dropout [22] (MCD, dropout rate = 0.5) was applied to each block in the bottom three levels of the 2D U-net. These settings were adopted for the following fine-tuning procedure.

Figure 1 provides the schematic of our proposed algorithm. Briefly, the trained U-net was applied to the 10 ACDC fine-tuning subjects. For each subject, test-time MCD was applied to generate 50 segmentation samples ($s_1(x), s_2(x), \ldots, s_{50}(x),\ x \in \Omega$); the mean of the associated probability maps were calculated to derive the "mean" segmentation $\bar{s}(x)$. In addition, the standard deviation of the 50 segmentation samples was calculated for each pixel and used as pixel-wise U-net segmentation uncertainty $\omega(x)$, i.e., $\omega(x) \propto \frac{1}{std(\{s1(x),s2(x)\ldots,s50(x)\})},\ x \in \Omega$. The derived "mean" segmentation $\bar{s}(x)$ was post-processed using a recently developed continuous kernel-cut (CKC) segmentation method, which demonstrated effectiveness in post-processing cardiac MRI CNN segmentation outputs [3,23,24]. The CKC segmentation algorithm employs normalized cut for balanced pair-wise feature clustering and continuous regularization on image grids to generate spatially smooth contours. In addition, we proposed to use the derived CNN "mean" segmentation as descent initialization of the CKC algorithm such that in regions with high U-net segmentation uncertainty (i.e., $\omega(x)$ is relatively low), the final segmentation $u(x)$ can be more different from the "mean" segmentation $\bar{s}(x)$ and vice versa. To this end, we derived the deep-learning uncertainty-guided CKC segmentation algorithm by minimizing the following function:

$$\sum_{l\in L} -\frac{u_l^T X u_l}{\mathbf{1} X u_l} + \int_\Omega g(x)|\nabla u_l(x)|dx + \int_\Omega \omega(x) \cdot |u_l(x) - \bar{s}_l(x)|dx, \quad u_l \in \{0,1\}, \tag{1}$$

subject to $\sum_{l\in L} u_l(x) = 1, \forall x \in \Omega$. In Equation (1), $u_l(x) \in \{0,1\}$ is decomposed from the final segmentation $u(x)$ and indicates if voxel x is in region $l \in L = \{LVM, LVC, background\}$, X is a matrix where each element $X(i,j)$ indicates if voxel j is within the K-nearest neighbor of voxel i, $\mathbf{1}$ is an all-ones matrix, $g(x)$ is a boundary weight function based on image contrast edges, and $\omega(x)$ enforces the similarity of CNN initial segmentation $\bar{s}_l(x)$ and CKC final segmentation $u_l(x)$ for each region l. Of note, $\bar{s}_l$ was decomposed from $\bar{s}_l$ similar to $u_l(x)$. The CKC algorithm in Equation (1) integrates the advantages of balanced portioning of image features in high-dimensional space and spatially smooth segmentation that mimics the behavior of manual delineation [3,24,25].

Direct optimization of the high-order and non-smooth function in Equation (1) is particularly challenging. Following the previous work [3,23,26], we adopted an upper bound optimization technique to simplify the optimization of Equation (1) by deriving and optimizing a series of upper bound functions of Equation (1), assuming that the upper bound function is easier to minimize than the original formulations. Briefly, for any given segmentation $\hat{u}_l, l \in L, x \in \Omega$, previous studies [26] showed that the following is an upper bound function of Equation (1):

$$\sum_{l \in L} \left\langle \frac{\mathbf{X1}\hat{u}_l^T X \hat{u}_l}{(1X\hat{u}_l)^2} - \frac{2X\hat{u}_l}{\mathbf{1}X\hat{u}_l}, u_l \right\rangle + \int_\Omega g(x)|\nabla u_l(x)|dx + \int_\Omega \omega(x) \cdot |u_l(x) - \bar{s}_l(x)|dx, \quad u_l \in \{0,1\}, \tag{2}$$

where $\langle , \rangle$ and ^{T} denote inner product and transpose, respectively; the first term in Equation (2) is linear with respect to u_l that we aim to solve. Through convex relaxation, i.e., by relaxing $u_l \in \{0,1\}$ to $u_l \in [0,1]$, we can derive a convex relaxed formulation of Equation (2), which can be efficiently and globally optimized using a continuous min-cut/max-flow algorithm on a graphics card [27,28]. We refer readers to the previous studies [27,28] for the details of the continuous min-cut/max-flow algorithm. Please note that Equation (1) was optimized iteratively; for each iteration, we derived an upper bound using Equation (2) and minimized the upper bound to generate a solution $\hat{u}_l$, which was used to update the upper bound for the next iteration. In particular, for the first iteration we used the derived CNN "mean" segmentation $\bar{s}_l$l as the given solution $\hat{u}_l$. This process was iterated several times until convergence (we observed convergence typically within five iterations in this study) to derive the final solution to Equation (1). We refer readers to previous studies [3,24] for the details of minimizing the CKC segmentation model in Equation (1). Upon CKC algorithm convergence, the final segmentation of the fine-tuning dataset (without manual labels) was saved and used to update the trained U-net model for another 20 epochs in ~10 min. This procedure was iterated until convergence and the final U-net model was tested on the remaining 90 ACDC subjects for LV indices quantification. We also implemented several commonly used methods for comparison, including:

1. A naive method (Naive): The trained U-net was used to segment the 90 ACDC test subjects directly.
2. A combined method (Combined) that integrated MCD, spatial augmentation, and style-intensity augmentation method. We explored the effects of MCD, spatial augmentation, and advanced style-intensity augmentation for U-net training; the optimal combination of the three components constitutes the combined method. A recent study [17] proposed style-intensity augmentation during network training to tackle the domain shift issue and demonstrated state-of-the-art performance in breast segmentation in MRI datasets from a different domain. Style-intensity augmentation comprises style transfer and intensity remapping, which produce non-realistic looking MR scans while preserving the image shapes. The style transfer procedure uses features extracted from style images to augment the training images, randomizing the color, texture and contrast but preserving the geometry [29]. The intensity remapping technique generates a random mapping function to map the original image signal intensities to new values. This method is based on the assumption that by considerably changing the appearance of training images, the network will focus on non-domain specific features, e.g., the geometric shape of breast that is preserved in different breast MR datasets [17]. The optimized combined method was applied to the ACDC test dataset.
3. DeepLab: DeepLabV3+ [30], a top performing neural network in several medical image segmentation challenges, was trained on the LVSC dataset and tested on the ACDC test dataset.

Of note, the proposed algorithm and the naive method were implemented based on the same settings, i.e., MCD+spatial augmentation, and the proposed algorithm incorporated the fine-tuning procedure. The proposed algorithm, the naive and the combined methods were implemented using TensorFlow 1.4.0; DeepLabV3+ was implemented with Keras 2.2.4. All were run on Python 2.7.14 platforms on a GPU (Tesla P100, NVIDIA Corp., Santa Clara, CA, USA). The CKC segmentation algorithm was implemented using MATLAB 2013a (MathWorks, Natick, MA, USA) and CUDA (CUDA v8.0, NVIDIA Corp., Santa Clara, CA, USA) on a Ubuntu 14.02 desktop with a GPU (GeForce, GTX TITAN X, Santa Clara, CA, USA).

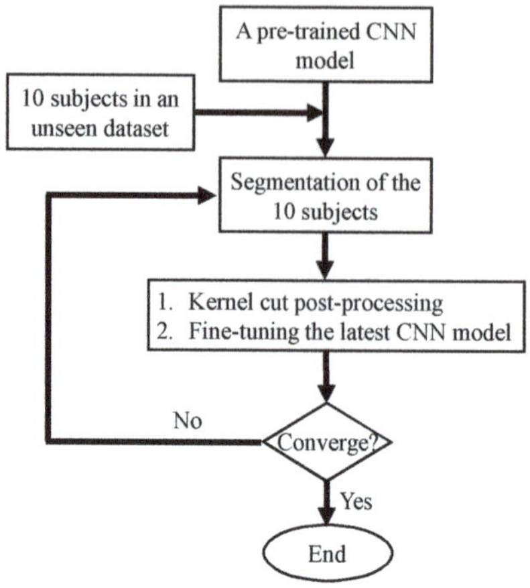

Figure 1. Schematic of the proposed algorithm pipeline for cardiac MR image segmentation using pre-trained CNNs. A trained CNN was applied to 10 previously unseen subjects; the initial segmentation was post-processed and post-processed using a kernel-cut algorithm. The resulting segmentation was used to update the trained CNN. This procedure was iterated till convergence to derive the final CNN model, which was applied to the unseen test dataset for LV function evaluation. Please note that no manual annotation of the unseen dataset was required in this procedure.

2.3. Evaluation Methods

Algorithm performance was evaluated for LV segmentation and function measurements. LV segmentation accuracy was evaluated using Dice coefficient (DSC) and average surface distance in a symmetric manner (ASSD) by comparing algorithm and manual segmentation masks [24,31]. We denote R_a and R_m the algorithm and manual segmentation, respectively. DSC measures the overlap of R_a and R_m and is calculated as: $\frac{2|R_a \cap R_m|}{|R_a|+|R_m|}$, where $|.|$ represents the size of a mask. ASSD evaluates the closeness between the algorithm and manual segmentation boundaries and is given as: $\frac{1}{2}\{\frac{1}{|\partial R_a|}\sum_{p\in\partial R_a} d(p,\partial R_m) + \frac{1}{|\partial R_m|}\sum_{p\in\partial R_m} d(p,\partial R_a)\}$, where ∂R_a represents the algorithm segmentation boundary and $d(p,\partial R_a)$ is the shortest Euclidean distance from a vertex p (e.g., a vertex from the manual segmentation boundary ∂R_m) to ∂R_a. ∂R_m and $d(p,\partial R_m)$ are defined the same way. Please note that traditional classification accuracy metrics, including true/false positives, true/false negatives and their combinations, can also be used to evaluate image segmentation accuracies [32] and DSC can be derived based on the four basic cardinalities when evaluating Boolean data. In fact, DSC, ASSD, and volume errors are widely used overlap, volume, and distance-based metrics for comprehensive evaluation of segmentation algorithms [33], and here we adopted the same or similar metrics consistent with most image segmentation studies.

In addition, the derived algorithm segmentation masks were used to determine LVMM, LVEDV, LVESV, LVSV, and LVEF For LVMM calculation, a density of 1.05 g/mL for myocardium [34] was used.

2.4. Statistical Analysis

Continuous variables were expressed as mean ± standard deviation (Mean ± SD). DSC provided by the proposed approach was compared with the other comparative methods using paired t-tests. Algorithm LV function measurement errors were compared using paired t-tests. Relationships and agreement for algorithm vs. manual LV indices were assessed using Pearson correlation coefficients (r, 95% confidence intervals [CI]) and Bland-Altman analyses (with 95% limits of agreement [LOA]). Fisher's z-transformation [35] was used to compare the correlation coefficients provided by each algorithm vs. manual analyses. Shapiro–Wilk tests were used to assess if the data can be modeled by a normal distribution and when data were not normally distributed, nonparametric tests were performed. We used GraphPad Prism v7.00 (GraphPad Software Inc., San Diego, CA, USA) for all the statistical analyses. Results were considered significant when the probability of making a two-tailed type I error was less than 5% ($p < 0.05$).

3. Results

Figure 2 shows segmentation of different regions of the heart at end-diastole and end-systole for three ACDC test subjects using the proposed algorithm (left) and the combined method (right).

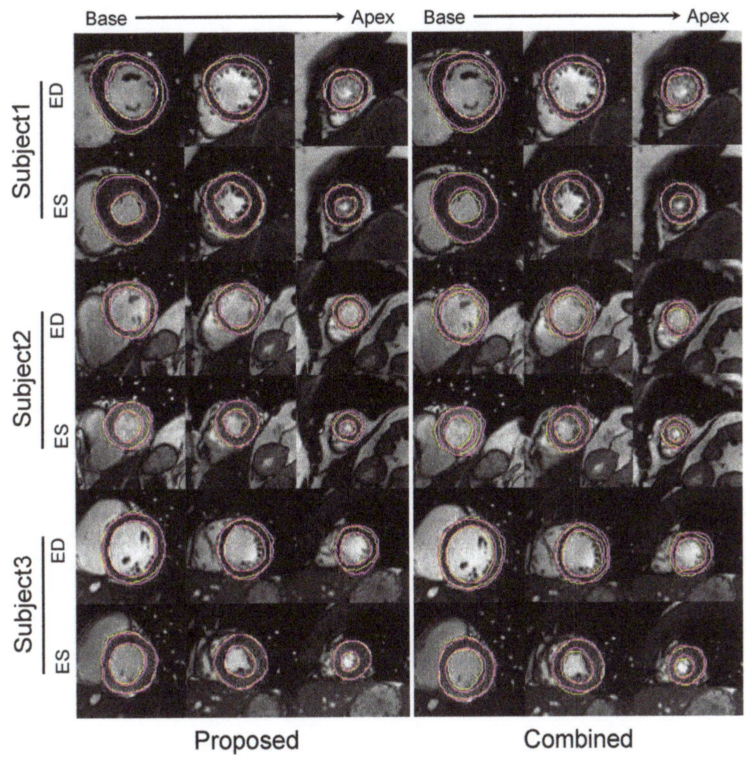

Figure 2. Representative segmentation of different regions of the heart at end-systole and end-diastole for three unseen ACDC test subjects (Subject1, Subject2, and Subject3) using the proposed (left) and the combined (right) methods. Algorithm and manual segmentation are shown in purple and yellow, respectively. ED: end-diastole; ES: end-systole.

We observed that direct application of the trained model to the 10 fine-tuning subjects yielded DSC of 0.77, 0.90 and ASSD of 2.32 mm, 2.88 mm for LVM, LVC; these accuracies were improved to 0.82, 0.92 for DSC and 1.97 mm, 1.91 mm for ASSD by the proposed CKC algorithm (data not shown), which were used to fine-tune the pre-trained model. As shown in Table 1, for the 90 test subjects the proposed algorithm yielded DSC of 0.81 ± 0.09 for LVM and 0.90 ± 0.09 for LVC. Meanwhile, the combined method generated DSC of 0.78 ± 0.08 and 0.87 ± 0.12 for the two regions, higher than the naive method and DeepLabV3+. Similarly, the proposed algorithm yielded substantially lower ASSD compared with the naive method, which further outperformed the combined method and DeepLabV3+. Of note, the DSC and ASSD provided by our approach were significantly different from each of the other algorithms ($p < 0.0001$), and the naive method demonstrated higher overall segmentation accuracy than the combined method and DeepLabV3+. As shown in Table A1 in Appendix A, MCD, spatial augmentation, and style-intensity augmentation each improved the segmentation accuracy and the combination of the three components, which constitutes the combined method, provided the highest segmentation accuracy among all the possible combinations.

Table 1. LV myocardium and cavity segmentation accuracy (mean ± SD) for n = 180 images from 90 previously unseen ACDC test subjects.

Methods	DSC ([0, 1])		ASSD (mm)	
	LVM	LVV	LVM	LVC
Proposed	0.81 ± 0.09	0.90 ± 0.09	2.04 ± 1.77	1.82 ± 2.18
Naive	0.74 ± 0.12 *	0.87 ± 0.12 *	2.43 ± 2.16 *	2.40 ± 2.58 *
Combined	0.78 ± 0.08 *	0.87 ± 0.12 *	2.71 ± 2.50 *	2.87 ± 2.61 *
DeepLab	0.26 ± 0.18 *	0.32 ± 0.27 *	18.60 ± 17.48 *	17.33 ± 12.37 *

DSC: Dice-similarity-coefficient, ASSD: average-symmetric-surfaced-distance; LVM: left ventricle myocardium, LVC: left ventricle cavity; *: $p < 0.0001$ when compared with the proposed algorithm.

Table 2 summarizes the LV functional parameters generated by manual and algorithm segmentation, illustrating that LV function measurements provided by the proposed approach are closer to manual results compared with the other methods. For example, paired t-tests showed that LV indices provided by the proposed approach were not significantly different from manual measurements, whereas the measurements generated by the naive and combined method were significantly different from manual LVMM (Proposed: $p = 0.1976$; Naive: $p < 0.0001$; Combined: $p = 0.0023$), LVEDV (Proposed: $p = 0.8015$; Naive: $p < 0.0001$; Combined: $p < 0.0001$), LVESV (Proposed: $p = 0.8631$; Naive: $p < 0.0001$; Combined: $p < 0.0001$), LVSV (Proposed: $p = 0.6617$; Naive: $p = 0.0734$; Combined: $p < 0.0001$), and LVEF (Proposed: $p = 0.2495$; Naive: $p = 0.0059$; Combined: $p < 0.0001$).

Table 2. Algorithm and manual LV function measurements (mean ± SD) for n = 180 images from 90 previously unseen ACDC test subjects.

	Manual	Proposed	Naive	Combined	DeepLab
LVMM (g) ¥	138.1 ± 54.3	$129.3 \pm 49.8_{0.1976}$	$110.8 \pm 48.2_{<0.0001}$	$154.4 \pm 83.6_{0.0023}$	$46.4 \pm 34.7_{<0.0001}$
LVEDV (mL)	163.8 ± 75.2	$162.9 \pm 72.0_{0.8015}$	$174.6 \pm 74.5_{<0.0001}$	$175.8 \pm 72.8_{<0.0001}$	$71.8 \pm 69.3_{<0.0001}$
LVESV (mL)	99.4 ± 80.4	$99.2 \pm 76.7_{0.8631}$	$108.2 \pm 80.0_{<0.0001}$	$118.4 \pm 76.2_{<0.0001}$	$58.3 \pm 62.1_{<0.0001}$
LVSV (mL)	64.4 ± 24.6	$63.7 \pm 25.8_{0.6617}$	$66.5 \pm 31.5_{0.0734}$	$57.4 \pm 25.9_{<0.0001}$	$13.5 \pm 24.2_{<0.0001}$
LVEF (%)	46.2 ± 20.4	$45.5 \pm 20.5_{0.2495}$	$43.0 \pm 23.6_{0.0059}$	$36.7 \pm 18.4_{<0.0001}$	$-4.4 \pm 142.2_{<0.0001}$

LVMM: LV myocardium mass (g); LVEDV: LV end-diastolic volume (mL); LVESV: LV end-systolic volume (mL); LVSV: LV stroke volume (mL); LVEF: LV ejection fraction (%); ¥: n = 180 images from 90 subject; p-values for comparison of algorithm vs. manual LV indices are shown in subscripts.

Table 3 and Figure 3 show that there were strong and significant correlations between the proposed algorithm and the naive method vs. manual analyses of LVMM (Proposed: r = 0.86, $p < 0.0001$; Naive: r = 0.79, $p < 0.0001$), LVEDV (Proposed: r = 0.99, $p < 0.0001$; Naive: r = 0.98, $p < 0.0001$), LVESV (Proposed: r = 0.99, $p < 0.0001$; Naive: r = 0.98, $p < 0.0001$), LVSV (Proposed: r = 0.92, $p < 0.0001$; Naive: r =0.84, $p < 0.0001$), and LVEF (Proposed: r = 0.93, $p < 0.0001$; Naive: r = 0.75, $p < 0.0001$). Please note that the correlations between the naive and combined methods vs. manual measurements were very similar for all the LV indices except for LVMM. Fisher's z-transformations showed that the correlations for the proposed algorithm and the naive method vs. manual measurements were significantly different for LVMM ($p = 0.0366$), LVEDV ($p = 0.0214$), LVESV ($p = 0.0214$), LVSV ($p = 0.0151$), and LVEF ($p < 0.0001$). Similar results were observed when comparing the correlations yielded by the proposed algorithm and the combined method.

Table 3. Relationships (Pearson r and []=95% CI) for algorithm vs. manual LV function measurements for n = 180 images from 90 previously unseen ACDC test subjects.

Pearson (r, 95% CI)	Proposed vs. Manual	Naive vs. Manual	Combined. vs. Manual	DeepLab vs. Manual
ine LVMM (g) ¥	0.86 ([0.80, 0.90])	0.79 ([0.73, 0.84])	0.41 ([0.28, 0.52])	0.47 ([0.35, 0.58])
LVEDV (mL)	0.99 ([0.98, 0.99])	0.98 ([0.97, 0.99])	0.99 ([0.99, 0.99])	0.57 ([0.41, 0.69])
LVESV (mL)	0.99 ([0.98, 0.99])	0.98 ([0.97, 0.99])	0.97 ([0.96, 0.98])	0.65 ([0.51, 0.75])
LVSV (mL)	0.92 ([0.88, 0.95])	0.84 ([0.76, 0.89])	0.83 ([0.75, 0.89])	0.13 ([−0.08, 0.33])
LVEF (%)	0.93 ([0.89, 0.95])	0.75 ([0.65, 0.83])	0.76 ([0.65, 0.83])	0.08 ([−0.13, 0.28])

LVMM: LV myocardium mass (g); LVEDV: LV end-diastolic volume (mL); LVESV: LV end-systolic volume (mL); LVSV: LV stroke volume (mL); LVEF: LV ejection fraction (%); ¥: n = 180 images from 90 subjects.

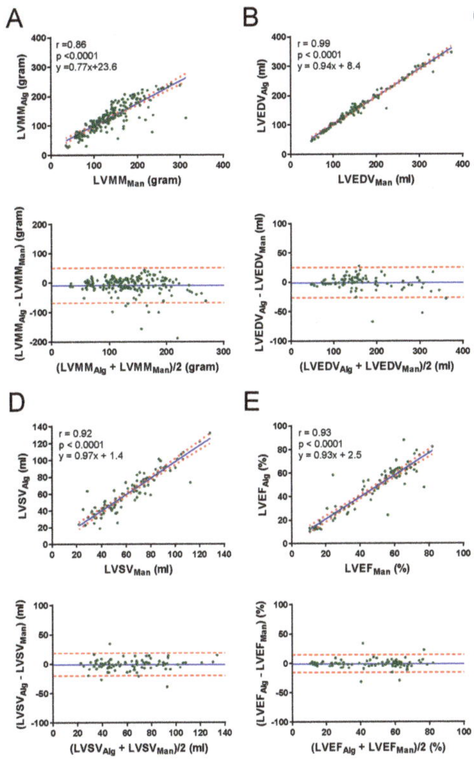

Figure 3. Relationships and agreement between the proposed algorithm vs. manual measurements of LVMM (**A**), LVEDV (**B**), LVESV (**C**), LVSV (**D**), and LVEF (**E**) (n = 180 images from 90 subjects). Linear regression and Bland-Altman analyses of algorithm vs. manual LV indices are shown in the top and bottom plots, respectively. $_{Alg}$: algorithm results; $_{Man}$: manual results. Solid lines (blue) indicate the biases and dotted lines (red) represent the 95% limits of agreement.

Figure 3 also shows the quantitative agreement between the proposed algorithm and manual LV indices. Bland-Altman analyses indicated that there was promising agreement between the proposed algorithm and manual LVMM (bias = −8.8 ± 30.3 g, 95% LOA = [−68.1, 50.5] g), LVEDV (bias = −0.9 ± 13.1 mL, 95% LOA = [−26.6, 24.8] mL), LVESV (bias = −0.2 ± 13.8 mL, 95% LOA = [−27.3, 26.9] mL), LVSV (bias = −0.7 ± 10.0 mL, 95% LOA = [−20.2, 18.9] mL), and LVEF (bias = −0.6 ± 7.8%, 95% LOA = [−15.9%, 14.6%]). In contrast, the naive and combined methods yielded greater biases and wider 95% LOAs for LVMM (Naive: bias = −27.3 ± 33.3 g, 95% LOA = [−92.5, 37.9] g; Combined:

bias =16.4 ± 79.0 g, 95% LOA = [−138.5, 171.3] g); LVEDV (Naive: bias=10.9 ± 14.5 mL, 95% LOA = [−17.5, 39.2] mL; Combined: bias = 12.0 ± 11.2 mL, 95% LOA = [−10.0, 34.0] mL), LVESV (Naive: bias = 8.8 ± 16.7 mL, 95% LOA = [−24.0, 41.5] mL; Combined: bias = 19.0 ± 18.1 mL, 95% LOA = [−16.4, 54.4] mL), LVSV (Naive: bias = 2.1 ± 17.4 mL, 95% LOA = [−31.9, 36.1] mL; Combined: bias = −7.0 ± 14.7 mL, 95% LOA = [−35.8, 21.8] mL), and LVEF (Naive: bias = −3.1 ± 15.8%, 95% LOA = [−34.0%, 27.7%]; Combined: bias = −9.5 ± 13.6%, 95% LOA = [−36.1%, 17.2%]).

For the proposed algorithm and the naive method, U-net training/pre-training was completed in approximately 5 h. The fine-tuning procedure required an additional ∼12 min, including 10 s to post-process each image using the CKC algorithm and 10 min to update the U-net parameters. The combined method required ∼15 h for training and DeepLabV3+ required ∼5 h. For all the trained/fine-tuned models, inference of a new 2D cine image stack required ∼1 s.

4. Discussion

Deep learning is emerging to potentially transform cardiac imaging workflow and clinical patient care. Here, we developed an approach to employing a trained CNN for LV segmentation and function evaluation in an independent cardiac cine MRI dataset. For a dataset of 180 cine MR images from 90 subjects with various cardiac disease, we made the following observations: (1) improved segmentation accuracy in the independent dataset; (2) strong correlations between the proposed approach and manual analyses of LV indices; and (3) rapid and fully automated fine-tuning procedure *without* needing manual labels for the independent dataset.

Cine MRI has been routinely performed for evaluation of LV structure and function in cardiovascular MR exams. Deep learning and machine leaning have demonstrated promise in several aspects of the cardiac research and clinical workflow, including but not limited to prediction of cardiac left ventricular kinematics and boundaries [36], classification of cardiac arrhythmias from electrocardiogram [37], and detection of cardiac structure and structural abnormalities [38]. However, direct application of a trained model to a previously unseen dataset often yields suboptimal performance [3,8,39]. For example, direct application of a CNN trained on a large cine MRI dataset of 4275 subjects [8] to 20 patients in a previously unseen ACDC dataset yielded DSC of 0.65 for LVM and 0.74 for LVC. Previous studies showed that DSC is usually sensitive to small differences when the segmented object is relatively small and not very sensitive to errors when the object is relatively large [40]. We note that the size of LVM is generally smaller than the LVC at end-diastole although the differences between the two regions are smaller at end-systole. A recent study [9] investigated the variabilities of intra and inter-observer manual segmentation. The authors reported greater DSC of 0.956–0.967 for LVC and 0.870–0.900 for LVM at end-diastole, and similarly, these were 0.898–0.941 and 0.891–0.917 at end-systole. The robustness of manual segmentation and the substantially lower algorithm DSC [8] than repeated manual analyses (0.65 vs. 0.870–917) suggest that manual segmentation errors have a minimal effect in this case. In addition, the training and testing datasets used by Bai et al. [8] differ substantially as the training dataset mainly consists of healthy volunteers whereas the testing dataset comprise patients with diverse cardiac pathologies, which affect the appearance of the myocardium in MR images. Based on the literature and our experience, we think that the relatively low DSC for LVM than LVC (0.65 vs. 0.74) reported by Bai et al. [8] is mainly caused by the combined effects of the large differences between training and testing datasets, the relatively small size, hollow shape, and image signal intensity inhomogeneity in the LVM compared with LVC. However, this warrants further investigation. Nonetheless, the initial suboptimal accuracies [8] were later improved by employing 80 manually segmented subjects in the ACDC dataset for fine-tuning. Previous studies [4,41] and our efforts have shown that manual segmentation of a 3D cardiac MR volume with 10–15 slices typically requires 20–30 min. This lengthy procedure requires experience and expertise from examiners, introduces user variability, and is not compatible with efficient research and clinical workflow. Similarly, another study [3] applied a pre-trained state-of-the-art CNN (1st place winner in the ACDC segmentation challenge) to 40 ACDC subjects and achieved DSC of 0.78 for LVM and 0.86 LVC. Compared with these previous works, our approach yielded greater DSC of 0.81 for LVM and 0.90 for LVC *without* requiring manual segmentation of the fine-tuning datasets. In our future work, we will compare the results from this study with that by fine-tuning the proposed algorithm framework using manually segmented unseen dataset in terms of segmentation accuracy and time. In addition, the derived LV function measurements provided by our approach were strongly correlated with expert manual analyses with no significant differences between the techniques (p = 0.1976–0.8631, Table 2). This is important because our approach implemented fully automated transfer learning to segment an independent cardiac cine MR dataset acquired using a different MR system at a different location in patients with different cardiac diseases *without* requiring manual segmentation of the target dataset, potentially enabling efficient clinical workflow and facilitating broader use of deep learning for a wide range of applications.

We also implemented a combined method that employed state-of-the-art style-intensity augmentation techniques [17] to address the domain shift issue, which had performed well for breast segmentation in different MRI datasets. Compared with our approach, the optimized implementation of the combined method yielded lower DSC of 0.78 for LVM and 0.87 for LVC with substantially greater ASSD, as shown in Table 1. In another study [18], the authors tackled a similar problem by developing a series of stacked transformations that performed extensive data augmentation (sharpening, blurring, adding noise, changing brightness/contrast, intensity perturbation, rotation, scaling, deformation) during network training. For eight public MRI/ultrasound datasets, the authors achieved improved segmentation accuracy with the use of the proposed data augmentation techniques. These studies [17,18] showed that advanced and extensive data augmentation techniques yielded higher segmentation performance than adversarial learning-based domain-adaptation techniques. Surprisingly, the naive method generally outperformed the combined method for segmentation accuracy measurements (except for DSC for LVM) but the correlations of LV function measurements with manual results were comparable. This warrants further investigation. Of note, the well-known DeepLabV3+ algorithm [30] performed poorly in this work (see Figure A1), further highlighting the challenges of domain shift for medical image segmentation. In fact, we previously trained the DeepLabV3+ model on 50 subjects from the UK Biobank dataset and applied the model to segment 50 previously unseen ACDC subjects. We achieved DSC of 0.437 for LVM and 0.568 for LVC. Similarly, we trained the DeepLabV3+ model on 50 ACDC subjects and tested the model on 50 subjects from the UK Biobank dataset. We obtained DSC of 0.745 for LVM AND 0.813 for LVC. Please note that these results are excluded in the final version of our previous paper [3] as suggested by the reviewers. In a recent study of lung MRI segmentation [23], we achieved DSC of 0.872 and 0.701 by training the DeepLabV3+ model on one dataset and testing the model on another different dataset. Collectively, these and other studies suggest the inability of deep learning, including DeepLabV3+ and other state-of-the-art models, to deal with the domain shift issue for medical image segmentation. Our approach outperformed the naive method and a combined method that used state-of-the-art data augmentation techniques [17] and differs from the other methods [18,31] in that in addition to comprehensive data augmentation, we implemented Monte-Carlo dropout to mitigate overfitting and a CKC algorithm to automatically update the "annotations" of the fine-tuning subjects. Previous studies [3,24] demonstrated the effectiveness of using CKC to improve CNN initial segmentation and here we substantially extended the previous work by demonstrating its utility in a new application, whereby the CKC post-processing results were incorporated to effectively tune the trained model to segment an independent cardiac cine MRI dataset. The proposed framework is relatively independent from commonly used domain adaptation and data augmentation techniques. Therefore, we think that our approach could be combined with these methods to address the domain shift issue, which represents an advantage of our approach that will be further investigated in future work.

Although our approach was based on a U-net implemented with Monte-Carlo dropout and a recently developed CKC algorithm, this work differs from other cardiac image segmentation methods developed to tackle a similar issue with higher performance for LV segmentation and biomarker quantification. In addition, the promise of our approach was demonstrated in the context of a U-net, which has been widely used for numerous applications, suggesting the generalizability of our framework for a broad range of segmentation tasks that involve a U-net. The improved segmentation generalizability may stem from the combination of the advantages of deep-learning and machine-learning methods without a deep architecture. As a result, both deep and shallow image features can be learned or employed, and the power of data-driven and rule-based segmentation methods was aggregated, potentially mitigating the limitations of the individual methods. However, further investigation of this is warranted. Efforts that can further improve the performance of our proposed approach including: (1) applying the CKC algorithm only to the fine-tuning subjects with problematic segmentation; (2) automatically selecting the datasets with acceptable CKC segmentation for CNNfine-tuning; and (3) adding a few more new unlabeled datasets for each iteration. We think that these strategies may be optimized and implemented in parallel to for potentially greater robustness. Regardless, the results realized here suggest that our approach provides a way to improve deep-learning segmentation generalizability without increasing the difficulties of algorithm interpretability, a major concern facing the community [42], and may facilitate broader use and translation of deep-learning techniques for research and clinical care.

We acknowledge several study limitations. First, the segmentation accuracy of our approach is lower than CNNs trained and tested on the same datasets. However, here we focus on adapting a trained CNN for segmentation of previously unseen cardiac MRI datasets, which is particularly challenging and requires urgent solution. Importantly, we achieved segmentation accuracies higher than two state-of-the-art segmentation methods (a combined method that employed style-intensity augmentation and DeepLabV3+) and LV function measurements that were strongly correlated with manual results. We note that the basal and apical slices of the heart are difficult to segment due to poor image qualities and the complexity of cardiac structures, which represent some of the major challenges facing the community. In addition, the proposed algorithm was validated on a retrospective dataset and the effectiveness of this approach warrants a prospective evaluation with datasets from different centers, MR scanners, imaging protocols, and disease phenotypes.

5. Conclusions

In conclusion, we developed a way to employ a pre-trained neural network to segment previously unseen cardiac MR datasets without requiring manual annotations of the unseen datasets for fine-tuning. For a clinical dataset of patients with diverse cardiac disease, we achieved LV segmentation and function evaluation accuracy and precision that may be suitable for research and clinical use. As such, our approach may facilitate the translation and use of deep learning in cardiac imaging workflow.

Appendix A

Table A1 shows that Monte-Carlo dropout, spatial augmentation, and style-intensity augmentation together led to the optimal performance of the combined method [17].

Table A1. Effects of Monte-Carlo dropout, spatial augmentation, and style-intensity augmentation on U-net training using the LVSC dataset. The three components were combined during U-net training (optimal implementation) and the trained U-net models were directly applied to the 90 ACDC test subjects (n = 180 images) for DSC and ASSD (mean ± SD) calculation.

MCD	Spa. Aug.	Sty.-Int. Aug.	DSC ([0, 1])		ASSD (mm)	
			LVM	LVC	LVM	LVC
ine ✗	✗	✗	0.33 ± 0.22	0.46 ± 0.30	16.85 ± 21.52	15.28 ± 18.02
✗	✗	✓	0.49 ± 0.18	0.68 ± 0.22	8.38 ± 7.33	8.55 ± 8.69
✗	✓	✗	0.73 ± 0.12	0.85 ± 0.14	2.92 ± 2.86	3.34 ± 3.48
✗	✓	✓	0.77 ± 0.07	0.87 ± 0.11	2.39 ± 2.05	2.80 ± 2.49
✓	✗	✗	0.34 ± 0.21	0.49 ± 0.29	11.30 ± 16.93	9.97 ± 15.36
✓	✗	✓	0.55 ± 0.17	0.71 ± 0.21	7.47 ± 7.00	7.37 ± 8.22
✓	✓	✗	0.75 ± 0.11	0.87 ± 0.12	2.30 ± 1.64	2.39 ± 1.85
✓	✓	✓	0.78 ± 0.08	0.87 ± 0.12	2.71 ± 2.50	2.87 ± 2.61

✗: a component is not used., ✓: a component is used. DSC: Dice-similarity-coefficient, ASSD: average-symmetric-surfaced-distance; LVM: left ventricle myocardium; LVC: left ventricle cavity; MCD: Monte-Carlo dropout; Spa. Aug.: spatial augmentation; Sty.-Int. Aug.: style-intensity augmentation.

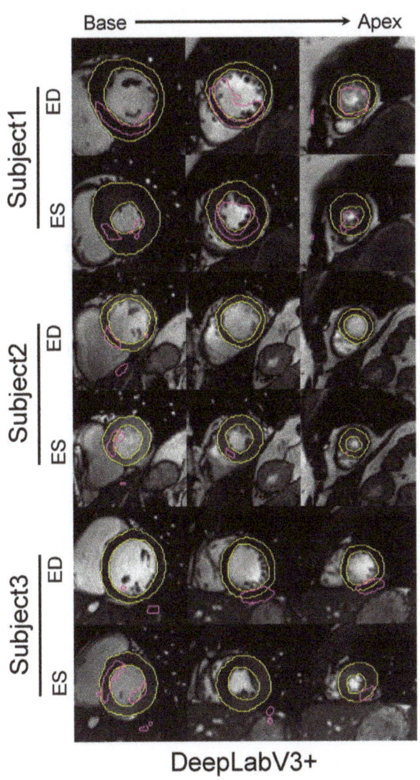

Figure A1. Representative segmentation of different regions of the heart at end-systole and end-diastole for the same three subjects as that in Figure 2 using DeepLabV3+. Algorithm and manual segmentation are shown in purple and yellow, respectively. ED: end-diastole; ES: end-systole.

Author Contributions: Conceptualization, F.G.; methodology, F.G. and M.N; software, F.G.; validation, F.G., M.N. and G.W.; formal analysis, F.G.; investigation, I.R. and G.W.; resources, I.R. and G.W.; writing—original draft preparation, F.G.; writing—review and editing, M.N., I.R. and G.W.; visualization, F.G.; supervision, G.W.; project administration, G.W.; funding acquisition, G.W. All authors have read and agreed to the published version of the manuscript.

Funding: This work was funded by Canadian Institutes of Health Research (CIHR) MOP: #93531, Ontario Research Fund and GE Healthcare. FG is supported by a Banting postdoctoral fellowship.

Institutional Review Board Statement: The datasets used here were published and made publicly available by previous studies; ethical review and approval were waived for this study.

Informed Consent Statement: Patients' consent was collected in the previous studies and were exempted in this study.

Data Availability Statement: The data used in this study is publicly available (https://www.cardiacatlas.org/studies/sunnybrookcardiac-data/, https://www.creatis.insalyon.fr/Challenge/acdc/, accessed 20 March 2021).

Acknowledgments: We acknowledge the use of the facilities of Compute Canada.

Conflicts of Interest: The authors declare no conflict of interest.

References

1. Flachskampf, F.A.; Biering-Sørensen, T.; Solomon, S.D.; Duvernoy, O.; Bjerner, T.; Smiseth, O.A. Cardiac imaging to evaluate left ventricular diastolic function. *JACC Cardiovasc. Imaging* **2015**, *8*, 1071–1093. [CrossRef] [PubMed]
2. Members, W.C.; Hundley, W.G.; Bluemke, D.A.; Finn, J.P.; Flamm, S.D.; Fogel, M.A.; Friedrich, M.G.; Ho, V.B.; Jerosch-Herold, M.; Kramer, C.M.; et al. ACCF/ACR/AHA/NASCI/SCMR 2010 expert consensus document on cardiovascular magnetic resonance: A report of the American College of Cardiology Foundation Task Force on Expert Consensus Documents. *Circulation* **2010**, *121*, 2462–2508. [CrossRef] [PubMed]
3. Guo, F.; Ng, M.; Goubran, M.; Petersen, S.E.; Piechnik, S.K.; Neubauer, S.; Wright, G. Improving cardiac MRI convolutional neural network segmentation on small training datasets and dataset shift: A continuous kernel cut approach. *Med. Image Anal.* **2020**, *61*, 101636. [CrossRef] [PubMed]
4. Petitjean, C.; Dacher, J.N. A review of segmentation methods in short axis cardiac MR images. *Med. Image Anal.* **2011**, *15*, 169–184. [CrossRef]
5. Peng, P.; Lekadir, K.; Gooya, A.; Shao, L.; Petersen, S.E.; Frangi, A.F. A review of heart chamber segmentation for structural and functional analysis using cardiac magnetic resonance imaging. *Magn. Reson. Mater. Phys. Biol. Med.* **2016**, *29*, 155–195. [CrossRef] [PubMed]
6. Shen, D.; Wu, G.; Suk, H.I. Deep learning in medical image analysis. *Annu. Rev. Biomed. Eng.* **2017**, *19*, 221–248. [CrossRef]
7. Leiner, T.; Rueckert, D.; Suinesiaputra, A.; Baeßler, B.; Nezafat, R.; Išgum, I.; Young, A.A. Machine learning in cardiovascular magnetic resonance: Basic concepts and applications. *J. Cardiovasc. Magn. Reson.* **2019**, *21*, 1–14. [CrossRef]
8. Bai, W.; Sinclair, M.; Tarroni, G.; Oktay, O.; Rajchl, M.; Vaillant, G.; Lee, A.M.; Aung, N.; Lukaschuk, E.; Sanghvi, M.M.; et al. Automated cardiovascular magnetic resonance image analysis with fully convolutional networks. *J. Cardiovasc. Magn. Reson.* **2018**, *20*, 65. [CrossRef]
9. Bernard, O.; Lalande, A.; Zotti, C.; Cervenansky, F.; Yang, X.; Heng, P.A.; Cetin, I.; Lekadir, K.; Camara, O.; Ballester, M.A.G.; et al. Deep learning techniques for automatic MRI cardiac multi-structures segmentation and diagnosis: Is the problem solved? *IEEE Trans. Med. Imaging* **2018**, *37*, 2514–2525. [CrossRef]
10. Chen, C.; Qin, C.; Qiu, H.; Tarroni, G.; Duan, J.; Bai, W.; Rueckert, D. Deep Learning for Cardiac Image Segmentation: A Review. *Front. Cardiovasc. Med.* **2020**, *7*, 25. [CrossRef]
11. Yan, W.; Wang, Y.; Gu, S.; Huang, L.; Yan, F.; Xia, L.; Tao, Q. The Domain Shift Problem of Medical Image Segmentation and Vendor-Adaptation by Unet-GAN. In *International Conference on Medical Image Computing and Computer-Assisted Intervention*; Springer: Cham, Switzerland, 2019; pp. 623–631.
12. Tzeng, E.; Hoffman, J.; Saenko, K.; Darrell, T. Adversarial discriminative domain adaptation. In Proceedings of the IEEE Conference on Computer Vision and Pattern Recognition, Honolulu, HI, USA, 21–26 July 2017; pp. 7167–7176.
13. Zhu, Y.; Fahmy, A.S.; Duan, C.; Nakamori, S.; Nezafat, R. Automated Myocardial T2 and Extracellular Volume Quantification in Cardiac MRI Using Transfer Learning—Based Myocardium Segmentation. *Radiol. Artif. Intell.* **2020**, *2*, e190034. [CrossRef] [PubMed]
14. Huo, Y.; Xu, Z.; Moon, H.; Bao, S.; Assad, A.; Moyo, T.K.; Savona, M.R.; Abramson, R.G.; Landman, B.A. Synseg-net: Synthetic segmentation without target modality ground truth. *IEEE Trans. Med. Imaging* **2018**, *38*, 1016–1025. [CrossRef] [PubMed]
15. Chen, C.; Dou, Q.; Chen, H.; Qin, J.; Heng, P.A. Unsupervised bidirectional cross-modality adaptation via deeply synergistic image and feature alignment for medical image segmentation. *IEEE Trans. Med. Imaging* **2020**, *39*, 2494–505. [CrossRef] [PubMed]

16. Krizhevsky, A.; Sutskever, I.; Hinton, G.E. Imagenet classification with deep convolutional neural networks. *Commun. ACM* **2017**, *60*, 84–90. [CrossRef]
17. Hesse, L.S.; Kuling, G.; Veta, M.; Martel, A. Intensity augmentation to improve generalizability of breast segmentation across different MRI scan protocols. *IEEE Trans. Biomed. Eng.* **2020**, *68*, 759–770. [CrossRef]
18. Zhang, L.; Wang, X.; Yang, D.; Sanford, T.; Harmon, S.; Turkbey, B.; Wood, B.J.; Roth, H.; Myronenko, A.; Xu, D.; et al. Generalizing deep learning for medical image segmentation to unseen domains via deep stacked transformation. *IEEE Trans. Med. Imaging* **2020**, *39*, 2531–2540. [CrossRef]
19. Guo, F.; Ng, M.; Roifman, I.; Wright, G. Cardiac MRI Left Ventricular Segmentation and Function Quantification Using Pre-trained Neural Networks. In *International Conference on Functional Imaging and Modeling of the Heart*; Springer: Cham, Switzerland, 2021; pp. 46–54.
20. Radau, P.; Lu, Y.; Connelly, K.; Paul, G.; Dick, A.; Wright, G. Evaluation framework for algorithms segmenting short axis cardiac MRI. *MIDAS J.-Card. MR Left Ventricle Segmentation Chall.* **2009**, *49*. [CrossRef]
21. Ronneberger, O.; Fischer, P.; Brox, T. U-net: Convolutional networks for biomedical image segmentation. In *International Conference on Medical Image Computing and Computer-Assisted Intervention*; Springer: Cham, Switzerland, 2015; pp. 234–241.
22. Gal, Y.; Ghahramani, Z. Dropout as a bayesian approximation: Representing model uncertainty in deep learning. In Proceedings of the 33rd International Conference on Machine Learning, New York, NY, USA, 20–22 June 2016; pp. 1050–1059.
23. Guo, F.; Capaldi, D.P.; McCormack, D.G.; Fenster, A.; Parraga, G. Ultra-short Echo-time Magnetic Resonance Imaging Lung Segmentation with Under-Annotations and Domain Shift. *Med. Image Anal.* **2021**, *72*, 102107. [CrossRef]
24. Guo, F.; Ng, M.; Wright, G. Cardiac cine MRI left ventricle segmentation combining deep learning and graphical models. In *Medical Imaging 2020: Image Processing*; International Society for Optics and Photonics: Bellingham, WA, USA, 2020; Volume 11313, p. 113130Z.
25. Guo, F.; Krahn, P.R.; Escartin, T.; Roifman, I.; Wright, G. Cine and late gadolinium enhancement MRI registration and automated myocardial infarct heterogeneity quantification. *Magn. Reson. Med.* **2021**, *85*, 2842–2855. [CrossRef]
26. Tang, M.; Ben Ayed, I.; Marin, D.; Boykov, Y. Secrets of grabcut and kernel k-means. In Proceedings of the IEEE International Conference on Computer Vision, Santiago, Chile, 7–13 December 2015; pp. 1555–1563.
27. Yuan, J.; Bae, E.; Tai, X.C. A study on continuous max-flow and min-cut approaches. In Proceedings of the 2010 IEEE Computer Society Conference on Computer Vision and Pattern Recognition, San Francisco, CA, USA, 13–18 June 2010; pp. 2217–2224.
28. Guo, F.; Yuan, J.; Rajchl, M.; Svenningsen, S.; Capaldi, D.P.; Sheikh, K.; Fenster, A.; Parraga, G. Globally optimal co-segmentation of three-dimensional pulmonary 1H and hyperpolarized 3He MRI with spatial consistence prior. *Med. Image Anal.* **2015**, *23*, 43–55. [CrossRef]
29. Jackson, P.T.; Abarghouei, A.A.; Bonner, S.; Breckon, T.P.; Obara, B. Style augmentation: Data augmentation via style randomization. In Proceedings of the CVPR Workshops, Long Beach, CA, USA, 16–21 June 2019; pp. 83–92.
30. Chen, L.C.; Zhu, Y.; Papandreou, G.; Schroff, F.; Adam, H. Encoder-decoder with atrous separable convolution for semantic image segmentation. In Proceedings of the European conference on computer vision (ECCV), Munich, Germany, 8–14 September 2018; pp. 801–818.
31. Guo, F.; Ng, M.; Wright, G. Cardiac MRI left ventricle segmentation and quantification: A framework combining U-Net and continuous max-flow. In *International Workshop on Statistical Atlases and Computational Models of the Heart*; Springer: Cham, Switzerland, 2018; pp. 450–458.
32. Nai, Y.H.; Teo, B.W.; Tan, N.L.; O'Doherty, S.; Stephenson, M.C.; Thian, Y.L.; Chiong, E.; Reilhac, A. Comparison of metrics for the evaluation of medical segmentations using prostate MRI dataset. *Comput. Biol. Med.* **2021**, *134*, 104497. [CrossRef]
33. Maier-Hein, L.; Eisenmann, M.; Reinke, A.; Onogur, S.; Stankovic, M.; Scholz, P.; Arbel, T.; Bogunovic, H.; Bradley, A.P.; Carass, A.; et al. Why rankings of biomedical image analysis competitions should be interpreted with care. *Nat. Commun.* **2018**, *9*, 1–13. [CrossRef] [PubMed]
34. Grothues, F.; Smith, G.C.; Moon, J.C.; Bellenger, N.G.; Collins, P.; Klein, H.U.; Pennell, D.J. Comparison of interstudy reproducibility of cardiovascular magnetic resonance with two-dimensional echocardiography in normal subjects and in patients with heart failure or left ventricular hypertrophy. *Am. J. Cardiol.* **2002**, *90*, 29–34. [CrossRef]
35. Kirby, M.; Svenningsen, S.; Owrangi, A.; Wheatley, A.; Farag, A.; Ouriadov, A.; Santyr, G.E.; Etemad-Rezai, R.; Coxson, H.O.; McCormack, D.G.; et al. Hyperpolarized 3He and 129Xe MR imaging in healthy volunteers and patients with chronic obstructive pulmonary disease. *Radiology* **2012**, *265*, 600–610. [CrossRef]
36. Damen, F.W.; Newton, D.T.; Lin, G.; Goergen, C.J. Machine Learning Driven Contouring of High-Frequency Four-Dimensional Cardiac Ultrasound Data. *Appl. Sci.* **2021**, *11*, 1690. [CrossRef]
37. Lee, H.; Yoon, T.; Yeo, C.; Oh, H.; Ji, Y.; Sim, S.; Kang, D. Cardiac Arrhythmia Classification Based on One-Dimensional Morphological Features. *Appl. Sci.* **2021**, *11*, 9460. [CrossRef]
38. Komatsu, M.; Sakai, A.; Komatsu, R.; Matsuoka, R.; Yasutomi, S.; Shozu, K.; Dozen, A.; Machino, H.; Hidaka, H.; Arakaki, T.; et al. Detection of cardiac structural abnormalities in fetal ultrasound videos using deep learning. *Appl. Sci.* **2021**, *11*, 371. [CrossRef]
39. Tao, Q.; Yan, W.; Wang, Y.; Paiman, E.H.; Shamonin, D.P.; Garg, P.; Plein, S.; Huang, L.; Xia, L.; Sramko, M.; et al. Deep learning–based method for fully automatic quantification of left ventricle function from cine MR images: A multivendor, multicenter study. *Radiology* **2019**, *290*, 81–88. [CrossRef]

40. Wong, K.C.; Moradi, M.; Tang, H.; Syeda-Mahmood, T. 3D segmentation with exponential logarithmic loss for highly unbalanced object sizes. In *International Conference on Medical Image Computing and Computer-Assisted Intervention*; Springer: Cham, Switzerland, 2018; pp. 612–619.
41. Wang, Y.; Zhang, Y.; Xuan, W.; Kao, E.; Cao, P.; Tian, B.; Ordovas, K.; Saloner, D.; Liu, J. Fully automatic segmentation of 4D MRI for cardiac functional measurements. *Med. Phys.* **2019**, *46*, 180–189. [CrossRef]
42. Rudin, C. Stop explaining black box machine learning models for high stakes decisions and use interpretable models instead. *Nat. Mach. Intell.* **2019**, *1*, 206–215. [CrossRef]

applied sciences

MDPI

Article

MUSIC: Cardiac Imaging, Modelling and Visualisation Software for Diagnosis and Therapy

Mathilde Merle [1], Florent Collot [1], Julien Castelneau [2], Pauline Migerditichan [2], Mehdi Juhoor [1], Buntheng Ly [3], Valery Ozenne [1], Bruno Quesson [1], Nejib Zemzemi [1,2], Yves Coudière [1,2], Pierre Jaïs [1], Hubert Cochet [1] and Maxime Sermesant [1,3,*]

[1] IHU Liryc, 33600 Pessac, France; mathilde.merle@ihu-liryc.fr (M.M.); florent.collot@ihu-liryc.fr (F.C.); mehdi.juhoor@ihu-liryc.fr (M.J.); valery.ozenne@ihu-liryc.fr (V.O.); bruno.quesson@u-bordeaux.fr (B.Q.); nejib.zemzemi@inria.fr (N.Z.); yves.coudiere@inria.fr (Y.C.); pierre.jais@ihu-liryc.fr (P.J.); hubert.cochet@chu-bordeaux.fr (H.C.)

[2] Inria Bordeaux, 33000 Bordeaux, France; julien.castelneau@inria.fr (J.C.); pauline.migerditichan@inria.fr (P.M.)

[3] Centre Inria d'Université Côte d'Azur, 06902 Sophia Antipolis, France; buntheng.ly@inria.fr

* Correspondence: maxime.sermesant@inria.fr

Abstract: The tremendous advancement of cardiac imaging methods, the substantial progress in predictive modelling, along with the amount of new investigative multimodalities, challenge the current technologies in the cardiology field. Innovative, robust and multimodal tools need to be created in order to fuse imaging data (e.g., MR, CT) with mapped electrical activity and to integrate those into 3D biophysical models. In the past years, several cross-platform toolkits have been developed to provide image analysis tools to help build such software. The aim of this study is to introduce a novel multimodality software platform dedicated to cardiovascular diagnosis and therapy guidance: MUSIC. This platform was created to improve the image-guided cardiovascular interventional procedures and is a robust platform for AI/Deep Learning, image analysis and modelling in a newly created consortium with international hospitals. It also helps our researchers develop new techniques and have a better understanding of the cardiac tissue properties and physiological signals. Thus, this extraction of quantitative information from medical data leads to more repeatable and reliable medical diagnoses.

Keywords: cardiac imaging; multimodal; electrophysiology; deep learning; biophysical modelling; inverse problems

Citation: Merle, M.; Collot, F.; Castelneau, J.; Migerditichan, P.; Juhoor, M.; Ly, B.; Ozenne, V.; Quesson, B.; Zemzemi, N.; Coudière, Y.; et al. MUSIC: Cardiac Imaging, Modelling and Visualisation Software for Diagnosis and Therapy. *Appl. Sci.* **2022**, *12*, 6145. https://doi.org/10.3390/app12126145

Academic Editor: Mihaela Pop and Cristian A. Linte

Received: 13 April 2022
Accepted: 2 June 2022
Published: 16 June 2022

1. Introduction

With the substantial progress in medical imaging and the boosting of multimodalities, it becomes increasingly challenging to obtain a comprehensive and integrated 3D view of a given heart. Unequivocally, most of these are consequences of larger size of acquired data (often in multiple centres), complex fusion of the imaging modalities, and variability in cardiac function.

MUSIC (Multimodality Platform for Specific Imaging in Cardiology) is a software platform developed by the *Multimodal Data Science* team (https://www.ihu-liryc.fr/en/technology-for-health-cluster/p/mds/, accessed on 1 June 2022) at Liryc IHU (https://www.ihu-liryc.fr accessed on 1 June 2022) and the *Epione* (https://team.inria.fr/epione/, accessed on 1 June 2022) and *Carmen* (https://team.inria.fr/carmen/, accessed on 1 June 2022) Inria (http://www.inria.fr, accessed on 1 June 2022) teams that aims to offer a large spectrum of functionalities and processing pipelines dedicated to cardiology applications for both diagnosis and therapy guidance.

The MUSIC project is led by a multidisciplinary team including researchers, developers, clinicians and medical imaging operators.

MUSIC is composed of a desktop application, a web portal, and an academic consortium. The desktop application is called MUSICardio and is built on the open-source software medInria (http://med.inria.fr, accessed on 1 June 2022), which comprises a large set of versatile tools for medical image segmentation, visualisation, filtering, histogram analysis, data reformatting, registration and mesh processing, all of which are being provided to the community with a user-friendly common framework for more efficiency. Furthermore, imaging data from various modalities, such as MDCT, MRI, PET and echography, are easily handled by the application. In addition, the cardiac models generated can be exported to clinical 3D electroanatomical mapping systems used by cardiologists during interventional procedures to repair electrical disorders (e.g., catheter-based ablations). Our research focuses on the following:

- Combined analysis of imaging (CT-scan, MRI, PET) and electrophysiology (invasive, non-invasive mapping or image-based simulation) data to improve knowledge regarding the structural substrate generating dangerous arrhythmia;
- The development of AI-based image processing tools for the automatic segmentation of cardiac structures and quantification of robust markers associated with the risk of arrhythmia;
- The development of customised image-based modelling methods to simulate electrophysiological tests with computation times compatible with clinical practice;
- The development of cardiac models for the optimal navigation of instruments in the virtual patient heart during interventional procedures;
- The clinical validation of these different tools for diagnosis, prognosis or real-time guided therapy in patients with cardiac electrical disorders.

Treatment improvement: MUSIC offers image-based algorithms for diagnosis and prognosis as well as pipelines dedicated to the guidance of either atrial or ventricular interventions through imaging. For instance, this application has been used to guide scar-related ventricular tachycardia (VT) ablation in over 500 consecutive procedures in Bordeaux University Hospital.

A booster for research: MUSICardio is an application allowing the analysis, in one environment, of multi-parametric data sets from different Liryc IHU teams (structural, mechanics, hemodynamics, electrical, etc.) and to interface with simulation platforms such as SOFA (https://www.sofa-framework.org/, accessed on 1 June 2022), CEPS (https://carmen.gitlabpages.inria.fr/ceps/, accessed on 1 June 2022), and CARP (https://carp.medunigraz.at/, accessed on 1 June 2022), in order to develop patient-specific modelling and simulation strategies. Our researchers can therefore use directly in MUSICardio our medical image analysis and processing algorithms to obtain quantitative information to improve the diagnosis of heart disease.

A VT MUSIC consortium has been created including more than 30 international hospitals with expertise in ventricular tachycardia from Europe, USA and Australia. Participating centres anonymise and upload their data to our MUSIC web portal, then our expert operators process the data in the MUSICardio software to obtain customised models of the heart. These new data are then sent back to the hospitals, which can immediately include them during operations to visualise catheters in real time in patient-specific heart modelling.

Some imaging platforms similar on some points to MUSIC allow us to situate ourselves: OsiriX (https://www.osirix-viewer.com, accessed on 1 June 2022) and Horos (https://horosproject.org, accessed on 1 June 2022) (OsiriX fork) are imaging applications allowing the reading of DICOM from sources or PACS systems, data visualisation in 2D/3D/4D, and have imaging tools and algorithms. They can be compared to our MUSICardio application within the MUSIC platform. There are differences: MUSICardio is dedicated to cardiology with dedicated tools, imports and exports several data types in addition to DICOM, and has a pipeline system of Python scripts allowing to chain tools and algorithms very easily for the users. OsiriX has two versions: one not free and FDA/CE certified, and the other free and non-certified. Horos is free and suggests a donation. Both applications are compatible only with macOS. MUSICardio is compatible with macOS, linux (Ubuntu) and Windows. Another similar application is 3DSlicer (https://www.slicer.org, accessed on 1 June 2022). It is a free and open-source medical imaging software. The common points with MUSICardio: 3DSlicer is compatible with macOS, linux and Windows; it allows the display of different types of medical data such as DICOM or meshes, offers algorithmic imaging treatments, AI-assisted segmentations and scripting in Python. For its part, MUSICardio also has algorithms and tools dedicated to cardiology. These tools are similar to our MUSICardio application, however they do not cover all the functionalities of our MUSIC platform dedicated to cardiology: the MUSICardio application, the web portal, as well as the dedicated consortium, in particular the expertise of our medical operators and clinicians within the consortium.

2. Architecture

MUSIC is a software platform allowing in a unique environment the analysis of multi-parametric data sets (e.g., structural, mechanic, hemodynamic, electrical, etc.) and to interface such data with simulation platforms in order to develop patient-specific modelling and simulation strategies. The MUSIC project is composed of the MUSICardio application and a data web portal, along with the VT MUSIC consortium.

2.1. Consortium

The VT MUSIC consortium brings together international centres with expertise in ventricular tachycardia and our institute. The data acquired in the hospitals are sent to us through a web portal and processed rapidly by our expert operators in our MUSICardio application. Thus, personalised models of the patient heart can be sent to the centres to be included in cardiac operations. Figure 1 shows the process of the consortium for the particular case of radiotherapy.

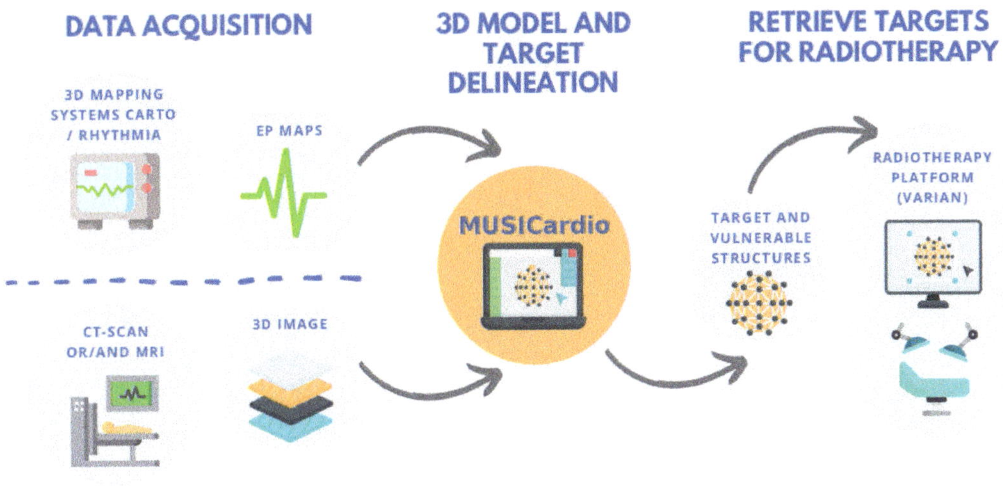

Figure 1. VT MUSIC consortium process (radiotherapy example).

VT centres interested in joining the consortium are invited to contact our team. Access to the consortium is free.

2.2. Web Portal

A web portal has been developed in Java through the Play framework (https://www.playframework.com, accessed on 1 June 2022). Users of the VT MUSIC consortium can anonymise their images through an anonymisation tool that we developed. Then, the web server allows users to upload these anonymised data and add relevant information for our operators and clinicians. The data are downloaded by our operators who process them in our MUSICardio application. Finally, the generated cardiac models are uploaded to the web portal and can be retrieved by hospital users for use during cardiac operations. In addition to the web portal, a mesh viewer developed by our team is provided so that clinicians from the VT MUSIC consortium hospitals can study the heart meshes before surgery.

2.3. *MUSICardio*

The MUSICardio application within the MUSIC project is based on the medInria open-source software developed by Inria. Our team participates in the improvement of medInria as the core of MUSICardio. We are part of the medInria consortium, created by Inria, which leads the development of the software. The modularity of the plugin system of the software allows easily adding new plugins dedicated to functionalities needed in the cardiac field.

2.3.1. Usage

There are three current ways of using MUSICardio. The application can be directly used by our researchers for instance to use algorithmic tools developed by our team. Researchers, post-doctoral fellows, PhD students or other members of our laboratory can also write Python scripts, with the help of our team, to create pipelines to process and analyse their data in the application. Python is an understandable language for non-developers, and it can be learnt quickly and easily. Pipelines allow complex suites of algorithms and tools to be processed in simple graphical steps in MUSICardio, making it easier for users, especially if they have to perform the tasks many times. Finally, our expert operators (who are trained medical imagery staff, and formed to use our VT pipeline script in MUSICardio) can process our pipeline in MUSICardio, using as input the data sent by partner hospitals in the VT MUSIC consortium, and they can send back the output heart models to these hospitals.

2.3.2. Libraries

The core of the application (i.e., medInria) is written in C++ and Qt, uses external libraries and includes Python wrappers. The main external libraries used are shown in the Table 1.

Table 1. External libraries used in medInria, core of MUSICardio.

Library	Description	Origin
DCMTK	DICOM management	OFFIS e.V
dtk	Tools for modular software development	Inria
ITK [1]	Scientific imaging management	Kitware
LCC LogDemons [2]	LCC Log Demons algorithms	Inria
QtDCM	Qt widgets to handle DICOM images	Inria
RPI	Image registration algorithms	Inria
TTK	Tensor algorithms	Inria
VTK	Image processing, 3D graphics, volume rendering and visualisation	Kitware

Our application gives access to a wide range of data tools, workspaces or toolboxes through a system of plugins that can be easily activated or deactivated, allowing a modularity of the application binaries. We use some additional external libraries in the MUSICardio application for our plugins (Table 2):

Table 2. External libraries used in MUSICardio plugins.

Library	Description	Origin
Eigen [3]	Linear algebra library	Inria
FFmpeg	Video tools	FFmpeg project
Mmg [4]	Remeshing tools	Inria
QuaZip	Qt zip/unzip tools	S.A. Tachenov
Qwt	Qt 2D plots library	U. Rathmann, J. Wilgen
TetGen [5]	Tetrahedral mesh generator, 3D Delaunay triangulator	Weierstrass Institute
zlib	Data compression library	J.-L. Gailly, M. Adler

These external libraries API can be accessed in plugins, allowing developers to access their functionalities and to develop more tools in the application.

2.3.3. Components

Figure 2 presents the architecture of the MUSICardio application through its medInria core, its import/export formats and functionalities plugins, and its external libraries.

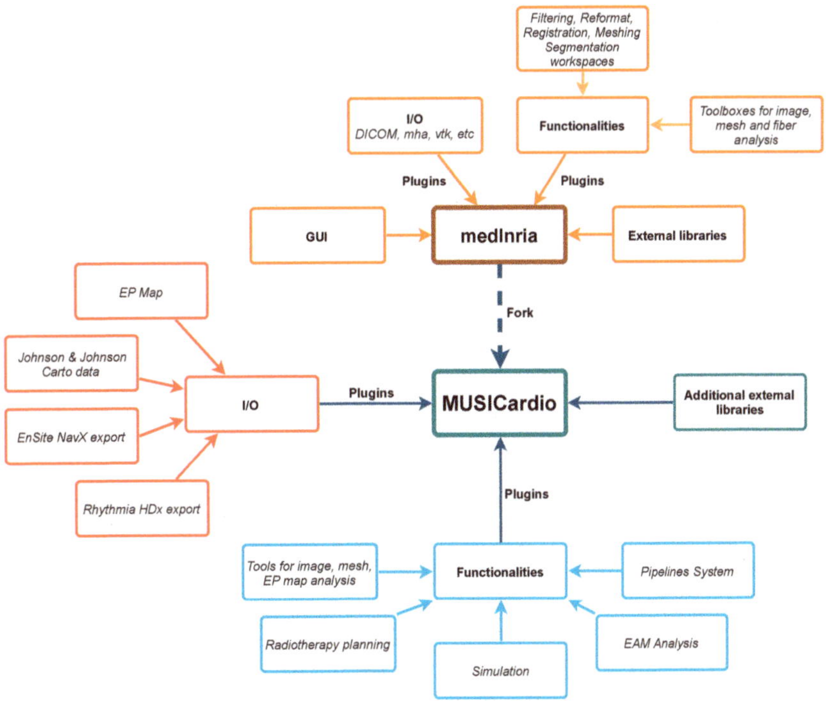

Figure 2. MUSICardio modules.

2.3.4. Python

Python 3 is provided in the software. A Python console has been added, which allows researchers to work on their data directly in Python in the application and further perform parameter testing in the console. The interfacing between Python and the application in C++/Qt is done with the Swig (https://www.swig.org, accessed on 1 June 2022) framework.

2.3.5. Pipelines

Processing workflows can be executed through a pipeline system included in the MUSICardio application. This allows us to script complex process sequences in order to simplify the work of the operators, make the results more robust and obtain the output data fast. The pipeline scripts were developed in Python, allowing researchers and operators to create and adapt their own customised pipelines.

For instance, within the VT MUSIC consortium, clinicians and expert operators at Liryc IHU use the pipeline system to segment CT and MR images, and to generate a 3D structural model comprising a detailed anatomy of the whole heart, myocardial scar maps from CT-derived wall thickness or late-enhancement MRI analysis, structures at risk during SBRT (stereotactic body radiotherapy) procedures, or other structures such as coronaries, left phrenic nerve, atrioventricular node, ICD (implantable cardioverter defibrillator) lead tip, GI (gastrointestinal) tract, etc. The 3D models are used to treat refractory ventricular tachycardia. EP maps (electrophysiology data) are registered in the same geometrical coordinates to allow the user to perform the 3D rendering of combined structural and EP data.

Pipeline scripts can handle automatic segmentation using Deep Learning-based algorithms that have been developed by our team. This allows us to obtain accurate and quick automatic segmentations of the anatomical heart structures for a more precise diagnosis and treatment of cardiac diseases.

3. Data Management

3.1. Data Format

The data management system in the MUSIC project includes local databases per user managing the data imported or generated by the MUSICardio application, access from the application to several data sources including PACS systems, and our online web portal that allows hospitals in the VT MUSIC consortium to send their data to our institute and to receive output data from us.

Within the application, users can import and export data in various formats, including VTK, ITK, DICOM, Nnrd, GIS, Nifti, Gipl, OBJ, STL, Medit, etc. MUSICardio can export data to electrophysiology mapping systems such as BioSense Webster, Johnson & Johnson (New Brunswick, NJ, USA) "CARTO", Boston Scientific (Natick, MA, USA) "RHYTHMIA", and Saint-Jude Medical (Saint Paul, MN, USA) "EnSite NavX". Within the VT MUSIC consortium, these compatible data are sent to VT centres around the world where clinicians can visualise their catheters in real time inside a virtual patient-specific heart created in the application through their navigation systems.

The application also allows us to export cardiac structures in DICOM RT-Struct format. These data are compatible with any radiation therapy (RT) planning software. Depending on the strategy and RT technology available onsite, these cardiac targets and associated segmentations may be registered onto a 4D planning CT and expanded to take into account the cardiac/respiratory motion and margins of errors for misregistration. Dose simulations are then performed to validate the final strategy, which is often the result of a trade-off between safety and efficacy of radiation delivery. Once validated, the actual SBRT (stereotactic body radiation therapy) treatment is delivered.

3.2. Data Sources

The MUSICardio application can connect to hospital PACS systems to access imaging data. This allows clinicians and hospital operators to easily work on data through a secure tool dedicated to hospitals. This avoids data import and export on physical media and travel across hospitals, and it allows for accurate data tracking. It also allows hospital users to filter data and easily find the information and data they need through metadata. Figure 3 shows the interface of the PACS access system.

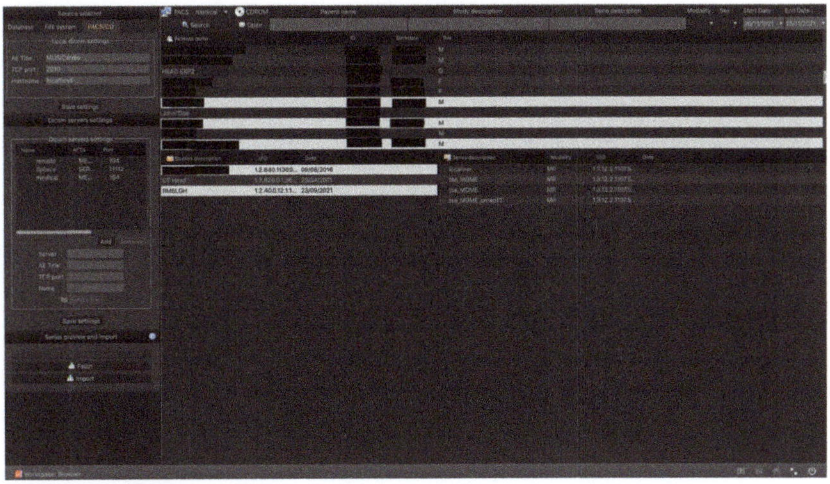

Figure 3. PACS data source.

4. Functionalities

4.1. Image and Mesh Workspaces

The MUSICardio application offers a wide range of tools and algorithms for segmenting, filtering, and reformatting, which are sorted into different workspaces to find them more easily.

The main image and mesh tools and algorithms in the application are outlined in the Table 3.

Figure 4 shows the interface of the thresholding process in the filtering workspace.

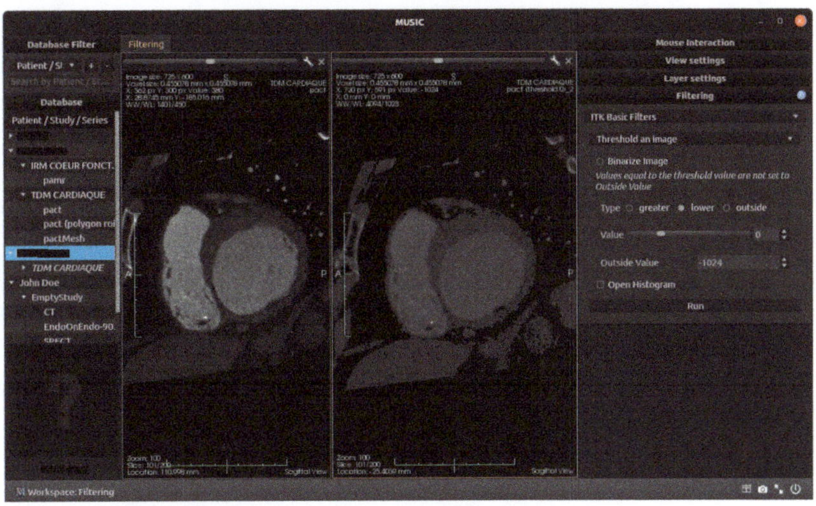

Figure 4. Example of the thresholding of CT data.

Table 3. List of the main imaging and meshing tools in MUSICardio.

Worspace	Description
Diffusion	Processing of diffusion and tractography.
EAM Analysis	Electroanatomical mapping analysis tools.
Filtering	Image binary operation (AND, OR, XOR, NOT), constant operation (addition, multiply, etc.), blurring, normalisation, thresholding, edge filters, mask application, morphological filters (dilate, erode, close, open), bias correction, etc.
Meshing	Convert mesh to and from mask, data array management, interpolation, projections, distance computations, mesh deformation, morphological mesh filters, binary operations on meshes, fiber computations, decimation, refine, etc.
Reformat	Crop data, extract 3D data from 4D ones, merge images and meshes, reslice (change orientation, pixel size, image dimensions), super resolution (compute accurate data from orthogonal ones), move an image to a new location.
Registration	Register two images with diffeomorphic demons [6], LCC Log demons, Optimus registration algorithm (gradient-free Mutual Information-based rigid registration), or manual registration.
Radiotherapy	Radiotherapy ablation planning tools with DICOM-RT-STRUCT data.
Segmentation	Mapping of myocardial depth isthmuses, histogram analysis, level set segmentation, paint and polygon segmentations, VOI (volume of interest) segmentation, variational segmentation through landmarks, as well as vessel segmentation.
Simulation	Simulation processing tools, forward problem, inverse problem, FEM and MFS simulation tools (see Section 4.5).

4.2. Segmentation

The Segmentation workspace handles multiple segmentation tools, one of the most important and used is the "Polygon ROI" toolbox. The "Polygon ROI" toolbox has been developed jointly by our team and radiologists. This tool allows us to manually segment images through polygon contours (ROI, region of interest), use a repulsor tool to refine these contours, automatically interpolate these segmentations between slices, or annotate them for AI. A focus in the development has been made on the production of contours with standardised labels. For now, eight different structures with different colours and names can be defined simultaneously. Figure 5 illustrates the ability to segment several anatomical zones and define a label name for each structure.

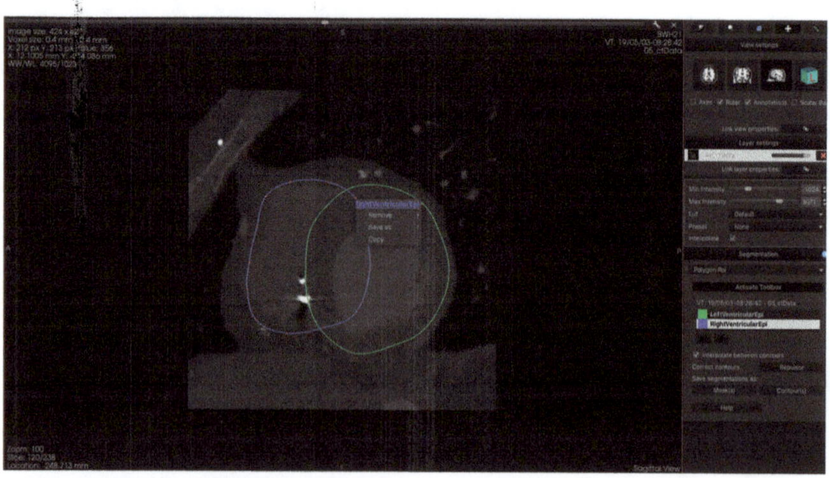

Figure 5. Contours segmented in polygon ROI toolbox. This example shows an ongoing segmentation of the left (green) and right (blue) ventricles of the heart.

The results of segmentation can be saved as binary masks or internal contours structures. A contour structure is an internal format of MUSICardio and medInria. It allows us to save the work in progress and start again later or reuse the result of a segmentation as the model on another dataset. Other useful features are available:

- Correction tools: each node of the polygon contour can be manipulated separately, or a repulsor tool is available allowing to push the segmentation boundary with a circle. See Figure 6a.
- Multiple Views: segmentations can be done on different series simultaneously in multiple views to keep the same label name for series, and we can use a cursor to mark a position in series. The data have to be first reformatted. Figure 6b shows two MRIs at different dates. A lesion, in green, is visible only on one series (the right one). The evolution of another lesion is showed on the second series in blue. The green cursor allows us to find the way through the images.

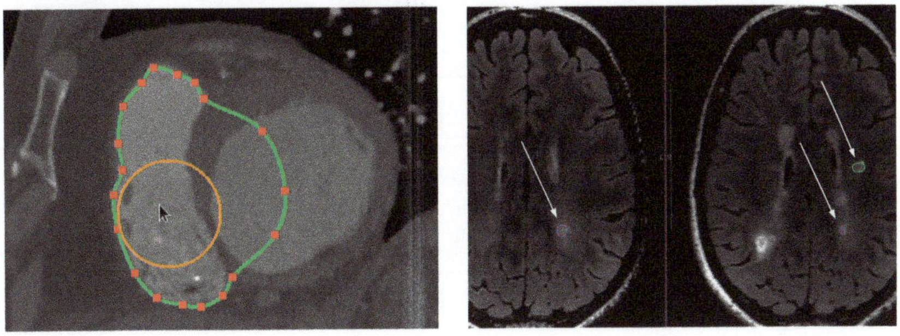

(a) (b)

Figure 6. Polygon ROI features. (**a**) Repulsor: the orange circle handled by the user allows to push the boundary of a previously segmented region of interest (**b**) Multiple views: a segmentation is done simultaneously in two views. The arrows show an ingoing segmentation of lesions in blue and green in two reformatted brain images, where the user can easily follow the differences.

4.3. Mesh Enhancement

The VTK library allows us to generate surface meshes from segmented masks, but these meshes are generally not of enough quality to be used for computer modelling and simulation. We decided to include the Mmg (http://www.mmgtools.org, accessed on 1 June 2022) library in order to improve the quality of the generated meshes. This library allows us to smooth, refine or decimate meshes without losing the original shape of the geometry. Volumetric meshes are then generated using the TetGen (https://wias-berlin.de/software/index.jsp?id=TetGen&lang=1, accessed on 1 June 2022) library, and if the quality of the volumetric mesh is not satisfactory, Mmg can also be used to improve the mesh quality. We developed some tools in MUSICardio to improve the mesh quality (Table 4).

4.4. Electroanatomical Map Analysis

ElectroAnatomical Mapping (EAM) systems are used to create a 3D model of the heart and to record the cardiac electrical activity at the mapped points [7]. More than two decades of device improvements and technological refinements have made the clinical EAM systems routinely used during invasive radio-frequency (RF) ablation procedures for treating dangerous cardiac arrhythmia [8].

An electroanatomical map (EAM) is a combination of a 3D heart geometry (i.e., a mesh) and electrophysiological data in the form of 1D electrical signals (e.g., uni/bipolar maps, local activation times, intracardiac electrograms). After the EP procedure, the study containing the maps can be exported from navigation systems. Our application is capable of importing EAMs from navigation systems used in the clinics: CARTO (Biosense Webster,

Johnson & Johnson) and RHYTHMIA (Boston Scientific), respectively. Once imported, these maps can be visualised and further processed using various toolboxes.

4.4.1. Review of Electrocardiograms

The recorded intracardiac waves and ECG signals can be reviewed. Users may create annotations to highlight important events. These annotations are given: name, colour, description. They can reference events that have a duration or just a specific time. Figure 7 illustrates the signal review and annotation window.

Table 4. List of tools dedicated to mesh enhancement.

Tool	Description
Binary operations	Intersection, union and difference between two meshes.
Convert mesh to mask	Convert mesh into mask to work on it and convert back.
Mesh cleaning	Filling holes in meshes, remove unused vertices, merge very close points.
Mesh interactors	Selecting and deleting selected cells.
Mesh manipulation	Apply transformation to mesh: translation, rotation, scaling.
Mesh registration	Registration of a mesh to another mesh using landmarks or the iterative closest point method. Registration of a mesh to an image using landmarks.
Merge volumes	Concatenate meshes.
Projection mapping and interpolation	Projection of data on meshes, scalar or vector field, linear and nonlinear interpolation.
Remeshing	Decimate, refine, smooth meshes.

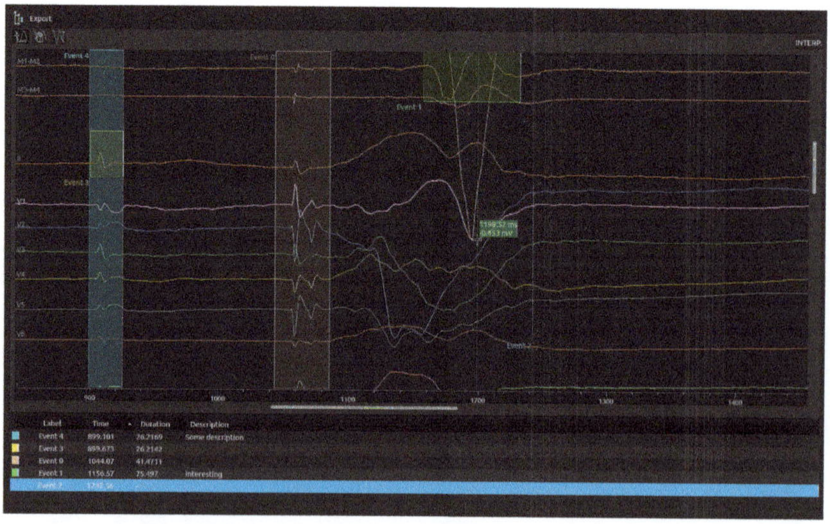

Figure 7. Annotations of 1D electrophysiological signals.

4.4.2. Radiotherapy Planning

Stereotactic body radiotherapy (SBRT) has recently been applied to treat refractory ventricular tachycardia [9,10]. MUSICardio provides an environment for an in-depth analysis and SBRT session planning.

CT and MR images are segmented, creating a 3D structural model comprising a detailed whole heart anatomy, myocardial scar maps generated from CT-derived wall thickness or late-enhancement MRI analysis, and the structures at risk during the interventions.

For patients who previously underwent catheter radio-frequency ablation, electroanatomical maps can be imported and co-registered with the CT and MR segmented meshes to allow the user to perform the 3D rendering of combined structural information and EP data.

The ablation targets and organs at risk can be delineated, creating contours to be exported in the DICOM RT-struct format. Figure 8 shows an example of such contours.

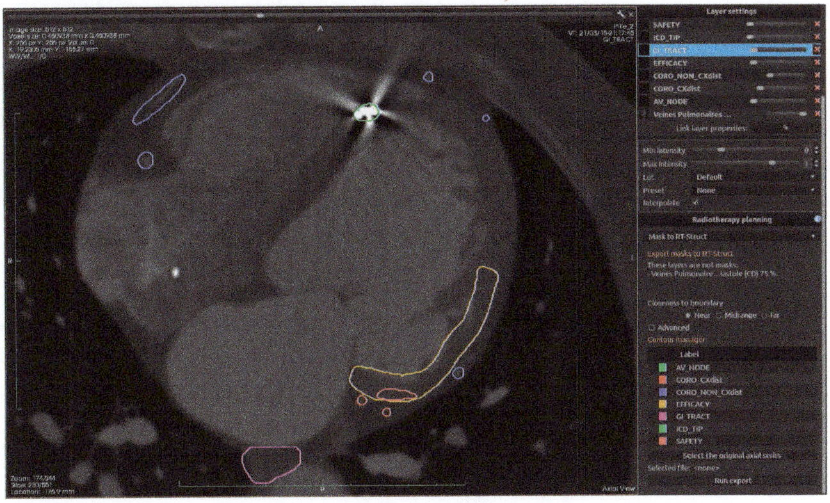

Figure 8. Automatic contouring of targets and organs at risk. The contours can be exported to the standard DICOM Rt-Struct format. Each ablation target and organ at risk is segmented with a color code: the list of contour names is displayed in a table in the right part of the application, as shown in this screenshot.

4.5. Electrocardiographic Imaging

In the Simulation workspace of MUSICardio, we developed different tools in order to enable the use of computational models for simulations along with their integration with multimodality imaging data. Of particular concern are the tools developed to improve the resolution of ElectroCardioGraphic Imaging (ECGI) inverse problems. Briefly, ECGI is a non-invasive technique that allows clinicians to construct electrical information on the heart surface or volume from electrical measurements on the body surface and patient-specific heart and torso geometries. The ECGI tools that are developed in MUSICardio are based on the formulation of Spach et al., who first related the epicardial and body surface electrical potentials by means of a transfer matrix that depends on the geometry of the heart and the torso of the patient [11]. Both surface and volume ECGI functionalities follow the following pipeline:

- Load the 3D image data of the patient (i.e., CT image scan or MR images).
- Segment body surface and heart surface or volume geometry and construct their corresponding masks. The user can also segment other organs such as lungs, bones and the liver. Figure 9a shows a 2D axial cross-section view of a segmented geometry showing the volume of the heart, lungs, liver and the body surface.
- Generate computational meshes of the segmented geometries. Figure 9b shows an example of a generated 3D computational geometry.
- Identify the position of electrodes in the body surface mesh (semi-automatic).
- Construct a transfer matrix based on finite elements method (FEM) or on the method of fundamental solutions (MFS) [12]. Details about both methods can be found in [13].
- Load the body surface electrical recordings.

- Construct the electrical potential on the heart surface or the current density on the heart volume by solving the ECGI inverse problem. Several approaches have been implemented using different regularisation methods, see [13].
- Visualise the electrical information constructed on the surface or the volume of the heart. Figure 10 illustrates an example of the electrical potential reconstruction using MFS (b) and FEM (c) as well as the ground truth of the corresponding solution (a).

For the post processing of the computed electrical potential, our ECGI pipeline script in Python allows one to compute activation maps.

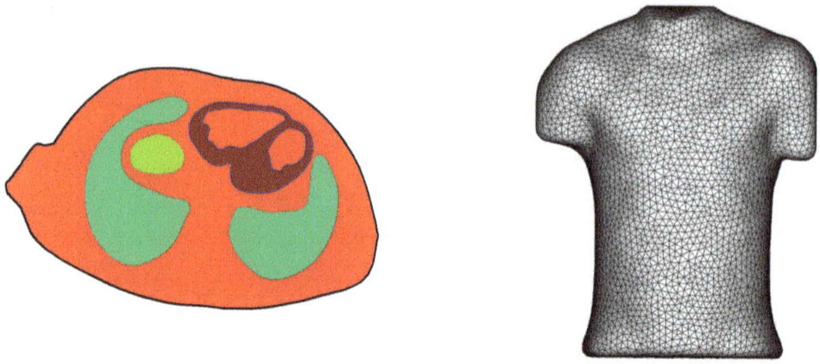

(a) 2D view (b) 3D view

Figure 9. Examples of ECGI-oriented segmentations and mesh generation. (**a**): A 2D cross-section view of a segmented geometry showing the heart (dark red), lungs (green), liver (yellow) and the remaining tissue (red). (**b**): A 3D mesh of a torso volume.

Figure 10 shows an example of typical results that can be obtained using MUSICardio.

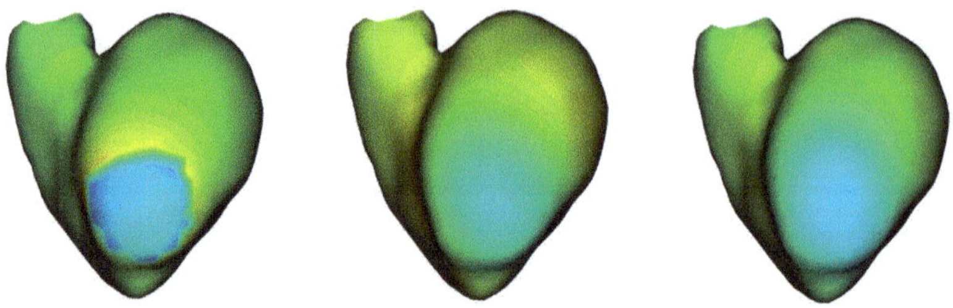

(a) Ground truth (b) MFS results (c) FEM results

Figure 10. Distribution of the electrical potential on the heart surface. Simulated ground truth solution (**a**), MFS solution (**b**) and FEM solution (**c**).

4.6. Artificial Intelligence

Deep Learning-based algorithms developed by our team have been recently included in Python script pipelines and further used in the MUSICardio application. These AI algorithms included in the scripts can improve the accuracy of cardiac segmentation and speed up the analysis and computations, resulting in a less invasive patient diagnostic, a more precise treatment, as well as a better therapy outcome.

The Python support in MUSICardio pipeline allows us to access an unlimited number of calculation libraries, including the AI libraries such as: TensorFlow, PyTorch, and scikit-learn. The pipeline script also has several extended front-end controls of the MUSICardio interface, which are able to facilitate the visualisation of processing steps undertaken during the AI project. The ability to display different type of images and resulted computations is crucial in building a coherent and easy-to-understand pipeline starting from the input image to the final Deep Learning (DL) model prediction. However, the majority of the AI project steps are already built in Python, and these can be easily integrated into the main pipeline.

One of the AI pipelines developed by our team automatically predicts the risk of ventricular arrhythmia (VA) from 3D CT image-based models, as proposed in [14]. This particular pipeline consists of several image processing steps including segmentation of CT images, visualisation of short-axis view orientation, wall thickness calculation, and 2D bullseye map flattening. The bullseye thickness map is then used as input to a VA classification network, after which the classification's attention map is calculated from the prediction score using a gradient back propagation algorithm. The automatic image segmentation is performed using a pre-trained Dual-UNet segmentation network, while the VA classifier network consists of a conditional variational autoencoder and a fully connected classifier model. The two DL networks and all matrix calculations can be integrated directly in the pipeline Python script. The complete image processing pipeline is presented in Figure 11.

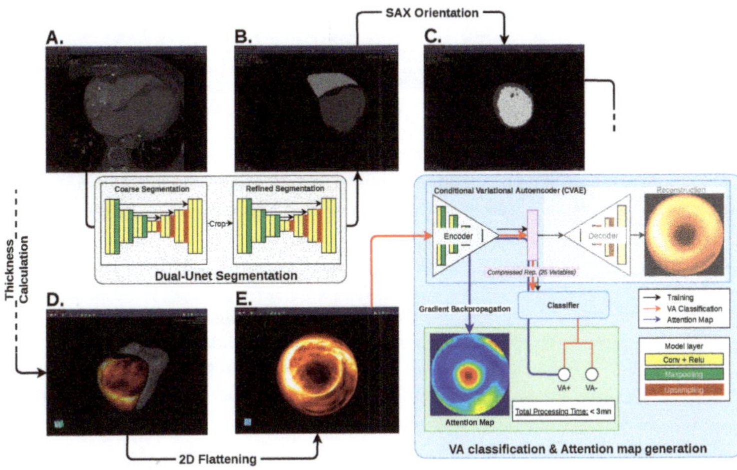

Figure 11. Automatic pipeline for LV (left ventricular) wall thickness map extraction and VA prediction. The arrows illustrate the background calculations and AI predictions. (**A**) Input 3D CT image, (**B**) segmentation masks, (**C**) short-axis oriented masks, (**D**) 3D wall thickness map, (**E**) 2D bullseye thickness map.

The pipeline ends with the VA classification from the 2D wall thickness map using a conditional variational autoencoder and a fully connected classifier model. A so-called "attention map" is also generated from the gradient back propagation of the classification score. The thickness map, attention map and classification score can be displayed at the end of the pipeline, as shown in Figure 12.

The pipeline that was implemented and designed to be used case by case was not adapted for the processing of multiple inputs/outputs required for the development stage of the AI models. Therefore, both AI models were built, trained and validated outside of MUSICardio using the Python environment and packages, including TensorFlow (Deep Learning algorithms) and SimpleITK, Numpy and VTK (image processing). The segmentation network was built using two consecutive 3D UNet models. The first model was used to coarsely segment the heart regions, which was used to crop the region of interest to be used as input in the second segmentation model for refine segmentation. The model achieved a Dice score of 0.90 for left ventricle wall segmentation on the validation dataset. Similarly, the post-infarct VA prediction was trained using a retrospective dataset of 600 patients (27.5% VA). The validation on the test dataset showed the prediction accuracy of 79.2%, sensitivity of 73.2% and specificity of 81.5%. The architectures of the AI networks used in the example pipeline are shown in Figure 11.

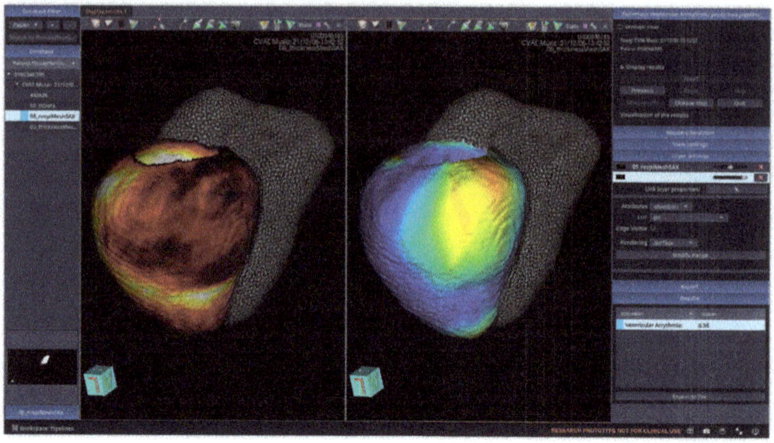

Figure 12. The output results at the end of the pipeline. The VA classification result is shown in the bottom right corner at 98%. The two views show the mesh of the LV wall thickness and classification attention map. The views could be linked together, showing that higher attention (red on left mesh) was placed within the thinning region of the LV (yellow/green on right mesh). The interactive side-by-side display of the thickness and attention map helps to explain the model prediction and increases the user's confidence in the AI tool.

Once a model such as this has been built, it can be implemented in MUSICardio pipelines for direct use: the compatibility of the application to run and install Python packages allows the direct integration of internal and external image processing tools, as well as the pre-trained AI models, into a single MUSICardio pipeline. Thus, MUSICardio proves to be a powerful tool to quickly prototype a fully functioning AI-based project.

4.7. MRI-Guided Radiofrequency Ablation of Cardiac Arrhythmias

Radiofrequency (RF) ablation procedures performed under MRI offer the advantage of a non-ionising imaging, providing that appropriate MR-compatible instrumentation (e.g., Vision-MR Ablation Catheter, Imricor Medical Systems, Burnsville, MN, USA) be used. MRI can provide 3D anatomical images of the heart at the beginning of the therapeutic session prior to real-time catheter navigation, online monitoring of therapy and immediate assessment of lesion transmurality. Therefore, such an integrated workflow does not require 3D registration between data acquired prior to and during the therapeutic procedure. Figure 13 shows an example of the use of MUSICardio in such a scenario on a preclinical study. Three-dimensional (3D) images (balanced ssfp sequence) acquired at the beginning of the session served as a road map and were loaded and processed in MUSICardio application (Figure 13a) in less than 5 min to segment (using the polygon ROI feature) right and left ventricle cavities. The resulting volumes were converted into meshes and then loaded in the prototype Monte Carlo platform provided by the MRI manufacturer (Siemens Healthcare, Erlangen, Germany). Real-time 3D catheter navigation was then performed using a dedicated tip-tracking MRI sequence (measuring X, Y and Z positions of MRI micro coils embedded into the catheter tip at an update rate of 40 ms) to precisely position the catheter at a desired location within the cavity (Figure 13b). Catheter position was displayed online in transparency over the original images and the previously computed 3D meshes. Then, RF energy was delivered under MRI thermometry ([15], data not shown) to monitor treatment progression on the fly. After several RF shots performed at different locations within the heart (positioned under 3D real-time tip tracking sequence), a 3D post-ablation T1-weighted MRI sequence (long TI without injection of contrast agent [16]) was run to visualise the lesion core (irreversible thermal coagulation) and surrounding oedema. Then, the wall thickness map of the left ventricle was computed from the epicardium and endocardium masks using a similar pipeline presented in Section 4.6. The segmentation was initially performed manually with MUSICardio, then a dual-UNet segmentation network was trained on the TI-weighted images to accelerate and automate the segmentation during the intervention (<1 min). An arbitrary threshold was applied on TI-weighted 3D images on the LV wall to automatically extract the mask of the RF lesions. This mask was then multiplied by the wall thickness map to create a measure of the lesion transmurality (expressed in percentage), which is a major parameter to assess the therapeutic success of RF ablation. The output result was finally displayed on the interface (Figure 13c), allowing the user to visualise native T1-weighted images and transmurality map within a single graphical interface.

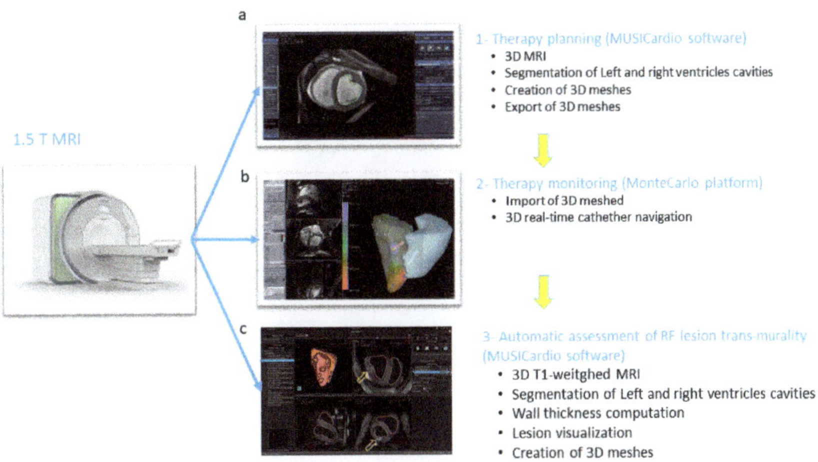

Figure 13. Example of preclinical RF ablation procedure performed under MRI. Three-dimensional (3D) balanced SSFP sequence acquired at the beginning of the session was processed within MUSICardio (**a**) to segment the cavities of the RV and LV. The resulting 3D meshes were then exported to the Montecarlo platform (**b**), allowing real-time 3D catheter navigation and visualisation. After completion of the procedure, 3D T1-weighted images were acquired and processed in MUSICardio (**c**) to automatically compute the transmurality map for each location of RF delivery. Orange arrows show RF lesions with a central thermal necrosis in the hyposignal surrounded by a hypersignal corresponding to oedema.

5. Discussion

The MUSIC project brings together various partners in the cardiology world: hospitals, research institutes, and therefore clinicians, researchers, developers, modelling engineers, database specialists, etc. This mixture of professions and objectives requires communication, adaptation and imagination in order to develop computer tools as well as interfaces that correspond to the greatest number of people. In particular, the tools must be easy to use and adapted to the needs of the medical world, which prefers stable and intuitive tools. They must also be modular enough to be used by the researchers who are less interested in stability than in rapid prototyping of the tools and algorithms, and who may want to change the parameters and the tools employed more frequently. The plurality of our users also requires cross-platform applications, including developing and testing on macOS, Windows and Linux. Thus, a continuous integration process had to be created and improved in order to obtain stable and efficient software tools.

Our quality approach is necessary because of the plurality of our users but also because we can promote our code through spin-off companies. The inHEART (https://www.inheart.fr, accessed on 1 June 2022) spin-off was created jointly by Liryc IHU and Inria, and it is based on a version of our software. It is thus required to have a controlled development processes in order to help with the regulatory process needed for any software as a medical device.

An important issue in the development of clinical and research applications is the long-term funding of personnel and equipment not only for software development and continuous improvement but also for database management, beta testing, management, etc. This is not always present in classical academic funding.

For now, the web portal is accessible only by the members of the VT MUSIC consortium. Our imaging application MUSICardio is accessible by members of our laboratory: researchers and engineers for research purposes, and expert operators and clinicians for the VT MUSIC consortium. The core of MUSICardio (medInria) is open-source and can be tested by the public; however, a lot of the functionalities that we developed are private for now.

Our tools must also manage more and more data, some of which are very heavy. This impacts the technologies we can use but also the minimal hardware that users must have to run our software: processors, graphics card, hard disk, etc. The balance is complicated to find between the power needed to visualise and apply treatments on very large data and the power of the computers that users can have, especially in hospitals or research centres.

6. Perspectives

The MUSIC project continues to be developed and improved. We are currently working on a future version of the MUSICardio 4 software in close collaboration with the Inria team that handles medInria. This new version will include new algorithms, tools, and will be based on medInria 4 on which we work also. New features will include an improvement of the global GUI of the software, the data management system as well as the management of plugins.

We have started an exploratory work of real-time simulation of catheters during a cardiac intervention in MUSICardio. This could allow us to display in real time a catheter in the application and to superimpose it with patient images or meshes to study their relative positions.

We also want to develop augmented reality tools to display and manipulate meshes in MUSICardio. This will allow us to study mesh in more detail as well as communicate about our tools in scientific events, and it is a first step to study the use of augmented reality in diagnostic or operative cardiology.

It would also be interesting to certify our binaries by Microsoft and Apple software platforms to simplify their installation. This would allow an official installation in hospitals where computer security is omnipresent: clinicians often do not have the computer rights to install unofficial software, so it can take a long time or a complicated route to install new software on these computers.

In addition, an online mesh viewer developed in Three.js (https://threejs.org, accessed on 1 June 2022) has been developed. It is currently deployed on the network of our laboratory for beta-testing, and it aims to replace the mesh viewer (desktop application) provided by our team to hospitals inside the VT MUSIC consortium. The online mesh viewer could eventually be included in the web portal for easier use by users. We also planned to include in the web portal the anonymisation tool, which is also a desk application for now. Concerning the web portal itself, we would like to improve it by updating or even changing the technology used and adding features for the users: pause/restart uploads, etc.

We will continue to develop the VT MUSIC consortium to welcome new hospital partners and to enable patients to receive the best possible cardiac treatments.

Our collaboration in the medInria consortium will allow us to organise the improvement of the core application used in MUSICardio and to be a driving force in the development of these innovative tools.

Finally, MUSIC is ideally suited to develop novel approaches combining artificial intelligence and biophysical modelling, being at the interface of these two scientific areas.

7. Conclusions

The exceptional improvement of the cardiac field in the past years has led to a critical need of better multimodal technologies. The MUSIC platform is an emerging and robust technology which allows the user to import, export and work on a large spectrum of imaging data from various modalities and modelling approaches used in the cardiac field, offering functionalities dedicated to cardiovascular diagnosis and therapy guidance. The VT MUSIC consortium brings international hospitals together around an innovative interface to perform patient-specific ablation procedures, helping doctors to improve the lives of many patients every day.

Author Contributions: Software, M.M., F.C., J.C., P.M., M.J., B.L. and V.O.; Supervision, B.Q., N.Z., Y.C., P.J., H.C. and M.S.; Writing—original draft, M.M., F.C., J.C., P.M., M.J., B.L. and V.O.; Writing—review & editing, M.S. All authors have read and agreed to the published version of the manuscript.

Funding: This work has benefited from funding provided by: ANR (French National Research Agency): IHU Liryc, L'Institut de RYthmologie et modélisation Cardiaque, reference 10-IAHU-0004, 2011/2024. ANR: EQUIPEX MUSIC, Plateforme multi-modale d'exploration en cardiologie, reference 11-EQPX-0030, 2012/2021. Inria: ADT IDAM, engineering for the integration of cardiac electrophysiology images, data and models, 2015/2017. Inria: ADT EPICARD-VIZ, electrocardiographic imaging tool: simulation and visualization, 2016/2018. Inria: ADT CIESCARD, combining electrical and structural information to help cardiologists better target cardiac therapy, 2018/2020. ERC: Starting Grant ECSTATIC, Electrostructural Tomography—Towards Multiparametric Imaging of Cardiac Electrical Disorders, reference 715093, 2017/2022 (https://www.ihu-liryc.fr/fr/ecstatic/ accessed on 1 June 2022).

Conflicts of Interest: The authors declare no conflict of interest.

References

1. McCormick, M.; Liu, X.; Ibanez, L.; Jomier, J.; Marion, C. ITK: Enabling reproducible research and open science. *Front. Neuroinform.* **2014**, *8*, 13. [CrossRef] [PubMed]
2. Lorenzi, M.; Ayache, N.; Frisoni, G.; Pennec, X. LCC-Demons: A robust and accurate symmetric diffeomorphic registration algorithm. *NeuroImage* **2013**, *81*, 470–483. [CrossRef] [PubMed]
3. Guennebaud, G.; Jacob, B.; Avery, P.; Bachrach, A.; Barthelemy, S. Eigen v3. Available online: http://eigen.tuxfamily.org (accessed on 1 June 2022).
4. Dapogny, C.; Dobrzynski, C.; Frey, P. Three-dimensional adaptive domain remeshing, implicit domain meshing, and applications to free and moving boundary problems. *J. Comput. Phys.* **2014**, *262*, 358–378. [CrossRef]
5. Si, H. TetGen, a Delaunay-Based Quality Tetrahedral Mesh Generator. *ACM Trans. Math. Softw.* **2015**, *41*, 11. [CrossRef]
6. Vercauteren, T.; Pennec, X.; Perchant, A.; Ayache, N. Diffeomorphic demons: Efficient non-parametric image registration. *NeuroImage* **2009**, *45*, S61–S72. [CrossRef] [PubMed]
7. Gepstein, L.; Hayam, G.; Ben-Haim, S.A. A Novel Method for Nonfluoroscopic Catheter-Based Electroanatomical Mapping of the Heart. *Circulation* **1997**, *95*, 1611–1622. [CrossRef]
8. Koutalas, E.; Rolf, S.; Dinov, B.; Richter, S.; Arya, A.; Bollmann, A.; Hindricks, G.; Sommer, P. Contemporary Mapping Techniques of Complex Cardiac Arrhythmias-Identifying and Modifying the Arrhythmogenic Substrate. *Arrhythm Electrophysiol. Rev.* **2015**, *4*, 19. [CrossRef]
9. Cuculich, P.S.; Schill, M.R.; Kashani, R.; Mutic, S.; Lang, A.; Cooper, D.; Faddis, M.; Gleva, M.; Noheria, A.; Smith, T.W.; et al. Noninvasive Cardiac Radiation for Ablation of Ventricular Tachycardia. *N. Engl. J. Med.* **2017**, *377*, 2325–2336. [CrossRef]
10. Gianni, C.; Rivera, D.; Burkhardt, J.D.; Pollard, B.; Gardner, E.; Maguire, P.; Zei, P.C.; Natale, A.; Al-Ahmad, A. Stereotactic arrhythmia radioablation for refractory scar-related ventricular tachycardia. *Heart Rhythm* **2020**, *17*, 1241–1248. [CrossRef] [PubMed]
11. Barr, R.C.; Ramsey, M.; Spach, M.S. Relating epicardial to body surface potential distributions by means of transfer coefficients based on geometry measurements. *IEEE Trans. Biomed. Eng.* **1977**, 1–11. [CrossRef] [PubMed]

12. Wang, Y.; Rudy, Y. Application of the method of fundamental solutions to potential-based inverse electrocardiography. *Ann. Biomed. Eng.* **2006**, *34*, 1272–1288. [CrossRef] [PubMed]
13. Karoui, A.; Bear, L.; Migerditichan, P.; Zemzemi, N. Evaluation of fifteen algorithms for the resolution of the electrocardiography imaging inverse problem using ex-vivo and in-silico data. *Front. Physiol.* **2018**, 1708. [CrossRef] [PubMed]
14. Ly, B.; Finsterbach, S.; Nuñez-Garcia, M.; Cochet, H.; Sermesant, M. Scar-Related Ventricular Arrhythmia Prediction from Imaging Using Explainable Deep Learning. In *Lecture Notes in Computer Science (Including Subseries Lecture Notes in Artificial Intelligence and Lecture Notes in Bioinformatics)*; Springer: Berlin, Germany, 2021; pp. 461–470. [CrossRef]
15. Yon, M.; Delcey, M.; Bour, P.; Grissom, W.; Quesson, B.; Ozenne, V. Continuous cardiac thermometry via simultaneous catheter tracking and undersampled radial golden angle acquisition for radiofrequency ablation monitoring. *Sci. Rep.* **2022**, *12*, 4006. [CrossRef] [PubMed]
16. Toupin, S.; Bour, P.; Lepetit-Coiffé, M.; Ozenne, V.; de Senneville, B.D.; Schneider, R.; Vaussy, A.; Chaumeil, A.; Cochet, H.; Sacher, F.; et al. Feasibility of real-time MR thermal dose mapping for predicting radiofrequency ablation outcome in the myocardium in vivo. *J. Cardiovasc. Magn. Reson.* **2017**, *19*, 1–12. [CrossRef] [PubMed]

applied sciences

MDPI

Article

Cardiac Radiofrequency Ablation Simulation Using a 3D-Printed Bi-Atrial Thermochromic Model

Shu Wang [1,†], Carlo Saija [1,†], Justin Choo [1], Zhanchong Ou [1], Maria Birsoan [1], Sarah Germanos [1], Joshua Rothwell [1], Behrad Vakili [1], Irum Kotadia [1], Zhouyang Xu [1], Adrian Rolet [1], Adriana Namour [1], Woo Suk Yang [1], Steven E. Williams [1,2,‡] and Kawal Rhode [1,*,‡]

1 Department of Surgical & Interventional Engineering, School of Biomedical Engineering & Imaging Sciences, King's College London, St. Thomas Hospital, London SE1 7EH, UK; shu.1.wang@kcl.ac.uk (S.W.); carlo.saija@kcl.ac.uk (C.S.); justin.choo@kcl.ac.uk (J.C.); zhanchong.ou@kcl.ac.uk (Z.O.); maria.birsoan@kcl.ac.uk (M.B.); sarah.germanos@kcl.ac.uk (S.G.); joshua.rothwell@kcl.ac.uk (J.R.); behrad.vakili@kcl.ac.uk (B.V.); irum.kotadia@kcl.ac.uk (I.K.); zhouyang.xu@kcl.ac.uk (Z.X.); adrian.rolet@kcl.ac.uk (A.R.); adriana.namour@kcl.ac.uk (A.N.); woo.s.yang@kcl.ac.uk (W.S.Y.); steven.e.williams@kcl.ac.uk (S.E.W.)

2 Centre for Cardiovascular Science, The University of Edinburgh, Edinburgh EH8 9YL, UK

* Correspondence: kawal.rhode@kcl.ac.uk

† These authors contributed equally to this work.

‡ These authors contributed equally to this work.

Featured Application: The simulator has applications in the training of electrophysiologists for cardiac radio-frequency ablation therapy and the evaluation of novel cardiac ablation devices.

Abstract: Radiofrequency ablation (RFA) is a treatment used in the management of various arrhythmias including atrial fibrillation. Enhanced training for electrophysiologists through the use of physical simulators has a significant role in improving patient outcomes. The requirements for a high-fidelity simulator for cardiac RFA are challenging and not fully met by any research or commercial simulator at present. In this study, we have produced and evaluated a 3D-printed, bi-atrial model contained in a custom-made enclosure for RFA simulation using a new soft tissue-mimicking polymer, Layfomm-40, combined with thermochromic pigment and barium sulphate in an acrylic paint carrier. We evaluated the conductive properties of Layfomm-40, its sensitivity to RFA, and its visibility in X-ray imaging, and carried a full simulation of RFA in the cardiac catheterization laboratory by an electrophysiologist. We demonstrated that a patient-specific 3D-printed Layfomm-40 bi-atrial model coated with a custom thermochromic/barium sulphate paint was compatible with the CARTO3 electroanatomic mapping system and could be effectively imaged using X-ray fluoroscopy. We demonstrated the effective delivery and visualization of radiofrequency ablation lesions in this model. The simulator meets nearly all the requirements for high-fidelity physical simulation of RFA. The use of such simulators is likely to have impact on the training of electrophysiologists and the evaluation of novel RFA devices.

Keywords: electrophysiology; cardiac radiofrequency ablation; 3D-printing; Layfomm-40; physical simulation; simulation training; thermochromic pigments

Citation: Wang, S.; Saija, C.; Choo, J.; Ou, Z.; Birsoan, M.; Germanos, S.; Rothwell, J.; Vakili, B.; Kotadia, I.; Xu, Z.; et al. Cardiac Radiofrequency Ablation Simulation Using a 3D-Printed Bi-Atrial Thermochromic Model. *Appl. Sci.* **2022**, *12*, 6553. https://doi.org/10.3390/app12136553

Academic Editors: Mihaela Pop and Cristian A. Linte

Received: 11 May 2022
Accepted: 27 June 2022
Published: 28 June 2022

1. Introduction

Cardiac ablation therapy is a minimally invasive interventional procedure used in the treatment of cardiac arrhythmias. The process involves the insertion of flexible catheters through peripheral blood vessels, which are guided to the site of abnormal electrical conduction in the myocardium. Here, the arrhythmia is terminated via destruction of the pathological tissue, most commonly using radiofrequency ablation (RFA). The procedure is performed in the cardiac catheterization laboratory under X-ray fluoroscopy guidance and often with the use of an electroanatomic mapping system (EAMS), which tracks the inserted catheters and allows measurement of the patient's cardiac anatomy and electrophysiology. Hong et al. describe the current strategies and technologies for the ablation of the most common arrhythmia, atrial fibrillation [1].

RFA procedures are complex and require considerable training, sometimes delivered via computer or physical simulators. Computers and physical simulators have an ethical and cost advantage but may lack fidelity. An example of a computer simulator is the Mentice VIST (Mentice AB, Göteborg, Sweden), which has been shown to improve electrophysiology-trainee performance [2]. The requirements for a high-fidelity physical simulator for cardiac RFA make it challenging to produce such a simulator. These requirements include representation of the cardiac anatomy, soft material properties, realistic appearance under imaging (particularly X-ray fluoroscopy), compatibility with EAMSs, sensitivity to RFA, and generation of electrophysiological signals. We are not aware of any physical simulator that meets all these criteria. However, several simulators have been developed that meet a subset of the criteria. Rossi et al. [3] developed a physical simulator using a 3D-printed whole heart model embedded into a custom torso. The heart was printed using thermoplastic polyurethane and the simulator was compatible with the CARTO3 EAMS (Biosense Webster, Irvine, CA, USA). The simulator was evaluated by 10 electrophysiologists and used to compare novel to experienced operators. Similar simulators are produced by Heartroid (JMC Corporation, Yokohama, Japan, https://www.heartroid.com/itemlist/ablation/, accessed on 1 December 2021) and Pangolin (Tel-Aviv, Israel, http://pangolin.co.il/en/gallery/simulators-endo-vascular/, accessed on 1 December 2021). However, none of these physical simulators have sensitivity to RFA, meaning that ablation lesions cannot be created in these simulators and as a result, effects of intended therapies delivered by trainees cannot be quantified. Lesion formation depends on the parameters of the RFA, such as power, duration, ablation temperature and importantly, the contact force between the ablation catheter and the tissue [4]. Therefore, a simulator that allows creation of lesions will add a valuable layer to assessing a trainee's progression.

Several attempts have been made to create a physical simulation medium that is sensitive to RFA and therefore able to demonstrate lesions. Bu-Lin et al. [5] used a polyacrylamide gel with bovine albumin which produced a noticeable color change after RFA due to coagulation between 50 and 60 °C. Negussie et al. [6] proposed the use of thermochromic pigments to create RFA-sensitive models that produced a permanent color change above 60 °C. However, there have been no attempts to incorporate ablation-sensitivity into a cardiac RFA simulator.

3D-printed models are widely used in the medical field for a variety of purposes. Current applications include, but are not limited to, implant and prosthetic design, biomedical device testing, and, notably, pre-operative/procedural planning and surgical/interventional simulation training [7]. Furthermore, advances in the field of additive manufacturing have facilitated the production of objects with complex geometries. Notably, the incorporation of patient scans as the basis for model design enables the fabrication of patient-specific simulation aids bearing greater anatomical accuracy. This is particularly relevant in the rehearsal of complex surgical/interventional procedures, also allowing for anatomical variation between patients. In our previous work, we evaluated the use of 3D-printed thermoplastics for creating patient-specific whole heart models that were multimodal-imaging-compatible [8]. We investigated a low-cost, soft-tissue-mimicking copolymer filament, known as Layfomm-40 from the Poro-Lay series (CC-Products, Köln, Germany) [9,10]. Layfomm-40 filament is rigid and consists of polyvinyl alcohol (PVA) and a thermoplastic elastomer (TPE) composite. This allows the material to be 3D printed using a fusion deposition modelling (FDM) printer at low cost. Once the 3D print is soaked in water, the PVA dissolves and leaves a spongey, microporous TPE composite which mimics soft tissue. In our work, we found Layfomm-40 to be an excellent material for creating cardiac models in terms of soft material properties and imaging properties [8].

The aim of this work was to evaluate the use of Layfomm-40 for creating a physical cardiac RFA simulator that could satisfy as many as possible of the requirements mentioned previously. We investigated the electrical and thermal conductivities of Layfomm-40 to ensure compatibility with RFA and EAMSs, investigated the use of thermochromic pigments for RFA-sensitivity, and investigated techniques for optimum visibility of Layfomm-40 under X-ray fluoroscopy. We integrated our findings to produce a bi-atrial, patient-specific model that was housed in a custom enclosure, and we tested this simulator in the cardiac catheterization laboratory environment to prove the overall concept.

2. Materials and Methods

2.1. Layfomm-40 Conductivity Analysis

For a 3D-printed Layfomm-40 model to be compatible with RFA and EAMSs, it must be both electrically and thermally conductive. We 3D printed five 10 mm cubes of Layfomm-40 and immersed these in saturated saline solution for 3 days. The concentration of the saline solution affects conductivity and using a saturated solution gave the maximum conductivity achievable. Electrical conductivity testing was completed using an AstroAI multimeter (AstroAI, Garden Grove, CA, USA) using the method shown in Figure 1a. The conductivity, σ, was calculated using Equation (1), where R denotes electrical resistance, A denotes the cross-sectional area, and L denotes the current path length.

$$\sigma = \frac{L}{R \times A} S \cdot m^{-1} \tag{1}$$

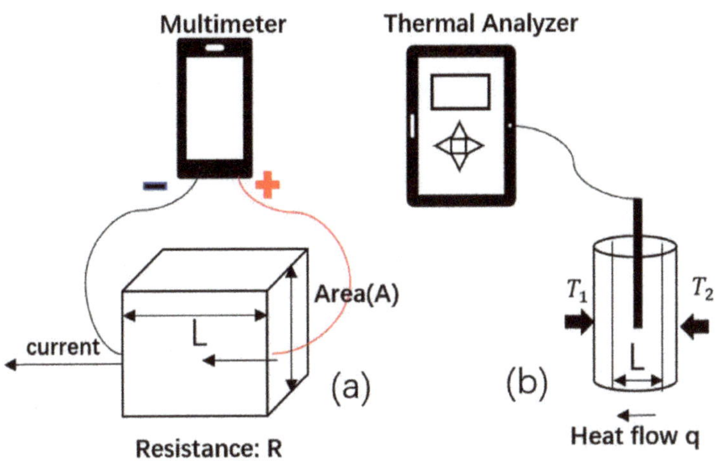

Figure 1. Illustration of (**a**) electrical conductivity measurement using a multimeter and (**b**) thermal conductivity measurement using a thermal property analyzer.

The thermal conductivity measurement was conducted with a KD2 Pro Thermal Analyzer (Decagon, Pullman, WA, USA) using the method shown in Figure 1b. The prepared samples of Layfomm-40 were 100 mm long, 15 mm thick, and had a 2.4 mm diameter lumen to insert the needle probe. Five samples were immersed in saturated saline solution for 3 days. The thermal conductivity was calculated using Equation (2), where q denotes the heat flow, κ denotes the thermal conductivity, $T_2 - T_1$ corresponds to the temperature difference, and L represents the travel distance of heat flow.

$$q = \kappa \cdot \frac{T_2 - T_1}{L} \tag{2}$$

2.2. Thermochromic Paint Formulation and RFA-Sensitivity

Thermochromic pigments display a color-changing effect, which is induced by heating or cooling [11,12]. In terms of whether the original color can be recovered, thermochromic materials can be classified as reversible or irreversible. Reversible pigments are usually composed of a color developer, a color former, and a co-solvent. On heating, the co-solvent changes from solid to liquid and allows the former to separate from the developer, leading to the color change. On cooling, the process is reversed (Figure 2a). Irreversible pigments have a volatile dye that evaporates when heated to the phase transition temperature, resulting in a permanent colorless state (Figure 2b).

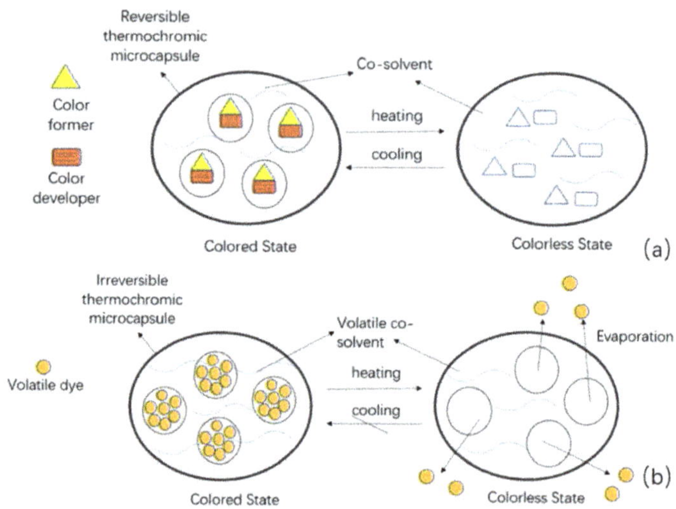

Figure 2. Mechanisms of action of thermochromic microcapsules during heating and cooling. (**a**) Reversible and (**b**) irreversible.

Since we want the color change to be permanent and occur at typical temperatures required for RFA lesion formation, we selected an irreversible thermochromic pigment with a transition temperature of 60 °C. This pigment was black below the transition temperature and colorless above this temperature (Special FX Creative, Newhaven, UK, https://www.sfxc.co.uk/products/sfxc-irreversible-thermochromic-pigment-60-c-black, accessed on 10 November 2021). The pigment was mixed with (white) unpigmented acrylic paint (UAP) (10:1 UAP:pigment mass ratio) to form a grey thermochromic paint that could be applied to the 3D-printed Layfomm-40.

Several discs (radius 20 mm and thickness 5 mm) of Layfomm-40 were printed and painted with the thermochromic paint. These were soaked in 0.9 w/w saline solution and then tested for RFA-sensitivity. The discs were immersed in a saline bath (~38 °C) and radiofrequency was delivered through a Stockert 70 Cardiac Ablation RF Generator (Biosense Webster, USA) and a non-irrigated ablation catheter. The catheter tip was kept orthogonal to the disc plane. Ablations of 60 s duration were applied while varying the power (30–90 W) and temperature (50–80 °C) settings. Only one ablation was applied for each combination of power and temperature. Ablations were also applied to a piece of chicken breast for comparison. The diameter of the lesions was measured for each setting with a digital caliper. Once parameters for consistent size lesion creation were identified, the experiment was repeated using a $50 \times 50 \times 2$ mm^3 sample of thermochromic-paint-coated Layfomm-40 which was divided into nine subsections. Nine ablations were applied keeping the power and temperature settings between 70–80 W and 70–80 °C, respectively. The lesion sizes were measured, and the average size computed for these power/temperature settings.

2.3. Inreasing X-ray Visibility

In our previous work [8], we found the X-ray attenuation of Layfomm-40 cardiac models was only marginally different to that of water. Therefore, since the models need to be immersed in a water tank as part of the simulation, these un-modified models are not easily visualized in X-ray fluoroscopy. We investigated the use of barium sulphate ($BaSO_4$) to increase the model visibility. $BaSO_4$ is a commonly used X-ray contrast agent and is often incorporated into polymers that need to be X-ray visible. We doped UAP with $BaSO_4$ in a range of mass ratios from 5:2 (UAP:$BaSO_4$) to 15:1. Hollow tubes of Layfomm-40 were prepared and coated with the different doped UAP formulations. These were soaked in saline for 24 h. These were then immersed in a tank filled with 0.9% w/w saline and imaged using our cardiac catheterization laboratory (Siemens Artis Q Biplane, Siemens Healthineers, Erlangen, Germany) using our standard electrophysiology X-ray protocol. We exported the captured image data and analyzed the maximum percentage image contrast between the walls of the sample (Layfomm-40 with doped UAP) and the background (saline) based on region-of-interest analysis. This analysis was performed with ImageJ (https://imagej.nih.gov/ij/index.html, accessed on 17 November 2021). For reference, we calculated the percentage contrast between the cardiac shadow and the surrounding lung tissue in representative clinical X-ray fluoroscopy images taken from patients in the same catheterization laboratory using the same protocol.

2.4. Atrial Model and Simulator Build-Up

Having verified the RFA-sensitivity and formulated the thermochromic and $BaSO_4$-doped paints, we proceeded to develop the complete simulator. To ensure reproducibility and cost-effectiveness of the simulator, most parts were manufactured using FDM or stereolithography (SLA) 3D-printing. The simulator comprises four main components: the cardiac model, its base, a transparent tank and its lid (with patch holders).

The computer model of the atria and associated great vessels was designed via medical image segmentation and processing. An adult male contrast-enhanced chest computer tomography (CT) scan was segmented using the semi-automatic segmentation feature of ITK-SNAP (University of Pennsylvania, Philadelphia, PA, USA). Following this initial step, the segmentation was refined using the smoothing tool (Level 2) in Seg3D (University of Utah, Salt Lake City, UT, USA). Subsequently, the dilation-erosion function was applied to create an inner mold of this segmentation. The final hollow model was the differential result between the segmentation and the inner mold, giving a wall thickness of 1 mm. This model was exported to Fusion360 (Autodesk, San Rafael, CA, USA) and extruded by 1.5 mm to give an overall wall thickness of 2.5 mm. The pulmonary veins were cut to a length of 20 mm. The venae cavae were cut and lofted to a standard-diameter tube fitting size (inner diameter: 16 mm, outer diameter: 23 mm). The model was divided into two sections for 3D printing to allow for the use and easy removal of support material. The exported meshes were sliced using Cura (Ultimaker, Utrecht, The Netherlands) and manufactured using a Chiron FDM 3D-printer (Anycubic, Shenzhen, China). A transeptal puncture was made in the atrial septum for right-to-left access. Paint was applied to the inner surface of model in four layers with a 3 h drying time between each layer. Following this, the two sections of the model were welded together using a digital soldering iron, ensuring not to transfer heat to the inner coats of paint. Finally, the outside of the model was coated with paint in a similar manner to the inside.

The simulator system was contained within a $48 \times 39 \times 31$ cm transparent box (Really Useful Plastic Box, Badford, UK) with catheter entry points connecting to each of the venae cavae via silicone tubing. The entry point for the inferior vena cava was extended outside the box using silicone tubing to simulate realistic pathlength from femoral venous access sites. The box was filled to a depth of 25 cm with 38 °C 0.9% w/w saline solution to simulate the human thorax and its conductive properties. The lid of the box was designed to allow compatibility with the CARTO3 EAMS. Six tubes were integrated into the lid to accommodate the six CARTO3 patches, with three patches above the heart model (simulating the patches on a patient's anterior chest wall) and three at the bottom (simulating the patches on the posterior chest wall). The tubes were constructed using standard polyvinylchloride (PVC) pipes and gasketed screw-on end-cap fittings with polytetrafluoroethene tape to ensure no leaks were present. The model base was designed using Fusion360 to hold the model in place during the simulation and 3D printed using a Photon SLA printer (Anycubic, China). The model rests on the base with a foam insert, custom clamps hold the venae cavae, and a hook-and-loop strap goes around the body of the model. This arrangement allows models with variable geometry to be firmly held in the simulator and to be inserted and removed easily.

2.5. X-ray Imaging, Mapping, and Ablation

The simulator was taken to the cardiac catheterization laboratory, placed on the patient table and connected to the CARTO3 system. X-ray fluoroscopy imaging was performed at several standard view angles using the standard electrophysiology protocol on the system. A cone beam CT scan was performed using the standard protocol.

An 8F ThermoCool SmartTouch SF (Biosense Webster, USA) ablation catheter was inserted via an 11 cm 8F introducer sheath (Cordis, Santa Clara, CA, USA) into the inferior vena cava. The ablation catheter was connected to the CARTO3 system, the irrigation system, and the RF generator. The right side of the model was mapped to generate the geometry of the right atrium and the venae cavae. The introducer sheath was replaced with a 60 cm 8.5F 55° Heartspan Transseptal Fixed Sheath (Biotronik, Berlin, Germany). This was manipulated under X-ray fluoroscopy guidance to pass the ablation catheter into the left atrium through the transseptal puncture in the heart model. Mapping was then performed of the left atrium.

Ablations were performed only in the right side of the heart model. A line of five ablations was performed at the superior aspect of the posterior intercaval line with increasing ablative power. From inferior to superior, the power settings for each ablation point were as follows: 15.9 W; 16.0 W; 22.0 W; 23.0 W; and 30.6 W. The ablation duration was fixed to 45 s. Subsequently, using a fixed power setting of 21.5 W and the same duration, a line of five ablations was performed in the posterior wall of the right atrium. The heart model was then removed from the simulator and cut open to examine and measure the lesions.

3. Results

3.1. Layfomm-40 Conductivity Analysis

The electrical conductivity of Layfomm-40 immersed in saturated saline solution was determined to be from 1.3×10^{-7} to 3.0×10^{-6} S/m. The thermal conductivity was determined to be from 0.34 to 0.45 W/m/K. In comparison, literature values for myocardium are 0.16 S/m [13] and 0.56 W/m/K [14], respectively. The electrical conductivity was much lower than the physiological value, but the thermal conductivity was similar.

3.2. Thermochromic Paint Formulation and RFA-Sensitivity

The Layfomm-40 discs with thermochromic paint coating were confirmed to be RFA-sensitive. Table 1 shows that lesions were formed on the discs with temperatures ≥60 °C and power settings ≥40 W. The diameter of the lesions best matched those in the chicken breast (Figure 3a) when the temperature and power were ≥70 °C and ≥70 W, respectively. Settings below these either produced no lesions or inconsistently sized lesions. Using temperatures and powers above 90 °C and 90 W could produce a maximal lesion diameter of approximately 7 mm (Figure 3b). Consistent lesions were produced with temperatures of 70–80 °C and powers of 70–80 W (Figure 3c). The average lesion diameter for these settings was measured to be 3.3 ± 0.3 mm (±1 SD, n = 9).

Table 1. Ablation lesion sizes (mm) on thermochromic Layfomm-40 discs using different ablation temperatures and power settings. The ablation duration was 60 s.

Ablation Temperature (°C)/Power (Watts)	50 °C	60	70	80
30 W	NA	NA	NA	NA
40	NA	NA	2.7	2.4
50	NA	2.3	1.3	1.6
60	NA	1.4	1.8	3.3
70	NA	2.4	3.2	3.3
80	NA	1.9	3.0	3.2
90	NA	2.5	3.2	3.5

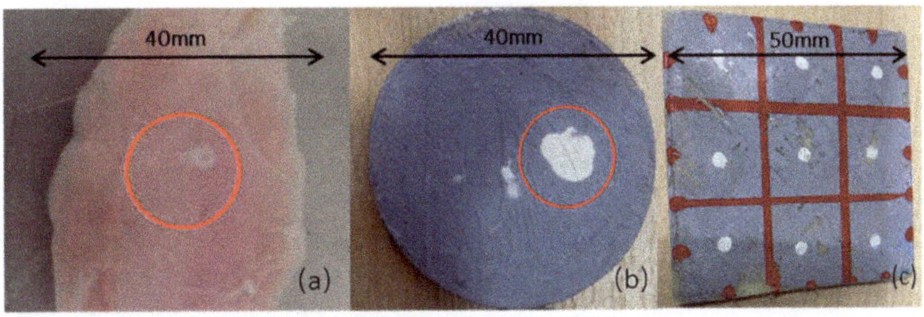

Figure 3. (**a**) Ablation lesion in chicken breast at 70 W, 70 °C, 60 s with a lesion diameter of ~3 mm, (**b**) Layfomm-40 disc with thermochromic paint coating showing the maximum size lesion that could be created at >90 W, >80 °C, 60 s with a lesion diameter of ~7 mm, and (**c**) Layfomm-40 square with thermochromic paint coating showing nine lesions at 70–80 W, 70–80 °C, 60 s with lesion diameters of ~3 mm.

3.3. *Increasing X-ray Visibility*

Figure 4 shows the effect of increasing the concentration of $BaSO_4$ in the acrylic paint on X-ray image contrast. Table 2 show the calculated maximum percentage image contrast for the different mass ratios with a value computed from clinical image data for comparison. It was found that the 5:1 mass ratio gave a contrast that best matched what was seen in clinical images.

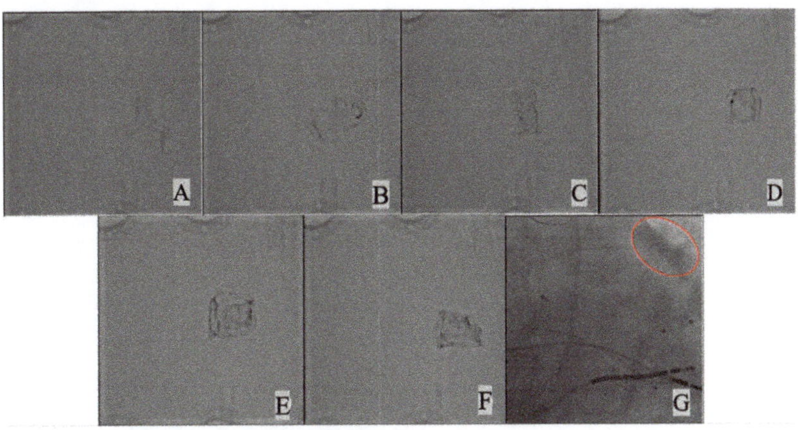

Figure 4. X-ray imaging of Layfomm-40 cylinders coated with UAP mixed with increasing amounts of $BaSO_4$. (**A**) mass ratio15:1 UAP:$BaSO_4$, (**B**) 15:2, (**C**) 5:1, (**D**) 15:2, (**E**) 3:1, (**F**) 5:2, and (**G**) clinical image showing the left heart border (red oval).

Table 2. The maximum percentage image contrast for $BaSO_4$–UAP coated Layfomm-40 cylinders at different UAP:$BaSO_4$ mass ratios compared to myocardial contrast achieved in clinical images.

UAP:$BaSO_4$	15:1	15:2	5:1	15:4	3:1	5:2	Myocardium
% Contrast	5.56	15.6	23.6	28.2	35.5	40.0	21.0

Combining the findings from 3.2 and 3.3, a strategy was formulated for the coating of Layfomm-40 for the heart model. Figure 5 illustrates the architecture of the coating. There are four layers of paint applied both externally and internally consisting of a sandwich of UAP, UAP doped with $BaSO_4$ and UAP doped with thermochromic pigment.

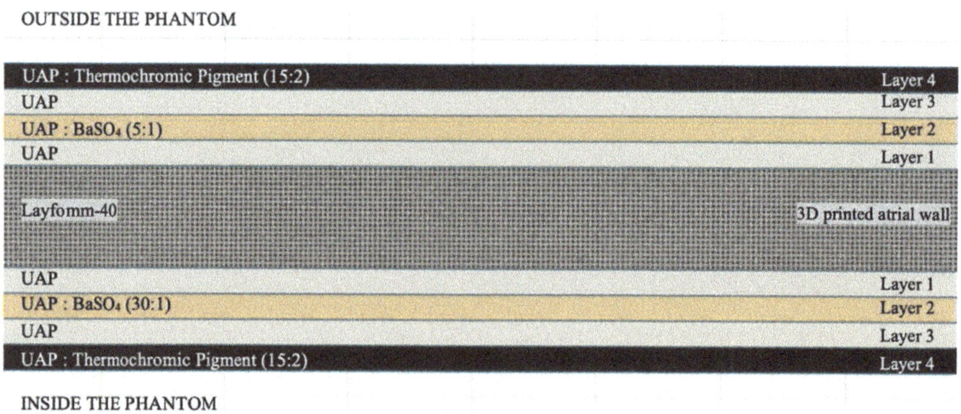

Figure 5. Architecture of custom paint coatings of the Layfomm-40 model.

3.4. Atrial Model and Simulator Build-Up

Figure 6 shows the solid and hollow computer model that was generated by image segmentation and processing of the patient CT data. Figure 7 shows the steps to produce the bi-atrial model from the computer model.

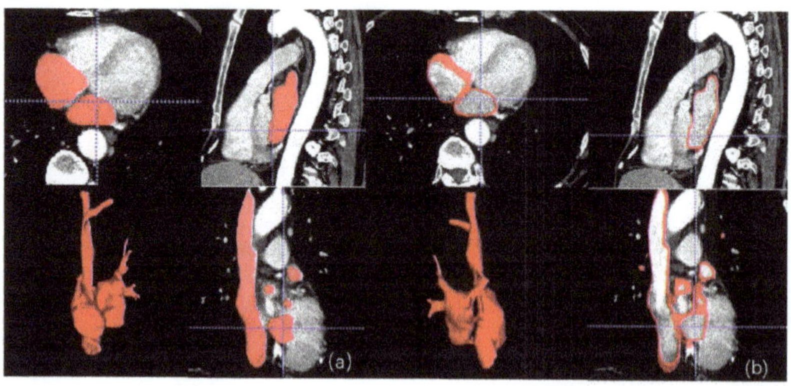

Figure 6. (**a**) Solid and (**b**) hollow bi-atrial computer model showing the segmentation in multiplanar views and a 3D rendering.

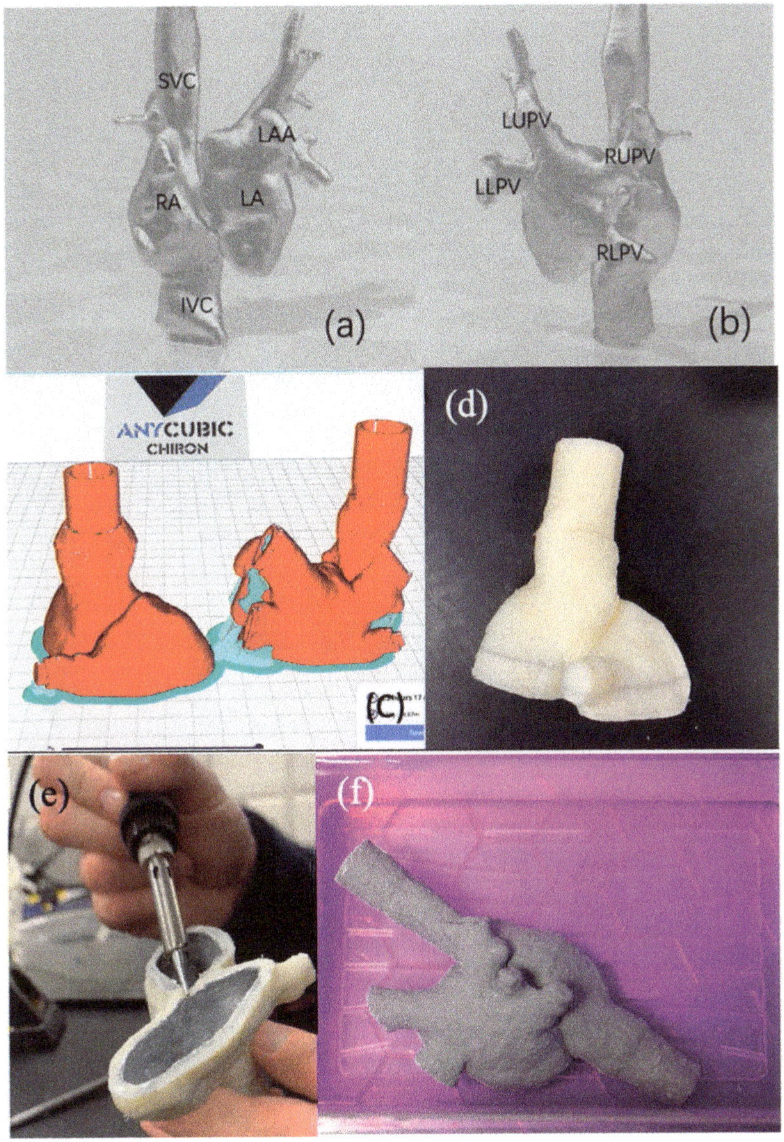

Figure 7. Bi-atrial model rendering from (**a**) anterior view and (**b**) posterior view. (**c**) Bi-atrial model cut into two sections for ease of printing and coating. (**d**) Lower section printed in Layfomm-40. (**e**) Lower section coated internally with paint layers and ready for plastic welding. (**f**) Fully welded and external-paint-coated bi-atrial model soaked in saline solution. SVC—superior vena cava, IVC—inferior vena cava, RA—right atrium, LA—left atrium, LAA—left atrial appendage, LUPV—left upper pulmonary vein, LLPV—left lower pulmonary vein, RUPV—right upper pulmonary vein, and RLPV—right lower pulmonary vein.

Figure 8 shows the features of the simulator base and enclosure. Arrangement of the CARTO3 patches can be seen as well as the methods used to allow entry of devices into the heart model.

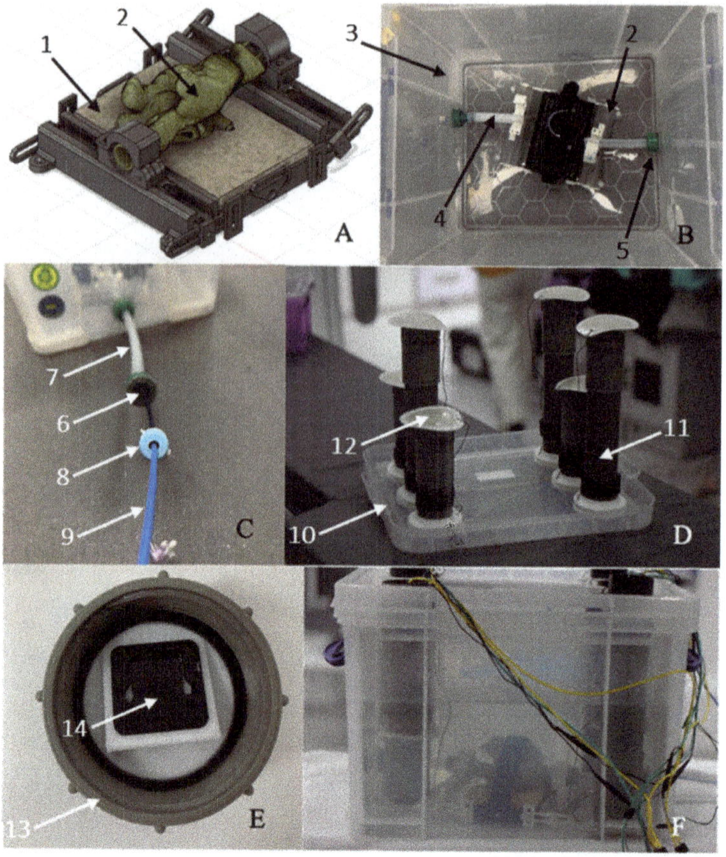

Figure 8. (**A**) Fusion360 model of simulator base (1) with inserted bi-atrial model (2). (**B**) Simulator base inside the Really Useful Box (3) with connecting silicone tubing (4) leading to standard hose connectors (5) with silicone plugs (6) to act as hemostatic valves. (**C**) Silicone tubing extension (7) to IVC entry point showing insertion of a sheath (8) and ablation catheter (9). (**D**) lid assembly (10) with PVC tubes (11) to accommodate the CARTO3 patches (12). (**E**) PVC tube end cap (13) with CARTO3 patch connector (14). (**F**) Complete simulator assembly with wiring to CARTO3 system and RF generator.

3.5. X-ray Imaging, Mapping, and Ablation

Figure 9 show examples of X-ray fluoroscopy images taken during the simulated procedure. The bi-atrial model is clearly visible and there is good epicardial contrast between the Layfomm-40 and the surrounding saline solution. The cone beam CT image (512 × 512 × 400 matrix size, 0.46 mm^3 voxel size) was used to verify the dimensions of the heart model. The overall model size was 72 × 66 × 160 mm. The LA dimensions were 38–44 mm, which fall within the typical range [8]. The RA dimensions were 29–35 mm, which also fall within the typical range [15]. Figure 10 shows the exported chamber geometries and ablation points from CARTO3 and the corresponding lesions formed on the heart model. During the first five ablations, the ablative power was increased gradually from inferior to superior and there was a clear corresponding effect on the lesion size. For the final five ablations, the ablative power was kept constant, and the mean lesion diameter was measured to be 3.4 ± 0.6 mm (±1 SD, n = 5). The results showed that the RFA only affected the internal coatings of the heart model and had no effect transmurally or on the outer coatings.

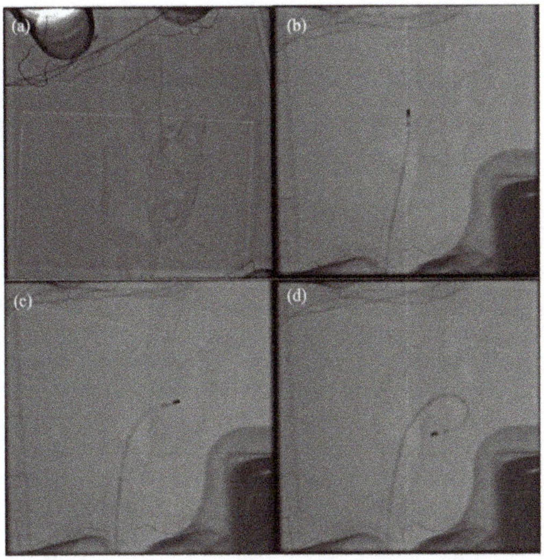

Figure 9. X-ray fluoroscopy imaging of the simulator. (**a**) With no catheters inserted, (**b**) ablation catheter in the RA, (**c**) ablation catheter across the septum, and (**d**) ablation catheter with transseptal sheath in the LA.

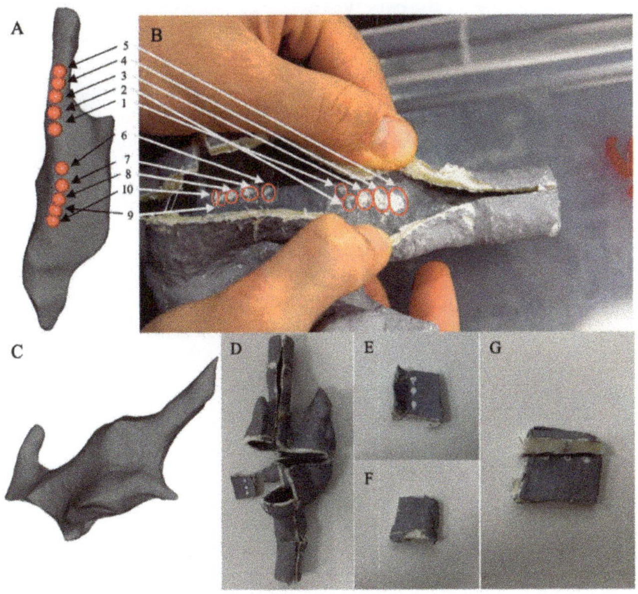

Figure 10. CARTO3 mapping and ablation. (**A**) Exported geometry and ablation points (numbered 1 to 10) from the CARTO3 system showing the RA and the venae cavae. (**B**) Cut-away of the heart model immediately after the ablation experiment showing the lesions and their correspondence to the ablation points. (**C**) Partial CARTO3 map of the LA. (**D**) Cut-away of the dry heart model. (**E**) Ablation points clearly visible on the internal surface. (**F**) External surface is not affected by the RFA. (**G**) Transmural cut that shows no effect inside the model wall.

4. Discussion

The aim of this study was to produce a novel 3D-printed, irreversible thermochromic bi-atrial model, meeting as many of the requirements for a high-fidelity physical simulator for cardiac RFA as possible. We combined Layfomm-40, a new tissue-mimicking 3D-printable polymer, with an irreversible thermochromic pigment and barium sulphate using acrylic paint as a carrier. We measured the electrical and thermal conductivities of Layfomm-40 soaked in saline solution. The thermal properties were found to be similar to those of the myocardium. Although the electrical conductivity was substantially lower it was sufficient to allow use of the material with RFA and EAM systems. We developed an irreversible thermochromic paint and demonstrated that Layfomm-40 coated with this paint was sensitive to RFA and produced visible lesions whose size was influenced by the power and temperature settings of the ablation system. Lesions with diameters similar to those produced in isolated muscle and in vivo could be generated. We overcame issues with the X-ray visibility of Layfomm-40 by formulating a barium-sulphate-doped acrylic paint that produced a similar tissue-to-background image contrast to that seen in clinical cardiac X-ray fluoroscopy images. The thermochromic paint and the barium-doped paint were combined into a four-layer coating for applying to Layfomm-40, both internally and externally. A bi-atrial model was constructed using these strategies, starting with a patient CT scan. A custom base and enclosure were designed and constructed that allowed insertion of catheters, stabilized the model during catheter manipulation and allowed compatibility with standard RF generators and the CARTO3 EAMS. Experimental validation of the simulator by a cardiologist demonstrated the successful ability to perform X-ray fluoroscopy and cone beam CT imaging, to insert and exchange interventional devices without leaks, to manipulate devices within the bi-atrial model without causing damage, to perform mapping using CARTO3, and to perform ablations. Inspection of the cut away model after the experiment clearly showed the ablation lesions.

The striking novel feature of our simulator is the ability to visualize delivered radiofrequency ablation lesions. To the best of our knowledge a common limitation of all prior physical simulators is the inability to visualize the delivered therapy. The state-of-the-art simulators developed by Rossi et al. [3], Heartroid, and Pangolin are limited by not being RFA-sensitive. Since the goal of electroanatomic mapping systems is to guide the delivery of radiofrequency ablation therapy we argue that visualization of this therapy is an essential requirement to support effective simulator-based training of electrophysiologists. In this work we have developed a simple but effective methodology supporting this visualization which can be easily applied to any 3D-printed cardiac chamber model which potentially could be of value during training or device evaluation scenarios. As mentioned previously, there are many factors that affect RFA lesion formation [4]. For lesions 1–5 shown in Figure 10B, we see that increasing power increased the lesion diameter while keeping all other parameters constant, which was as expected. One parameter which is difficult to keep constant is the contact force and this ranged from an average (over the time of each ablation) of 11.1 g to 21.7 g during these five ablations. Masnok and Watanabe conducted experiments to investigate the effect of varying contact force on RFA lesion formation in an in vitro set up [16]. They measured the surface lesion diameter, the intramural width, and the intramural depth. Although they used different power and temperature settings (30 W, 30 °C) for their irrigated ablation catheter compared to our experiments, the average surface lesion diameters that they measured were 4.1 mm (for 2 g force) to 6.9 mm (40 g), with our measurement being 3.4 mm (19.7 g average, 14.4–23.4 g range for lesions 6–10 (Figure 10B)). One limitation of our current model is that we cannot see lesion formation intramurally. In fact, the inner layer of doped paint was unaffected by the RFA. It is desirable to create a lesion that extends transmurally but one that does not lead to risk of perforation of the myocardium [16]. Since the doped paint does not penetrate the wall of our model, we cannot see the intramural effects. This is a limitation that could be addressed in future work.

Physical simulators, such as the one presented, are a way of implementing the 3Rs principle—Replacement, Reduction, and Refinement. By performing more humane animal experiments [17], this simulator falls into the Replacement category. Furthermore, although we have not performed a detailed cost analysis, we estimate that the cost of parts to construct our simulator is less than USD 1000 (including suitable SLA and FDM printers and not including labor costs). This compares favorably with the simulator of Rossi et al. which was estimated to have a parts cost of circa USD 7500 [3]. The cost of using animal models or cadavers would also be several thousands of dollars per unit, not taking into account the cost of the specialist facilities that are required to support this type of work. Therefore, our proposed solution is not only cost-effective but also has an ethical advantage.

5. Conclusions

Our novel simulator meets many of the requirements for a fully functional physical simulator for cardiac RFA. We believe that it is currently the most comprehensive example of such a simulator. Anatomically accurate, 3D-printed tissue-mimicking thermochromic models, as presented in this paper, may prove to be a reliable, inexpensive, and clinically useful tool in simulating cardiac catheter ablation for either training of healthcare professionals or evaluation of novel RF ablation devices. These provide a valuable alternative to computer simulations, animal models, or cadavers.

6. Future Work

In this work we focused on the technical aspects of the simulator and reached a proof-of-concept stage. Future work will focus on evaluating the simulator using a cohort of electrophysiologists and performing standard ablation strategies such as pulmonary vein isolation or wide area circumferential ablation for atrial fibrillation. Several simulators mentioned in the introduction have flow capability. We have not tested this feature in our simulator but there is no reason to believe that the simulator would not be flow-compatible and we aim to develop this feature. Electrophysiologists rely on electrophysiology signals to guide their treatment strategies and currently our simulator is not capable of simulating this. This feature could also be incorporated either via computer-based simulation or physically and this is another area for future investigation.

Author Contributions: All authors contributed to conceptualization of methods, carrying out experiments, analysis of results, and writing of the manuscript. A.R. was responsible for the photography. All authors have read and agreed to the published version of the manuscript.

Funding: This work was supported by the National Institute for Health Research Biomedical Research Centre at Guy's and St. Thomas' NHS Foundation Trust and King's College London, the Wellcome/EPSRC Centre for Medical Engineering (WT 203148/Z/16/Z) and the KCL-China Scholarship Scheme. The views expressed are those of the authors and not necessarily those of the NHS, the NIHR, or the Department of Health. SEW is supported by the British Heat Foundation (FS/20/26/34952).

Informed Consent Statement: Not applicable.

Data Availability Statement: Not applicable.

Conflicts of Interest: The authors declare no conflict of interest.

References

1. Hong, K.L.; Borges, J.; Glover, B. Catheter Ablation for the Management of Atrial Fibrillation: Current Technical Perspectives. *Open Heart* **2020**, *7*, e001207. [CrossRef] [PubMed]
2. De Ponti, R.; Marazzi, R.; Doni, L.A.; Tamborini, C.; Ghiringhelli, S.; Salerno-Uriarte, J.A. Simulator Training Reduces Radiation Exposure and Improves Trainees' Performance in Placing Electrophysiologic Catheters during Patient-based Procedures. *Heart Rhythm.* **2012**, *9*, 1280–1285. [CrossRef] [PubMed]
3. Rossi, L.; Penela, D.; Doni, L.; Marazzi, R.; Napoli, V.; Napoli, L.; Vilotta, M.; Villani, G.Q.; De Ponti, R. Development of Simulation Combining a Physical Heart Model and Three-dimensional System for Electrophysiology Training. *Pacing Clin. Electrophysiol.* **2018**, *41*, 1461–1466. [CrossRef] [PubMed]

4. Kumar, S.; Barbhaiya, C.R.; Balindger, S.; John, R.M.; Epstein, L.M.; Koplan, B.A.; Tedrow, U.B.; Stevenson, W.G.; Michaud, G.F. Better Lesion Creation and Assessment during Catheter Ablation. *J. Atr. Fibrillation* **2015**, *8*, 1189. [CrossRef] [PubMed]
5. Bu-Lin, Z.; Bing, H.; Sheng-Li, K.; Huang, Y.; Rong, W.; Jia, L. A Polyacrylamide Gel Phantom for Radiofrequency Ablation. *Int. J. Hyperth.* **2008**, *24*, 568–576. [CrossRef] [PubMed]
6. Negussie, A.H.; Partanen, A.; Mikhail, A.S.; Xu, S.; Abi-Jaoudeh, N.; Maruvada, S.; Wood, B.J. Thermochromic Tissue-mimicking Phantom for Optimisation of Thermal Tumour Ablation. *Int. J. Hyperth.* **2016**, *32*, 239–243. [CrossRef] [PubMed]
7. Yap, Y.L.; Tan, Y.S.; Tan, H.K.; Peh, Z.K.; Low, X.Y.; Yeong, W.Y.; Tan, C.S.; Laude, A. 3D Printed Bio-Models for Medical Applications. *Rapid Prototyp. J.* **2017**, *23*, 227–235. [CrossRef]
8. Wang, S.; Noh, Y.; Brown, J.; Roujol, S.; Li, Y.; Wang, S.; Housden, R.; Ester, M.C.; Al-Hamadani, M.; Rajani, R.; et al. Development and Testing of an Ultrasound-Compatible Cardiac Phantom for Interventional Procedure Simulation Using Direct Three-Dimensional Printing. *3D Print. Addit. Manuf.* **2020**, *7*, 269–278. [CrossRef]
9. Talalwa, L.; Natour, G.; Bauer, A.; Drzezga, A.; Gordji-Nejad, A.; Beer, S. T 1-mapping and Dielectric Properties Evaluation of A 3D Printable Rubber-Elastomeric Polymer as Tissue Mimicking Materials for MRI Phantoms. *Mater. Res. Express* **2020**, *7*, 115306. [CrossRef]
10. Talalwa, L.; Gordji-Nejad, A.; Natour, G.; Drzezga, A.; Bauer, A.; Beer, S. Evaluation of 3D Printable Rubber-Elastomeric Polymer as Phantom Material for Hybrid PET/MRI. In Proceedings of the IEEE Nuclear Science Symposium and Medical Imaging Conference (NSS/MIC), Manchester, UK, 26 October–2 November 2019; pp. 1–3. [CrossRef]
11. Nguyen, D.K.; Bach, Q.-V.; Lee, J.-H.; Kim, I.-T. Synthesis and Irreversible Thermochromic Sensor Applications of Manganese Violet. *Materials* **2018**, *11*, 1693. [CrossRef] [PubMed]
12. Fu, F.; Hu, L. Temperature Sensitive Colour-changed Composites. In *Advanced High Strength Natural Fibre Composites in Construction*; Fan, M., Fu, F., Eds.; Woodhead Publishing: Cambridge, UK, 2017; pp. 405–423. [CrossRef]
13. Raghavan, K.; Porterfield, J.E.; Kottam, A.T.G.; Feldman, M.D.; Escobedo, D.; Valvano, J.W.; Pearce, J.A. Electrical conductivity and permittivity of murine myocardium. *IEEE Trans. Bio-Med. Eng.* **2009**, *56*, 2044–2053. [CrossRef] [PubMed]
14. Thermal Conductivity » IT'IS Foundation. Available online: https://itis.swiss/virtual-population/tissue-properties/database/thermal-conductivity (accessed on 8 November 2021).
15. Maceira, A.M.; Cosín-Sales, J.; Roughton, M.; Prasad, S.K.; Pennell, D.J. Right Atrial Dimensions and Volume Estimation by Steady State Free Precession Cardiovascular Magnetic Resonance. *J. Cardiovasc. Magn. Reson.* **2013**, *15*, 29. [CrossRef] [PubMed]
16. Masnok, K.; Watanabe, N. Role of Catheter Contact Force on Biophysical Properties of the Ablation Lesion Formation in Radiofrequency Catheter Cardiac Ablation. In Proceedings of the IEEE Region 10 Symposium (TENSYMP), Jeju-si, Korea, 23–25 August 2021; pp. 1–4. [CrossRef]
17. Russell, W.M.S.; Burch, R.L. *The Principles of Humane Experimental Technique*; Methuen: London, UK, 1959.

applied sciences

MDPI

Article

Meshless Electrophysiological Modeling of Cardiac Resynchronization Therapy—Benchmark Analysis with Finite-Element Methods in Experimental Data

Carlos Albors [1,*], Èric Lluch [2], Juan Francisco Gomez [3], Nicolas Cedilnik [4], Konstantinos A. Mountris [5,6], Tommaso Mansi [7], Svyatoslav Khamzin [8], Arsenii Dokuchaev [8], Olga Solovyova [8], Esther Pueyo [5,6], Maxime Sermesant [4], Rafael Sebastian [9], Hernán G. Morales [10] and Oscar Camara [1,*]

1 Sensing in Physiology and Biomedicine (PhySense), Department of Information and Communication Technologies, Universitat Pompeu Fabra, 08018 Barcelona, Spain
2 Digital Technology and Innovation, Siemens Healthineers, 91052 Erlangen, Germany; eric.lluch@gmail.com
3 Valencian International University, 46002 Valencia, Spain; juanfrancisco.gomez@campusviu.es
4 INRIA, Université Côte d'Azur, Epione Team, 06902 Sophia Antipolis, France; nicoco@nicoco.fr (N.C.); maxime.sermesant@inria.fr (M.S.)
5 Aragón Institute of Engineering Research, IIS Aragón, University of Zaragoza, 50018 Zaragoza, Spain; konstantinos.mountris@gmail.com (K.A.M.); epueyo@unizar.es (E.P.)
6 CIBER in Bioengineering, Biomaterials & Nanomedicine (CIBER-BBN), 50018 Zaragoza, Spain
7 Digital Technology and Innovation, Siemens Healthineers, Princeton, NJ 08540, USA; tmansi@its.jnj.com
8 Institute of Immunology and Physiology, Ural Branch of the Russian Academy of Sciences (UB RAS), 620049 Yekaterinburg, Russia; svyatoslav.khamzin@gmail.com (S.K.); zodelheim@gmail.com (A.D.); o.solovyova@iip.uran.ru (O.S.)
9 Computational Multiscale Simulation Lab (CoMMLab), Universitat de València, 46100 Burjassot, Spain; rafael.sebastian@uv.es
10 Dassault Systèmes, 78140 Vélizy-Villacoublay, France; hmorales81@gmail.com
* Correspondence: carlos.albors@upf.edu (C.A.); oscar.camara@upf.edu (O.C.)

Citation: Albors, C.; Lluch, È.; Gomez, J.F.; Cedilnik, N.; Mountris, K.A.; Mansi, T.; Khamzin, S; Dokuchaev, A.; Solovyova, O.; Pueyo, E.; et al. Meshless Electrophysiological Modeling of Cardiac Resynchronization Therapy—Benchmark Analysis with Finite-Element Methods in Experimental Data. *Appl. Sci.* **2022**, *12*, 6438. https://doi.org/10.3390/app12136438

Academic Editors: Mihaela Pop and Cristian A. Linte

Received: 30 April 2022
Accepted: 21 June 2022
Published: 24 June 2022

Abstract: Computational models of cardiac electrophysiology are promising tools for reducing the rates of non-response patients suitable for cardiac resynchronization therapy (CRT) by optimizing electrode placement. The majority of computational models in the literature are mesh-based, primarily using the finite element method (FEM). The generation of patient-specific cardiac meshes has traditionally been a tedious task requiring manual intervention and hindering the modeling of a large number of cases. Meshless models can be a valid alternative due to their mesh quality independence. The organization of challenges such as the CRT-EPiggy19, providing unique experimental data as open access, enables benchmarking analysis of different cardiac computational modeling solutions with quantitative metrics. We present a benchmark analysis of a meshless-based method with finite-element methods for the prediction of cardiac electrical patterns in CRT, based on a subset of the CRT-EPiggy19 dataset. A data assimilation strategy was designed to personalize the most relevant parameters of the electrophysiological simulations and identify the optimal CRT lead configuration. The simulation results obtained with the meshless model were equivalent to FEM, with the most relevant aspect for accurate CRT predictions being the parameter personalization strategy (e.g., regional conduction velocity distribution, including the Purkinje system and CRT lead distribution).

Keywords: electrophysiology; parameter optimisation; smoothed particle hydrodynamics; meshless model; cardiac resynchronization therapy; CRT-EPiggy19 challenge

1. Introduction

Cardiovascular diseases (CVDs) are one among the leading causes of death worldwide, accounting for 32% of all global deaths [1,2]. The high prevalence of CVD leads to substantial health and economic expenses, as it is one of the most critical challenges in healthcare. Heart failure (HF) is a cardiac pathology that causes CVDs; a non-negligible number of HF patients have left ventricle (LV) heart's dyssynchrony [3] induced by a left bundle branch block (LBBB) [4,5]. LBBB patients exhibit an abnormal His–Purkinje system, which produces a delay of activation between the interventricular septum and LV-free wall [6].

Cardiac resynchronization therapy (CRT) has demonstrated in randomized clinical trials to be an effective treatment for patients having (i) symptomatic HF; (ii) depressed left ventricular ejection fraction (EF < 35%); (iii) evidence of ventricular dyssynchrony by a prolonged QRS complex (>120 ms). CRT enhances cardiac structure and function through reverse remodeling [4,7,8]. The most consolidated methodology to deliver CRT, biventricular pacing (BiV-CRT), creates an artificial pacemaker in both ventricles and right atrium to resynchronize the electrical activation and, thus, the mechanical contraction between the LV septal and lateral walls at every cardiac beat [9,10]. Nevertheless, more than 30% of patients fulfilling the criteria for CRT implantation do not respond to the therapy (non-responders, NR), although ratios differ according to the applied definition and criteria [11,12]. One of the main reasons for the high rate of CRT non-responders is the use of too simple indices for patient selection (e.g., EF, QRS, New York Heart Association class). Beyond optimization of patient selection, the correct electrode placement is a key factor to reduce the number of CRT-negative responses. Potential therapeutic alternatives to traditional BiV-CRT are emerging based on optimization of lead placement and number [13] or on new physiological stimulation modalities [14].

Computational electrophysiological models can be valuable tools for a better understanding of pacing-based therapies such as CRT, providing additional information to physicians and device manufacturers to improve therapy efficacy. The interested reader is referred to Niederer et al. [15] and Lee et al. [16] for comprehensive reviews on computational models in cardiology and specific to LBBB and CRT, respectively. More recently, some studies have focused on CRT response optimization through electromechanical models including coronary perfusion [17], or myocardial strains with a complete cardiovascular system, adding both atria as well as systemic and pulmonary circulations [18]. Other studies particularly investigate lead placement. For instance, Albatat et al. [19] analyzed the benefits of multi-site pacing in CRT patients with myocardial infarction. Carpio et al. [20] explored RV lead optimization in a complete simulated torso, while Oomen et al. [21] used fast electro-mechanical simulations to study the role of post-infarction ischemia in reverse LV remodelling following CRT.

Patient-specific personalization plays an important role to make computational models more realistic. However, detailed electrical and mechanical information of the heart is needed, often only available from invasive techniques [16]. Due to the difficulties of obtaining the required in vivo data in humans at different stages of the disease (e.g., from healthy to LBBB and with a CRT device), the validation of CRT computational models is challenging.

Cardiac computational modeling can be improved by translating pre-clinical data into patient-specific models, linking animal and clinical research. For example, as a result of the participation in the Cardiac Electrophysiological Simulation Challenge (CESC'10) MICCAI-STACOM workshop (https://stacom.github.io (accessed on 26 April 2022)), several research groups [22] developed a pipeline integrating different modeling approaches to predict depolarization isochrones from optical mapping data of a perfused ex vivo porcine heart with different pacing conditions [23], acquired at the Sunnybrook Health Sciences Centre, Toronto, Canada. However, experimental data were available for two cases.

Some years later, Rigol et al. [24] developed a swine model of LBBB to study the link between electrical and mechanical dyssynchrony, and their correction with CRT. The authors generated a unique dataset with signal, multi-modal images and electro-anatomical maps at different stages of the disease in tens of infarcted and non-infarcted animals. Soto Iglesias et al. [25] proposed advanced visualization techniques and metrics to quantify the differences in electrical activation patterns at baseline, LBBB and CRT stages. A subset of the database was the foundation for the organization of the CRT-EPiggy19 challenge (https://crt-epiggy19.surge.sh/ (accessed on 15 April 2022)) at the MICCAI-STACOM19 workshop, which is available open access in a public repository (https://zenodo.org/record/3249511 (accessed on 18 April 2022)). More recently, Ramirez et al. [26] also developed a swine model of the heart that was coupled with electrophysiological models to study advanced biomaterial injection therapies for ischemic heart failure.

Participants at the CRT-EPiggy19 challenge adopted different modeling approaches to predict the electrical activation after CRT. Khamzin et al. [27] and Cedilnik and Sermesant [28] developed personalization strategies based on genetic algorithms to estimate regional conduction velocities with simple but fast phenomenological Eikonal-based models. Meanwhile, Gomez and Sebastian [29] used a more detailed Ten Tusscher–Panfilov [30] for cellular electrophysiology, considering transmural heterogeneity and electrical propagation by a monodomain model. After the challenge, other researchers have used the provided data to better understand cardiac physiology and pacing-based therapies [31].

All the aforementioned approaches are based on solving the electrophysiological model equations with the finite-element method (FEM) as a numerical technique based on a mesh discretization of the biventricular heart geometry, as it is the common choice in cardiac modeling [32]. In FEM, the computational domain is divided into discrete subsets of interconnected nodes as elements. However, the explicit connectivity required in the domain leads to great difficulty in generating the irregular patient-specific cardiac meshes, which then becomes a tedious, manual, highly interactive, and time-consuming process. Moreover, the reliability of the simulation results is highly dependent on the quality of the built geometrical mesh [33]. Additionally, mesh distortion that can occur during large cardiac deformations enforces the use of remeshing algorithms to restore mesh shape and numerical accuracy, thereby increasing the computational cost and efforts [34]. Meshless methods are an interesting alternative to avoid meshing difficulties, since the spatial domain is composed of an unstructured particle cloud without connectivity. Therefore, the meshless domain construction procedure can be used for any type of complex geometry. In addition, large deformations or the linking of meshes with different spatial resolution, often necessary in cardiac electromechanics, can be better handled with meshless methods than with FEM, as FEM-based connectivity does not need to be satisfied. For instance, authors in [35] have shown the potential of meshless methods for fluid–structure interaction (FSI) applications, which are extremely time-consuming for mesh-based methods.

Meshless approaches have already been applied to cardiac modeling. For example, Wong et al. [36] used an element-free Galerkin meshless method for modeling cardiac mechanics. On the other hand, Lluch et al. [37] developed meshless methods based on smoothed particle hydrodynamics (SPH) meshless technique for modeling cardiac mechanics. The same authors later [38] employed genetic algorithms to calibrate a SPH-based fully coupled electro-mechanical model of the heart with high-resolution imaging and invasive in vivo measurements from a healthy canine heart. Recently, Mountris and Pueyo [39] proposed a meshfree mixed collocation method with interpolating trial functions to solve the monodomain equations for cardiac electrophysiology and the O'Hara ventricular cell model [40], which was applied to one of the CRT-EPiggy19 challenge dataset under healthy and LBBB conditions.

In this manuscript, we present a benchmark analysis of a meshless SPH method with finite-element methods for the prediction of cardiac electrical patterns in CRT, based on a subset of the CRT-EPiggy19 dataset, including infarcted and non-infarcted cases. A data assimilation strategy was designed to personalize the most relevant parameters of the electrophysiological simulations and identify the optimal CRT lead configuration.

2. Materials and Methods

2.1. CRT-EPiggy19 Data and Experiments

The experiments to create the CRT-EPiggy19 data were performed at Hospital Clínic de Barcelona, Spain, after animal handling approval of the Institutional Review Board and Ethics Commitee of the hospital. In the animals, radiofrequency ablations were carried out to induce LBBB, where half of them presented a myocardial infarction located at the septal wall with different levels of transmurality. Then, a CRT device was implanted to later study the effects of the therapy. More details of the experimental protocol can be found in Rigol et al. [24].

A subset of the CRT-EPiggy19 data was used in our study, including three cases for training and testing, respectively (two non-infarcted and one infarcted dataset in each group). In the dataset provided by the challenge organizers, image segmentation and biventricular finite-element mesh reconstruction were performed using an in-house Siemens algorithm applied on cine sequences of Magnetic Resonance Imaging (MRI), acquired from the swines during the experimental studies. The scar was manually segmented and quantified from delay-enhancement MR images. In terms of electrophysiological data, anatomical point-based reconstructions from CARTO XP of epicardial and endocardial layers were obtained at baseline, LBBB and CRT phases. The electro-anatomical map (EAM) clouds of points were then interpolated onto the MRI biventricular FEM meshes through a quasi-conformal mapping method [25]. Finally, a rule-based method [41] was used for the generation of the cardiomyocyte orientation in each mesh. In addition, regional labels (AHA regions, ventricle definition, endo- and epi-cardial wall distinction) and scarred AHA segments were also included in the models. In the training set, each porcine model was reported in two distinct pathologic stages: with a block in the left bundle branch of the purkinje system and after CRT. For the testing dataset, only the LBBB stage was provided to personalize the electrophysiological models and used them to predict CRT electrical patterns. The RV endocardium was not acquired in the EAM data; therefore, the analysis was centered on the endocardial LV layer and biventricular epicardial layer.

2.2. Meshless Model Based on Smoothed Particle Hydrodynamics

The total Lagrangian meshless method (TL-SPH) developed by Lluch et al. [38] was used in our experiments. As a meshless model, SPH is easy to parallelize, and memory efficient. Additionally, it is mathematically rigorous since it satisfies the Kronecker's delta property. Figure 1 illustrates the developed meshless SPH-based modeling pipeline to predict CRT electrical patterns in the experimental data. The first step of the pipeline consisted on discretizing the continuous domain provided by the biventricular meshes of the porcine hearts with a cloud of particles without connectivity, where each particle had the following individual properties: three-dimensional position, cardiomyocyte orientation, tissue type, initial impulse, conduction velocity, area and volume.

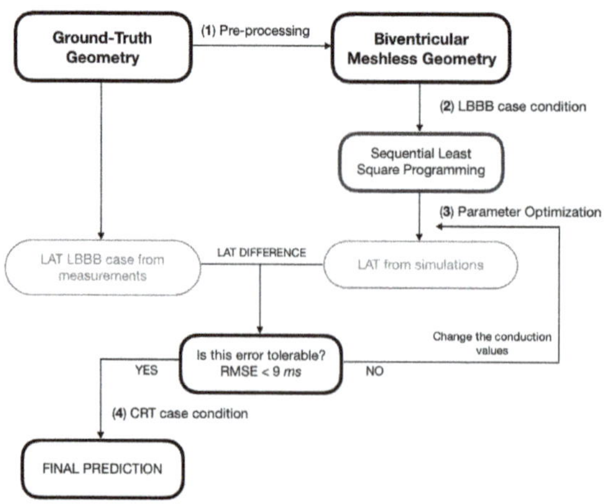

Figure 1. Scheme of the developed meshless modeling pipeline to predict the electrical patterns in experimental data after cardiac resynchronization therapy. LBBB: left bundle branch block. RMSE: root mean square error. LAT: local activation time.

To determine the particle properties a Gaussian Kernel function was enforced, defining the number of neighboring particles around each particle of interest that is then employed to estimate the area and volume indices. Parameters such as the kernel size (i.e., smoothness length of the Gaussian kernel) and the geometry discretization according to the number of particles were key factors to determine when setting up the simulation. To define the optimal values of the Gaussian kernel function, sensitivity analyses of particle resolution and kernel size were performed. Several simulations were carried out fixing the kernel size and increasing the particle numbers; configurations with 15 $\times 10^3$ (2 h simulation), 20 $\times 10^3$, 80 $\times 10^3$ and 100 $\times 10^3$ (11 h simulation) particles were studied in one of the analyzed geometries. The kernel size was inversely proportional to the number of particles to avoid excessive computational cost; kernels from 3.5 to 9 mm were tested in intervals of 0.5 mm. A kernel size of 6.5–8.5 mm was finally defined, as a function of the swine model morphology and required conduction velocities (i.e., larger kernels for higher conduction velocities and morphologies with higher curvature), in combination with geometries of 15 $\times 10^3$ particles. As shown in [42], configurations with higher number of particles (e.g., 50 $\times 10^3$) and smaller kernel sizes (e.g., 3 mm), computational costs exponentially increases without a substantial accuracy gain, which will hamper the parameter optimization process. Furthermore, we also analyzed the effect of the time-step, testing values of 10^{-3}, 10^{-4}, and 10^{-5} in one of the studied cases for LBBB simulations. The computational cost associated with each time-step value was of >42 min, around 20 min and around 7 min, providing RMSE of 6.2 ms, 6.8 ms, and 7.9 ms, respectively. A time-step value of 10^{-4} was finally chosen as a trade-off between computational cost and result accuracy. The total simulated time was of 0.15 s, based on the total activation times of the available EAM dataset (i.e., most cases with TAT < 0.1 s).

2.3. Electrophysiological Model

The simplified reaction–diffusion Mitchell–Schaeffer (MS) electrophysiological model [43], together with a diffusion term [42], was used at the cellular level. The MS method allowed us to simulate the electrical activation sequence of the swine hearts with an ionic model of the ventricular action potential duration (APD) composed of only two currents: one inward and one outward. The computation of the voltage and depolarization phase over time is performed with the following partial differential equations:

$$\begin{cases} \dfrac{\partial v}{\partial t} = div(\boldsymbol{D}\nabla v) + \dfrac{wv(1-v)}{\tau_{in}} - \dfrac{v}{\tau_{out}} + \boldsymbol{I}_{app} \\ \dfrac{\partial w}{\partial t} = \begin{cases} \dfrac{1-w}{\tau_{open}} \text{if} v s. < v_{gate} \\ \dfrac{-w}{\tau_{close}} \text{if} v s. > v_{gate} \end{cases}, \end{cases} \tag{1}$$

where $\boldsymbol{I}_a pp \in \mathbb{R}$ describes the initial stimulus of the transmembrane potential $v \in \mathbb{R}$, $w \in \mathbb{R}$ controls the depolarization phase, and $v_{gate} \in \mathbb{R}$ determines where the APD starts. Furthermore, τ_{open}, τ_{close}, τ_{in}, and $\tau_{out} \in \mathbb{R}$ govern the duration of the four stages of the APD (i.e., initiation, plateau, decay, and recovery). The diffusion term, $div(\boldsymbol{D}\nabla v)$, includes cardiomyocyte orientation, with the diffusion tensor, $\boldsymbol{D} \in \mathbb{R}^{3\times3}$, defined as in [44]:

$$\boldsymbol{D} = (\boldsymbol{f} \otimes \boldsymbol{f}(1-ar) + \boldsymbol{I} \cdot ar) \cdot \boldsymbol{d} \tag{2}$$

There are three main parameters in Equation (2) to take into account: the cardiomyocyte orientation vector, $\boldsymbol{f} \in \mathbb{R}^3$; the diffusion coefficient, $d \in \mathbb{R}$, which controls the action potential propagation speed; and the anisotropic ratio, $ar \in \mathbb{R}$, which determines the relation between conduction velocities and cardiomyocyte orientation (e.g., $ar = 1$ will define an isotropic behavior). We tested different values for ar (from 0.01 to 0.5), finally fixing to 0.01 (i.e., giving more weight to cardiomyocyte orientation) for all cases. The cardiomyocyte orientation was provided in all studied biventricular meshes by the CRT-Epiggy19 organizers from the rule-based model proposed by Doste et al. [41], which is adapted to replicate histological data of both left and right ventricles. Finally, $\boldsymbol{I} \in \mathbb{R}^{3\times3}$ defines the identity matrix and ⊗ the tensor product. Overall, only six parameters are necessary in the MS model, which is convenient for model personalization.

2.4. Left Bundle Branch Block Simulation with Personalized Parameters

An initial stimulus was set in the atrio-ventricular (AV) node, with an average of 60 particles, being identified from the earliest activated points in the EAM data of each case, to initiate the simulated electrical pattern over the two ventricles. The Purkinje (PK) system, which has fast conduction velocities, needs to be incorporated in the model for simulating a LBBB and disrupt the normal electrical propagation in the LV branch. Therefore, particles located in the lower (i.e., closer to the apex) half of the endocardial RV (around 500 particles) and the lower third of the LV (around 300 particles), if no scar was present, were labeled as Purkinje (see Figure 2), following the distribution of PK–myocardial junctions found in PK-based simulation studies [45].

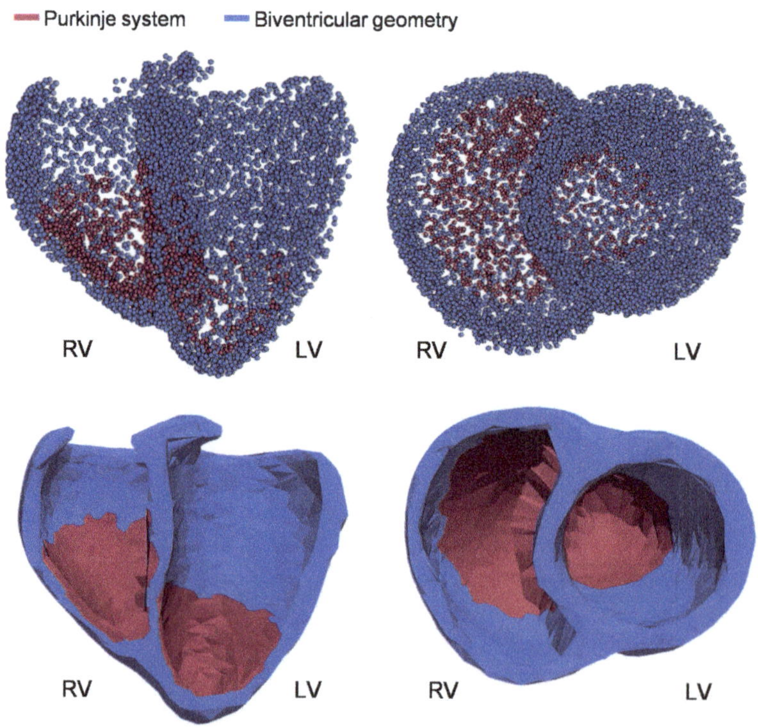

Figure 2. Biventricular geometries with particles labelled as regular myocardial tissue (in blue) and Purkinje system (in red), used for the meshless solver. Top: point cloud representation. Bottom: mesh-based triangulation from applying the Delaunay algorithm to the cloud of points, for visualization purposes.

Including PK particles in the LV, could seem contradictory for simulating LBBB electrical patterns. However, it was necessary to consider the LV retrograde activation due to the transmurality of the PK system in pigs [46], leading to latest activation points being located at the basal LV in several cases.

As other participants at the CRT-EPiggy19 challenge [28,29], we also added a second impulse in the electrophysiological simulations to replicate an early activation of the RV epicardium observed in the EAM data (see RV_{epi} initial activation point in Figure 3). Besides the expected activation induced by the AV node, the septomarginal trabecula may have a role in the fast activation of the RV that needs to be incorporated to replicate the electrophysiological measurements. The dynamics of the electrical pattern in LBBB cases are displayed in Figure 3, starting from the two stimulus (1 in the Figure 3), followed by the propagation to the biventricular apex (2 in the Figure 3), and the propagation from the LV apex to the base (3 in the Figure 3), with fast activation in the endocardium and slow in the epicardium.

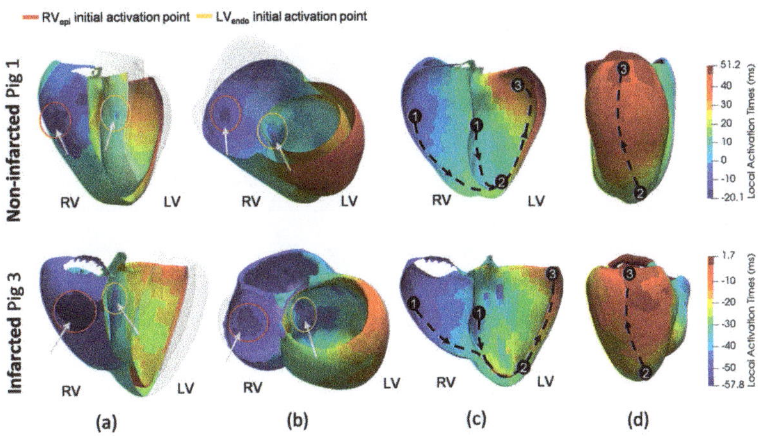

Figure 3. Electroanatomical maps of an infarcted and a non-infarcted training case in left bundle branch block condition. In (**a**,**b**), both initial activation points (1) in the RV epicardial layer (red circle) and LV endocardial one (yellow circle) are shown. At (**c**,**d**), the numbering sequences describe the followed electrical pattern to fully activate the biventricular geometries: starting from the initial stimulus (1), following to the biventricular apex (2), and propagating to the left ventricular base (3). The colourscale represents the local activation times, from earliest to the latest activation points, in blue and red, respectively. RV_{epi}: right ventricle epicardium. LV_{endo}: left ventricle endocardium.

The local conduction velocity (CV) values defined in each geometry, guiding the wave propagation speed in the direction established by the modeled cardiomyocyte orientation, was one of the main parameters affecting the simulated electrical pattern. However, it is not simple to set up the number of heterogeneous conductivity regions: different values at each voxel would both be impractical (too many parameters to optimise) and does not make sense in relation with the sparsity of the available electroanatomical data; too few regions would not consider the existing CV heterogeneity (e.g., faster CV in PK system, complex electrical propagation in the septum due to discontinuities in cardiomyocyte orientation [41], presence of scar, etc.). Consequently, we performed a sensitivity analysis to determine the optimal number of different regions with local conductivities to optimize, from only a single region considering the whole biventricular geometry, to 21 regions including the 17 AHA segments. In total, the following seven regional CV configurations were tested:

- 1 region (*whole biventricular geometry*).
- 2 regions ($RV - LV$).
- 3 regions ($RV - Purkinje\ system - LV$).
- 4 regions ($RV_{epi} - RV_{endo} - LV_{epi} - LV_{endo}$).
- 5 regions ($RV_{epi} - RV_{endo} - Purkinje\ system - LV_{epi} - LV_{endo}$).
- 6 regions ($RV_{epi} - RV_{endo} - Purkinje\ system - Septum - LV_{epi} - LV_{endo}$).
- 21 regions ($RV_{epi} - RV_{endo} - Purkinje\ system - 17\ LV_{AHA_{segments}} - LV_{endo}$).

The optimization of the CV distribution in each analyzed case was performed with the constrained non-linear Sequential Least Squares Programming algorithm. The cost function was based on minimizing the root mean square error of each particle activation time between simulation results and the EAM-based electrical patterns. An iterative method was used for parameter optimization, updating the five regional CVs until the best possible fit was obtained. The choice of a constrained algorithm was made so that: (1) conductivity values always were positive; (2) a purkinje system always being the fastest regional layer; (3) the lowest conductivity value always was in the necrotic/scar zone for infarcted cases. In the end, an average of 70 simulations were performed for each analyzed case, mainly for the optimization of the CV configuration.

2.5. Simulation of Cardiac Resynchronization Therapy

Once model parameters were personalized with the SLSQP optimization algorithm to better replicate the electrical pattern of the LBBB data, the next step was to simulate CRT using the same personalized parameters (see Figure 1). Additional initial stimulus were incorporated in the model, simulating the LV and RV leads of CRT. The position of the CRT leads in the training cases was determined by identifying the earliest activated points in the provided electroanatomical maps. In the testing cases, as EAM data were not available, several lead configurations were evaluated to find the one furnishing better evaluation metrics, as described below.

2.6. Evaluation Metrics and Experiments

As mentioned above, the root mean square error difference between the local activation time (e.g., time when each particle activates, with the initial stimulus as reference) given by the simulations and the EAM measurements, integrated over each particle of the biventricular geometries, was used in the parameter optimization in the training cases. As for testing, global and regional metrics were used to evaluate the prediction accuracy of different modeling strategies in each analyzed case.

First, the total activation time (TAT) required to activate the whole biventricular geometry from the initial impulses, was employed as a general metric. Additionally, as proposed by Soto Iglesias et al. [25], we computed some activation delays to better characterize regional patterns, specifically, the inter-ventricular delay (IVD), which is time difference between earliest activation points of both ventricles (LV and RV) in the epicardial layer; and the left ventricular transmural delay (LV-TD), defined as time difference between LV layers (epi- and endocardium) first activated points. Finally, we also estimated the recovery as follows:

$$Recovery = \frac{TAT_{baseline} - TAT_{LBBB}}{TAT_{LBBB} - TAT_{CRT}} * 100, \quad (3)$$

which indicates the percentage of how close the TAT is to the baseline after applying CRT. Finally, we created histograms of the percentage of activated tissue over time for the right and left epicardial regions, which provides an intuitive visualization of the different intra- and inter-ventricular delay differences between LBBB and CRT conditions (including distinct lead configurations).

After the sensitivity analyses of different modeling choices of the SPH-based solution (e.g., number of particles, kernel size, time step), as explained in Section 2.2, the initial experiments in our study consisted on personalizing model parameters (e.g., regional conduction velocities) with the EAM data in the three studied training cases in LBBB condition. Subsequently, the resulting regional conduction velocity distribution was used for modeling CRT, using the lead position provided by the challenge organizers in the training dataset. The initial stimulus characteristics (e.g., location, depolarization times) were maintained in all models).

The meshless simulation results were qualitatively and quantitatively compared with the ones provided by an FEM-based method [29] presented in the CRT-EPiggy19 challenge, since it was the only participant processing the three analyzed training cases. Additionally, metric comparisons are already made with the mesh-based methodology presented by another challenge participant [27], which reduced the biventricular geometry mesh resolution to 12 $\times 10^3$ elements to facilitate the exploration of a larger range of parameters (>30 different conduction velocity regions) in an optimization process using the L-BFGS optimization algorithm.

In the CRT-EPiggy19 challenge, EAM data of testing cases after CRT were not provided, thus lead location was unknown. For this reason, we tested four different lead locations (Figure 4) in each testing geometry to determine the one providing the best recovery: (1) RV apex and LV basal region ($RV_{apx} - LV_{bas}$); (2) RV and LV Apex ($RV - LV_{apx}$); (3) earliest and latest point activation from the LBBB cases in the EAM ($Early - Late_L AT$); (4) RV Outflow track (RVOT) septal and LV basal region ($RVOT_{sep} - LV_{bas}$). If recovery was similar in different lead configurations, TAT and delay values were analyzed to choose the final lead configuration.

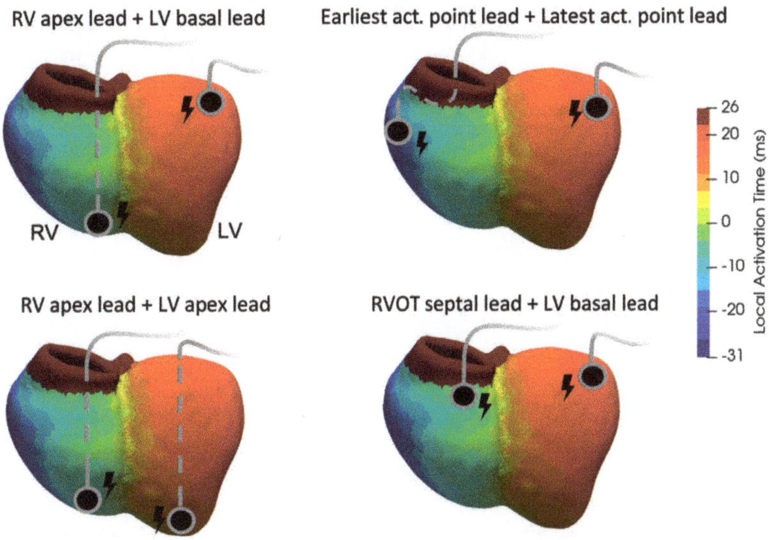

Figure 4. Different configuration of cardiac resynchronization leads analysed in the testing cases. LV/RV: left and right ventricle, respectively. Earliest/Latest act: Earliest/Latest activation. RVOT: right ventricular outflow tract.

For comparison purposes, colourmaps representing the electrical activation patterns of the figures have been adjusted by setting the initial depolarization of the RV of each model (local activation values) as the initial times and dividing them into several isochrones. For visualization, the Open Source Paraview (ParaView, v.5.8) (https://www.paraview.org (accessed on 1 April 2022)) software tool was used. Computational resources for the meshless electrophysiological models consisted of a Nvidia RTX 2080 Ti GPU and an i9-9900k CPU executed in Code::Blocks software (Code::Blocks IDE, v.16.01).

3. Results

3.1. Training Data

The sensitivity analysis to determine the best regional distribution of conduction velocities in the LBBB condition resulted in best fittings of simulations with EAM data when increasing the number of regions, with a RMSE of 6.4 ms and 5.3 ms in the non-ischemic and ischemic models, respectively, for 21 regions (vs. 6.7 ms and 5.8 ms in the non-ischemic and ischemic models, respectively, for 5 regions). However, when applied to CRT data, electrophysiological simulations with 6 and 21 regions produced larger errors than with 5: 10.2 ms and 9.8 ms in the non-ischemic and ischemic cases, respectively, for 21 regions, and 9.3 ms and 7.7 ms in the non-ischemic and ischemic cases, respectively, for 5 regions. Therefore, five regions were finally chosen for the conduction velocity distribution in the remaining simulations. The optimization process took between 10 and 25 h to converge (15–25 min per simulation), depending on the studied case.

3.1.1. Left Bundle Branch Block Simulations

Table 1 summarizes the accuracy obtained with the meshless SPH-based in the training dataset, as quantified by the metrics detailed above. Equivalent results were obtained for both LBBB and CRT conditions. Figure 5 shows the local activation time maps for a non-infarcted and an infarcted case of the training database at LBBB and CRT conditions, provided by the EAM, and from FEM- and SPH-based simulations.

Table 1. Metrics characterizing the electrical activation maps in training cases from measurements and meshless simulations. EAM: electroanatomical maps. SPH-Sim: Simulation with smoothed particle hydrodynamics meshless method. LBBB: left bundle branch block. CRT: cardiac resynchronization therapy. TAT: total activation time. LAT-RMSE: local activation time root mean square error. IVD: inter-ventricular delay. LV-TD: left ventricle transmural delay. (*) indicates an infarcted pig.

	Pig 1				**Pig 2**				**Pig 3 (*)**			
	EAM		**SPH-Sim**		**EAM**		**SPH-Sim**		**EAM**		**SPH-Sim**	
	LBBB	CRT	LBBB	CRT	LBBB	CRT	LBBB	CRT	LBBB	CRT	LBBB	CRT
TAT (ms)	72.0	70.0	70.0	66.0	66.0	45.0	78.0	39.0	59.0	35.0	49.7	46.0
LAT RMSE (ms)			6.8	7.9			9.4	7.7			5.1	6.6
17 IVD (ms)	18.0	7.3	18.6	11.4	19.8	−3.3	14.3	0.0	17.7	−12.5	16.9	−4.8
LV-TD (ms)	7.2	0.0	9.0	2.0	9.9	−6.6	13.0	0.4	11.8	−5.2	18.4	0.6
Recovery (%)	−2.6		−8.0		47.7		69.6		342.9		185.0	

The initial impulses set for the non-ischemic case shown in Figure 5 (Pig 1) were established with a difference of 12 ms between them. The first, located in the RV endocardium, was set at 2 ms, and the second, the septal one, at 14 ms. Table A1 and Figure 5 revealed similar activation patterns in the biventricular epicardium with an average regional LAT RMSE of 5.7 ms compared to the EAM data. However, larger differences were observed in the LV endocardial layer, increasing the regional error to 9.1 ms (Table A1). The remaining metrics (e.g., TAT, IVD and LV-TD) were similar between meshless simulations and measurements, with a difference <3 ms, while the overall error was 6.8 ms, as reported in Table 1.

For the infarcted case shown in Figure 5 (Pig 3), the initial stimulus was placed as in Pig 1, but at 2.5 ms in the RV endocardial layer, and in the septal area at 12 ms to better match the EAM data. Figure A3 in the Appendix shows simulation results obtained with only one stimulus. The scar in Pig 3, which had a transmurality of 86%, was located in the septo-apical and antero-septal LV regions. In the scar region, a 75% conduction velocity reduction with respect to the Purkinje system was established from the best simulation result. Although the electrical activation pattern provided by the SPH simulations was very close to the EAM measurements (5.1 ms average LAT error), differences in conductivity were appreciated between both ventricles. The posterior basal region of the RV epicardial layer had a slower activation than the ground-truth data, whereas LV layers (endo- and epicardium) had a higher conductivity. The metrics in Table 1 show differences greater than 6 ms for LV-TD, 9 ms for TAT, and less than 1 ms for IVD.

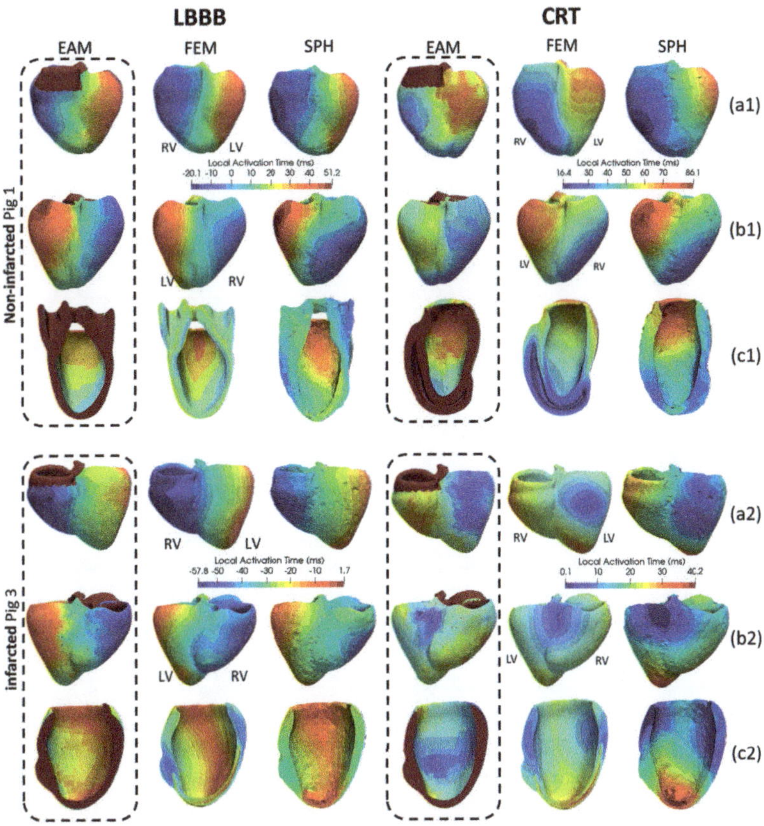

Figure 5. Local activation time maps for a non-infarcted and an infarcted (top and bottom panels corresponding to Pig 1 and Pig 3, respectively) case of the training database in left bundle branch block (LBBB) and cardiac resynchronization therapy (CRT) conditions, provided by the electroanatomical measurements (EAM) and electrophysiological simulations performed with a finite-element method (FEM) and a meshless (SPH) model. (**a1**,**a2**) and (**b1**,**b2**) correspond to anterior and posterior biventricular epicardial visualizations. (**c1**,**c2**) show endocardial view of the left ventricle (LV) lateral wall. RV: right ventricle.

Table 2 shows the conduction velocity values estimated by the SPH-based model in the five selected regions for an ischemic and a non-ischemic cases of the LBBB training dataset. Additionally, the corresponding parameters obtained with the FEM-based approach of Gomez and Sebastian [29] on the same cases are also included for comparison purposes. The reader can be referred to Figure A2 for a visual representation.

Table 2. Conduction velocity values (m/s) estimated by the SPH- and FEM-based solvers [29] for an ischemic and non-ischemic training cases at LBBB scenario. SPH: Smoothed particle hydrodynamics meshless method. FEM: Finite element method. RV epi: Right ventricle epicardium. RV endo: Right ventricle endocardium. LV: Left ventricle. PK: Purkinje system.

	SPH-Based							**FEM-Based**	
	RV endo	RV epi	LV endo	LV epi	Scar	Average heart tissue	PK	Average heart tissue	PK
Ischemic	1.53	1.40	1.36	1.62	0.49	1.30	1.69	1.78	1.30
Non-ischemic	0.83	0.65	0.63	0.51	-	0.65	2.40	0.50	2.60

3.1.2. Cardiac Resynchronization Therapy Simulations

The configuration of the CRT leads was initially positioned close to the apical regions of both ventricles in the non-infarcted training case shown in Figure 5, following the information provided by the organizers of the CRT-EPiggy19. In the EAM data, the mid-apical lead location on the lateral wall of the RV endocardial layer resulted in fast epicardial conduction, specifically in the posterior part. In contrast, LV lead apicality with weak access to the PK system implied slower activation of its endocardial layer than of the epicardial one, with the former presenting the last activation point. The SPH-based simulation produced the largest differences (9.7 ms of LAT error in Table A1) in the RV. As for the LV, an 8.5 ms LAT error was found, since it was not possible to fully capture the conduction velocity change between the endocardium and epicardium. The metrics summarized in Table 1 present differences between simulations and observations of 2 ms for LV-TD, 4 ms in TAT, and 4.1 ms in IVD with an error of 9.2 ms in the overall LAT.

For the infarcted testing case shown in Figure 5 (pig 3), non-physiological conduction velocities above 2 m/s, specifically in the ischemic zone, were required to match the fast electrical patterns observed in the EAM data (35 ms), with both CRT leads located in the LV epicardial layer (anterior and posterior regions). The parameter optimization process in SPH-based simulations did not capture these high conduction velocities due to the physiological constrains, providing slower values and exhibiting large differences with EAM in the apex, reflected in the 11 ms of TAT and in delay metrics over 6–7 ms. However, the overall LAT error was not large (6.6 ms).

3.2. *Testing Data*

3.2.1. Left Bundle Branch Block Simulations

Table 3 summarizes the accuracy obtained with the meshless SPH-based in the testing dataset. The initial impulses were fixed at 0 ms in the endocardial RV layer and at 9 ms in the septal area for Pig 4, one of the non-infarcted testing cases. The SPH-based simulation correctly replicated the conduction velocities of the LBBB EAM at different layers showing a low error of 5.1 ms, and specifically the RV epicardium with a 4 ms regional error (Table A2). Nevertheless, the anterior part of the LV endocardium showed a greater number of variations, corroborated by a regional error above the mean (6.4 ms in Table A2). The LV epicardial sequence was also similar (5.5 ms regional LAT RMSE in Table A2) in measurements and simulations, with the same latest activation point.

Table 3. Metrics characterizing the electrical activation maps in testing cases from measurements and meshless simulations, including the best lead configuration in the cardiac resynchronization therapy scenario. EAM: electroanatomical maps. SPH-Sim: Simulation with smoothed particle hydrodynamics meshless method. LBBB: left bundle branch block. CRT: cardiac resynchronization therapy. TAT: total activation time. LAT-RMSE: local activation time root mean square error. IVD: inter-ventricular delay. LV-TD: left ventricle transmural delay. $RV_{apx} - LV_{bas}$: CRT leads in the right ventricular apex and basal left ventricle. (*) indicates an infarcted pig.

	Pig 4				Pig 5				Pig 6 (*)			
	EAM		SPH-Sim		EAM		SPH-Sim		EAM		SPH-Sim	
	LBBB	CRT	LBBB	CRT ($RV_{apx} - LV_{bas}$)	LBBB	CRT	LBBB	CRT ($RV_{apx} - LV_{bas}$)	LBBB	CRT	LBBB	CRT ($RV_{apx} - LV_{bas}$)
TAT (ms)	61	59	56	36.8	92	70	76	55	67	49	73.7	47
LAT RMSE (ms)			5.1	13.2			8.2	14.6			5.9	10.7
IVD (ms)	19.61	−9.04	14.7	0.5	25.58	7.83	21.5	1.2	12.75	2.63	18.3	0.8
LV-TD (ms)	16.55	−12.95	5.8	−0.7	32.68	−8.45	9.8	-0.3	14.22	−1.97	6.8	0.6
Recovery (%)	17.95		100.84		46.09		68.2		637		666	

The EAM of the infarcted testing case (Pig 6) had an initial impulse at the RV endocardium lateral wall, inducing a rapid RV activation, while the LV one was much more gradual. The scar in Pig 6 was located in over the whole septum, with 57% of transmurality. In the SPH-based simulations, the initial stimulus were placed at similar regions of the non-infarcted case but at 2.6 ms and 14 ms for the RV endocardial layer and septal area, respectively. To faithfully represent the ischemic region in the simulations, a reduction of over 87% in the conduction velocity with respect to the PK system was determined by the SLSQP optimization algorithm. The epicardial layer depicted the highest LAT regional error (see Table A2), specifically in the posterior part for the RV and in the LV anterior part. With a LAT-RMSE of 5.9 ms for Pig 6 (Table 3), the LV endocardial layer showed the best fitting (5.7 ms regional LAT-RMSE) for a delayed basal activation of the LV (Table A2).

3.2.2. Cardiac Resynchronization Therapy Simulations

In the three analyzed testing cases, the optimal configuration consisted in leads located in the RV apex and the basal LV ($RV_{apx} - LV_{bas}$ configuration), providing the best recovery metric values and overall cardiac resynchronization. The CRT leads were activated at practically the same time in the three testing cases, with a time interval under 0.5 ms between them. The RV was typically triggered prior to the LV (see histograms in Figure 6 for the infarcted testing case), where it was always located the last activation point.

Figure 6 shows the local activation time maps for a an infarcted case (Pig 6) of the testing database in CRT condition, provided by the EAM, and for different simulated lead configurations. Furthermore, histograms of the percentage of electrically activated heart tissue for the right and left epicardial layers are displayed to better represent inter-ventricular delays with different lead configurations. It can be easily appreciated the better inter-ventricular synchronization provided by the $RV_{apx} - LV_{bas}$ lead configuration when analyzing the histograms, which was also confirmed by a low TAT and recovery discrepancy between the SPH-based simulation and EAM data, as shown in Table 3. We can also see in Figure 6 the impact of changing the LV lead from the basal (or latest activated point) to the apex, increasing the inter-ventricular delay compared to the remaining configurations. Additionally, placing the RV lead in the RVOT was better than in the earliest activated point (i.e., lateral wall), as can be seen in Figure 6 (fourth and fifth row, respectively), the former having less IVD.

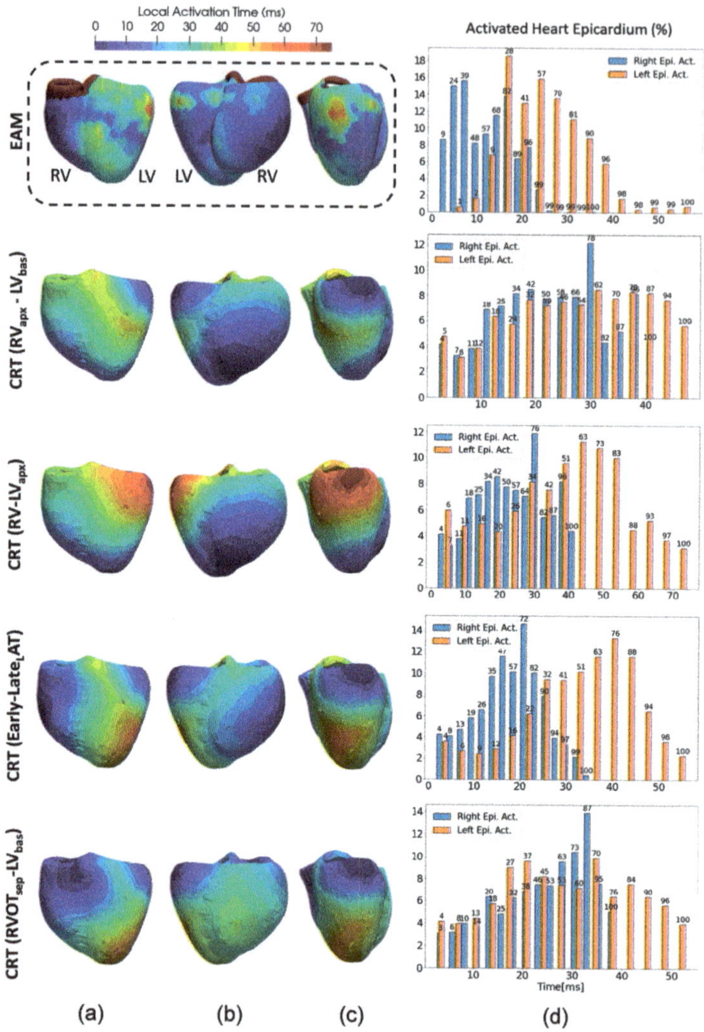

Figure 6. Local activation time maps ((**a–c**) showing anterior/posterior of biventricular epicardium and LV lateral wall epicardium, respectively) after cardiac resynchronization therapy (CRT) from the electro-anatomical (EAM) data and meshless simulations in the infarcted testing case, Pig 6. Histograms of the percentage of electrically activated heart tissue for the right and left epicardial layers are in the right column (**d**). From the second to the fifth row, different simulation results obtained with different CRT lead locations are displayed. $RV_{apx} - LV_{bas}$: leads on right ventricle (RV) apex and basal left ventricle (LV). $RV - LV_{apx}$: both leads are located in the biventricular apex. $Early - Late_L AT$: leads located at the the earliest and latest EAM ventricular activated points, respectively. $RVOT_{sep} - LV_{bas}$: leads in the septal RV outflow track and in the basal LV.

The non-infarcted cases of the testing database (Pig 4 and Pig 5) had an identical overall behavior with respect to optimal lead configuration. However, they presented large TAT errors between SPH-based simulations and EAM data (22.2 ms and 15 ms for Pig 4 and Pig 5, respectively), with CRT simulations providing lower TAT values and, consequently, larger recovery than EAM data (see Table 3). The difference between optimal SPH-based simulations and EAM measurements was due to a different lead location. Figure 7 shows the histograms of the percentage of electrically activated heart tissue for both ventricles under LBBB and CRT conditions for the three analyzed testing cases, showing the large synchronization recovery achieved for all cases.

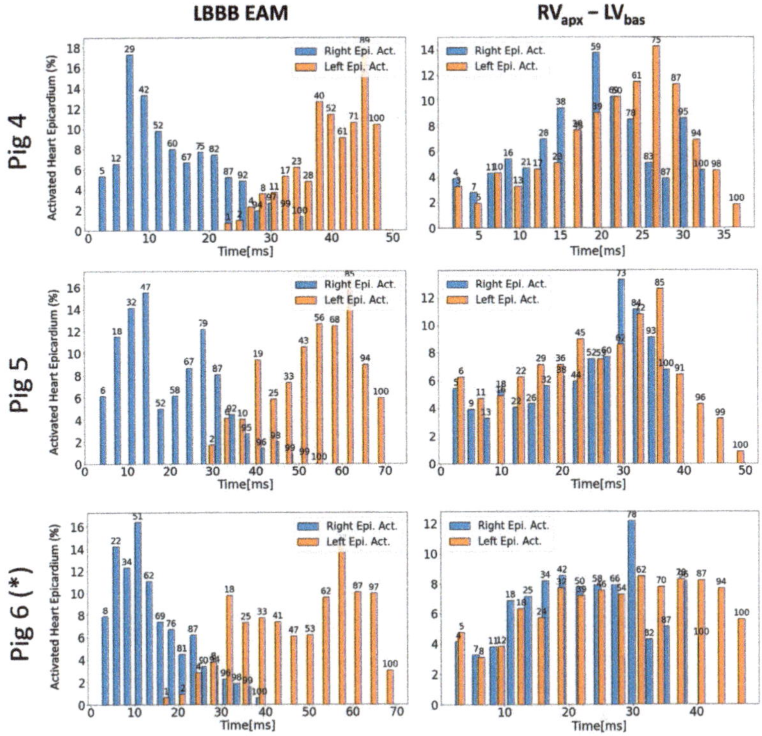

Figure 7. Percentage of electrically activated heart epicardium with a left bundle branch block (LBBB) and with the best cardiac resynchronization (CRT) lead configuration in the three testing cases. $RV_{apx} - LV_{bas}$: RV apex and LV basal stimulation regions. Epi: epicardium. act: Activation. (*) indicates an infarcted pig.

4. Discussion

Computational models of the heart can provide useful insight on the pathophysiological mechanisms and device options in CRT, contributing to reduce the high rate of non-responders. However, computational models need to be personalized and validated (after verification) with data coming from different sources, following standards such as the V&V40, to build the required credibility to be part of the device design and regulatory evaluation pipelines in silico trials [47]. Regrettably, it is not straightforward to acquire rich in vivo human data in clinical applications such as CRT. However, researchers (e.g., Rigol et al. [24,26]) have developed realistic experimental models, generating animal data that can be used to test and personalized the developed computational models of the heart.

The CRT-EPiggy19 challenge provided as open access multi-modal data of swine models under healthy, LBBB and CRT conditions, for model benchmark purposes. Three research teams [27–29] participated in the challenge, running different FEM-based electrophysiological models in patient-specific biventricular meshes that were provided by the organizers. In practice, mesh generation from patient-specific data of complex organs such as the heart often involves tedious manual interactions that hinder the application of computational models to large patient databases. Meshless models are an interesting alternative that have already been applied in cardiac electromechanics [33,36,38,39], but they have not been benchmarked with FEM-based approaches in LBBB and CRT experimental data. Meshless methods are obviously independent of the generation of patient-specific cardiac meshes, thus solving one of the main bottleneck steps of mesh-based alternatives for translating computational models into clinical environment.

In this manuscript we present a meshless modeling pipeline, based on the SPH-based approach developed by Lluch et al. [38], where the most relevant parameters have been optimized to fit a subset of the CRT-EPiggy19 dataset. Basically, training data of three cases with LBBB were used to estimate the model parameters minimizing the differences between local activation times provided by meshless simulation results and EAM measurements, consequently predicting CRT electrical activation patterns with known lead location. Several metrics, proposed by Soto Iglesias et al. [25], were used, beyond the common global TAT parameter, to better quantify the local electrical heterogeneity in the ventricles. Although computational times could be further reduced, the meshless method could provide CRT predictions and lead configuration optimal strategies in around 20 min, once the LBBB pattern has been assimilated. Timings which are compatible with the clinical routine workflow. Moreover, meshless methods make the potential coupling with other physical models very easy, compared with FEM alternatives, with electromechanical models allowing large deformations without the risk of convergence issues due to mesh element quality degeneration.

The most relevant parameters related to the SPH-based model were the number of particles and the kernel size, which were set up to different values (15×10^3 and 6.5–8.5 mm, respectively) than the original SPH formulation in [42] (51×10^3 and 3 mm, respectively). The main reason was to decrease the computational cost for each simulation, without compromising result accuracy, so that the meshless method could be embedded into an parameter optimization framework. On the other hand, Mountris and Pueyo [39] employed a higher number of particles (240×10^3) and a fixed neighbourhood size (150 particles) in their meshless model applied to CRT-EPiggy19 data, which would be prohibitive in our application due to the exponential growth in computational cost of the SPH-based solution.

4.1. Benchmark Analysis of Meshless and Finite-Element Method Solutions on Training Data

Despite the complex pipeline to process EAM data and the variability of the analyzed cases, the SPH-based model provided low LAT errors in LBBB (6.75 ± 1.59 ms) and CRT (10.38 ± 3.80 ms) cases. The meshless simulation results were generally similar to FEM-based ones from CRT-EPiggy19 participants (see Table 1 and Figure 5), when qualitatively analyzing the electrical activation patterns, and with the quantitative metrics (e.g., TAT, LAT, delays). However, some methodological differences were found that could explain small variations in the obtained results. For instance, the approach by Gomez and Sebastian [29], using a biophysical Ten Tusscher–Panfilov model, with a larger number of parameters and a more personalized Purkinje system differentiating between RV and LV, will certainly be more appropriate than simplified phenomenological Eikonal models in some cases. On the other hand, the low computational cost of Eikonal-based solutions allow running a lot of simulations and a larger exploration of the parameter space to match EAM data. Computational times for both meshless and FEM-based methods depend on the domain resolution (i.e., number of points/elements), the complexity of the electrophysiological model, and the number of parameters to estimate in the optimization procedure. Moreover, there is also a variety of IT resources involved. Independently of these factors, the main advantage of the meshless methods is the time saved to prepare the simulation domain compared to FEM alternative, which can be a matter of hours for complex geometries.

The key parameters related to the electrophysiological modeling for better fitting the EAM data were (1) the initial stimuli (number and position) of the electrical activation, (2) the modeling of the PK system, (3) and the regional conduction velocity distribution, which was optimized for each analyzed case. For instance, most participants [27,29] and ourselves adapted their modeling solutions to consider a possible retrograde activation of the PK system, via an extra-stimulus, to replicate the rapid activation from the apex to the LV base observed in the EAM data. Figure A3 in the Appendix shows how using only one stimulus provided simulation results farther from the EAM data (error of 15.3 ms vs. 5.1 ms for two stimuli), demonstrating the dependence of the simulated activation patterns on the stimulation protocol. Potential causes for the PK retrograde activation might be the more transmural PK system in pigs compared to humans, which lead to incomplete LBBB such as in Pig 3 (anterior PK branch being functional while posterior branch being damaged, affecting the epicardial propagation). Modeling solutions with dedicated PK models such as in [29,39] could explain their better performance in these cases, justifying the use of more detailed and personalized PK system estimation algorithms [46,48].

The most important parameter to optimize in all electrophysiological modeling solutions to match EAM data were the regional distribution of conduction velocities, with computational costs directly linked to the chosen number of regions. We performed a sensitivity analysis that resulted in the use of 5 regions (RV/LV endocardium/epicardium, PK system), which avoided overfitting of LBBB-estimated results when applied to CRT cases (effect seen with a larger number of regions) and reasonable computational times (e.g., around 20 min per simulation). The CV distribution provided by the SPH-based model (see Table 2) are physiologically meaningful (e.g., PK being the fastest region, endocardial regions faster than epicardial ones, the scar having the lowest CV values), due to the imposed constraints in the optimization step. Cedilnik and Sermesant [28] used the same regions without PK in the only case they processed (Pig 3), however obtaining similar qualitative results in CRT simulations to Gomez and Sebastian [29] and ourselves. The regional strategies selected by Khamzin et al. [27] and Gomez and Sebastian [29] were the opposite, personalizing 30 and 34 (17 $AHA_{segments}$ division in both ventricles) regional parameters of conduction velocities, respectively, which gave them a lot of flexibility to match EAM data at the expense of risk of overfitting, as could be the reason of non-physiological CV distribution in some cases, compared with the SPH-based results (see ischemic case in Table 2, with conduction velocity slower in PK than in heart tissue).

Aiming at a perfect matching of simulation results to EAM data is not a simple task due to the variability of electrical patterns and the data uncertainty coming from the nature of EAM acquisitions and the post-processing (e.g., interpolation) required to create the biventricular meshes with local activation time maps. For instance, the sequential way (point-to-point) for acquiring the EAM data made the measurements dependent on the heart's anatomy and the number of EAM points, which was relatively low since an old system (CARTO XP) was used. Unexpected electrical activation patterns in the EAM of some cases could be explained by EAM interpolation effects. For instance, Pig 2 and Pig 4 non-infarcted cases had a significantly smaller amount of anatomical point-based acquisitions from CARTO XP in the LV anterior epicardial layer, leading to a 20 ms slower activation in the posterior vs. the anterior epicardial LV. We could not capture such heterogeneity in the SPH-based modeling pipeline since a single conduction velocity parameter was used for the entire LV epicardium, leading to the highest regional error in this area (see Tables A1 and A2 in the Appendix A).

Additionally, data uncertainty can lead to unrealistic and non-physiological parameters providing a better fitting between simulations and observations. For example, similar to Gomez and Sebastian [29], we needed high conduction velocities in areas near the scar in the infarcted cases (Pig 3 and Pig 6) to better fit EAM data with LBBB. Additionally, some participants included a second stimulus in the RV to better replicate the available electrophysiological measurements, which could correspond to the influence of the RV septomarginal trabecula but it could also be an interpolation artefact due to the sparsity of the EAM data. In the SPH-based modeling pipeline, we chose a constrained optimization algorithm to impose certain physiological requirements, at the expense of having less degrees of freedom, unlike approaches taken by other participants [27] that help them to achieve better fitting with EAM data (3–4.5% of LAT error in both LBBB and CRT training cases).

Another source of uncertainty is the position of the CRT leads, which justifies some differences between simulation results from all participants and EAM data. In Pig 1 and Pig 3 of the training dataset, the sub-optimal lead configuration was remarkable. In Pig 1, the apicality of the leads in both ventricles reduced the benefit of biventricular pacing, thus being a CRT non-responder (low recovery metric) both in the meshless simulation and in real data. A different CRT configuration with the LV lead placed at the LBBB latest activation point improved the recovery for both meshless and FEM-based simulations. In Pig 3, the meshless and FEM-based simulations managed to improve the recovery percentage after a reduction of the TAT due to the estimated high conduction velocities. However, the basal configuration of the leads also both located in the epicardial layer of the LV for the high extension in the apical zone and transmurality of the scar, determined the ineffectiveness of biventricular resynchronisation therapy reflected in the delay metrics such as the IVD. An analogous behavior was found by Cedilnik and Sermesant [28] with a practically identical CRT prediction LAT error (6.5 ms and 6.6 ms for them and us, respectively) to SPH (6.6 ms).

The simulation protocol designed by CRT-EPiggy19 organizers asked to personalize model parameters with the LBBB data and use them to predict CRT measurements. However, some participants [27,29] applied correction strategies to better fit EAM data after CRT. Gomez and Sebastian [29] recalculated the conduction velocities for Pig 3, allowing a better match in the LV apical part than with the SPH-based model without corrections. Khamzin et al. [27] estimated a weight to adapt LBBB regional conduction velocities to CRT using Montecarlo random sampling and simulating 1000 different electrical activation patterns for each sample due to the low computational cost of their Eikonal-based model. Additionally, we did not use warming-up cardiac cycles to establish robust initial boundary conditions in the SPH-based model, while Gomez and Sebastian [29] had 10 cardiac cycles for stabilization purposes (taking 36 h), following the pipeline they previously optimized for arrhythmia simulation [49]. Although initial boundary conditions should not have a large influence for predicting activation maps, a rigorous study should be performed to confirm this assumption.

4.2. Validation of Meshless Method Results on Testing Data

Three testing cases of the CRT-EPiggy19 challenge were also processed with the SPH-based modeling pipeline. Lamentably, FEM-based simulation results were not available for benchmarking. As the CRT lead location was not provided in the testing cases, four different lead configurations based on literature [11,50–52] were evaluated. In the three analyzed testing cases, the optimal configuration was with a RV apical lead and the LV one placed at the epicardial lateral wall ($RV_{apx} - LV_{bas}$ in Figures 6, A4 and A5). The $RV_{apx} - LV_{bas}$ lead configuration not only provided better recovery percentages, but also had a smaller LAT error with CRT EAM data (see Table A2 in the Appendix A). This is in agreement with multiple clinical studies and guidelines [11,53], although different alternatives (e.g., different RV location [20]) are still being proposed. For instance, some studies suggest that RVOT pacing may be more beneficial than standard one, specifically in cases with a decreased left ventricular ejection fraction [11,50]. In our study, RVOT pacing was the second best lead configuration, but still with slightly worse overall efficiency compared to $RV_{apx} - LV_{bas}$. The worst scenario was when both leads were in apical locations, as in the case of Pig 1, where the benefits of bi-ventricular pacing are reduced to only one lead due to an overlap of the electrical breakthrough waves.

4.3. Limitations and Future Work

The presented study has several limitations at different levels. First, the available data from the CRT-EPiggy19 challenge ertr useful to identify and better understand key aspects of different CRT models. However, several factors associated with EAM acquisition and processing induced a non-negligible data uncertainty that can limit the conclusions from the study. As well, hemodynamic descriptors, e.g., based on Doppler-derived measurements, were not available from the experimental study in Rigol et al. [24], preventing the optimization of important CRT parameters such as the AV delay, which has been found a potential non-responder factor [54]. Moreover, even in the case of better animal experiments, models should also be tested on in vivo human data to investigate its added value in the CRT clinical pipeline. Furthermore, the processing and modeling of each case, including a large number of simulations for parameter optimization, is very time consuming. The consequence is that only a few cases could be processed in our study and by the CRT-EPiggy19 participants, limiting the impact and generalizability of the benchmark analysis. A more comprehensive comparison with other FEM-based and meshless models in common data would be beneficial.

The proposed SPH-based modeling pipeline provided simulation results comparable to the state-of-the-art alternatives, but several improvements could be incorporated. Firstly, the inclusion of the anisotropic ratio and the myocardial layer for each ventricle in the optimization pipeline could give more degrees of freedom to match EAM data. Additionally, the parameter optimization schemes used by all participants of the CRT-Epiggy19 challenge were not taking advantage of recent technological advances such as the use of deep learning algorithms [55,56], variational approaches [57], reduced-order models [58,59] or GPU-based architectures [60], which allows for the exploration of a larger space of parameter solutions at reduced computational times. Moreover, cardiac multi-physical models should provide more realistic simulations, allowing for the inclusion of hemodynamic factors and improving the adjustment of CRT configuration through flow ratios [61], perfusion models [17], lumped models of the whole cardiovascular circulation [18] or with a complete torso [20].

5. Conclusions

A meshless modeling pipeline to simulate cardiac electrical patterns in CRT was compared to FEM-based alternatives, providing equivalent results on fitting experimental data available from the CRT-EPiggy19 challenge. The main advantage of the meshless model is the independence from the usually arduous patient-specific meshing process, one of the most important bottlenecks of translating computational models into a clinical environment. However, the most relevant aspect for accurate CRT predictions was the chosen parameter personalization strategy rather than the geometrical discretization. In particular, the regional conduction velocity distribution was key, requiring at least five different regions and ideally including a PK label. A larger number of regions was associated with better data fitting but higher computational costs and more risk of overfitting. Additionally, the optimal CRT configuration was found with apical RV and basal LV leads, as reported in the literature. Despite the uniqueness of the CRT-EPiggy19 challenge dataset, data uncertainty was high in some cases due to challenging EAM acquisition and processing, which could lead to the estimation of non-physiological parameters and the requirement of prior constraints in the optimization algorithm. Nevertheless, having several teams of modeling researchers working on the same data have been beneficial for each challenge participant, jointly improving the different modeling solutions.

Author Contributions: Conceptualization, C.A., J.F.G., N.C., K.A.M., T.M., S.K., A.D., O.S., E.P., M.S., R.S., H.G.M. and O.C.; Data curation, T.M.; Formal analysis, C.A., J.F.G., N.C., K.A.M., S.K., A.D., O.S., E.P., M.S. and R.S.; Investigation, C.A., J.F.G., N.C., K.A.M., S.K., A.D., O.S., E.P., M.S., R.S. and O.C.; Methodology, C.A., J.F.G., N.C., K.A.M., S.K., A.D., O.S., E.P., M.S. and R.S.; Project administration, O.C.; Software, È.L. and H.G.M.; Supervision, O.S., E.P., M.S., R.S. and O.C.; Writing—original draft, C.A., J.F.G., N.C., K.A.M., S.K., A.D., O.S., E.P., M.S., R.S. and O.C.; Writing—review & editing, C.A. and O.C. All authors have read and agreed to the published version of the manuscript.

Funding: This work was supported in part by the Spanish Ministry of Science and Innovation (TIN2011-28067, REDINSCOR RD06/003/008, PID2019-105674RB-I00), the Spanish Industrial and Technological Development Center (cvREMOD-CEN-20091044), the Spanish Ministry of Economy and Competitiveness under the Maria de Maeztu Units of Excellence Programme (MDM-2015-0502), the Government of Aragón (Spain) (LMP94_21), the Directorate General of Science Policy of the Generalitat Valenciana (Spain) (PROMETEU 2016/088), the Seventh Framework Programme (FP7/2007-2013) for research, technological and demonstration under grant agreement VP2HF (no. 611823), by the RSF grant no. 19-14-00134, and by European Regional Development Fund (ERDF) DPI2016-75799-R (AEI/ERDF, EU).

Institutional Review Board Statement: The present manuscript was based on a subset of the data from the experimental study conducted by Rigol et al. [24] at Hospital Clinic de Barcelona in which animal handling was approved by the Institutional Review Board and Ethics Committee (approval reference number: DMAH 5648). Additional information can be found at: Rigol M, Solanes N, Fernandez-Armenta J, Silva E, Doltra A, Duchateau N, Barcelo A, Gabrielli L, Bijnens B, Berruezo A, Brugada J, Sitges M. Development of a swine model of left bundle branch block for experimental studies of cardiac resynchronization therapy. J Cardiovasc Transl Res. 2013 Aug;6(4):616-22. doi: 10.1007/s12265-013-9464-1.

Conflicts of Interest: The authors declare no conflict of interest.

Appendix A

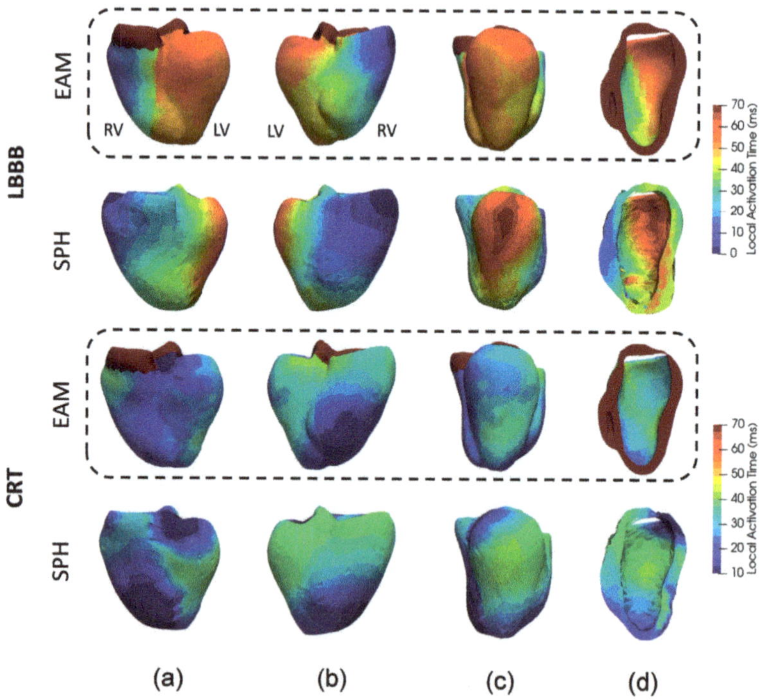

Figure A1. Local activation time maps for a non-infarcted case, Pig 2, of the training database in left bundle branch block (LBBB) and cardiac resynchronization therapy (CRT) conditions, provided by the electroanatomical measurements (EAM) and a meshless (SPH) model. (**a**,**b**) correspond to anterior and posterior biventricular epicardial visualizations. (**c**,**d**) show epicardial and endocardial view of the left ventricle (LV) lateral wall, respectively. RV: right ventricle.

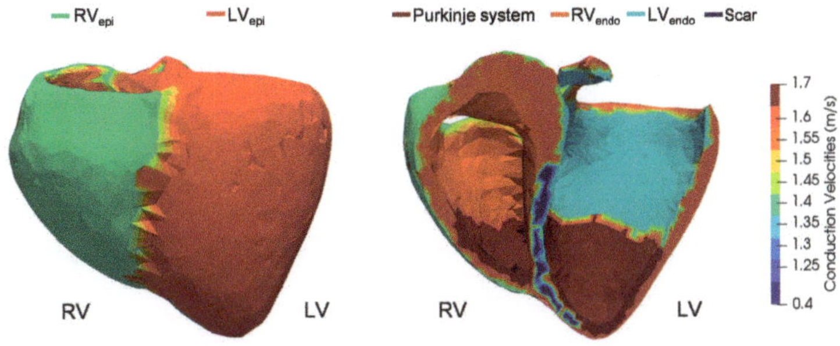

Figure A2. Conduction velocity map for one of the ischemic cases analyzed in our study.

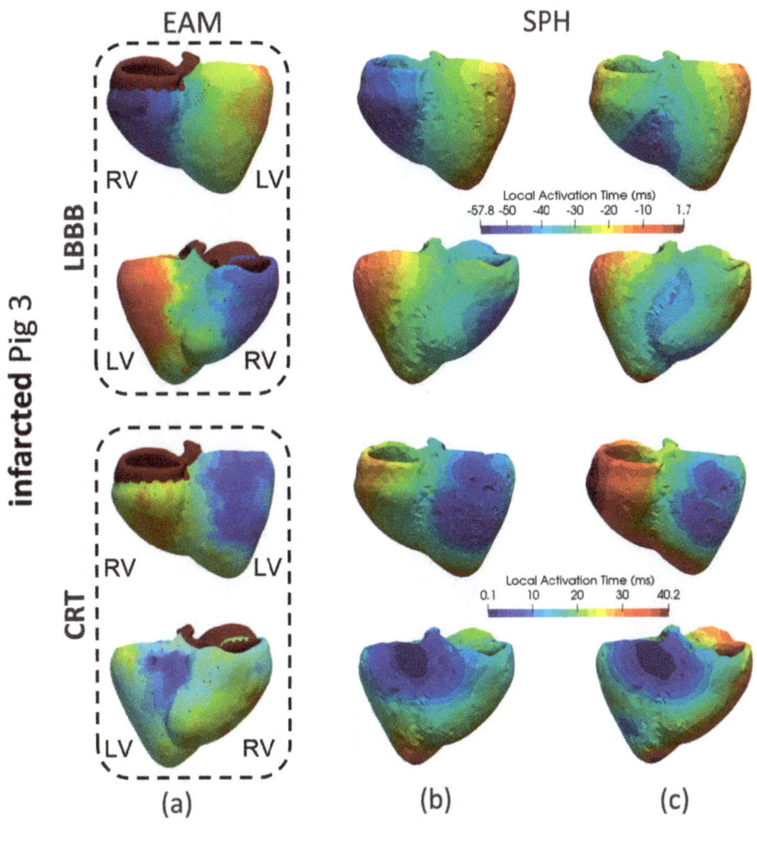

Figure A3. Local activation time maps for the infarcted case, Pig 3, of the training database in left bundle branch block (LBBB) and cardiac resynchronization therapy (CRT) conditions, provided by the (**a**) electroanatomical measurements (EAM) and the (**b**,**c**) meshless (SPH) model. From top to bottom in each condition, the anterior and posterior biventricular epicardial visualizations are shown, respectively. The electrical activation patterns acquired by maintaining the strategy of two initial stimuli (Right Ventricle (RV) and septal) are represented in (**b**) and disregarding only the initial RV stimulus in (**c**). LV: left ventricle.

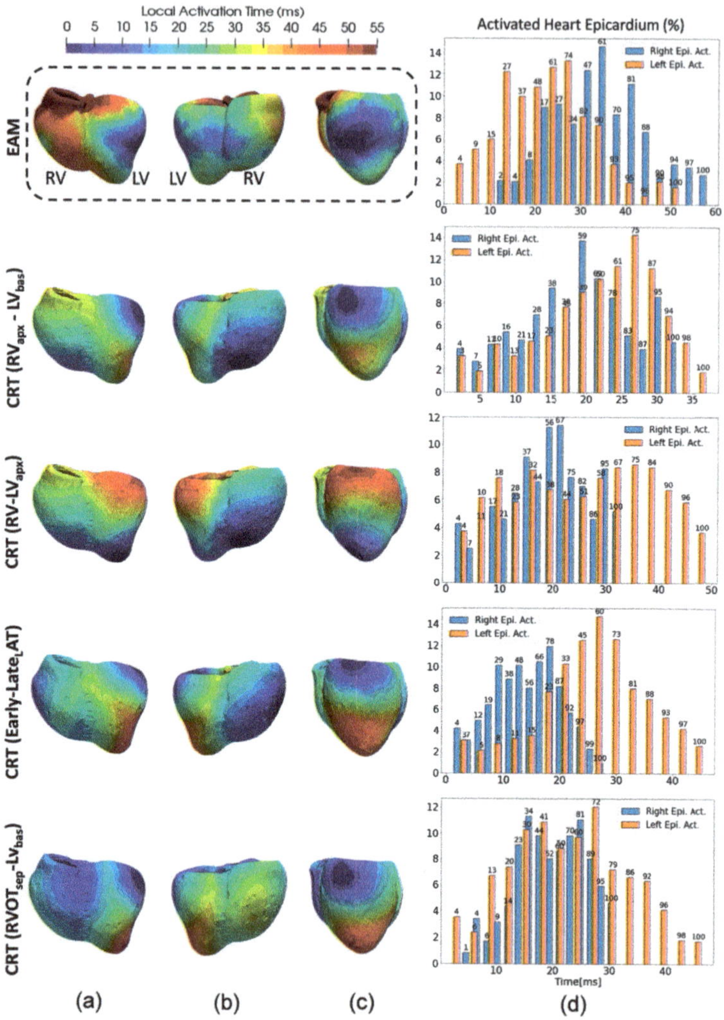

Figure A4. Local activation time maps ((**a**–**c**) showing anterior/posterior of biventricular epicardium and LV lateral wall epicardium, respectively) after cardiac resynchronization therapy (CRT) from the electro-anatomical (EAM) data and meshless simulations in the non-infarcted testing case, Pig 4. Histograms of the percentage of electrically activated heart tissue for the right and left epicardial layers are in the right column (**d**). From the second to the fifth row, different simulation results obtained with different CRT lead locations are displayed. $RV_{apx} - LV_{bas}$: leads on right ventricle (RV) apex and basal left ventricle (LV). $RV - LV_{apx}$: both leads are located in the biventricular apex. $Early - Late_L AT$: leads located at the the earliest and latest EAM ventricular activated points, respectively. $RVOT_{sep} - LV_{bas}$: leads in the septal RV outflow track and in the basal LV.

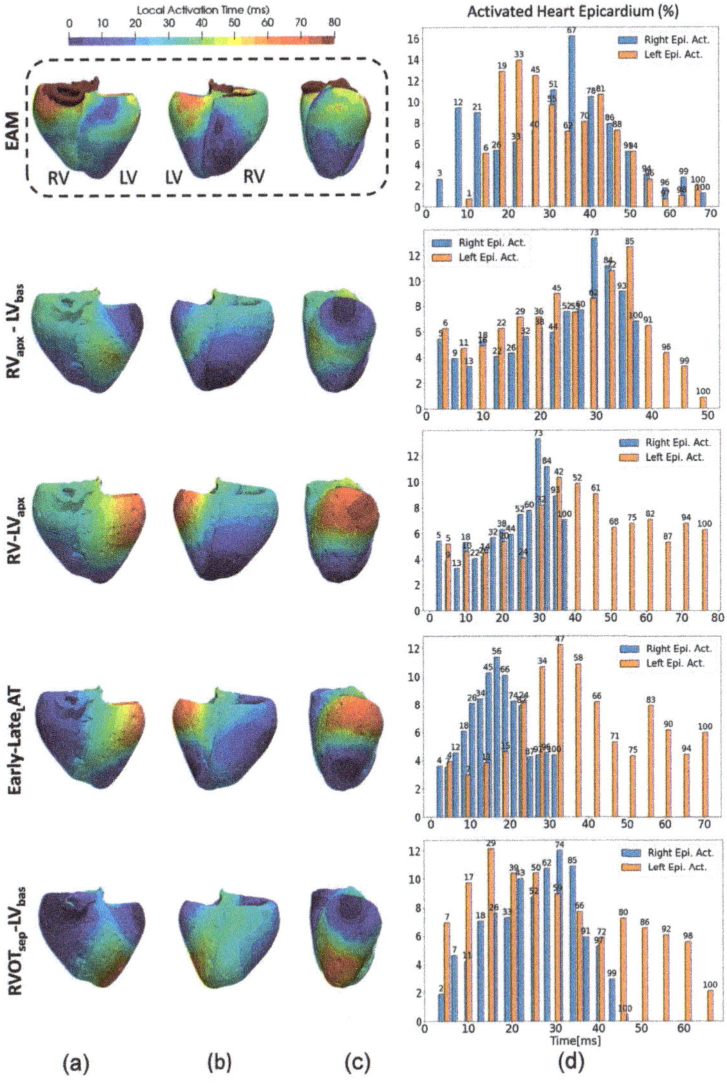

Figure A5. Local activation time maps ((**a**–**c**) showing anterior/posterior of biventricular epicardium and LV lateral wall epicardium, respectively) after cardiac resynchronization therapy (CRT) from the electro-anatomical (EAM) data and meshless simulations in the non-infarcted testing case, Pig 5. Histograms of the percentage of electrically activated heart tissue for the right and left epicardial layers are in the right column (**d**). From the second to the fifth row, different simulation results obtained with different CRT lead locations are displayed. $RV_{apx} - LV_{bas}$: leads on right ventricle (RV) apex and basal left ventricle (LV). $RV - LV_{apx}$: both leads are located in the biventricular apex. $Early - Late_L AT$: leads located at the the earliest and latest EAM ventricular activated points, respectively. $RVOT_{sep} - LV_{bas}$: leads in the septal RV outflow track and in the basal LV.

Table A1. Quantitative measures characterising the regional local activation time error in the electrical activation maps for the training cases from meshless simulations. LBBB: left bundle branch block. CRT: cardiac resynchronization therapy. RMSE: root mean square error. LV: left ventricle. RV: right ventricle. Epi: epicardium. Endo:endocardium. (*) indicates an infarcted pig.

RMSE (ms)	Pig 1		Pig 2		Pig 3 (*)	
RV Epi	LBBB	CRT	LBBB	CRT	LBBB	CRT
	5.6	8.73	7.95	7.04	5.6	4.3
LV Epi	5.9	7.37	10.17	8.53	4	6.2
LV Endo	9.1	7.45	9.76	7.48	6.4	9.7

Table A2. Quantitative measures characterising the regional local activation time error in the electrical activation maps for the testing cases from meshless simulations. LBBB: left bundle branch block. CRT: cardiac resynchronization therapy. RMSE: root mean square error. Epi: epicardium. Endo:endocardium. $RV_{apx} - LV_{bas}$: right ventricle apex and basal region of the left ventricle. $RV - LV_{apx}$: RV apex and LV apex. $Early - Late_L AT$: earliest and latest EAM activation points. $RVOT_{sep} - LV_{bas}$: RV outflow track septal and LV basal region. (*) indicates an infarcted pig.

	RMSE (ms)	RV Epi	LV Epi	LV Endo
Pig 4	LBBB	3.9	5.47	6.42
	CRT ($RV_{apx} - LV_{bas}$)	17.14	11.6	10.71
	CRT ($RV - LV_{apx}$)	17.23	14.56	13.16
	CRT ($Early - Late_L AT$)	20.67	13.4	11.86
	CRT ($RVOT_{sep} - LV_{bas}$)	15.46	13.17	12.33
Pig 5	LBBB	4.92	7.68	13.4
	CRT ($RV_{apx} - LV_{bas}$)	10.98	16.26	17.85
	CRT ($RV - LV_{apx}$)	10.98	14.74	19.27
	CRT ($Early - Late_L AT$)	20.87	10.82	14.73
	CRT ($RVOT_{sep} - LV_{bas}$)	19.82	21.88	20.8
Pig 6 (*)	LBBB	5.97	6.11	5.7
	CRT ($RV_{apx} - LV_{bas}$)	11.94	12.21	7.98
	CRT ($RV - LV_{apx}$)	12.03	18.95	14.01
	CRT ($Early - Late_L AT$)	5.91	14.48	12.7
	CRT ($RVOT_{sep} - LV_{bas}$)	13.78	12.85	11.93

References

1. Timmis, A.; Vardas, P.; Townsend, N.; Torbica, A.; Katus, H.; De Smedt, D.; Gale, C.P.; Maggioni, A.P.; Petersen, S.E.; Huculeci, R.; et al. European Society of Cardiology: Cardiovascular disease statistics 2021. *Eur. Heart J.* **2022**, *43*, 716–799. [CrossRef] [PubMed]
2. World Health Organization. Cardiovascular Diseases (CVDs). June 2021. Available online: https://www.who.int/news-room/fact-sheets/detail/cardiovascular-diseases-(cvds) (accessed on 11 May 2022).
3. Zhang, F.; Wang, Y. Left ventricular mechanical dyssynchrony in patients with heart failure: What is the next step? *J. Nucl. Cardiol.* **2021**. [CrossRef] [PubMed]
4. McDonagh, T.A.; Metra, M.; Adamo, M.; Gardner, R.S.; Baumbach, A.; Böhm, M.; Burri, H.; Butler, J.; Čelutkienė, J.; Chioncel, O.; et al. 2021 ESC Guidelines for the diagnosis and treatment of acute and chronic heart failure: Developed by the Task Force for the diagnosis and treatment of acute and chronic heart failure of the European Society of Cardiology (ESC) With the special contribution of the Heart Failure Association (HFA) of the ESC. *Eur. Heart J.* **2021**, *42*, 3599–3726. [PubMed]
5. Scherbak, D.; Hicks, G.J. Left Bundle Branch Block (LBBB). In *StatPearls*; StatPearls Publishing: Treasure Island, FL, USA, 2019.
6. Strauss, D.G.; Selvester, R.H.; Wagner, G.S. Defining left bundle branch block in the era of cardiac resynchronization therapy. *Am. J. Cardiol.* **2011**, *107*, 927–934. [CrossRef] [PubMed]
7. Healey, J.S.; Hohnloser, S.H.; Exner, D.V.; Birnie, D.H.; Parkash, R.; Connolly, S.J.; Krahn, A.D.; Simpson, C.S.; Thibault, B.; Basta, M.; et al. Cardiac resynchronization therapy in patients with permanent atrial fibrillation: Results from the Resynchronization for Ambulatory Heart Failure Trial (RAFT). *Circ. Heart Fail.* **2012**, *5*, 566–570. [CrossRef] [PubMed]

8. Stockburger, M.; Moss, A.J.; Klein, H.U.; Zareba, W.; Goldenberg, I.; Biton, Y.; McNitt, S.; Kutyifa, V. Sustained clinical benefit of cardiac resynchronization therapy in non-LBBB patients with prolonged PR-interval: MADIT-CRT long-term follow-up. *Clin. Res. Cardiol.* **2016**, *105*, 944–952. [CrossRef]
9. Bozyel, S.; Ağır, A.A.; Şahin, T.; Çelikyurt, U.; Aktaş, M.; Argan, O.; Yılmaz, İ.; Karaüzüm, K.; Derviş, E.; Vural, A.; et al. Improvement in left ventricular intrinsic dyssynchrony with cardiac resynchronization therapy. *Anatol. J. Cardiol.* **2017**, *17*, 298. [CrossRef]
10. Moss, A.J.; Brown, M.W.; Cannom, D.S.; Daubert, J.P.; Estes, M.; Foster, E.; Greenberg, H.M.; Hall, W.J.; Higgins, S.L.; Klein, H.; et al. Multicenter automatic defibrillator implantation trial–cardiac resynchronization therapy (MADIT-CRT): Design and clinical protocol. *Ann. Noninvasive Electrocardiol.* **2005**, *10*, 34–43. [CrossRef]
11. Daubert, C.; Behar, N.; Martins, R.P.; Mabo, P.; Leclercq, C. Avoiding non-responders to cardiac resynchronization therapy: A practical guide. *Eur. Heart J.* **2017**, *38*, 1463–1472. [CrossRef]
12. Sieniewicz, B.J.; Gould, J.; Porter, B.; Sidhu, B.S.; Teall, T.; Webb, J.; Carr-White, G.; Rinaldi, C.A. Understanding non-response to cardiac resynchronisation therapy: Common problems and potential solutions. *Heart Fail. Rev.* **2019**, *24*, 41–54. [CrossRef]
13. Antoniadis, A.P.; Behar, J.M.; Claridge, S.; Jackson, T.; Sohal, M.; Rinaldi, C.A. Multisite pacing for cardiac resynchronization therapy: Promise and pitfalls. *Curr. Cardiol. Rep.* **2016**, *18*, 64. [CrossRef] [PubMed]
14. Zhang, S.; Shan, Q. Discussion of LBBP synchronization effects in HF patients with LBBB and comparison with BiV-CRT. *Heart Fail. Rev.* **2022**, 1–6. [CrossRef] [PubMed]
15. Niederer, S.A.; Lumens, J.; Trayanova, N.A. Computational models in cardiology. *Nat. Rev. Cardiol.* **2019**, *16*, 100–111. [CrossRef] [PubMed]
16. Lee, A.W.; Costa, C.M.; Strocchi, M.; Rinaldi, C.A.; Niederer, S.A. Computational modeling for cardiac resynchronization therapy. *J. Cardiovasc. Transl. Res.* **2018**, *11*, 92–108. [CrossRef]
17. Fan, L.; Choy, J.S.; Raissi, F.; Kassab, G.S.; Lee, L.C. Optimization of cardiac resynchronization therapy based on a cardiac electromechanics-perfusion computational model. *Comput. Biol. Med.* **2022**, *141*, 105050. [CrossRef]
18. Owashi, K.; Taconné, M.; Courtial, N.; Simon, A.; Garreau, M.; Hernandez, A.; Donal, E.; Le Rolle, V.; Galli, E. Desynchronization Strain Patterns and Contractility in Left Bundle Branch Block through Computer Model Simulation. *J. Cardiovasc. Dev. Dis.* **2022**, *9*, 53. [CrossRef]
19. Albatat, M.; Bergsland, J.; Arevalo, H.; Odland, H.H.; Wall, S.; Sundnes, J.; Balasingham, I. Multisite pacing and myocardial scars: a computational study. *Comput. Methods Biomech. Biomed. Eng.* **2020**, *23*, 248–260. [CrossRef]
20. Carpio, E.F.; Gomez, J.F.; Sebastian, R.; Lopez-Perez, A.; Castellanos, E.; Almendral, J.; Ferrero, J.M.; Trenor, B. Optimization of lead placement in the right ventricle during cardiac resynchronization therapy. A simulation study. *Front. Physiol.* **2019**, *10*, 74. [CrossRef]
21. Oomen, P.J.; Phung, T.K.N.; Weinberg, S.H.; Bilchick, K.C.; Holmes, J.W. A rapid electromechanical model to predict reverse remodeling following cardiac resynchronization therapy. *Biomech. Model. Mechanobiol.* **2022**, *21*, 231–247. [CrossRef]
22. Camara, O.; Sermesant, M.; Lamata, P.; Wang, L.; Pop, M.; Relan, J.; De Craene, M.; Delingette, H.; Liu, H.; Niederer, S.; et al. Inter-model consistency and complementarity: Learning from ex vivo imaging and electrophysiological data towards an integrated understanding of cardiac physiology. *Prog. Biophys. Mol. Biol.* **2011**, *107*, 122–133. [CrossRef]
23. Pop, M.; Sermesant, M.; Lepiller, D.; Truong, M.; McVeigh, E.; Crystal, E.; Dick, A.; Delingette, H.; Ayache, N.; Wright, G. Fusion of optical imaging and MRI for the evaluation and adjustment of macroscopic models of cardiac electrophysiology: A feasibility study. *Med. Image Anal.* **2009**, *13*, 370–380. [CrossRef] [PubMed]
24. Rigol, M.; Solanes, N.; Fernandez-Armenta, J.; Silva, E.; Doltra, A.; Duchateau, N.; Barcelo, A.; Gabrielli, L.; Bijnens, B.; Berruezo, A.; et al. Development of a swine model of left bundle branch block for experimental studies of cardiac resynchronization therapy. *J. Cardiovasc. Transl. Res.* **2013**, *6*, 616–622. [CrossRef] [PubMed]
25. Soto Iglesias, D.; Duchateau, N.; Kostantyn Butakov, C.; Andreu, D.; Fernandez-Armenta, J.; Bijnens, B.; Berruezo, A.; Sitges, M.; Camara, O. Quantitative Analysis of Electro-Anatomical Maps: Application to an Experimental Model of Left Bundle Branch Block/Cardiac Resynchronization Therapy. *IEEE J. Transl. Eng. Health Med.* **2017**, *5*, 1900215. [CrossRef] [PubMed]
26. Ramírez, W.A.; Gizzi, A.; Sack, K.L.; Guccione, J.M.; Hurtado, D.E. In-silico study of the cardiac arrhythmogenic potential of biomaterial injection therapy. *Sci. Rep.* **2020**, *10*, 12990. [CrossRef] [PubMed]
27. Khamzin, S.; Dokuchaev, A.; Solovyova, O. Prediction of CRT Response on Personalized Computer Models. In *International Workshop on Statistical Atlases and Computational Models of the Heart*; Springer: Berlin/Heidelberg, Germany, 2019; pp. 352–363.
28. Cedilnik, N.; Sermesant, M. Eikonal Model Personalisation using Invasive Data to Predict Cardiac Resynchronisation Therapy Electrophysiological Response. In *Statistical Atlases and Computational Models of the Heart. Multi-Sequence CMR Segmentation, CRT-EPiggy and LV Full Quantification Challenges*; Springer: Berlin/Heidelberg, Germany, 2019.
29. Gomez, J.F.; Trenor, B.; Sebastian, R. Prediction of CRT Activation Sequence by Personalization of Biventricular Models from Electroanatomical Maps. In *International Workshop on Statistical Atlases and Computational Models of the Heart*; Springer: Berlin/Heidelberg, Germany, 2019; pp. 342–351.
30. Ten Tusscher, K.H.; Panfilov, A.V. Alternans and spiral breakup in a human ventricular tissue model. *Am. J. -Physiol.-Heart Circ. Physiol.* **2006**, *291*, H1088–H1100. [CrossRef] [PubMed]
31. Lluch, E.; Mihalef, V.; Vizitiu, A.; Passerini, T.; Audigier, C.; Halperin, H.; Haschemi, M.; Ashikaga, H.; Mansi, T. Is Personalized Computational Model of Atrial Fibrillation Really Personalized? *Circulation* **2021**, *144*, A11195. [CrossRef]

32. Chabiniok, R.; Wang, V.Y.; Hadjicharalambous, M.; Asner, L.; Lee, J.; Sermesant, M.; Kuhl, E.; Young, A.A.; Moireau, P.; Nash, M.P.; et al. Multiphysics and multiscale modelling, data–model fusion and integration of organ physiology in the clinic: Ventricular cardiac mechanics. *Interface Focus* **2016**, *6*, 20150083. [CrossRef]
33. Zhang, L.; Ademiloye, A.; Liew, K. Meshfree and particle methods in biomechanics: Prospects and challenges. *Arch. Comput. Methods Eng.* **2019**, *26*, 1547–1576. [CrossRef]
34. Zhang, H.; Gao, Z.; Xu, L.; Yu, X.; Wong, K.C.; Liu, H.; Zhuang, L.; Shi, P. A meshfree representation for cardiac medical image computing. *IEEE J. Transl. Eng. Health Med.* **2018**, *6*, 1–12. [CrossRef]
35. Liu, M.; Zhang, Z. Smoothed particle hydrodynamics (SPH) for modeling fluid–structure interactions. *Sci. China Phys. Mech. Astron.* **2019**, *62*, 984701. [CrossRef]
36. Wong, K.C.; Wang, L.; Zhang, H.; Liu, H.; Shi, P. Meshfree implementation of individualized active cardiac dynamics. *Comput. Med. Imaging Graph.* **2010**, *34*, 91–103. [CrossRef] [PubMed]
37. Lluch, È.; De Craene, M.; Bijnens, B.; Sermesant, M.; Noailly, J.; Camara, O.; Morales, H.G. Breaking the state of the heart: meshless model for cardiac mechanics. *Biomech. Model. Mechanobiol.* **2019**, *18*, 1549–1561. [CrossRef] [PubMed]
38. Lluch, È.; Camara, O.; Doste, R.; Bijnens, B.; De Craene, M.; Sermesant, M.; Wang, V.Y.; Nash, M.P.; Morales, H.G. Calibration of a fully coupled electromechanical meshless computational model of the heart with experimental data. *Comput. Methods Appl. Mech. Eng.* **2020**, *364*, 112869. [CrossRef]
39. Mountris, K.A.; Pueyo, E. Cardiac electrophysiology meshfree modeling through the mixed collocation method. *arXiv* **2021**, arxiv:2110.06671.
40. O'Hara, T.; Virág, L.; Varró, A.; Rudy, Y. Simulation of the undiseased human cardiac ventricular action potential: Model formulation and experimental validation. *PLoS Comput. Biol.* **2011**, *7*, e1002061. [CrossRef]
41. Doste, R.; Soto-Iglesias, D.; Bernardino, G.; Alcaine, A.; Sebastian, R.; Giffard-Roisin, S.; Sermesant, M.; Berruezo, A.; Sanchez-Quintana, D.; Camara, O. A rule-based method to model myocardial fiber orientation in cardiac biventricular geometries with outflow tracts. *Int. J. Numer. Methods Biomed. Eng.* **2019**, *35*, e3185. [CrossRef]
42. Lluch, E.; Doste, R.; Giffard-Roisin, S.; This, A.; Sermesant, M.; Camara, O.; de Craene, M.; Morales, H.G. Smoothed Particle Hydrodynamics for Electrophysiological Modeling: An Alternative to Finite Element Methods. FIMH 2017. In Proceedings of the FIMH 2017—9th International Conference on Functional Imaging and Modelling of the Heart, Toronto, ON, Canada, 6 June 2017; pp. 333–343.
43. Mitchell, C.C.; Schaeffer, D.G. A two-current model for the dynamics of cardiac membrane. *Bull. Math. Biol.* **2003**, *65*, 767–793. [CrossRef]
44. Talbot, H.; Marchesseau, S.; Duriez, C.; Sermesant, M.; Cotin, S.; Delingette, H. Towards an interactive electromechanical model of the heart. *Interface Focus* **2013**, *3*, 20120091. [CrossRef]
45. Camara, O.; Pashaei, A.; Sebastian, R.; Frangi, A. *Personalization of Fast Conduction Purkinje System in Eikonal-Based Electrophysiological Models with Optical Mapping Data*; Springer: Berlin/Heidelberg, Germany, 2010; Volume 6364 LNCS.
46. Cárdenes, R.; Sebastian, R.; Soto-Iglesias, D.; Berruezo, A.; Camara, O. Estimation of Purkinje trees from electro-anatomical mapping of the left ventricle using minimal cost geodesics. *Med. Image Anal.* **2015**, *24*, 52–62. [CrossRef]
47. Viceconti, M.; Pappalardo, F.; Rodriguez, B.; Horner, M.; Bischoff, J.; Musuamba Tshinanu, F. In silico trials: Verification, validation and uncertainty quantification of predictive models used in the regulatory evaluation of biomedical products. *Methods* **2021**, *185*, 120–127. [CrossRef]
48. Barber, F.; Langfield, P.; Lozano, M.; Garcia-Fernandez, I.; Duchateau, J.; Hocini, M.; Haissaguerre, M.; Vigmond, E.; Sebastian, R. Estimation of Personalized Minimal Purkinje Systems from Human Electro-Anatomical Maps. *IEEE Trans. Med. Imaging* **2021**, *40*, 2182–2194. [CrossRef]
49. Lopez-Perez, A.; Sebastian, R.; Izquierdo, M.; Ruiz, R.; Bishop, M.; Ferrero, J.M. Personalized cardiac computational models: from clinical data to simulation of infarct-related ventricular tachycardia. *Front. Physiol.* **2019**, *10*, 580. [CrossRef] [PubMed]
50. Shimony, A.; Eisenberg, M.J.; Filion, K.B.; Amit, G. Beneficial effects of right ventricular non-apical vs. apical pacing: A systematic review and meta-analysis of randomized-controlled trials. *Europace* **2012**, *14*, 81–91. [CrossRef] [PubMed]
51. Leclercq, C.; Sadoul, N.; Mont, L.; Defaye, P.; Osca, J.; Mouton, E.; Isnard, R.; Habib, G.; Zamorano, J.; Derumeaux, G.; et al. Comparison of right ventricular septal pacing and right ventricular apical pacing in patients receiving cardiac resynchronization therapy defibrillators: The SEPTAL CRT Study. *Eur. Heart J.* **2016**, *37*, 473–483. [CrossRef] [PubMed]
52. Huntjens, P.R.; Walmsley, J.; Ploux, S.; Bordachar, P.; Prinzen, F.W.; Delhaas, T.; Lumens, J. Influence of left ventricular lead position relative to scar location on response to cardiac resynchronization therapy: a model study. *Europace* **2014**, *16*, iv62–iv68. [CrossRef]
53. Sharma, S.P.; Dahal, K.; Dominic, P.; Sangha, R.S. Clinical and echocardiographic response of apical vs. nonapical right ventricular lead position in CRT: A meta-analysis. *J. Arrhythmia* **2018**, *34*, 185–194. [CrossRef] [PubMed]
54. Brabham, W.W.; Gold, M.R. The role of AV and VV optimization for CRT. *J. Arrhythmia* **2013**, *29*, 153–161. [CrossRef]
55. Treml, L.M.; Bartocci, E.; Gizzi, A. Modeling and analysis of cardiac hybrid cellular automata via GPU-accelerated Monte Carlo simulation. *Mathematics* **2021**, *9*, 164. [CrossRef]
56. Shahi, S.; Fenton, F.H.; Cherry, E.M. Prediction of chaotic time series using recurrent neural networks and reservoir computing techniques: A comparative study. *Mach. Learn. Appl.* **2022**, *8*, 100300. [CrossRef]
57. Barone, A.; Gizzi, A.; Fenton, F.; Filippi, S.; Veneziani, A. Experimental validation of a variational data assimilation procedure for estimating space-dependent cardiac conductivities. *Comput. Methods Appl. Mech. Eng.* **2020**, *358*, 112615. [CrossRef]

58. Barone, A.; Carlino, M.G.; Gizzi, A.; Perotto, S.; Veneziani, A. Efficient estimation of cardiac conductivities: A proper generalized decomposition approach. *J. Comput. Phys.* **2020**, *423*, 109810. [CrossRef]
59. Fresca, S.; Manzoni, A.; Dedè, L.; Quarteroni, A. Deep learning-based reduced order models in cardiac electrophysiology. *PLoS ONE* **2020**, *15*, e0239416.
[CrossRef] [PubMed]
60. Kaboudian, A.; Cherry, E.M.; Fenton, F.H. Real-Time Interactive Simulations of Complex Ionic Cardiac Cell Models in 2D and 3D Heart Structures with GPUs on Personal Computers. In *2021 Computing in Cardiology (CinC)*; IEEE: Piscataway, NJ, USA, 2021; Volume 48, pp. 1–4.
61. Santiago, A.; Aguado-Sierra, J.; Zavala-Aké, M.; Doste-Beltran, R.; Gómez, S.; Arís, R.; Cajas, J.C.; Casoni, E.; Vázquez, M. Fully coupled fluid-electro-mechanical model of the human heart for supercomputers. *Int. J. Numer. Methods Biomed. Eng.* **2018**, *34*, e3140. [CrossRef] [PubMed]

applied sciences

MDPI

Article

Structure (Epicardial Stenosis) and Function (Microvascular Dysfunction) That Influence Coronary Fractional Flow Reserve Estimation †

Jermiah J. Joseph [1,2], Clara Sun [1], Ting-Yim Lee [2,3], Daniel Goldman [2], Sanjay R. Kharche [1,2,*] and Christopher W. McIntyre [1,2,*]

1 Lawson Health Research Institute, Western University, 750 Baseline Road East, London, ON N6C 2R5, Canada; jjosep56@uwo.ca (J.J.J.); clara.sun@lhsc.on.ca (C.S.)
2 Department of Medical Biophysics, Western University, 1151 Richmond Street North, London, ON N6A 5C1, Canada; tlee@rri.ca (T.-Y.L.); dgoldma2@uwo.ca (D.G.)
3 Robarts Research Institute, Western University, 1151 Richmond Street North, London, ON N6A 5C1, Canada
* Correspondence: sanjay.kharche@lhsc.on.ca (S.R.K.); christopher.mcintyre@lhsc.on.ca (C.W.M.)
† This paper is an extended version of paper published in Functional Imaging and Modelling of the Heart 2021 in 11th International Conference, held virtually in Stanford, CA, USA, 21–25 June 2021.

Citation: Joseph, J.J.; Sun, C.; Lee, T.-Y.; Goldman, D.; Kharche, S.R.; McIntyre, C.W. Structure (Epicardial Stenosis) and Function (Microvascular Dysfunction) That Influence Coronary Fractional Flow Reserve Estimation. *Appl. Sci.* **2022**, *12*, 4281. https://doi.org/10.3390/app12094281

Academic Editors: Mihaela Pop and Cristian A. Linte

Received: 28 February 2022
Accepted: 22 April 2022
Published: 23 April 2022

Abstract: *Background.* The treatment of coronary stenosis is decided by performing high risk invasive surgery to generate the fractional flow reserve diagnostics index, a ratio of distal to proximal pressures in respect of coronary atherosclerotic plaques. Non-invasive methods are a need of the times that necessitate the use of mathematical models of coronary hemodynamic physiology. This study proposes an extensible mathematical description of the coronary vasculature that provides an estimate of coronary fractional flow reserve. *Methods.* By adapting an existing computational model of human coronary blood flow, the effects of large vessel stenosis and microvascular disease on fractional flow reserve were quantified. Several simulations generated flow and pressure information, which was used to compute fractional flow reserve under several conditions including focal stenosis, diffuse stenosis, and microvascular disease. Sensitivity analysis was used to uncover the influence of model parameters on fractional flow reserve. The model was simulated as coupled non-linear ordinary differential equations and numerically solved using our implicit higher order method. *Results.* Large vessel stenosis affected fractional flow reserve. The model predicts that the presence, rather than severity, of microvascular disease affects coronary flow deleteriously. *Conclusions.* The model provides a computationally inexpensive instrument for future in silico coronary blood flow investigations as well as clinical-imaging decision making. A combination of focal and diffuse stenosis appears to be essential to limit coronary flow. In addition to pressure measurements in the large epicardial vessels, diagnosis of microvascular disease is essential. The independence of the index with respect to heart rate suggests that computationally inexpensive steady state simulations may provide sufficient information to reliably compute the index.

Keywords: coronary vasculature; lumped parameter model; fractional flow reserve; computational cardiology

1. Introduction

This manuscript is an extension of work originally presented in Functional Imaging and Modelling of the Heart, 2021 [1].

Clinical relevance of and potential sources of uncertainty in fractional flow reserve estimation: Coronary vessel severity of stenosis is clinically quantified using a quantity called fractional flow reserve (FFR) [2,3]. Quantities such as FFR allow objective clinical decision making, especially when computed tomography subjectively indicates intermediate coronary stenosis. Several clinical trials have led to a partial clinical acceptance of FFR for objective diagnostics [4–6]. FFR is clinically measured by determination of the ratio of blood flow through a stenosed vessel to that in the same vessel in the absence of stenosis [7]. In recent times, non-invasive computed tomography angiography combined with computational fluid dynamics (CFD) have become increasingly prevalent in estimating FFR, and aim at reducing the significant risks associated with invasive pressure wire measurements [8]. However, multiple complex physiological processes render uncertain FFR estimation [9]. In particular, the clinical literature suggests that micro-vascular dysfunction and stenosis morphology play a significant role in the estimated FFR. In addition, surgical and pharmacological sensitivity remains limited where adverse events often occur in critically ill patients, such as those with renal failure [10], where diagnostics are sub-optimal. As such, the use of FFR to determine clinical intervention depends on quantification of the vascular structure and function.

A brief overview of FFR modelling: Computed tomography angiography-driven computational estimation of FFR is now an advanced technology [11]. Combining imaging with computational fluid dynamics assessment of FFR is known to increase the specificity of diagnosing lesion-specific ischemia [12]. It is facilitated by ready availability of open source advanced scientific platforms [13–16], including those developed in house (Virtual Cardiac Physiology Laboratory) [17,18]. Typically, computation of FFR combines an imaging-generated 3D coronary geometry coupled to models of coronary hemodynamic physiology. Others have used the approach to study a spectrum of processes involving FFR estimation refinement [19], interplay among multiple stenosis complexes [20], and perioperative treatment assessment [21], among several other applications. Computer modelling can be performed at a simple lumped parameter or detailed 3D spatial scales. As a specialized high performance computing application, 3D modelling cannot be performed onsite by the clinician. Due to the large variety of data collection resulting in the need to explore parameter spaces [22,23], large scale computations remain unwarranted in a clinical environment. Recent studies demonstrate the deployable nature of lumped parameter (0D) modelling in a clinical environment. Our recent study, where the role of peripheral arterial disease in hypertension was addressed, illustrates a 0D model deployable nature [24]. We also used 0D hemodynamic modelling of the whole human model to test the effects of treatments such as hypothermia and exercise on systemic circulation [17]. The debilitating effects of atrial fibrillation on cerebral circulation were illuminated recently by Hunter et al. [25]. However, the availability of computationally efficient coronary blood flow models remains limited [26]. It was therefore relevant to develop an open source and extensible coronary model.

Study aims: In this work, an existing lumped parameter (0D) model of the coronary vasculature [27] was further developed and used to demonstrate important factors that regulate FFR. Specifically, the dependence of FFR on the nature of stenosis (focal or diffuse) and on micro-vascular status was investigated. Further, a partial rank correlation coefficient (PRCC)-based sensitivity analysis [1,28,29] was performed to determine the impact of model parameters on FFR. For this purpose, a 0D modelling approach was found to be suitable as the study's goal was to understand coronary flow in the presence of pathological conditions. It can be appreciated that model identification (personalization), although highly desirable, was not essential in this theoretical study. As such, the presented model is theoretical in nature, in which a better understanding of pathophysiological processes was prioritized over model personalization.

2. Methods

Model development: A recent model of the coronary circulation [27] was adapted. It consists of 16 epicardial coronary artery segments, including the left anterior descending (LAD), left circumflex artery (LCX), right coronary artery (RCA), and several of their clinically significant daughter segments. The closed loop connectivity of the structured tree network is illustrated in Figure 1 and the names of all arteries are elaborated in Table 1. Each artery segment is characterized by the Windkessel time-independent parameters that consist of a hydraulic resistance (R_n), the inertia to flow of blood represented by an inductance (L_n), and the elastic capacity of the vessel, C_n [30]. The Windkessel parameters are determined using vessel lengths, vessel wall thickness, diameters, elasticity, blood viscosity, and blood density. In this study blood viscosity was taken to be 4×10^{-3} kg/(m-s) and density to be 1.06×10^3 kg/m^3, and Young's modulus (inverse of elasticity) to be 2×10^5 Pa, in agreement with current knowledge [31–33]. Vessel wall thickness was estimated as h = 0.08 D [30]. Each artery segment entering a capillary bed leading into the venous circulation was further assumed to experience a microvasculature terminal impedance (Z_i) that was estimated using a structured tree model by Olufsen [34] as

$$Z_i = \frac{8\mu\lambda((2\gamma^3)^{-(N+1)} - 1)}{\pi r_0{}^3(0.5\gamma^{-3} - 1)}, \; i = 1, \ldots, 9. \tag{1}$$

where $\gamma = 2^{-\frac{1}{\epsilon}}$ and ϵ represents the daughter vessel radius taper exponent, λ is the ratio of microvascular length to its diameter, and r_0 is the root vessel radius of the structured tree. N represents the number of generations for each structured tree [27,30,34]. The lumped coronary system was further developed by incorporating a detailed four chamber heart description (Figure 1A) [35]. For simplicity, this model does not account for the phase altering effects of cardiac contractility on microvascular coronary flow.

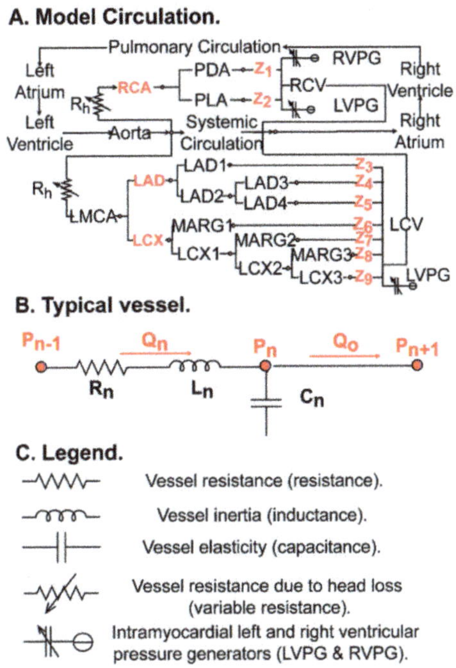

Figure 1. The modelled lumped parameter coronary vasculature tree network. (**A**) Closed loop vascular structure including tree network and functional components. See Table 1 for vessel names. Z_i (i = 1 to 9) represent terminal vessel impedances. Vessels as well as impedances shown in red were used in the simulation experiments. (**B**) Typical blood vessel represented by a resistance (R_n), inductance (L_n), and a capacitance (C_n). P_{n-1}: vessel inlet pressure; Q_n: flow through vessel; P_n: pressure in vessel; Q_o: outlet flow; P_{n+1}: outlet pressure, or pressure in distal vessel. (**C**) Symbols used in panels (**A**, **B**), and elsewhere in this work.

Table 1. Model parameter values. See Figure 1 for vessel connectivity. The rows are colour-coded to suggest the major epicardial coronaries, either LAD, LCX, or RCA.

Vessel	R (mmHg-s/mL)	C (ml/mmHg $\times 10^{-3}$)	L (mmHg-s^2/mL)
LMCA	0.2299	2.9	0.00228
LAD	0.4662	1.6	0.0298
LAD1	0.5729	1.6	0.0342
LAD2	1.7077	3.4	0.0916
LAD3	3.7484	1.3	0.1115
LAD4	3.2930	0.4	0.0716
LCX	0.3929	1.2	0.0241
LCX1	0.4730	0.7	0.0231
LCX2	1.0264	0.7	0.0380
LCX3	3.2342	1.1	0.0944
MARG1	1.7351	1.2	0.0655
MARG2	2.9195	0.8	0.0787
MARG3	3.0683	1	0.0896
RCA	1.8302	6.3	0.1171
PLA	2.4412	1.1	0.0799
PDA	1.2571	1.8	0.0596

Using the parameters given in Tables 1 and 2, summarized from the parent model [27] and microvascular impedances calculated using Equation (1), pressure at each node of the model (Figure 1) was computed as

$$\frac{dP_n}{dt} = \frac{Q_n - Q_0}{C_n} \quad (2)$$

and the flow through each vessel (resistance) was calculated as

$$\frac{dQ_n}{dt} = \frac{P_{n-1} - P_n - R_n Q_n}{L_n} \quad (3)$$

Table 2. Parameters used to compute microvascular impedances.

Z (Figure 1).	Root Vessel Radius, r_0 (mm).		N.	Control Z Values (mmHg-s/mL).
Z_1	PDA	0.108	19	134.100
Z_2	PLA	0.130	20	083.710
Z_3	LAD1	0.146	20	059.095
Z_4	LAD3	0.103	19	154.592
Z_5	LAD4	0.088	18	227.185
Z_6	MARG1	0.116	19	108.224
Z_7	MARG2	0.098	19	179.482
Z_8	MARG3	0.102	19	159.184
Z_9	LCX3	0.102	19	159.184

Legend. Z: terminal impedance; N: number of generations in microvasculature.

Further, the flow through each of the terminals was calculated as

$$Q_{z,n} = \frac{P_{n-1} - P_n}{Z_n} \quad (4)$$

Simulation experiments: In all simulations, fractional flow reserve (FFR) was computed as the average of the ratio of the time-dependent distal pressure, P_d (pressure downstream from stenosis) to the time-dependent proximal (aortic) pressure, P_a:

$$FFR_{vessel} = \frac{1}{M}\sum_{n=1}^{n=M}\frac{P_{vessel,n}}{P_{aorta,n}} \tag{5}$$

where M represents the total number of fractions over a given time T, which consisted of M time step recordings. T was taken to be 100 heart beats and the final 20 were analyzed. Simulations were designed to explore the effects of stenosis severity in the largest epicardial vessels (either LAD, LCX, or RCA; see Figure 1) or microvascular disease, or both. A sensitivity analysis was performed as described below.

Stenosis in three large vessels, namely the left anterior descending artery (LAD), the left circumflex artery (LCX), and the right coronary artery (RCA), was investigated. Simulations were performed by imposing focal or diffuse stenosis in a given large vessel.

To simulate focal stenosis, the blood vessel was divided into two and its biophysical parameters (Table 1) were revised using

$$\begin{aligned} R_s &= R_o\alpha^{-2} \\ C_s &= C_o\alpha^{3/2} \\ L_s &= L_o\alpha^{-1} \end{aligned} \tag{6}$$

where the stenosis severity, α, is given by the parameter

$$\alpha = \frac{A_s}{A_o}. \tag{7}$$

which is always between 0 and 1 by definition. To simulate diffuse stenosis extended through a certain length percentage x_s $(0 \le x_s \le 1)$ of a vessel, the revised parameters were calculated as

$$\begin{aligned} R &= R_s x_s + R_o(1-x_s) \\ L &= L_s x_s + L_o(1-x_s) \\ C &= C_s x_s + C_o(1-x_s) \end{aligned} \tag{8}$$

and used in Equations (2)–(4). Microvascular disease was simulated by decreasing the terminal vessel radius by a predefined amount in all terminals. In this model, radius regulated microvascular impedance was increased by decreasing the ϵ in Equation (1)'s γ parameter.

Sensitivity analysis: Sensitivity of multiple model parameters, including stenosis lengths, focal stenosis severity, heart rate, terminal vessel impedances, microvascular vessel taper parameter (ε), and number of downstream vasculature generations to FFR, was computed. To do so, we used our implementation of partial ranked correlation coefficients (PRCC) [17,36]. The coefficients were used to rank the parameters in descending order of significance, and the most relevant results reported.

Numerical methods: The model is a system of 36 coupled stiff ordinary differential equations. Pressures and flows were computed as state variables according to governing ordinary differential equations, Equations (2)–(4), for each vessel. The system was solved using our robust implicit solver available in our simulation software [18,24]. The method used in the solver is based on implicit backward difference formulae that provides $O(\mathrm{dt}^6)$ accuracy. A maximum user time step of 0.005 s gave stable solutions which remained unaffected when the maximum time step was halved and doubled. Each instance generated 500 s of simulated dynamics from which the final 10 s of activity were used to generate results. Simulations were performed on local and national clusters. Each instance of the model is a serial run that took 15 s. To construct results in the presented work, a large number of model instances (10^6) for predefined values of physiologically relevant parameters were executed within 4 h using 48 processors. The trivially parallel simulations were performed using GNU Utilities [37]. The simulation outputs were post-processed using a combination of UNIX and MATLAB scripts.

3. Results

Model FFR during the cardiac cycle: Time-dependent FFR in the three major coronary arteries (LAD, LCX, and RCA) under predefined large vessel stenosis and microvascular disease is illustrated in Figure 2. The control simulation (Figure 2, top row) devoid of stenosis or microvascular disease shows that FFR is high (more than 0.8) during the complete cardiac cycle in all three vessels. Due to flow distribution from the aorta to the smaller coronary network, the time dependent FFR was seen to reduce during systole. The time dependent FFR when either LAD, LCX, or RCA were focally stenosed by 90% ($\alpha = 0.9$) is shown in Figure 2, middle row. When there was a full vessel length stenosis the FFR values reduced to 0.56 for the LAD, 0.52 for the LCX, and 0.5 for the RCA. Whereas the overall FFR was observed to reduce significantly in all three simulations, large vessel stenosis led to minimal FFR during the cardiac cycle's diastole. Simulated microvascular disease, simulated by augmenting all terminal impedances by 50% ($\varepsilon = 2.55$, a reduction of ε increases impedance, Z), led to amplifying the difference between the aortic and respective distal pressures and gave a minimal FFR estimate during the systole (Figure 2, bottom row). When microvascular disease was simulated, the maximum time dependent FFR value was calculated to be 1 and minimum to be 0.7 in all three blood vessels.

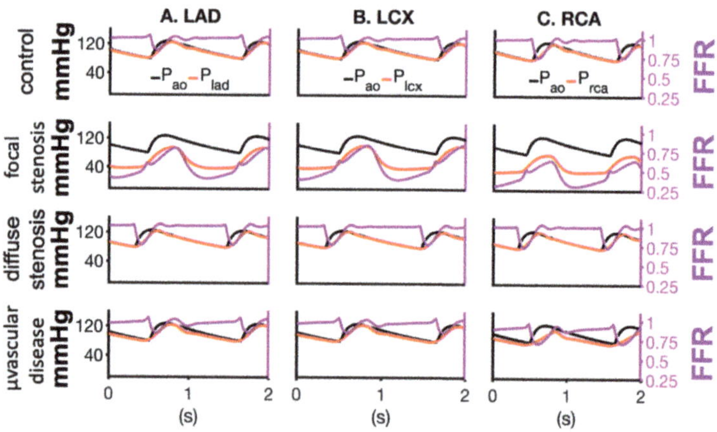

Figure 2. Pressure profiles and FFR in the LAD (column **A**), LCX (column **B**), and RCA (column **C**). In all columns, top row shows non-stenosed model behavior, second row shows the result of focal stenosis ($\alpha = 90\%$), and third row shows the result of downstream microvascular disease in the absence of focal stenosis ($\alpha = 100\%$; $\epsilon = 2.33$). In all panels, black lines and axis represent aortic pressure (proximal pressure) while red lines and axis represent the pressure of vessel of interest (distal pressure). Time dependent FFR is shown as orange dashed lines.

The coronary flow in the control coronary model (Figure 3, top row) and its reduction due to focal stenosis (Figure 3, middle row) and microvascular disease (Figure 3, bottom row) was computed. Relative to the control case (Figure 3, top row), focal stenosis (Figure 3, second row) restricted flow significantly in all three blood vessels. When microvascular disease was implemented, the maximum flow and overall flow in the network decreased. Further, the impact of individual artery resistances, inertances, and compliances were blunted as reflected in the flow profiles (Figure 3, bottom row).

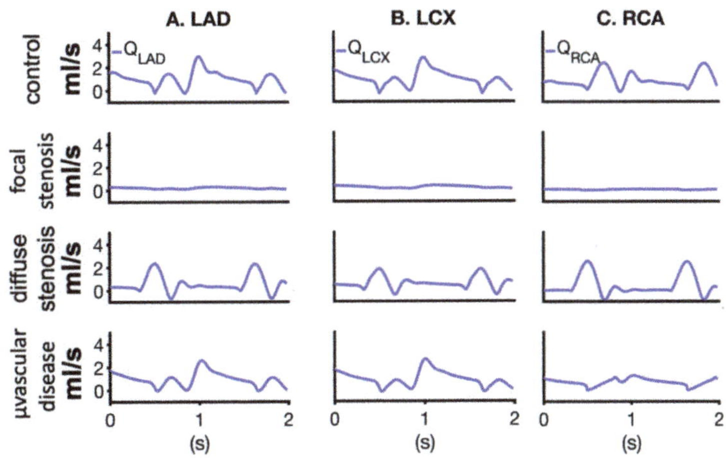

Figure 3. Flow profiles in the LAD (column **A**), LCX (column **B**), and RCA (column **C**). In all columns, top row shows non-stenosed (control) model flow, second row shows flow under focal stenosis ($\alpha = 90\%$) and the third row shows the flow under microvascular disease in the absence of focal stenosis ($\alpha = 100\%$; $\epsilon = 2.33$).

Focal and diffuse stenosis interplay: The dependence of average flow (flow), maximum flow, and FFR on simultaneous presence of reduced vessel diameters (focal stenosis) and diffuse stenosis (reduction of diameters along a predefined length) were quantified (Figure 4). In all vessels, the detrimental effects of stenosis on flow (Figure 4, top row) and maximum flow (Figure 4, middle row) were impacted by the severity of focal stenosis (horizontal axis) to a greater extent than the severity of diffuse stenosis (vertical axis). Progressive focal stenosis alone was found to minimally impact the estimated FFR (Figure 4, bottom row) due to the model formulation (see above). As such, a reduction of FFR was observed when the stenosis was diffuse to a certain extent. Conversely, diffuse stenosis in the absence of focal stenosis (vertical axis in Figure 4) also did not reduce FFR. Progressive focal stenosis in the RCA caused the largest reduction in FFR (Figure 4, third row) as compared to focal stenosis in the LAD and LCX in the presented model. In the presented model, the RCA was more susceptible to FFR reduction due to stenosis in comparison to the LAD and LCX. Simultaneous presence of focal and diffuse stenosis caused the most severe reduction of FFR in the RCA, followed by the LAD and LCX.

Role of microvascular disease in the modelled FFR: The average flow (flow), maximum flow, and FFR values of simultaneous focal stenosis and microvascular disease are shown in Figure 5. Microvascular disease was simulated by varying the daughter vessel's radius taper exponent ϵ (Equation (1)) from 2.76 (0% microvascular disease, control) to 2.33 (100% microvascular disease) which represents turbulent flow [22]. At diameter reductions below 70%, the flow in each blood vessel (Figure 5, top row) is significantly restricted by up to half of the control flow with the increase in severity of microvascular disease. At similar diameter reductions in the LAD and LCX however, the peak reduction in max flow values (near 0.5 of the control values) occur at 50% microvascular disease and returns to near control values at maximal microvascular disease. At diameter reductions above 80%, microvascular exacerbates the effect of the stenosis on FFR values. However, an almost unique value of diameter reduction for each, LAD, LCX, and RCA, was observed to characterize a clinically significant FFR transition to below 0.8 in the presence of an arbitrarily severe microvascular disease. While the diameter reduction was 0.7 for LAD and LCX, it was seen to be a much lower 0.55 in the case of RCA.

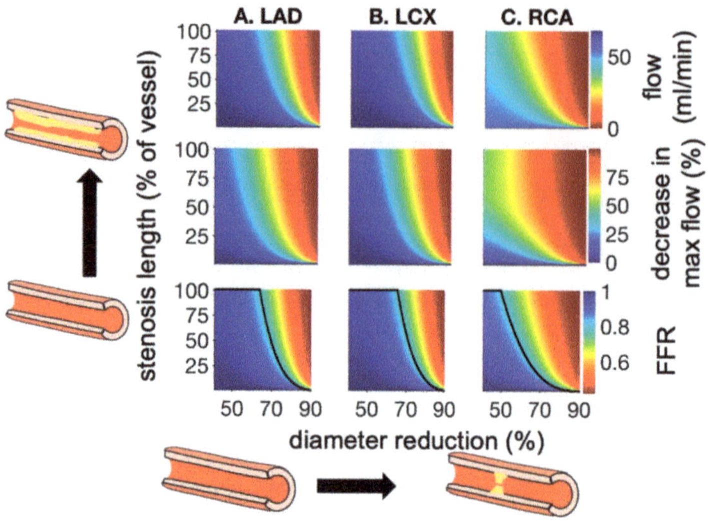

Figure 4. Dependence of flow rate (top row), maximum flow (middle row), and mean FFR (bottom row) on stenosis length (vertical axis, all panels) and vessel diameter (horizontal axis, all panels). Columns (**A–C**) show LAD, LCX, and RCA results, respectively. The black line in the bottom row demarcates the FFR = 0.8 threshold.

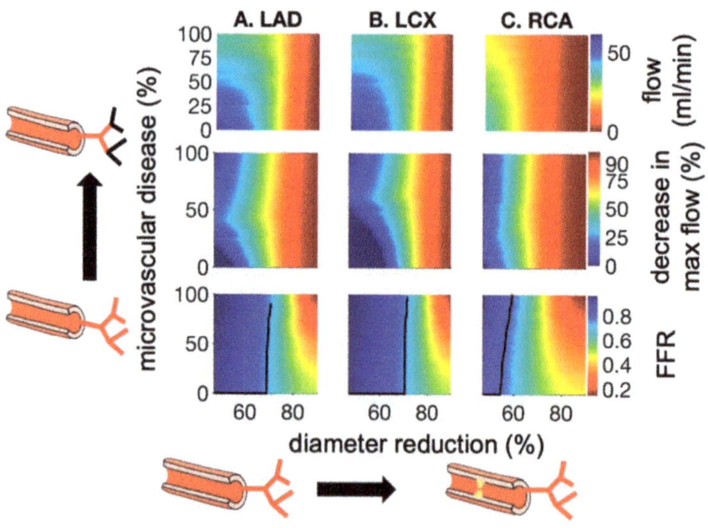

Figure 5. Dependence of flow rate (top row), maximum flow (middle row), and mean FFR (bottom row) on microvascular resistance increase (microvascular disease, vertical axis in all panels) and vessel diameter (horizontal axis, all panels). Columns (**A–C**) show LAD, LCX, and RCA results, respectively. The black line in the bottom row demarcates the FFR = 0.8 threshold.

Sensitivity analysis to stratify FFR impacting parameters: The results of the sensitivity analysis are shown in Figures 6 and 7. The histograms of FFR values obtained during the PRCC calculation are shown in Figure 6. As can be seen, the model did not produce any instances with FFR less than 0.3 due to the ranges of parameters considered. The model appears to produce FFR values centered around 0.54. Further, in all three coronaries, the FFR values appear to be distributed in a left-skewed Gaussian manner.

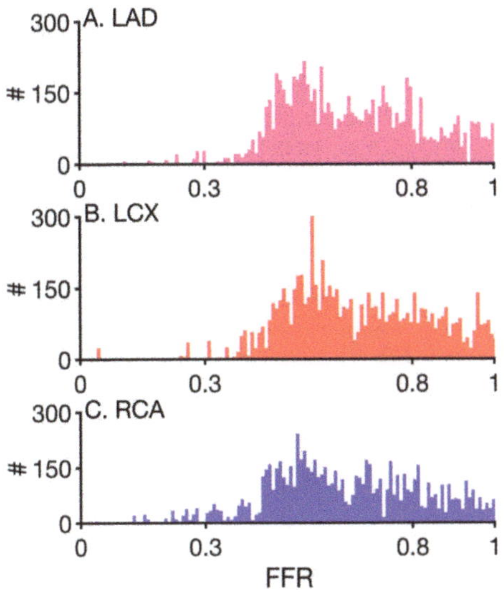

Figure 6. Histograms of FFR obtained from PRCC simulations (see Figure 7). Panel (**A**) shows the data for LAD, panel (**B**) for LCX, and panel (**C**) for RCA.

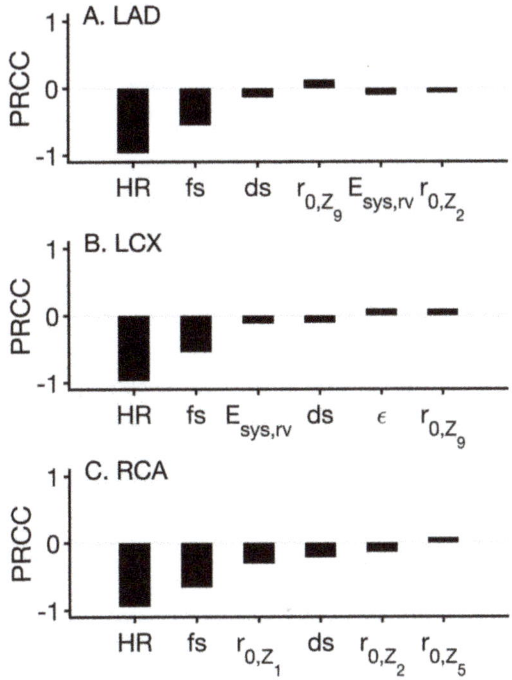

Figure 7. PRCC sensitivity of FFR to model control parameters. In all panels, the sensitivity of FFR to the six most relevant parameters are shown. Panel (**A**) shows the PRCC for LAD, panel (**B**) for the LCX, and panel (**C**) for the RCA. In all panels, HR: heart rate; fs: focal stenosis; ds: diffuse stenosis; r_0: root radius of microvascular bed; Z_i: microvascular impedance; ε: microvasculature taper exponent; and $E_{sys,rv}$: systolic elastance of the right ventricle.

The sensitivity analysis generated PRCC coefficients are shown in Figure 7. Heart rate (HR) is the most impactful model parameter regulating the FFR. Consistently, focal stenosis (fs) is also a significant regulator of PRCC. Both HR and fs negatively regulate FFR. Diffuse stenosis (ds) and the right ventricular systolic elastance ($E_{sys,rv}$) also negatively regulate FFR. The microvascular parameters (microvascular root radius r_0 and tapering factor ε) also affect FFR according to our sensitivity analysis.

4. Limitations and Future Directions

Whereas blood is known to be a non-Newtonian liquid [38] whose rheology depends on blood vessel size, especially at special scales, from coronary epicardial vessels to capillaries, the extant literature appears to use standard blood viscosity and density values [31–33]. Inclusion of detailed blood rheology into the model is planned but is not expected to alter presented results.

Further development of the presented model will lead to its clinical applicability. The sensitivity of FFR to heart rate requires further investigation. Although the sensitivity analysis presented heart rate as a primary regulator of FFR, the results of past work indicate its significance is unsettled. Kwasiborski et al. [39] found a significant correlation between FFR and heart rate in the LAD yet no correlation in the RCA in their porcine model. However, an investigation by Kolli et al. [40] found no statistically significant effect on mean FFR due to fluctuations in HR. Due to the increasing prominence of FFR regarding revascularization procedure design, further investigation into the measurement of heart rate as a potential FFR affecting factor is necessary. Patient specific model identification will increase the applicability of the model and reduce its prediction uncertainty. The inclusion of vessel-specific biomechanical properties and inclusion of a reactive vascular tone module [41–43] is expected to allow simulation of clinical parameters such as pulse wave velocities and residence times [44–49]. The inclusion of autoregulatory processes will further assist making the model's FFR estimates quantitatively reliable [50].

Although lumped parameter models for clinical bedside patient-specific hemodynamic simulation have potential, a significant limitation is the identification of initial parameters to ensure accuracy. The vessel parameters and microvascular impedances in this study were summarized from the literature [27,30]. Estimating the resistances, compliances, and inductances, for each three-element Windkessel model representing each blood vessel requires measurements of blood vessel diameters and lengths. Patient-specific model identification will increase the applicability of the model and reduce its prediction uncertainty. Time domain or frequency domain methods for parameter investigations from pressure and flow profiles have been developed [51,52]. Recent work demonstrated the capability of the unscented Kalman filter to personalize parameters for lumped parameter models using iterative simulations between 0D and 3D [53]. However, this method extinguishes the advantage of time and resource use by resorting to multi-scale simulations. Advancements in image processing algorithms for the visualization and quantification of vessel morphometry can be used to calculate the necessary parameters [13,54,55]. In addition, this study demonstrates that further investigation into the influence of cardiac parameters is permitted. Whereas a detailed heart model [56] was incorporated into the lumped parameter description [27], the simulated aortic root inflow to the coronary vasculature remains generic. Upon availability, patient-specific aortic root blood flow profiles will alleviate the limitation. Furthermore, the parameters of the heart model, such as ventricular and atrial elastances, require personalization. The development of high-resolution echocardiography and magnetic resonance imaging have demonstrated potential in estimating heart chamber volume [57,58]. Patient-specific estimation of blood rheology parameters such as blood viscosity will require clinical measurements [38]. By measurement of blood pressure and heart rate, the blood flow into the coronaries can be personalized. Using routine hematocrit blood tests, the viscosity can be personalized to a certain extent [20]. Also with potential, 4D flow MRI data can provide subject-specific temporal inlet flow information. Using all the temporal signals and spatial imaging data, a large number of modelling parameters can be estimated [59] using sophisticated methods such as steepest gradient algorithms and particle swarm techniques [60]. Multi-electrode noninvasive electrocardiogram may act as a confirmative test for existing ischemia, i.e., the presence of downstream sub-perfused myocardium [61]. Autoregulatory processes can be personalized using a combination of ultrasound, transcranial doppler and similar noninvasive routine clinical measurements [62].

Although the model is theoretical in nature, the presented results will guide our future work. As such, the findings of the study remain informative for deeper lumped parameter modelling and will inform our spatially extended modelling.

5. Conclusions

Focal and diffuse coronary stenosis were both observed to modulate FFR (Figure 3). However, our simulations indicate that FFR estimation must consider other conditions, such as AF and microvascular disease, both of which are routinely diagnosed among patients using non-invasive techniques. Furthermore, it appears that blood flow to the right ventricle is more severely affected due to extra-coronary and RCA stenosis conditions (Figures 4 and 5).

As seen in Figure 4, focal as well as diffuse stenosis reduces FFR relative to the control case. However, it can also be seen that extra-coronary conditions such as microvascular disease also affect FFR estimates. It is therefore clear that consideration of the effects of co-morbidities is essential in FFR estimation. The result also indicates that our approach is suitable for ranking the severity of co-morbidities. Figure 3, especially, indicates that microvascular disease alone does affect FFR estimation (see definition of FFR). Furthermore, the left and right heart's coronary are affected differentially. Whereas imaging studies are optimized to provide information regarding left coronaries, the model suggests that the right coronaries should also be considered. Our model suggests that stenosis may not be an exclusive focal or diffuse phenomenon. As Figure 4 shows, consideration of a combination of the two natures of stenosis is essential, especially in our future higher dimensional modelling (see Figure 8). In future studies, the 0D models in this detailed investigation will be useful as boundary conditions to 3D model computational fluid dynamics [63]. In addition to detailed geometry, Figure 5 indicates that *a priori* knowledge of microvascular health status will permit 3D models to provide better FFR estimates. Within the confines of the presented model, the sensitivity analysis (Figure 7) suggests that heart rate and severity of the large vessel occlusion are prime regulators of FFR. In addition, systolic heart function was found to be relevant.

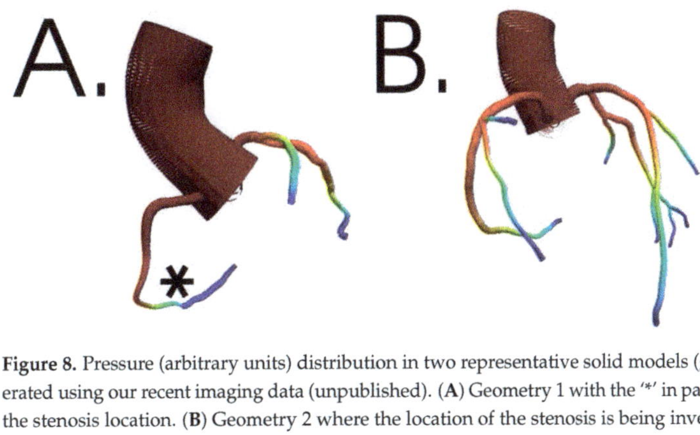

Figure 8. Pressure (arbitrary units) distribution in two representative solid models (geometries) generated using our recent imaging data (unpublished). (**A**) Geometry 1 with the '*' in panel (**A**) indicates the stenosis location. (**B**) Geometry 2 where the location of the stenosis is being investigated.

6. Discussion

We appreciate that model clinical testing routinely acquires immense amounts of data specific to the subject/patient. This includes the special organ that is the heart. The pulsatility provided by the heart is important in measurements such as FFR. However, modelling is an essential complement of CTA-driven FFR. In addition, modelling is essential due to existing heterogeneity among clinical providers.

The ready availability of high performance computing combined with high resolution clinical imaging modalities have augmented the application of computational fluid dynamics for in silico modelling and simulation-based investigation of complex biological processes [64]. However, due to the time and resource-intensive nature of large, multi-scale hemodynamic simulations, the clinical uptake of 3D modelling remains limited as it presently cannot be performed in real time. In contrast, the predictive capability of reduced order surrogates such as 1D and lumped parameter models have shown promise in their reliability relative to 3D models [17,56].

The wide use and reputation of FFR as the gold standard for coronary artery disease diagnosis motivates an investigation into the factors affecting FFR. Bearing in mind the utility and credibility of reduced order models for CFD simulation, a lumped parameter model of the human coronary vasculature [27] was further developed in this study. The model is capable of personalization based on clinical measurements of aortic pressure waves, imaging based vascular geometry (lengths, radii, and morphometry), as well as cardiac wall motion kinematics [65]. As such, the model permits imaging-clinical data assessment as a computationally efficient instrument, prior to detailed 3D computational fluid dynamics simulations. Novel imaging protocols that account for cardiac chamber to chamber diastole will further fortify refinement of the diagnostic instrument. This theoretical study illuminates the relative relevance of focal and diffuse stenosis. It also suggests that knowledge of co-morbidities will improve our clinical diagnostics. Furthermore, it informs our upcoming 3D investigation regarding the clinical data that will permit both validation as well as prediction.

Author Contributions: Study design: J.J.J., S.R.K. and C.W.M.; model development, T.-Y.L. provided the CTA data; J.J.J.; simulations: J.J.J. and C.S.; data analysis and figures, J.J.J.; draft writing, J.J.J. and S.R.K.; final manuscript, S.R.K., J.J.J., C.S., T.-Y.L., D.G. and C.W.M.; funding secured by S.R.K., D.G. and C.W.M. All authors have read and agreed to the published version of the manuscript.

Funding: This research was funded by Canada Canarie Inc. (RS3-111), Canadian NSERC grant (R4081A03) and Canada Heart and Stroke Foundation (G-20-0028717).

Institutional Review Board Statement: This study did not use sensitive data, patient information, or any other identity data. This computational study did not require ethics approvals of any nature.

Informed Consent Statement: This computational study did not require consent from patients or otherwise.

Data Availability Statement: The authors agree to make the model data openly available upon the publication of the manuscript. The model code is available in our laboratory's GITHUB, available online (accessed on 25 February 2022): https://github.com/mccsssk2/MDPI2022_JermiahsCoronary.

Acknowledgments: This work was supported by Canada Canarie Inc. (project: RS3-111), partly by Canadian Heart and Stroke Foundation grant (G-20-0028717, PI: CWM), and partly by Canadian NSERC grant (R4081A03, PI: DG). This work is part of CS's research-based undergraduate training in Western University. SRK also thanks Lawson Health Research Institute, Western University, MITACS Globalink (Canada), and Compute Canada for funding, compute resources, labor, and other resources. We thank the FIMH 2021 organizers for the opportunity to disseminate our work. We thank our research intern Shambhavi Malik (IIT BHU, shambhavimalik.bme18@itbhu.ac.in) for assistance with the spatially extended modelling of Figure 8.

Conflicts of Interest: The authors declare no conflict of interest.

References

1. Joseph, J.J.; Lee, T.-Y.; Goldman, D.; McIntyre, C.W.; Kharche, S.R. *The Role of Extra-Coronary Vascular Conditions that Affect Coronary Fractional Flow Reserve Estimation*; Springer: Cham, Switzerland, 2021; pp. 595–604.
2. Pijls, N.H.; Fearon, W.F.; Tonino, P.A.; Siebert, U.; Ikeno, F.; Bornschein, B.; van't Veer, M.; Klauss, V.; Manoharan, G.; Engstrom, T.; et al. Fractional flow reserve versus angiography for guiding percutaneous coronary intervention in patients with multivessel coronary artery disease: 2-year follow-up of the FAME (Fractional Flow Reserve Versus Angiography for Multivessel Evaluation) study. *J. Am. Coll. Cardiol.* **2010**, *56*, 177–184. [CrossRef]
3. Pijls, N.H.; Van Gelder, B.; Van der Voort, P.; Peels, K.; Bracke, F.A.; Bonnier, H.J.; el Gamal, M.I. Fractional flow reserve. A useful index to evaluate the influence of an epicardial coronary stenosis on myocardial blood flow. *Circulation* **1995**, *92*, 3183–3193. [CrossRef]
4. Fearon, W.F.; Tonino, P.A.; De Bruyne, B.; Siebert, U.; Pijls, N.H. Rationale and design of the Fractional Flow Reserve versus Angiography for Multivessel Evaluation (FAME) study. *Am. Heart J.* **2007**, *154*, 632–636. [CrossRef]
5. Fearon, W.F.; Bornschein, B.; Tonino, P.A.; Gothe, R.M.; Bruyne, B.D.; Pijls, N.H.; Siebert, U.; Fractional Flow Reserve Versus Angiography for Multivessel Evaluation Study Investigators. Economic evaluation of fractional flow reserve-guided percutaneous coronary intervention in patients with multivessel disease. *Circulation* **2010**, *122*, 2545–2550. [CrossRef]
6. Pijls, N.H.; van Schaardenburgh, P.; Manoharan, G.; Boersma, E.; Bech, J.W.; van't Veer, M.; Bar, F.; Hoorntje, J.; Koolen, J.; Wijns, W.; et al. Percutaneous coronary intervention of functionally nonsignificant stenosis: 5-year follow-up of the DEFER Study. *J. Am. Coll. Cardiol.* **2007**, *49*, 2105–2111. [CrossRef]
7. Tonino, P.A.; De Bruyne, B.; Pijls, N.H.; Siebert, U.; Ikeno, F.; van't Veer, M.; Klauss, V.; Manoharan, G.; Engstrom, T.; Oldroyd, K.G.; et al. Fractional flow reserve versus angiography for guiding percutaneous coronary intervention. *N. Engl. J. Med.* **2009**, *360*, 213–224. [CrossRef]
8. Ball, C.; Pontone, G.; Rabbat, M. Fractional Flow Reserve Derived from Coronary Computed Tomography Angiography Datasets: The Next Frontier in Noninvasive Assessment of Coronary Artery Disease. *Biomed. Res. Int.* **2018**, *2018*, 2680430. [CrossRef]
9. Jeremias, A.; Stone, G.W. Fractional flow reserve for the evaluation of coronary stenoses: Limitations and alternatives. *Catheter. Cardiovasc. Interv.* **2015**, *85*, 602–603. [CrossRef]
10. Odudu, A.; Francis, S.T.; McIntyre, C.W. MRI for the assessment of organ perfusion in patients with chronic kidney disease. *Curr. Opin. Nephrol. Hypertens.* **2012**, *21*, 647–654. [CrossRef]
11. Taylor, C.A.; Fonte, T.A.; Min, J.K. Computational fluid dynamics applied to cardiac computed tomography for noninvasive quantification of fractional flow reserve: Scientific basis. *J. Am. Coll. Cardiol.* **2013**, *61*, 2233–2241. [CrossRef]
12. Coenen, A.; Lubbers, M.M.; Kurata, A.; Kono, A.; Dedic, A.; Chelu, R.G.; Dijkshoorn, M.L.; Gijsen, F.J.; Ouhlous, M.; van Geuns, R.J.; et al. Fractional flow reserve computed from noninvasive CT angiography data: Diagnostic performance of an on-site clinician-operated computational fluid dynamics algorithm. *Radiology* **2015**, *274*, 674–683. [CrossRef]
13. Updegrove, A.; Wilson, N.M.; Merkow, J.; Lan, H.; Marsden, A.L.; Shadden, S.C. SimVascular: An Open Source Pipeline for Cardiovascular Simulation. *Ann. Biomed. Eng.* **2017**, *45*, 525–541. [CrossRef]
14. Arthurs, C.J.; Khlebnikov, R.; Melville, A.; Marcan, M.; Gomez, A.; Dillon-Murphy, D.; Cuomo, F.; Silva Vieira, M.; Schollenberger, J.; Lynch, S.R.; et al. CRIMSON: An open-source software framework for cardiovascular integrated modelling and simulation. *PLoS Comput. Biol.* **2021**, *17*, e1008881. [CrossRef]
15. McCullough, J.W.S.; Richardson, R.A.; Patronis, A.; Halver, R.; Marshall, R.; Ruefenacht, M.; Wylie, B.J.N.; Odaker, T.; Wiedemann, M.; Lloyd, B.; et al. Towards blood flow in the virtual human: Efficient self-coupling of HemeLB. *Interface Focus* **2021**, *11*, 20190119. [CrossRef]
16. Randles, A.; Draeger, E.W.; Bailey, P.E. Massively parallel simulations of hemodynamics in the primary large arteries of the human vasculature. *J. Comput. Sci.* **2015**, *9*, 70–75. [CrossRef]
17. Joseph, J.J.; Hunter, T.J.; Sun, C.; Goldman, D.; Kharche, S.R.; McIntyre, C.W. Using a Human Circulation Mathematical Model to Simulate the Effects of Hemodialysis and Therapeutic Hypothermia. *Appl. Sci.* **2022**, *12*, 307. [CrossRef]
18. Kharche, S.R.; Lemoine, S.; Tamasi, T.; Hur, L.; So, A.; McIntyre, C.W. Therapeutic Hypothermia Reduces Peritoneal Dialysis Induced Myocardial Blood Flow Heterogeneity and Arrhythmia. *Front. Med.* **2021**, *8*, 700824. [CrossRef]
19. Ghorbanniahassankiadeh, A.; Marks, D.S.; LaDisa, J.F. Correlation of Computational Instantaneous Wave-Free Ratio With Fractional Flow Reserve for Intermediate Multivessel Coronary Disease. *J. Biomech. Eng.* **2021**, *143*, 051011. [CrossRef]
20. Vardhan, M.; Gounley, J.; Chen, S.J.; Chi, E.C.; Kahn, A.M.; Leopold, J.A.; Randles, A. Non-invasive characterization of complex coronary lesions. *Sci. Rep.* **2021**, *11*, 8145. [CrossRef]
21. Chandola, G.; Zhang, J.M.; Tan, R.S.; Chai, P.; Teo, L.; Allen, J.C.; Low, R.; Huang, W.; Leng, S.; Fam, J.M.; et al. Computed Tomography Coronary Angiography and Computational Fluid Dynamics Based Fractional Flow Reserve Before and After Percutaneous Coronary Intervention. *Front. Bioeng. Biotechnol.* **2021**, *9*, 739667. [CrossRef]
22. Jonasova, A.; Vimmr, J. On the relevance of boundary conditions and viscosity models in blood flow simulations in patient-specific aorto-coronary bypass models. *Int. J. Numer. Method Biomed. Eng.* **2021**, *37*, e3439. [CrossRef]
23. Vignon-Clementel, I.E.; Figueroa, C.A.; Jansen, K.E.; Taylor, C.A. Outflow boundary conditions for 3D simulations of non-periodic blood flow and pressure fields in deformable arteries. *Comput. Methods Biomech. Biomed. Eng.* **2010**, *13*, 625–640. [CrossRef]

24. Altamirano-Diaz, L.; Kassay, A.D.; Serajelahi, B.; McIntyre, C.W.; Filler, G.; Kharche, S.R. Arterial Hypertension and Unusual Ascending Aortic Dilatation in a Neonate With Acute Kidney Injury: Mechanistic Computer Modeling. *Front. Physiol.* **2019**, *10*, 1391. [CrossRef]
25. Hunter, T.J.; Joseph, J.J.; Anazodo, U.; Kharche, S.R.; McIntyre, C.W.; Goldman, D. Atrial Fibrillation and Anterior Cerebral Artery Absence Reduce Cerebral Perfusion: A De Novo Hemodynamic Model. *Appl. Sci.* **2022**, *12*, 1750. [CrossRef]
26. Olufsen, M.S.; Nadim, A. On deriving lumped models for blood flow and pressure in the systemic arteries. *Math. Biosci. Eng. MBE* **2004**, *1*, 61–80. [CrossRef]
27. Duanmu, Z.; Yin, M.; Fan, X.; Yang, X.; Luo, X. A patient-specific lumped-parameter model of coronary circulation. *Sci. Rep.* **2018**, *8*, 874. [CrossRef]
28. Kharche, S.R.; Mironova, G.Y.; Goldman, D.; McIntyre, C.W.; Welsh, D.G. Sensitivity Analysis of a Smooth Muscle Cell Electrophysiological Model. In *Functional Imaging and Modeling of the Hear*; Ennis, D.B., Perotti, L.E., Wang, V.Y., Eds.; Springer International Publishing: Cham, Switzerland, 2021; pp. 540–550.
29. Britton, O.J.; Bueno-Orovio, A.; Van Ammel, K.; Lu, H.R.; Towart, R.; Gallacher, D.J.; Rodriguez, B. Experimentally calibrated population of models predicts and explains intersubject variability in cardiac cellular electrophysiology. *Proc. Natl. Acad. Sci. USA* **2013**, *110*, E2098–E2105. [CrossRef]
30. Pietrabissa, R.; Mantero, S.; Marotta, T.; Menicanti, L. A lumped parameter model to evaluate the fluid dynamics of different coronary bypasses. *Med. Eng. Phys.* **1996**, *18*, 477–484. [CrossRef]
31. Nader, E.; Monedero, D.; Robert, M.; Skinner, S.; Stauffer, E.; Cibiel, A.; Germain, M.; Hugonnet, J.; Scheer, A.; Joly, P.; et al. Impact of a 10 km running trial on eryptosis, red blood cell rheology, and electrophysiology in endurance trained athletes: A pilot study. *Eur. J. Appl. Physiol.* **2020**, *120*, 255–266. [CrossRef]
32. Wang, S.H.; Lee, L.P.; Lee, J.S. A linear relation between the compressibility and density of blood. *J. Acoust. Soc. Am.* **2001**, *109*, 390–396. [CrossRef]
33. Wu, D.; Wang, S.; Xie, J.; Mao, B.; Li, B.; Jin, C.; Feng, Y.; Li, G.; Liu, Y. Hemodynamic Mechanism of Coronary Artery Aneurysm High Occurrence on Right Coronary Artery. *Front. Physiol.* **2020**, *11*, 323. [CrossRef]
34. Olufsen, M.S. Structured tree outflow condition for blood flow in larger systemic arteries. *Am. J. Physiol. Heart Circ. Physiol.* **1999**, *276*, H257–H268. [CrossRef]
35. Heldt, T.; Shim, E.B.; Kamm, R.D.; Mark, R.G. Computational modeling of cardiovascular response to orthostatic stress. *J. Appl. Physiol.* **2002**, *92*, 1239–1254. [CrossRef]
36. Marino, S.; Hogue, I.B.; Ray, C.J.; Kirschner, D.E. A methodology for performing global uncertainty and sensitivity analysis in systems biology. *J. Theor. Biol.* **2008**, *254*, 178–196. [CrossRef]
37. Tange, O. GNU Parallel—The Command-Line Power Tool. *USENIX Mag.* **2011**, *36*, 42–47.
38. Chien, S.; Usami, S.; Taylor, H.M.; Lundberg, J.L.; Gregersen, M.I. Effects of hematocrit and plasma proteins on human blood rheology at low shear rates. *J. Appl. Physiol.* **1966**, *21*, 81–87. [CrossRef]
39. Kwasiborski, P.J.; Czerwinski, W.; Kowalczyk, P.; Buksinska-Lisik, M.; Horszczaruk, G.; Aboodi, M.S.; Derbisz, K.; Hochul, M.; Janas, A.; Cwetsch, A.; et al. Influence of heart rate on FFR measurements: An experimental and clinical validation study. *Int. J. Cardiol.* **2020**, *317*, 13–17. [CrossRef]
40. Kolli, K.K.; Banerjee, R.K.; Peelukhana, S.V.; Helmy, T.A.; Leesar, M.A.; Arif, I.; Schneeberger, E.W.; Hand, D.; Succop, P.; Gottliebson, W.M.; et al. Influence of heart rate on fractional flow reserve, pressure drop coefficient, and lesion flow coefficient for epicardial coronary stenosis in a porcine model. *Am. J. Physiol. Heart Circ. Physiol.* **2011**, *300*, H382–H387. [CrossRef]
41. Arciero, J.C.; Carlson, B.E.; Secomb, T.W. Theoretical model of metabolic blood flow regulation: Roles of ATP release by red blood cells and conducted responses. *Am. J. Physiol. Heart Circ. Physiol.* **2008**, *295*, H1562–H1571. [CrossRef]
42. Lucker, A.; Secomb, T.W.; Barrett, M.J.P.; Weber, B.; Jenny, P. The Relation Between Capillary Transit Times and Hemoglobin Saturation Heterogeneity. Part 2: Capillary Networks. *Front. Physiol.* **2018**, *9*, 1296. [CrossRef]
43. Pries, A.R.; Secomb, T.W. Microcirculatory network structures and models. *Ann. Biomed. Eng.* **2000**, *28*, 916–921. [CrossRef]
44. Clavica, F.; Alastruey, J.; Borlotti, A.; Sherwin, S.J.; Khir, A.W. One-dimensional computational model of pulse wave propagation in the human bronchial tree. *Conf. Proc. IEEE Eng. Med. Biol. Soc.* **2010**, *2010*, 2473–2476. [CrossRef]
45. Poleszczuk, J.; Debowska, M.; Dabrowski, W.; Wojcik-Zaluska, A.; Zaluska, W.; Waniewski, J. Patient-specific pulse wave propagation model identifies cardiovascular risk characteristics in hemodialysis patients. *PLoS Comput. Biol.* **2018**, *14*, e1006417. [CrossRef]
46. Womersley, J.R. Oscillatory flow in arteries. II. The reflection of the pulse wave at junctions and rigid inserts in the arterial system. *Phys. Med. Biol.* **1958**, *2*, 313–323. [CrossRef]
47. Butty, V.D.; Gudjonsson, K.; Buchel, P.; Makhijani, V.B.; Ventikos, Y.; Poulikakos, D. Residence times and basins of attraction for a realistic right internal carotid artery with two aneurysms. *Biorheology* **2002**, *39*, 387–393.
48. Esmaily-Moghadam, M.; Hsia, T.Y.; Marsden, A.L. A non-discrete method for computation of residence time in fluid mechanics simulations. *Phys. Fluids* **2013**, *25*, 110802. [CrossRef]
49. Hashemi, J.; Patel, B.; Chatzizisis, Y.S.; Kassab, G.S. Study of Coronary Atherosclerosis Using Blood Residence Time. *Front. Physiol.* **2021**, *12*, 625420. [CrossRef]
50. Carlson, B.E.; Arciero, J.C.; Secomb, T.W. Theoretical model of blood flow autoregulation: Roles of myogenic, shear-dependent, and metabolic responses. *Am. J. Physiol. Heart Circ. Physiol.* **2008**, *295*, H1572–H1579. [CrossRef]

51. Shim, Y.; Pasipoularides, A.; Straley, C.A.; Hampton, T.G.; Soto, P.F.; Owen, C.H.; Davis, J.W.; Glower, D.D. Arterial windkessel parameter estimation: A new time-domain method. *Ann. Biomed. Eng.* **1994**, *22*, 66–77. [CrossRef]
52. Pochet, T.; Gerard, P.; Marnette, J.M.; D'Orio, V.; Marcelle, R.; Fatemi, M.; Fossion, A.; Juchmes, J. Identification of three-element windkessel model: Comparison of time and frequency domain techniques. *Arch. Int. Physiol. Biochim. Biophys.* **1992**, *100*, 295–301. [CrossRef]
53. Pant, S.; Fabreges, B.; Gerbeau, J.F.; Vignon-Clementel, I.E. A methodological paradigm for patient-specific multi-scale CFD simulations: From clinical measurements to parameter estimates for individual analysis. *Int. J. Numer. Method Biomed. Eng.* **2014**, *30*, 1614–1648. [CrossRef]
54. Ward, E.P.; Shiavazzi, D.; Sood, D.; Marsden, A.; Lane, J.; Owens, E.; Barleben, A. Computed Tomography Fractional Flow Reserve Can Identify Culprit Lesions in Aortoiliac Occlusive Disease Using Minimally Invasive Techniques. *Ann. Vasc. Surg.* **2017**, *38*, 151–157. [CrossRef]
55. Boskamp, T.; Rinck, D.; Link, F.; Kummerlen, B.; Stamm, G.; Mildenberger, P. New vessel analysis tool for morphometric quantification and visualization of vessels in CT and MR imaging data sets. *Radiographics* **2004**, *24*, 287–297. [CrossRef]
56. Heldt, T. Continuous blood pressure-derived cardiac output monitoring–should we be thinking long term? *J. Appl. Physiol.* **2006**, *101*, 373–374. [CrossRef]
57. Schiavazzi, D.E.; Baretta, A.; Pennati, G.; Hsia, T.Y.; Marsden, A.L. Patient-specific parameter estimation in single-ventricle lumped circulation models under uncertainty. *Int. J. Numer. Method Biomed. Eng.* **2017**, *33*, e02799. [CrossRef]
58. Laser, K.T.; Horst, J.P.; Barth, P.; Kelter-Klopping, A.; Haas, N.A.; Burchert, W.; Kececioglu, D.; Korperich, H. Knowledge-based reconstruction of right ventricular volumes using real-time three-dimensional echocardiographic as well as cardiac magnetic resonance images: Comparison with a cardiac magnetic resonance standard. *J. Am. Soc. Echocardiogr.* **2014**, *27*, 1087–1097. [CrossRef]
59. Casas, B.; Lantz, J.; Viola, F.; Cedersund, G.; Bolger, A.F.; Carlhall, C.J.; Karlsson, M.; Ebbers, T. Bridging the gap between measurements and modelling: A cardiovascular functional avatar. *Sci. Rep.* **2017**, *7*, 6214. [CrossRef]
60. Kharche, S.; Ludtke, N.; Panzeri, S.; Zhang, H. A Global Sensitivity Index for Biophysically Detailed Cardiac Cell Models: A Computational Approach. *LNCS* **2009**, *5528*, 10.
61. Nørgaard, B.L.; Hjort, J.; Gaur, S.; Hansson, N.; Bøtker, H.E.; Leipsic, J.; Mathiassen, O.N.; Grove, E.L.; Pedersen, K.; Christiansen, E.H.; et al. Clinical Use of Coronary CTA–Derived FFR for Decision-Making in Stable CAD. *JACC Cardiovasc. Imaging* **2017**, *10*, 541–550. [CrossRef]
62. Chen, J.; Martin, C.; Ball, I.M.; McIntyre, C.W.; Slessarev, M. Impact of Graded Passive Cycling on Hemodynamics, Cerebral Blood Flow, and Cardiac Function in Septic ICU Patients. *Front. Med.* **2020**, *7*, 569679. [CrossRef]
63. Marsden, A.; Moghadam, M.E. Multiscale Modeling of Cardiovascular Flows for Clinical Decision Support. *Appl. Mech. Rev.* **2015**, *67*, 030804. [CrossRef]
64. Bluestein, D. Utilizing Computational Fluid Dynamics in Cardiovascular Engineering and Medicine-What You Need to Know. Its Translation to the Clinic/Bedside. *Artif. Organs* **2017**, *41*, 117–121. [CrossRef]
65. Crowley, L.E.; McIntyre, C.W. Remote ischaemic conditioning therapeutic opportunities in renal medicine. *Nat. Rev. Nephrol.* **2013**, *9*, 739–746. [CrossRef]

applied sciences

MDPI

Article

Atrial Fibrillation and Anterior Cerebral Artery Absence Reduce Cerebral Perfusion: A De Novo Hemodynamic Model †

Timothy J. Hunter [1,2], Jermiah J. Joseph [1,2], Udunna Anazodo [1,2], Sanjay R. Kharche [1,2,*], Christopher W. McIntyre [1,2,*] and Daniel Goldman [1,*]

1 Department of Medical Biophysics, Western University, 1151 Richmond Street North, London, ON N6A 5C1, Canada; thunte27@uwo.ca (T.J.H.); jjosep56@uwo.ca (J.J.J.); uanazodo@lawsonimaging.ca (U.A.)
2 Lawson Health Research Institute, 750 Baseline Road East, London, ON N6C 2R5, Canada
* Correspondence: sanjay.kharche@lhsc.on.ca (S.R.K.); christopher.mcintyre@lhsc.on.ca (C.W.M.); dgoldma2@uwo.ca (D.G.)

† This paper is an extended version of paper published in Functional Imaging and Modelling of the Heart 2021 in 11th International Conference, FIMH 2021, held in Stanford, CA, USA, 21–25 June 2021.

Abstract: Background: Atrial fibrillation is a prevalent cardiac arrhythmia and may reduce cerebral blood perfusion augmenting the risk of dementia. We hypothesize that geometric variations in the cerebral arterial structure called the Circle of Willis (CoW) play an important role in influencing cerebral perfusion. The objective of this work was to develop a novel cardio-cerebral lumped parameter hemodynamic model to investigate the role of CoW variants on cerebral blood flow dynamics under atrial fibrillation conditions. Methods: A computational blood flow model was developed by coupling whole-body and detailed cerebral circulation descriptions, modified to represent six common variations of the CoW. Cerebral blood flow dynamics were simulated in common CoW variants, under control and imposed atrial fibrillation conditions. Risk was assessed based on the frequency of beat-wise hypoperfusion events, and sensitivity analysis was performed with respect to this model output. Results: It was found that the geometry of the CoW influenced the frequency of hypoperfusion events at different heart rates, with the variant missing a P1 segment having the highest risk. Sensitivity analysis revealed that intrinsic heart rate is most associated with the considered outcome. Conclusions: Our results suggest that CoW geometry plays an important role in influencing cerebral hemodynamics during atrial fibrillation. The presented study may assist in guiding our future clinical-imaging research.

Keywords: reduced order model; atrial fibrillation; Circle of Willis variants; cerebral blood flow; sensitivity analysis

Citation: Hunter, T.J.; Joseph, J.J.; Anazodo, U.; Kharche, S.R.; McIntyre, C.W.; Goldman, D. Atrial Fibrillation and Anterior Cerebral Artery Absence Reduce Cerebral Perfusion: A De Novo Hemodynamic Model. *Appl. Sci.* **2022**, *12*, 1750. https://doi.org/10.3390/app12031750

Academic Editors: Mihaela Pop and Cristian A. Linte

Received: 31 December 2021
Accepted: 30 January 2022
Published: 8 February 2022

1. Introduction

This paper is an extension of work originally presented in Functional Imaging and Modelling of the Heart 2021 [1]. Atrial fibrillation (AF) currently affects a large part of the population. In addition to commonly known risks such as strokes and transient ischemic attacks, AF has been associated with increased cognitive decline and early dementia [2]. AF is known to reduce cerebral perfusion [3], and silent cerebral ischemia is thought to be a key mechanism in the increased cognitive risk [2,4]. Ongoing imaging research strongly suggests that a disrupted cerebral blood flow promotes debilitating early dementia [5]. The effects of AF on cerebral perfusion may be modulated by cerebral vascular geometry, and specifically by common congenital Circle of Willis (CoW) variants [6,7]. The function of a complete CoW is to ensure consistent distribution of blood flow to all regions of the brain. In cases with missing segments in the CoW, regions of the brain may be more susceptible to harmful altered hemodynamics. The aim of this work is to investigate whether structural variants of the CoW behave differently with respect to cerebral perfusion in AF conditions.

Multi-scale hemodynamic modelling has been used to study cerebral circulation and gain insight into patient-specific hemodynamics [8]. 3D modelling is a useful tool, which provides realistic and accurate patient-specific insight into patient hemodynamics. It has increasingly been used as the gold standard in computational hemodynamic studies as computational fluid dynamics platforms become more accessible [9]. We have recently found a close relationship between cardiac arrhythmia and systemic perfusion (Kharche et al., 2021; Frontiers in Medicine). However, current 3D methods remain computationally resource intensive, require high-definition vascular imaging, and are therefore unsuitable for applications studying large population hemodynamics.

In contrast to 3D hemodynamic models, lumped parameter (0D) models are known to provide clinically relevant information using significantly less time and computational resources [8]. 0D models are particularly useful in studies where there are poorly understood outcomes of diseases with well understood mechanisms because of their ability to assess the impact of a range of parameters or cases on a particular outcome. This lab has previously used 0D models to gain insight into the causes of pediatric hypertension [10] and investigate therapeutic hypothermia [11]. Anselmino et al. [4] have previously used 0D modelling to investigate the interplay between AF and cerebral hemodynamics. They determined that AF does indeed expose the brain to the risk of ischemia via low blood flow, or so-called hypoperfusion events. Saglietto et al. [12] have also used 0D modelling to predict that the optimal goal for a heart rate control strategy should be around 60 bpm, considered strict rate control.

The findings by Saglietto et al. [12] are in contrast to the common practice of lenient rate control (<110 bpm), which is based on findings from the RACE II trial, a large, randomized control trial [13]. The RACE II trial was a consequential study, which found that, compared to lenient rate control (<110 bpm), strict rate control (<80 bpm) was not more effective in reducing mortality in persistent AF patients. These findings have informed treatment strategies for persistent AF patients; however, they do not consider the increased risk for dementia, later confirmed by de Brujin et al. [2] in a longitudinal study. Modelling studies following de Brujin et al. [2] have aimed at elucidating the mechanism behind the increased risk and finding potential treatment strategies that mitigate it.

Previous modelling studies form the basis for the present work; however, the role of cerebral vascular structural variants, i.e., CoW variants, in the AF-cerebral perfusion relationship remains underexplored. As the CoW is known to play an important role in the distribution of blood flow to the brain, common variants should be considered while studying the interplay between AF and cerebral hemodynamics. In this study, a composite 0D model of human circulation with detailed cerebral vasculature was developed to discover the effects of AF on cerebral perfusion in cases with common CoW variants. Model composition is described in Section 2.1. The model is used to assess cerebral hemodynamics during AF in all six common CoW variants, the strategy for which is described in Section 2.4. Finally, the model itself is also assessed using sensitivity analysis as described in Section 2.5, which details the parameter values that are correlated to model outcomes.

2. Methods

This study is a modelling study that examines the role of varying blood vessel geometries in AF-related cerebral hypoperfusion. A previously developed composite 0D model [1] (Figure 1) was used to simulate cerebral hemodynamics under control and AF conditions. Six common variants of CoW geometry were modelled as separate cases, and the results for each were presented. Sensitivity analysis was also performed on the model to assess model parameters that had the greatest impact on simulated outcomes.

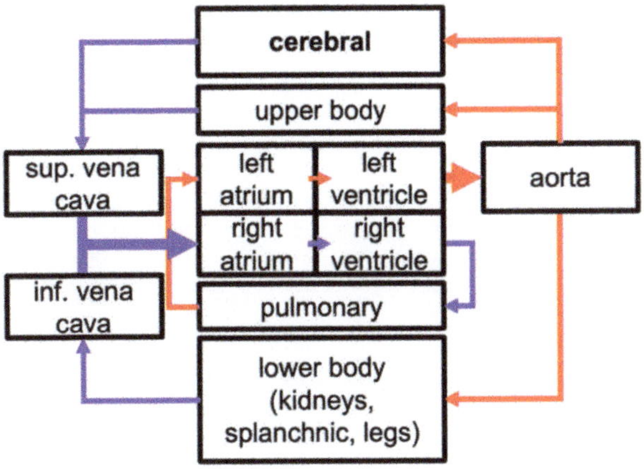

Figure 1. Caricature of the whole-body blood flow model. The cerebral model (black box, bold, top) is expanded in Figure 2.

Figure 2. Cerebral arterial architecture consisting of all Circle of Willis arteries. RA: Right anterior artery; LA: Left anterior artery; RM: Right middle artery; LM: Left middle artery; RP: Right posterior artery; LP: Left posterior artery; ACA1: Pre-communicating anterior cerebral artery; PCoA: Posterior communicating artery; and PCA1: Pre-communicating posterior cerebral artery.

2.1. Model Components

The 0D model is a composite model that consists of whole-body circulation, a blood-pressure modulated baroreflex control mechanism, and detailed cerebral circulation with an autoregulation function. All model parameters were inherited from the literature values, unless otherwise stated.

The whole-body circulation model was adapted from the model published by Heldt [14]. It consists of a network of blood containing elastic Windkessel compartments, which represent individual, or networks of, blood vessels. The time-dependent change in pressure within each compartment is a function of the change in volume (i.e., flow in or out) divided by the compliance of the compartment shown by the equation:

$$\frac{dP}{dt} = \frac{q_{in} - q_{out}}{C}, \tag{1}$$

where P denotes compartment pressure, t denotes time, q denotes flow, and C denotes compliance. The flow between connected compartments is calculated using the following equation:

$$q = \frac{P_p - P_d}{R}, \tag{2}$$

where P_p and P_d denote proximal and distal pressure, respectively, and R denotes resistance.

The pumping heart is represented as four compartments with variable elastance (inverse of compliance), representing the four chambers of the heart. The time-dependent elastances of ventricles and atria were calculated using activation terms. The equation for atrial activation is:

$$\begin{array}{cc} 0 < t_{loc} \le t_{asys}: & act_a = 1 - \cos\left(\pi \frac{t_{loc}}{t_{asys}}\right), \\ t_{asys} < t_{loc} \le 1.5t_{asys}: & act_a = 1 + \cos\left(2\pi \frac{t_{loc} - t_{asys}}{t_{asys}}\right) \\ otherwise: & act_a = 0 \end{array} \tag{3}$$

in which act_a is the activation term, t_{loc} is the time since the initiation of the cardiac cycle, and t_{asys} is a contraction timing parameter. Similarly, ventricular activation is calculated using:

$$\begin{array}{cc} t_{av} < t_{loc} \le t_{av} + t_s: & act_v = 1 - \cos\left(\pi \frac{t_{loc} - t_{av}}{t_s}\right) \\ t_{av} + t_s < t_{loc} \le t_{av} + 1.5t_s: & act_v = 1 + \cos\left(2\pi \frac{t_{loc} - t_{av} - t_s}{t_s}\right) \\ otherwise: & act_v = 0, \end{array} \tag{4}$$

where t_{av} is the atrioventricular time delay and t_s is a contraction timing parameter. The activation constants are applied to each heart compartment using the equation:

$$E = E_{dias} + 0.5\left(E_{sys} - E_{dias}\right) \times act, \tag{5}$$

where E_{sys} and E_{dias} are systolic and diastolic elastances, respectively.

Additionally, backflow is prevented in the heart and systemic veins by setting flow between compartments equal to 0 if distal pressure is greater than proximal pressure. A simplified caricature of the circulation model is presented in Figure 1, in which compartments are represented by boxes, and connections are indicated by arrows. It should be noted that the boxes labeled "cerebral" and "lower body" each represents multiple compartments.

The baroreflex is a feedback mechanism, which works to maintain hemodynamic homeostasis. It modulates peripheral vascular resistance, heart rate, and heart contractility to maintain systemic blood pressure and flow at healthy levels. The baroreceptor mechanism is implemented according to the model proposed by Lin et al. [15]. The model dynamically calculates sympathetic nervous activity (SNA) and parasympathetic nervous activity (PNA) based on the mean arterial pressure, as well as arterial PCO_2, which is assigned a constant value of 40 mmHg. Values for SNA and PNA are then used to dynamically modulate peripheral vascular resistance, the intrinsic heart rate, as well as heart contractility via modulation terms [15].

The cerebral circulation model is comprised of a network of elastic compartments with compliances and resistances similar to the systemic circulation. While pressure and flow are governed by the same equations as the systemic model, the formulation is more complicated and is beyond the scope of this article. Readers may refer to Ursino and Gianessi [16] for further details. The model also implements cerebral autoregulation, which is a physiological mechanism that alters vascular resistance and compliance in order to maintain blood flow within healthy ranges in the case of widely varying cerebral perfusion pressure. Each downstream region (Figure 2, RA, LA, RM, LM, RP, LP) is regulated by its own autoregulation function comprised of two integrated signals: Blood flow rate in the region, which is calculated dynamically, and arterial PCO_2, which is assigned at 40 mmHg. These two signals are applied to a first-order filter with time constants of 20 s for autoregulation and 40 s for CO_2 control, and the resulting values are used to modulate compliance and resistance within the corresponding vascular region. Blood flow from the whole-body model to the cerebral model was allowed by connecting the basilar and internal carotid arteries to the aortic compartment, and by connecting the cerebral outlet vein to the superior vena cava compartment. A caricature of the arterial segments of the cerebral model is shown in Figure 2.

This work considers the six common variants of the Circle of Willis found in the cerebral vasculature, represented in Figure 3 [7]. All variants, aside from the complete variant, are characterized by one or multiple missing segments of the CoW. To model the absence of the relevant cerebral vessel, its inlet and outlet flow was assigned a nil value.

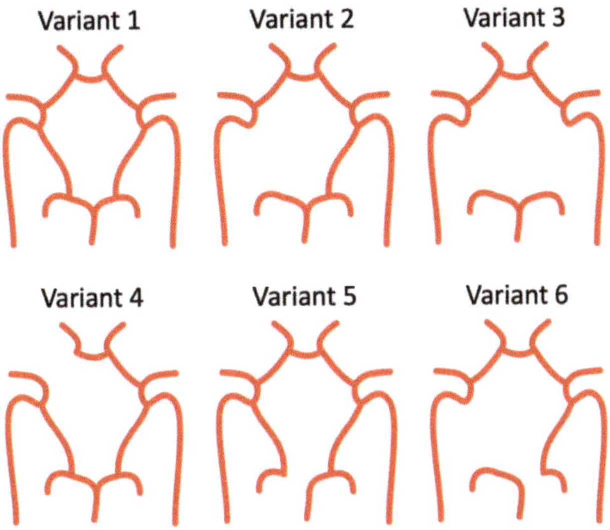

Figure 3. Caricature representations of all the common CoW variants. Variant 1 has all CoW vessels. Variant 2 has a missing posterior communicating artery (PCoA). Variant 3 has both missing PCoAs. Variant 4 has a missing precommunicating anterior cerebral artery, ACA1, segment. Variant 5 has a missing precommunicating posterior cerebral artery, PCA1, segment. Variant 6 has a missing PCoA and contralateral PCA1 segment.

2.2. *Atrial Fibrillation*

Each instance was simulated under AF and control conditions. The control was defined as having normal sinus rhythm (NSR) with stochastic RR intervals sampled from a normally distributed pink noise generator [17]. AF was modelled by assigning stochastic RR intervals sampled from an exponentially modified Gaussian distribution around a mean heart rate, modifying ventricular elastances (contractility), and assigning nil atrial contractility [4,17–19]. Pink noise and exponential samples were generated using in-house MATLAB scripts.

2.3. *Computational Methods*

The model used in this study has 57 coupled stiff ordinary differential equations (ODEs). An in-house ODE solver [20] was deployed to generate stable and accurate numerical solutions. The maximum integration timestep in the adaptive and implicit solver was 0.001 s, which was found to provide the same solution when the timestep was halved. The solutions were obtained using a relative tolerance of 10^{-6}, with an accuracy of $O(dt^6)$. Each instance of the simulation could be processed by available computing resources running Red Hat Linux within 60 s. Instances were trivially parallelized using GNU Parallel [21] in order to run a large number of instances (10^4) on multi-core compute nodes.

2.4. Hemodynamic Differences in CoW Variants

Blood flow was simulated in each variant at nine different intrinsic heart rates (50 to 130 bpm) in accordance with clinical practice [22]. The probability distribution functions underlying the RR intervals and the representative RR interval time series are illustrated in Figure 4. In each simulation, the number of hypoperfusion events was recorded to represent cerebral perfusion deficit. Derived measurements were the number of hypoperfusion events in each vascular bed over the 5000 beats of the simulation. A hypoperfusion event in any vascular bed was defined as a heartbeat in which the mean blood flow through the vascular bed fell below the 5th percentile of blood flow in the corresponding NSR experiment.

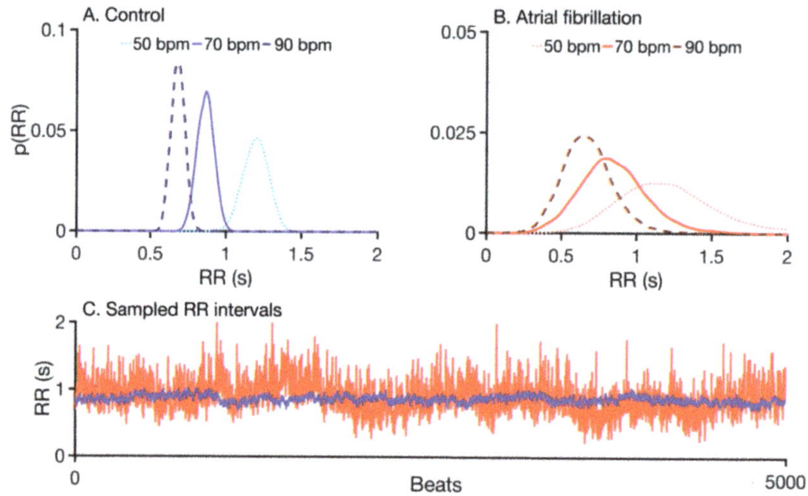

Figure 4. Stochastic RR interval assignment. Top row: Probability distribution functions for sampled RR intervals in NSR (**A**) and AF (**B**) at shown heart rates. (**C**) Sampled RR intervals with mean of 70 bpm over the span of 5000 beats under NSR (blue line) and AF (red line) conditions.

2.5. Sensitivity Analysis

Sensitivity analysis is a tool that provides a comprehensive understanding of the workings of a computational model with respect to its parameters and a specified modelling outcome [23]. The model has 95 parameters, which include all resistances, compliances, vessel geometry attributes, time constants, and scaling factors. Parameters' descriptions and acronyms, as well as their control values relevant to this work, are provided in Table 1. Model behavior was defined as the total number of hypoperfusion events in the distal cerebral circulation over a 5000-heartbeat simulation.

To permit sensitivity analysis, a control model population of 10^4 instances was constructed. To generate the population, 95 modelling parameters were each randomly sampled simultaneously from uniform distributions using a non-repetitive Mersenne Twister random number generator [24]. The sampling was constrained using Latin Hypercube Sampling [25]. The lower and upper limits adopted for each parameter's uniform distribution were obtained by multiplying the literature value by 0.5. for the lower limit, and by 2.0 for the upper limit. The adopted limits provided a large range sampling for each parameter. The model parameters and model outputs were stored for further analysis. Sensitivity analysis, which ranked parameters according to their impact on model behavior, was performed using partial rank correlation coefficients (PRCC) [26].

Table 1. Relevant model parameters.

Parameter	Description	Baseline Value
	Whole-body circulation	
HR_0	Intrinsic heart rate	75 bpm
$Edias_{rv}$	Right ventricular diastolic elastance.	0.07 (mmHg ml^{-1})
$Esys_{rv}$	Right ventricular systolic elastance.	1.3 (mmHg ml^{-1})
$Esys_{ra}$	Right atrial systolic elastance.	0.74 (mmHg ml^{-1})
$Edias_{ra}$	Right atrial diastolic elastance.	0.3 (mmHg ml^{-1})
$Edias_{lv}$	Left ventricular diastolic elastance.	0.13 (mmHg ml^{-1})
R_{pv}	Pulmonary venous resistance.	0.01 (mmHg s ml^{-1})
	Cerebral circulation	
G_{aut}	Autoregulation function gain.	0.9 (unitless)
tau_{aut}	Autoregulation function time constant.	20 (s)
C_d	Distal cerebral arterial compliance.	200 (ml mmHg^{-1})
k_R	Distal cerebral resistance scaling term.	13,100 (mmHg^{-3} s ml^{-1})

To compute PRCC, the normally distributed parameters (x_i), as well as the observed outputs (y_i), were rank transformed. Then, the linear effects of other additional variables were accounted for by expressing each as a linear regression of the inputs:

$$\hat{x}_j = a_0 + \sum_{\substack{k=1 \\ k \neq j}}^{N} a_k x_k, \ and\ \hat{y}_j = b_0 + \sum_{\substack{k=1 \\ k \neq j}}^{N} b_k x_k. \tag{6}$$

Using residuals defined as $r_{xi} = x_j - \hat{x}_j$ and $r_{yi} = y_j - \hat{y}_j$, PRCC is defined as the correlation among these residuals normalized using their respective variances:

$$PRCC(x_i, y_j) = \frac{Cov(r_{xi}, r_{yj})}{Var(r_{xi})Var(r_{yj})}. \tag{7}$$

As evident in Equation (6), PRCC assumes an underlying statistical model that is linear (regression), and the assumption of monotonicity provides the strength of the linear relationship between a given pair of a parameter and an output [26,27]. The PRCC indices range from −1 to +1.

3. Results

Model output statistics are presented from a single simulation instance with CoW variant 1 (complete CoW) at 80 bpm in Table 2. The statistics from the AF case are shown to be similar to those in the NSR case. Median systemic blood pressures of 117.44/77.81 mmHg (systolic/diastolic) for NSR and 119.51/78.95 mmHg for AF are shown to be similar to physiological levels. Additionally, total cerebral blood flow is 12.54 mL s^{-1} for the NSR case and 12.31 mL s^{-1} for the AF case.

The model has demonstrated that large variations in blood pressure are propagated through the large arterial circulation and have a high impact on small vessels in the distal cerebral circulation, annotated as RA, LA, RM, LM, RP, and LP in Figure 2. This effect is demonstrated in Figure 5 where a drop in aortic blood pressure due to a long RR interval is associated with two consecutive hypoperfusion events. The example shows aortic blood pressure and simultaneous blood flow into the LM in a control and AF case, colored in blue and red, respectively. On panel A, the dip in blood pressure can be seen in the AF case between seconds 1 and 3. Corresponding with this dip, two hypoperfusion events are annotated with black in panel B, with horizontal lines indicating the mean blood flow value during the heartbeat to show that it is indeed below the fifth percentile of normal blood flow.

Table 2. Model outputs under NSR conditions.

Output Name	Output Values	
	NSR	AF
$P_{a,sys}$ (mmHg)	117.44 ± 21.35	119.51 ± 17.45
$P_{a,dias}$ (mmHg)	77.81 ± 15.85	78.95 ± 16.64
Q_{ACA} (ml s^{-1})	0.99 ± 0.37	0.95 ± 0.45
Q_{MCA} (ml s^{-1})	3.68 ± 1.21	3.64 ± 1.37
Q_{PCA} (ml s^{-1})	1.47 ± 0.52	1.44 ± 0.59
CBF (ml s^{-1})	12.54 ± 4.24	12.31 ± 4.78

Model output statistics for a simulation run with CoW variant 1 (complete CoW), at an HR of 80 BPM under AF and NSR conditions. Systemic pressure and cerebral blood flow statistics are shown to be similar in both NSR and AF cases. Values are shown as median ± standard deviation. $P_{a,sys}$: Arterial systolic pressure; $P_{a,dias}$: Arterial diastolic pressure; Q_{ACA}: Anterior cerebral artery flow rate; Q_{MCA}: Middle cerebral artery flow rate; Q_{PCA}: Posterior cerebral artery flow rate; CBF: Cerebral blood flow.

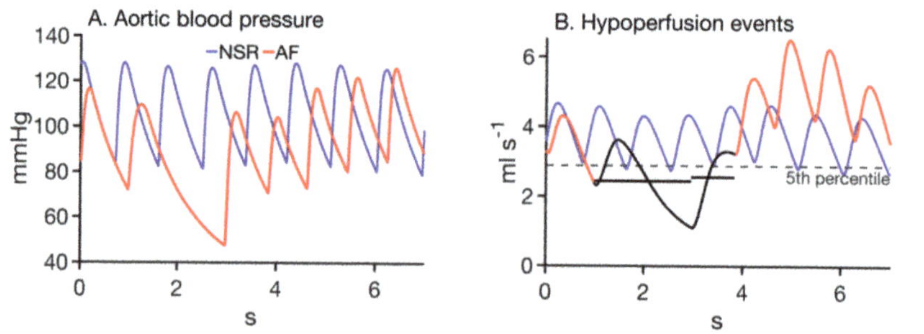

Figure 5. Hemodynamic outputs of a simulation of AF (red) and NSR (blue) at 70 bpm in the normal CoW. (**A**) Aortic blood pressures. (**B**) Blood flow through the left middle distal artery with hypoperfusion events shown in black.

The heart rate and vascular geometry dependence of hypoperfusion events is illustrated in Figure 6. For each of the six common variants of the CoW, total hypoperfusion event counts are shown for simulations at imposed heart rates ranging from 50 to 130 bpm.

All variants displayed similar behavior within the range of heart rates examined, with some differences in the number of counts, as well as the point at which they have the highest hypoperfusion count. Variant 1, with a complete CoW, is represented in Figure 6A. This variant had a minimum count occurring at a heart rate of 50 bpm with 178 total events, and a maximum count at 120 bpm with 2048 total events. Variant two, with a missing PCoA, is represented in Figure 6B. This variant had a minimum count at 50 bpm with 264 events, and a maximum count at 120 bpm with 2248 events. Variant number three, with both PCoAs missing, is shown in Figure 6C. It had a minimum count at 50 bpm with 268 events, and a maximum count at 130 bpm with 2120 events. Variant four, with a missing ACA1, is represented in Figure 6D. This variant had a minimum count at 50 bpm with 675 events, and a maximum count at 110 bpm with 2861 events. Variant five, with a missing PCA1, is represented in Figure 6E. This variant had a minimum count at 50 bpm with 211 events, and a maximum count at 120 bpm with 2458 events. Variant six, with a missing PCoA and contralateral PCA1, is represented in Figure 6F. This variant had a minimum count at 50 bpm with 97 events, and a maximum count at 110 bpm with 2386 events. All variants had minimum counts at a 50 bpm heart rate. Maximum points varied between different variants, although all were within 110 to 130 bpm. Variant 3 is notable in that there is no count drop off at 130 bpm as there is in all other variants. Overall, all variants exhibit similar behavior, increasing count with bpm, up to a maximum around 120 bpm.

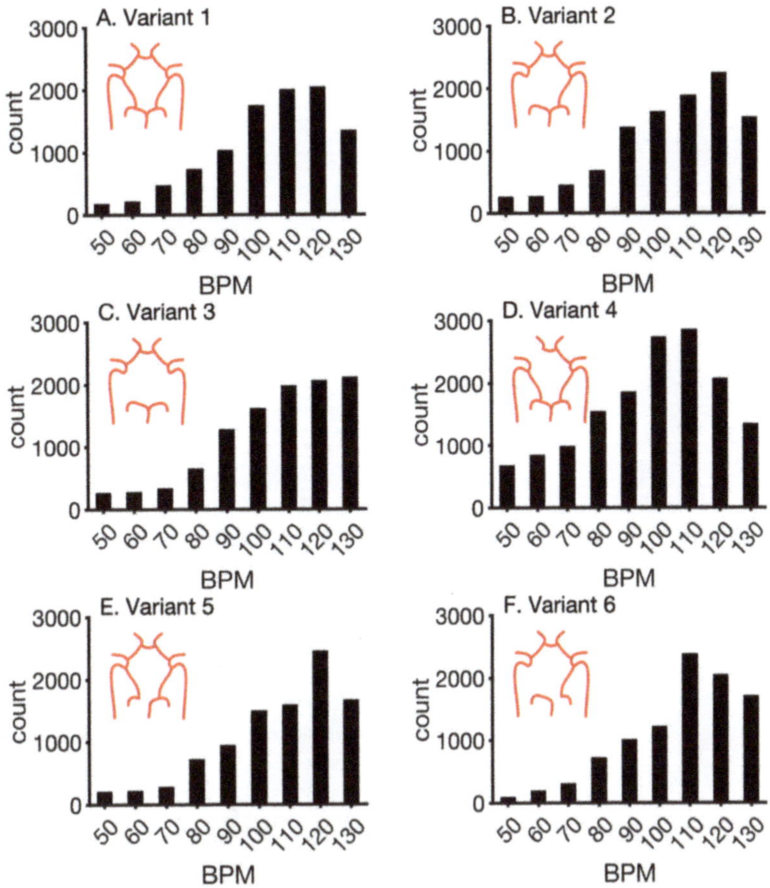

Figure 6. Absolute frequencies of hypoperfusion events in the distal cerebral circulations at varying heart rates under AF conditions. Count is the sum of all hypoperfusion events in each of the six distal circulation regions over a 5000 heart-beat simulation. Each panel shows hypoperfusion counts for heart rates from 50 bpm to 130 bpm for a particular CoW variant. (**A**) Complete CoW. (**B**) Missing PCoA. (**C**) Missing both PCoAs. (**D**) Missing ACA. (**E**) Missing PCA. (**F**) Missing PCoA and PCA.

Figure 7 illustrates alterations in cerebral blood flow heterogeneity between the six variants. Under AF conditions, the left middle, left anterior, and left posterior regions experience a balanced outflow in variants 1 through 4 (Figure 7A–D), indicating virtually uniform cerebral perfusion. Alternatively, variants 5 and 6 show flow patterns that are more irregular. Both these variants also have either out-of-phase or negative-flow amplitude in the LP region relative to the other two regions shown. Additionally, flow oscillations in the LP region for these two variants have much larger amplitudes than flow to the other regions and compared to all flow in the other variants.

As illustrated in Figure 8, the maximal PRCC values regarding hypoperfusion count are the intrinsic heart rate (HR_0), $Edias_{rv}$, $Esys_{rv}$, G_{aut}, and $Edias_{ra}$. Notably, HR_0 ranked the highest for each variant, with varying amplitudes across the variants. Additionally, mechanical characteristics of the right ventricle and atria have high PRCC values, i.e., $Edias_{rv}$, $Esys_{rv}$, $Edias_{ra}$, and $Esys_{ra}$. G_{aut} and tau_{aut}, which are both parameters that play a role in the cerebral autoregulation mechanism, also have high PRCC values for all variants.

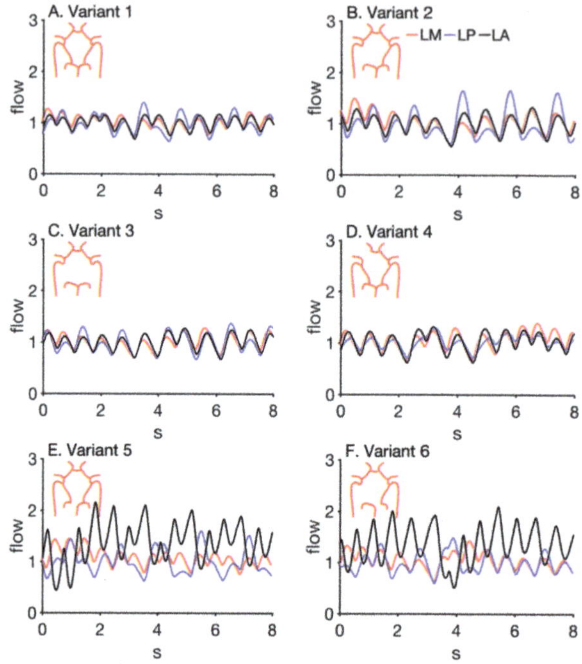

Figure 7. Perfusion to distal regions of the brain, represented by outflow at three distinct vessel terminals. Color coding for all panels is provided in (**B**). In all panels, red line represents flow at the terminal of left middle artery, blue lines represent flow at the terminal of left posterior artery, and black line represents flow at the terminal of left anterior artery. Each panel illustrates blood flow in a particular CoW variant. (**A**) Complete CoW. (**B**) Missing PCoA. (**C**) Missing both PCoAs. (**D**) Missing ACA. (**E**) Missing PCA. (**F**) Missing PCoA and PCA.

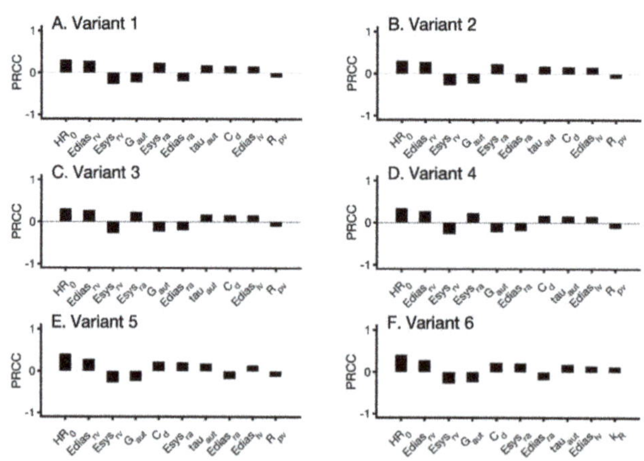

Figure 8. PRCCs for hypoperfusion count for each of the six considered CoW variants. The 10 PRCC values with the greatest magnitude are shown for each case and are ordered from greatest to least magnitude. Symbols are described in Table 1. Each panel illustrates PRCC analysis for a particular CoW variant. (**A**) Complete CoW. (**B**) Missing PCoA. (**C**) Missing both PCoAs. (**D**) Missing ACA. (**E**) Missing PCA. (**F**) Missing PCoA and PCA.

4. Discussion

While current treatment methods for AF, such as heart rate control and atrial ablation, are assessed based on treatment mortality, there is growing evidence that other factors, such as the impact on cognitive function, should be considered [2]. As research continues in this field, the results of the present study suggest that the cerebrovascular structure should be considered in treatment planning to ensure better clinical outcomes.

The present model is a composite of previously published models. It is based on established biophysical modelling techniques, i.e., lumped-parameter modelling using windkessel compartments. The components have been used previously to model a variety of disease cases, including AF. While direct model validation with in vivo data was not within the scope of the study, model outputs were presented for comparison with published values. Median arterial blood pressures (systolic/diastolic) were 117.44/77.81 mmHg and 119.51/78.95 mmHg for NSR and AF, respectively, which are considered to be within healthy ranges. Additionally, blood flow in major cerebral arteries is presented for comparison with measured values published by Zarrinkoob et al. [28]. Zarrinkoob reports blood flow in the ACA, MCA, and PCA to be 12%, 21%, and 8% of total CBF, respectively. The model shows corresponding values of 8%, 29%, and 12% for the NSR case, and 8%, 30%, and 12% for the AF case. Therefore, the model reflects clinically measured blood flow distribution, with predominant blood flow occurring in the MCA.

Variations from regular blood pressure in large arteries due to AF were shown to be associated with large changes in blood flow in the distal circulation of the brain (Figure 5). These changes lead to occurrences of critical hypoperfusion events in the brain, which may lead to silent cerebral ischemia, damaging brain tissue over time. The present modelling of this phenomenon is in agreement with previous works [1,4,12], and is the primary motivation for further investigation into the impacts of AF with respect to the cerebral circulation. Additionally, in Figure 5, it can be observed that the initial hypoperfusion seen at 2–4 s is followed by hyperperfusion from 4–7 s. This is to be expected because of the reflexive nature of the autoregulation mechanism. The autoregulatory function modulates the resistance and compliance of the downstream cerebral vessels within which the blood flow is being observed. The autoregulation function acts on a time scale of approximately 20 s, therefore there is a small delay between the drop in blood flow and the response of decreased resistance and increased compliance. This small delay in autoregulation function is thought to be the reason spontaneous drops in arterial pressure due to irregular heartbeats can cause transient hypoperfusion in the brain.

The result illustrated in Figure 6 shows that all considered CoW variants follow largely the same pattern with respect to the effect of heart rate on hypoperfusion frequency. All variants had a minimum hypoperfusion count at 50 bpm (in the heart rates considered), with the maximum occurring around 120 bpm. The most consequential result from this section is the result from variant 4, shown in Figure 6D. Variant 4 has a minimum hypoperfusion count of 675 at 50 bpm, which is over 2.5 times higher than variant 3, which has the next highest minimum. This demonstrates that although patients with CoW variant 4 may respond to a heart rate control strategy, it may not be sufficient to protect against hypoperfusion in the distal circulation of the brain. Based on this result, it is recommended that for patients with variant 4 of the CoW, alternative treatment methods be used in addition to, or instead of, heart rate control, in order to avoid ischemic cerebral damage.

It should be noted that this finding, along with previous modelling results [12], contradicts the recommendation made based on the RACE II trial [13]. The study found that relative to strict rate control, lenient rate control was as effective in preventing mortality and other outcomes, and was easier to achieve. This finding has informed clinicians on rate control strategies in relation to preventing mortality in recent years. However, cognitive impairment/dementia was not considered to be outcomes of this study, and heart rate had not yet been linked to hypoperfusion events associated with AF. Therefore, there is now growing evidence supporting strict rate control for preventing deleterious cognitive outcomes.

It was shown that certain variants could lead to increased heterogeneity in cerebral blood flow, with increased blood flow in some regions, and decreased in others (Figure 7). In particular, both variants with a missing PCA1 segment (variants 5 and 6) displayed heterogeneous flow patterns, as well as having larger amplitudes of the oscillatory flow rate than the other variants. This indicates that the PCA1 segment plays a key role in the distribution of blood flow with respect to homogeneity among the distal cerebral vessels. Although the large oscillations in blood flow to the left posterior circulation present in these variants are not considered harmful by the metric of hypoperfusion events, which is the primary focus of this study, they may lead to detrimental outcomes via other mechanisms, such as abnormal wall shear stress or acute hypertension. These phenomena will be further investigated in future work.

Sensitivity analysis, as shown in Figure 8, shows the model parameters that have the largest impact on modelling outcomes, namely the hypoperfusion event frequency. It was shown that in all cases of variant CoWs, HR_0 had the highest PRCC value, meaning that it is the parameter that most influences the hypoperfusion event frequency. This was expected, as heart rate control has been shown to be an effective method for decreasing hypoperfusion events [12,22]. In all variants, elastance values for the right heart were among the parameters with the largest PRCC values. This is an indication that the function of the right heart is strongly related to cerebral hypoperfusion outcomes, and warrants further study.

The present work is an investigation into the impact of AF on cerebral circulation considering common cases of congenital variations to the CoW. The presented model considers AF in the absence of other common cardiovascular conditions such as hypertension or atherosclerosis and represents simple cases of missing arterial segments, for the purposes of direct comparison. The model components have previously been used to study such conditions as hypertension, atherosclerotic lesions, and arterial occlusions. Additionally, small variations in cerebrovascular structure can be trivially modelled by assigning modified resistances to blood vessels. Future work will focus on incorporating these common conditions into our modelling, to further understand the impact of AF on cerebral circulation. Previously used techniques for representing populations using 0D models will be employed to elucidate the impacts of varied cerebrovascular structures [11].

In a clinical environment, it is critical for computational models to be applicable on a patient-specific basis. Methods for the incorporation of imaging data into 0D blood flow models are currently under development and will be used to further assess the impact of variant vascular structures using patient-specific data [8,29]. Such methods will also be effective in the clinic, opening up the possibility of patient-specific assessments for persistent AF patients. The presented model is extensible and personalizable, which will permit patient-specific risk stratification [30]. Further investigation will be conducted using spatially resolved 1D modelling to investigate the impacts of these phenomena on the blood vessels as well as the surrounding tissue in greater detail.

Author Contributions: Conceptualization, T.J.H., S.R.K. and D.G.; methodology, T.J.H. and S.R.K.; software, T.J.H. and J.J.J.; formal analysis, T.J.H.; original draft, T.J.H.; final manuscript, T.J.H., U.A., C.W.M. and D.G.; visualization, T.J.H.; supervision, D.G. and C.W.M.; project administration, S.R.K.; funding acquisition, C.W.M., D.G., and S.R.K.; All authors have read and agreed to the published version of the manuscript.

Funding: This research was funded by Canarie Inc. (#RS3-111, SRK, CWM, DG), the Canada Heart and Stroke Foundation (G-20-0028717, CWM), and the Canada NSERC operational grant (#R4081A03, DG).

Institutional Review Board Statement: Not applicable.

Informed Consent Statement: Not applicable.

Data Availability Statement: The source code to reproduce all results presented in this work is openly provided in the lab's github: https://github.com/mccsssk2/MDPI2021_TH_CircleofWillis (accessed on 31 December 2021).

Acknowledgments: We thank Compute Canada for unrestricted compute access to S.R.K. We thank Lawson and Western University I.T. services for their support. We thank members of the McIntyre lab, Goldman labx, Kharche lab, and many others for many in-depth discussions.

Conflicts of Interest: The authors declare no conflict of interest. The funders had no role in the design of the study; in the collection, analyses, or interpretation of data; in the writing of the manuscript, or in the decision to publish the results.

References

1. Hunter, T.J.; Joseph, J.J.; Anazodo, U.; Kharche, S.R.; McIntyre, C.W.; Goldman, D. Computational modelling of the role of atrial fibrillation on cerebral blood perfusion. In *Functional Imaging and Modeling of the Heart*; Ennis, D.B., Perotti, L.E., Wang, V.Y., Eds.; Springer International Publishing: Cham, Switzerland, 2021; Volume 12738 LNCS, pp. 679–686.
2. De Bruijn, R.F.A.G.; Heeringa, J.; Wolters, F.J.; Franco, O.H.; Stricker, B.H.C.; Hofman, A.; Koudstaal, P.J.; Ikram, M.A. Association Between Atrial Fibrillation and Dementia in the General Population. *JAMA Neurol.* **2015**, *72*, 1288–1294. [CrossRef] [PubMed]
3. Gardarsdottir, M.; Sigurdsson, S.; Aspelund, T.; Rokita, H.; Launer, L.J.; Gudnason, V.; Arnar, D.O. Atrial fibrillation is associated with decreased total cerebral blood flow and brain perfusion. *EP Eur.* **2018**, *20*, 1252–1258. [CrossRef] [PubMed]
4. Anselmino, M.; Scarsoglio, S.; Saglietto, A.; Gaita, F.; Ridolfi, L. Transient cerebral hypoperfusion and hypertensive events during atrial fibrillation: A plausible mechanism for cognitive impairment. *Sci. Rep.* **2016**, *6*, 8635. [CrossRef]
5. Anazodo, U.C.; Shoemaker, J.K.; Suskin, N.; St. Lawrence, K.S. An investigation of changes in regional gray matter volume in cardiovascular disease patients, pre and post cardiovascular rehabilitation. *Neuroimage* **2013**, *3*, 388. [CrossRef]
6. Steinman, D.A.; Poepping, T.L.; Tambasco, M.; Rankin, R.N.; Holdsworth, D.W. Flow Patterns at the Stenosed Carotid Bifurcation: Effect of Concentric versus Eccentric Stenosis. *Ann. Biomed. Eng.* **2000**, *28*, 415–423. [CrossRef]
7. Alastruey, J.; Parker, K.H.; Peiró, J.; Byrd, S.M.; Sherwin, S.J. Modelling the circle of Willis to assess the effects of anatomical variations and occlusions on cerebral flows. *J. Biomech.* **2007**, *40*, 1794–1805. [CrossRef] [PubMed]
8. Antiga, L.; Piccinelli, M.; Botti, L.; Ene-Iordache, B.; Remuzzi, A.; Steinman, D.A. An image-based modeling framework for patient-specific computational hemodynamics. *Med. Biol. Eng. Comput.* **2008**, *46*, 1097–1112. [CrossRef] [PubMed]
9. Lan, H.; Updegrove, A.; Wilson, N.M.; Maher, G.D.; Shadden, S.C.; Marsden, A.L. A Re-Engineered Software Interface and Workflow for the Open-Source SimVascular Cardiovascular Modeling Package. *J. Biomech. Eng.* **2018**, *140*, 024501. [CrossRef]
10. Altamirano-Diaz, L.; Kassay, A.D.; Serajelahi, B.; McIntyre, C.W.; Filler, G.; Kharche, S.R. Arterial Hypertension and Unusual Ascending Aortic Dilatation in a Neonate with Acute Kidney Injury: Mechanistic Computer Modeling. *Front. Physiol.* **2019**, *10*, 1391. [CrossRef]
11. Joseph, J.J.; Hunter, T.J.; Sun, C.; Goldman, D.; Kharche, S.R.; McIntyre, C.W. Using a Human Circulation Mathematical Model to Simulate the Effects of Hemodialysis and Therapeutic Hypothermia. *Appl. Sci.* **2021**, *12*, 307. [CrossRef]
12. Saglietto, A.; Scarsoglio, S.; Ridolfi, L.; Gaita, F.; Anselmino, M. Higher ventricular rate during atrial fibrillation relates to increased cerebral hypoperfusions and hypertensive events. *Sci. Rep.* **2019**, *9*, 1–9. [CrossRef]
13. Van Gelder, I.C.; Groenveld, H.F.; Crijns, H.J.G.M.; Tuininga, Y.S.; Tijssen, J.G.P.; Alings, A.M.; Hillege, H.L.; Bergsma-Kadijk, J.A.; Cornel, J.H.; Kamp, O.; et al. Lenient versus Strict Rate Control in Patients with Atrial Fibrillation. *N. Engl. J. Med.* **2010**, mboxemph362, 1363–1373. [CrossRef]
14. Heldt, T. Computational Models of Cardiovascular Response to Orthostatic Stress. Ph.D. Dissertation, Massacheusetts Institute of Technology, Cambridge, MA, USA, 2004.
15. Lin, J.; Ngwompo, R.F.; Tilley, D.G. Development of a cardiopulmonary mathematical model incorporating a baro-chemoreceptor reflex control system. *Proc. Inst. Mech. Eng. Part H J. Eng. Med.* **2012**, *226*, 787–803. [CrossRef] [PubMed]
16. Ursino, M.; Giannessi, M. A model of cerebrovascular reactivity including the circle of Willis and cortical anastomoses. *Ann. Biomed. Eng.* **2010**, *38*, 955–974. [CrossRef]
17. Hennig, T.; Maass, P.; Hayano, J.; Heinrichs, S.; Hennig, T.; Maass, P.; Hayano, J.; Heinrichs, S. Exponential Distribution of Long Heart Beat Intervals During Atrial Fibrillation and Their Relevance for White Noise Behaviour in Power Spectrum. *J. Biol. Phys.* **2006**, *32*, 383–392. [CrossRef] [PubMed]
18. Scarsoglio, S.; Guala, A.; Camporeale, C.; Ridolfi, L. Impact of atrial fibrillation on the cardiovascular system through a lumped-parameter approach. *Med. Biol. Eng. Comput.* **2014**, *52*, 905–920. [CrossRef] [PubMed]
19. Anselmino, M.; Scarsoglio, S.; Saglietto, A.; Gaita, F.; Ridolfi, L. A computational study on the relation between resting heart rate and atrial fibrillation hemodynamics under exercise. *PLoS ONE* **2017**, *12*, e0169967. [CrossRef] [PubMed]
20. Hindmarsh, A.C.; Brown, P.N.; Grant, K.E.; Lee, S.L.; Serban, R.; Shumaker, D.E.; Woodward, C.S. SUNDIALS: Suite of nonlinear and differential/algebraic equation solvers. *ACM Trans. Math. Softw.* **2005**, *31*, 363–396. [CrossRef]
21. Tange, O. GNU Parallel 2018. Available online: https://zenodo.org/record/1146014#.YgJqUerMKUl (accessed on 31 December 2021). [CrossRef]

22. Pianelli, M.; Scaglione, M.; Anselmino, M.; Caponi, D.; Garcia, P.; Cesarani, F.; Toso, E.; Raimondo, C.; Halimi, F.; Leclercq, J.F.; et al. Delaying cardioversion following 4-week anticoagulation in case of persistent atrial fibrillation after a transcatheter ablation procedure to reduce silent cerebral thromboembolism: A single-center pilot study. *J. Cardiovasc. Med.* **2011**, *12*, 785–789. [CrossRef]
23. Kharche, S.R.; Mironova, G.Y.; Goldman, D.; McIntyre, C.W.; Welsh, D.G. Sensitivity analysis of a smooth muscle cell electrophysiological model. In *Functional Imaging and Modeling of the Heart*; Ennis, D.B., Perotti, L.E., Wang, V.Y., Eds.; Springer International Publishing: Cham, Switzerland, 2021; Volume 12738 LNCS, pp. 540–550.
24. Matsumoto, M.; Nishimura, T. Mersenne twister. *ACM Trans. Model. Comput. Simul.* **1998**, *8*, 3–30. [CrossRef]
25. Malone, B.P.; Minansy, B.; Brungard, C. Some methods to improve the utility of conditioned Latin hypercube sampling. *PeerJ* **2019**, *7*, e6451. [CrossRef] [PubMed]
26. Kharche, S.; Lüdtke, N.; Panzeri, S.; Zhang, H. A global sensitivity index for biophysically detailed cardiac cell models: A computational approach. In *Functional Imaging and Modeling of the Heart*; Ayache, N., Delingette, H., Sermesant, M., Eds.; Springer: Berlin/Heidelberg, Germany, 2009; pp. 366–375.
27. Hamby, D.M. A comparison of sensitivity analysis techniques. *Health Phys.* **1995**, *68*, 195–204. [CrossRef] [PubMed]
28. Zarrinkoob, L.; Ambarki, K.; Wåhlin, A.; Birgander, R.; Eklund, A.; Malm, J. Blood flow distribution in cerebral arteries. *J. Cereb. Blood Flow Metab.* **2015**, *35*, 648–654. [CrossRef] [PubMed]
29. Joseph, J.J.; Lee, T.-Y.; Goldman, D.; McIntyre, C.W.; Kharche, S.R. The role of extra-coronary vascular conditions that affect coronary fractional flow reserve estimation. In *Functional Imaging and Modeling of the Heart*; Ennis, D.B., Perotti, L.E., Wang, V.Y., Eds.; Springer International Publishing: Cham, Switzerland, 2021; Volume 12738 LNCS, pp. 595–604.
30. Grande Gutierrez, N. Hemodynamic Based Thrombotic Risk Stratification in Kawasaki Disease Patients with Coronary Artery Aneurysms. Ph.D. Dissertation, Stanford University, Stanford, CA, USA, 2019.

applied sciences

MDPI

Review

Patient-Specific Inverse Modeling of In Vivo Cardiovascular Mechanics with Medical Image-Derived Kinematics as Input Data: Concepts, Methods, and Applications

Johane H. Bracamonte [1], Sarah K. Saunders [1], John S. Wilson [2], Uyen T. Truong [3] and Joao S. Soares [1,*]

[1] Department of Mechanical and Nuclear Engineering, Virginia Commonwealth University, Richmond, VA 23284, USA; bracamontej@vcu.edu (J.H.B.); saunderssk@vcu.edu (S.K.S.)
[2] Department of Biomedical Engineering and Pauley Heart Center, Virginia Commonwealth University, Richmond, VA 23219, USA; jswilson3@vcu.edu
[3] Department of Pediatrics, School of Medicine, Children's Hospital of Richmond at Virginia Commonwealth University, Richmond, VA 23219, USA; uyen.truong@vcuhealth.org
* Correspondence: jsoares@vcu.edu

Abstract: Inverse modeling approaches in cardiovascular medicine are a collection of methodologies that can provide non-invasive patient-specific estimations of tissue properties, mechanical loads, and other mechanics-based risk factors using medical imaging as inputs. Its incorporation into clinical practice has the potential to improve diagnosis and treatment planning with low associated risks and costs. These methods have become available for medical applications mainly due to the continuing development of image-based kinematic techniques, the maturity of the associated theories describing cardiovascular function, and recent progress in computer science, modeling, and simulation engineering. Inverse method applications are multidisciplinary, requiring tailored solutions to the available clinical data, pathology of interest, and available computational resources. Herein, we review biomechanical modeling and simulation principles, methods of solving inverse problems, and techniques for image-based kinematic analysis. In the final section, the major advances in inverse modeling of human cardiovascular mechanics since its early development in the early 2000s are reviewed with emphasis on method-specific descriptions, results, and conclusions. We draw selected studies on healthy and diseased hearts, aortas, and pulmonary arteries achieved through the incorporation of tissue mechanics, hemodynamics, and fluid–structure interaction methods paired with patient-specific data acquired with medical imaging in inverse modeling approaches.

Keywords: inverse models; data assimilation; cardiovascular imaging; image-based kinematics; biomechanics; tissue mechanics; hemodynamics; patient-specific models

Citation: Bracamonte, J.H.; Saunders, S.K.; Wilson, J.S.; Truong, U.T.; Soares, J.S. Patient-Specific Inverse Modeling of In Vivo Cardiovascular Mechanics with Medical Image-Derived Kinematics as Input Data: Concepts, Methods, and Applications. *Appl. Sci.* **2022**, *12*, 3954. https://doi.org/10.3390/app12083954

Academic Editors: Mihaela Pop and Cristian A. Linte

Received: 8 January 2022
Accepted: 30 March 2022
Published: 14 April 2022

1. Introduction

The primary role of numerical modeling in cardiovascular biomechanics has been to predict the performance of medical devices and to estimate physiological and mechanical cues acting on tissues, such as pressure and flow-driven stresses. Given the vast experimental evidence of mechanical factors producing effects on cellular differentiation, signaling, communication, and function [1–5], in silico experiments have explored the role of mechanical stimuli on normal and pathological tissue growth and remodeling [6]. From a clinical research standpoint, the development of patient-specific biomechanical models could provide more accurate and detailed data leading to a better understanding of the onset and progression of cardiovascular disease [7]. In addition, computational modeling has also been proposed as a supporting tool for medical practice on a patient-specific basis, which could provide non-invasive assessments of tissue properties, structure, and mechanical loads as physiologically meaningful risk stratification factors. Such patient-specific analyses have the potential to bring immense benefits to clinical practice by supporting diagnosis, treatment planning, and predictions of the outcome of surgical procedures with minimum associated costs and risk to the patients [8].

However, for biomechanical models to provide low-risk patient-specific solutions, personalized non-invasive clinical studies must be readily available to quantify regional cardiovascular function. Current medical imaging technology, namely echocardiography and magnetic resonance imaging (MRI), offers not only anatomical information but also high-resolution kinematics data of tissue motion and blood flow [9–12]. Kinematic-derived quantities, such as peak and average strain on the myocardium and aortic walls, have shown a good correlation with clinical risk markers [13]. Nevertheless, kinematic information alone cannot provide insights about mechanical forces, stresses, and tissue material properties, which are necessary for a full understanding of healthy and pathophysiological phenomena [14].

The inverse method, or data assimilation method, is an approach that allows solving classic mechanics problems "backwards"; that is, retrieving material properties and dynamic information (stress and forces) using measured kinematic information and loading boundary conditions as input [15]. Several research groups have coupled computational-mechanics tools with medical imaging technology to retrieve relevant biomechanical and hemodynamic markers from normal and pathological human tissues and organs [16], including diverse cardiovascular components [17]. Inverse modeling approaches have the potential to become a valuable tool for the non-invasive assessment of patient-specific cardiovascular health by providing quantitative physiological metrics that cannot be directly measured in vivo but may be derived entirely from clinical evaluations and the application of biomechanical principles.

The relevance of patient-specific modeling and its potential impact on the future of personalized and predictive health care has been acknowledged by several funding agencies. In 2003, the Interagency Modeling and Analysis Group was formed from the collaboration of nine institutes of the National Institutes of Health (NIH) and three directorates of the National Science Foundation (NSF). This group released its first funding opportunity in 2004 under the title "Interagency Opportunities in Multiscale Modeling in Biomedical, Biological, and Behavioral Systems Solicitation", which has been regularly reissued since then, and led to the creation of the Multiscale Modeling Consortium which includes over 100 projects on multiscale modeling of biological systems. The European Union initiated the "Structuring the Europhysiome" Consortium in 2006, which led to the Virtual Physiological Human project, an ongoing initiative that aims to bring together policymakers, regulatory agencies, funding bodies, industry, and research organizations towards the development of integrated computer models of the mechanical, physical, and biochemical functions of the living human body [7]. These initiatives have motivated the integration of multidisciplinary teams of biologists, physicians, and engineers who are faced with the challenge of bringing together field-specific nomenclatures, techniques, and analytical approaches.

Inverse modeling of biomechanical systems requires the confluence of state-of-the-art techniques from several disciplines including clinical care, medical imaging, simulation engineering, data analysis, and computer science (Figure 1). Inverse methods have been developed on a highly specific basis and tailored to the available clinical data, tissue/pathology of interest, and available resources; thus, inverse modeling developments only share a general data processing pipeline, while differing on the clinical data source, imaging technique, and modeling approach. Therefore, the task of designing an inverse method pipeline requires a comprehensive understanding of the process at all stages, for which familiarity with fundamental concepts and terminology is a prerequisite. The latter can be challenging due to the multidisciplinary nature not only of the method itself but also of the clinically relevant phenomena to model.

Inverse modeling analyses can also be applied to in vitro experimental setups. The main advantage of this approach is that input data is not limited by the available clinical tests, mechanical loads can be precisely controlled, and kinematics can be measured with high-resolution instruments. Furthermore, the target of the inverse method can be defined not only in terms of kinematic information but also in terms of controlled mechanical loads and stress measurements [18]. Moreover, the outputs of inverse modeling can be validated with controlled experimental parameters. Inverse analyses of in vitro setups have been applied to explanted animal and human tissue, and to engineered tissue constructs. Notably, inverse modeling has been applied to resolve mechanics at a cellular level. The traction force microscopy (TFM) technique was introduced by Butler et al., to estimate the force that adherent cells exert on their surroundings by solving the traction field in a hydrogel of known properties, cultured with cells, and with embedded beads as visual markers [19]. By tracking the bead displacements through microscopy, and setting known boundary conditions, the traction field is resolved by an exact solution assuming a semi-infinite medium. Further development was introduced by Tambe et al., with the monolayer stress microscopy (MSM) technique, which allowed the inverse estimation of stress fields across monolayer cellular constructs under static and dynamic conditions by inducing controlled displacements of the boundary under a motorized microscope [20]. These, and other similar techniques, have been used to explore the response of cardiovascular cells (endothelial cells, cardiomyocytes, smooth muscle cells, etc.), cellular layers, and engineered tissue to mechanical stimuli in terms of cell proliferation, migration, expression, and synthesis of extracellular matrix components [21–23]. The detailed and accurate results that can be retrieved from inverse analyses of in vitro experiments can provide valuable insights into cardiovascular mechanobiology. These insights contribute to the understanding of how macroscopic biomechanical factors affect the healthy or pathological growth and remodeling of cardiovascular and engineered tissues. However, the replication of in vivo physiological conditions in vitro is cumbersome, and the results of in vitro experimentation can be challenging to extrapolate to patient-specific situations. As a result, the clinical application of inverse analyses of in vitro experiments remains limited.

This article aims first to serve as a referential document for concepts and methods from all involved disciplines on patient-specific in vivo inverse modeling; and secondly, to highlight the potential clinical application of patient-specific inverse modeling in the cardiovascular research field. Specifically, we review the fundamentals of cardiovascular tissue and blood biomechanics, modeling and simulation, and medical imaging, as they relate to the inverse modeling approach and its applications in cardiovascular medicine. In Section 2, we review the application of the principles of classical continuum mechanics to the study of blood and tissue motion, with special emphasis on the constitutive equations that have been proposed to describe the mechanical behavior of cardiovascular tissue and blood. Section 3 briefly summarizes the fundamentals and main features of the finite element method (FEM) and finite volume method (FVM), as the most popular formulations for the numerical solution of biomechanical models. Next, we review the general definition of inverse problems and the available alternatives to solve inverse mechanics problems in Section 4. In Section 5, we review the working principles and main features of ultrasound (US), magnetic resonance imaging (MRI), and computational tomography (CT) imaging, giving special attention to the available techniques for the resolution of tissue and blood kinematics. Finally, in Section 6, we present a comprehensive review of the applications of imaging-based inverse modeling approaches to patient-specific human cardiovascular mechanics, including the resolution of the unloaded configuration and the estimation of tissue properties and stresses. Reviewed applications include healthy and diseased heart valves, cardiac and arterial walls, and hemodynamics of large arteries. To highlight the potential application of the inverse-modeling approach in cardiovascular medicine, we focus herein mostly on developments made in human studies, with a few mentions of relevant and pioneering studies in animals.

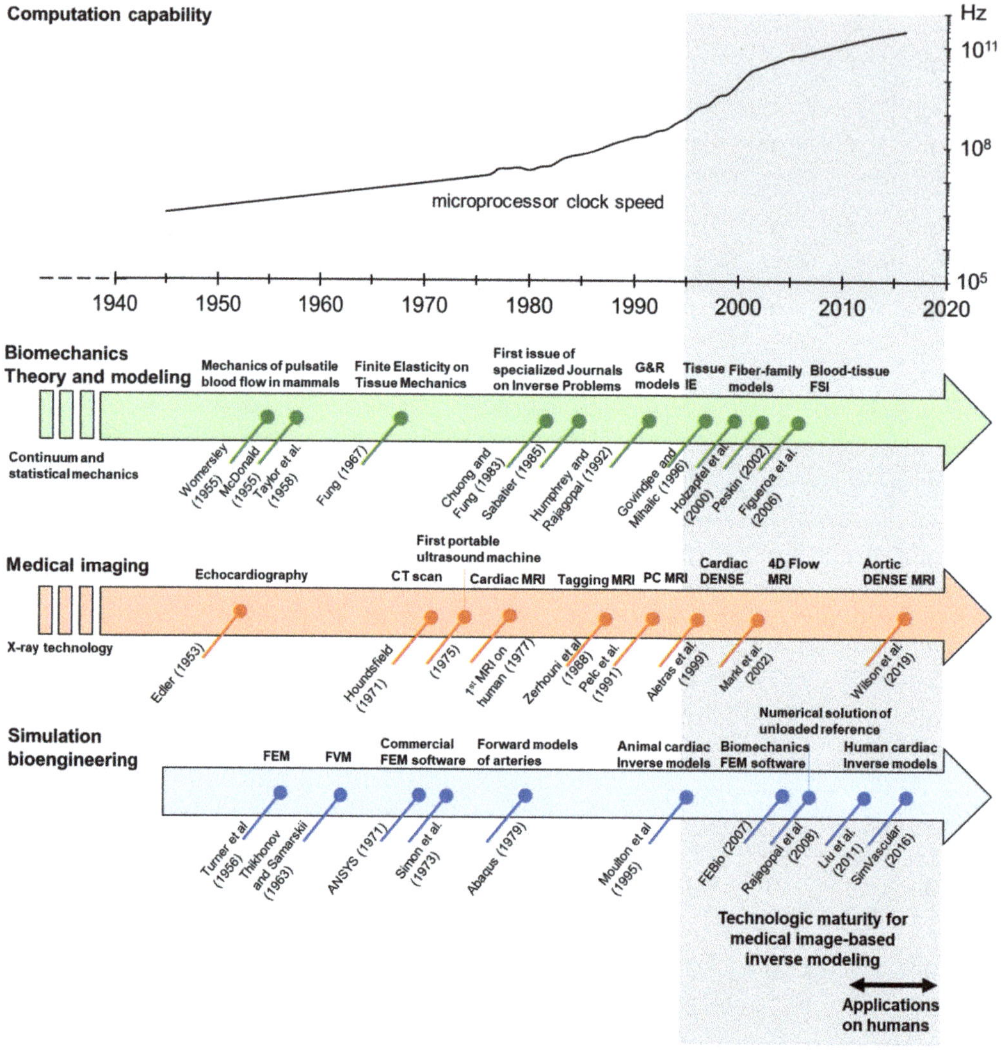

Figure 1. Timeline of microprocessor speed as a measure of computation capability, and relevant landmarks on the fields of biomechanics theory, medical imaging and simulation that make possible modern patient-specific image-based inverse modeling of the cardiovascular system. Acronyms: CT, computerized tomography; DENSE, displacement encoding with stimulated echoes; FEM, finite element method; FVM, finite volume method; IE, inverse elastostatics; FSI, fluid–structure interactions; PC, phase contrast; MRI, magnetic resonance imaging.

2. Governing Principles of Biomechanics

Modern biomechanics consists of the formulation of governing equations describing balances of mass, linear and angular momentum, and energy to biological systems and physiological processes. The human body maintains a uniform and stable temperature through homeostatic thermoregulation. Thus, contributions due to temperature change in the internal energy of the material, heat fluxes, and heat supply are typically negligible to the energy balance, which is in turn reduced to the balance between deformational energy and stress power (thermodynamic work). In classical continuum mechanics of purely mechanical processes, the balance of angular momentum directly translates to the symmetry of the stress tensor, and therefore, the relevant governing equations for most cardiovascular mechanics applications consist only of the balances of linear momentum and mass. However, most biological systems are open, and continuously interact with their surroundings, and thus the conservation principles must be handled carefully, especially with respect to tissue growth and atrophy within relevant timescales.

Given that the resolution of most in vivo medical imaging is on the scale of millimeters, only phenomena occurring at the tissue level can be directly associated with these measurements. The assumption of material continuity is reasonable for the formulation of the governing principles at this scale, leaving any additional considerations dealing with the extracellular and intracellular micro-environments to be included ad hoc with additional modeling formulations and constitutive equations.

To apply these principles, it is necessary to relate the stress tensor to kinematic measures, which is in essence the description of the mechanical behavior of the material under study. This information is provided by a constitutive equation; these can be either phenomenological equations "arbitrarily" formulated to reproduce experimental observations, or analytical expressions inspired by theoretical interactions of the material constituents at the micro or molecular scale. The selection of adequate models to describe the phenomena of interest is key to the success of any engineering analysis. The selected model must be complex enough to describe the most salient observable features at the scale of interest, while ideally being simple enough to provide a rational interpretation of its parameters and results and render a computationally tractable numerical problem. After fitting these parametrized models to experimental data, the constitutive equation can provide an additional understanding of the underlying mechanisms associated with the mechanical response of the material. Models of increased complexity usually require a larger number of parameters to be fitted, and overparametrized models can lead to solution multiplicity which obscures its interpretation and validity. For the sake of generality, the constitutive equation must also be independent of the frame of reference, comply with the second principle of thermodynamics, and yield amenable mathematical treatment and systems of equations that are solvable [24].

After formulating the constitutive relation as a function of specific unknowns (e.g., displacement or velocity fields), the resulting system of governing equations is then particularized into specific problems by the definition of temporal and spatial domains of interest and the imposition of appropriate boundary and initial conditions. To obtain a unique solution, it is necessary to constrain the problem by assigning a first-order boundary condition on at least one of the boundaries (e.g., by prescribing a known displacement or velocity). On the rest of the boundaries, higher-order boundary conditions can be applied to impose distributed forces such as a known pressure. Once solved, the result of this forward formulation is the transient spatial distribution of displacement or velocity throughout the domain as a result of the specified loads and material properties. From this kinematic information, strain and strain-rate distributions can be numerically derived, and the stress distributions retrieved via the constitutive equation.

In the following subsections, we present a review of the main features of the mechanical behavior and composition of cardiovascular tissues and blood, as well as the constitutive equations that have been developed and applied to model such behaviors. Then, we review the algorithms implemented to model fluid–structure interactions and their importance in cardiovascular mechanics simulations. Finally, we briefly discuss the modeling of biological tissue growth and remodeling by the application of the constrained mixture theory.

2.1. Structural Mechanics of Cardiovascular Tissue

The study of soft biological tissues under the framework of finite elasticity was initiated by Y.C. Fung and others in the late 1960s, setting the basis of modern biomechanics [25,26]. As in classical solid mechanics, the mechanical analyses of cardiovascular tissues are usually performed with a Lagrangian formulation of the governing principles (Figure 2).

Applied forces are imposed as boundary conditions. For blood vessels, these are prescribed as transmural pressure differences that often assume a traction-free condition on the adventitial surface. More recently, however, growing attention to the role of perivascular and pericardial support and tethering has promoted the inclusion of restrictions to the displacement of the outer surface of the heart and vasculature [14,27,28]. In addition, more complex formulations of cardiovascular tissue mechanics which departs from classical elastic solids have been proposed to account for complex microstructural compositions, the inclusion of pre-stress/strains, chemically activated muscular tone, and viscous energy dissipation.

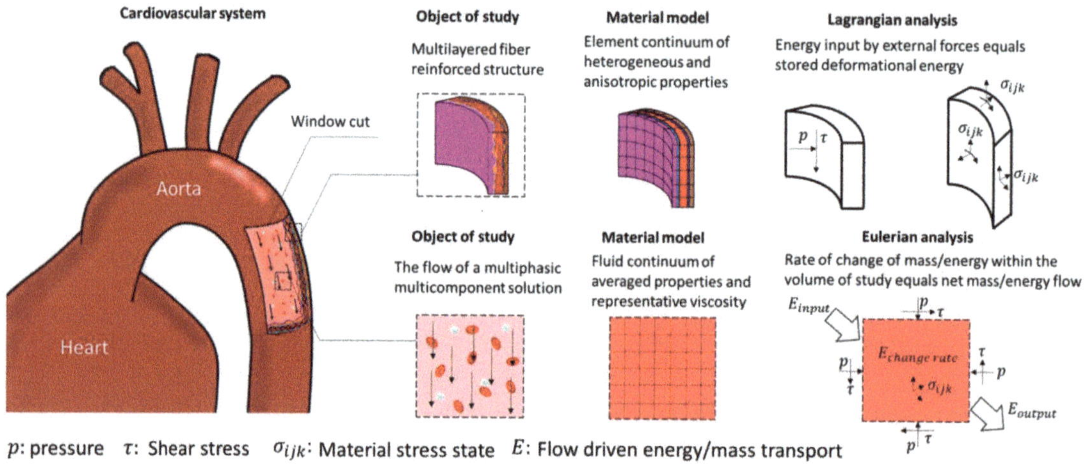

Figure 2. Representation of the modeling process of structural mechanics of cardiovascular tissue and fluid mechanics of the blood flow with a continuum mechanics approach. Structural mechanics of cardiovascular tissue are usually analyzed with a Lagrangian formulation that follows the deformation of a given portion of the tissue. Blood flow mechanics is usually analyzed with an Eulerian formulation, that is, analyzing the mass and energy balances on a fixed volume of interest through which the fluid flows.

Cardiovascular tissues are comprised of multiple layers of cells and extracellular matrix (ECM) components. The ECM is a network of macromolecules that is continuously synthesized and degraded by active cells and functionally provides them with structural and biochemical support. Typically, collagen, elastin, and fibrillin are regarded as the main structural constituents responsible for the macroscopic mechanical behavior of cardiovascular tissues [29]. Healthy cardiovascular tissue retains residual stress even when unloaded (i.e., the tissue is pre-strained relative to a reference state of zero transmural pressure). Circumferential and axial pre-stress/strain in vascular conduits have been widely established by measuring how much these tissues recoil to an open configuration when excised and cut transversely and longitudinally to relieve the residual stress [30,31]. It has been hypothesized that pre-strain plays a relevant role in balancing higher stresses on the luminal surface of blood vessels and promoting a homogenized transmural stress distribution and homeostatic equilibrium of the vascular tissue [32]. Notably, residual strain is heterogeneous and has been shown to vary with patient age and health, likely as a consequence of heterogeneous growth and remodeling and/or damage.

Additionally, cardiovascular tissue is muscular in nature and actively contracts/distends. Thus, its mechanical behavior is affected by the activation of actin-myosin sliding filaments, which depends on ion-based chemical signaling and determines the muscular tone. In the myocardium, striated muscle activation is responsible for cardiac contraction. In large arteries, contraction of smooth muscle cells regulates downstream vascular resistance, blood flow, and propagation of the pressure-pulse wave along the vascular tree.

Of note, cardiovascular tissue also exhibits viscoelastic behavior, which has been established with stress relaxation, creep, and strain-rate experiments. It has been argued that viscous energy dissipation of healthy tissue, functioning at a regular physiological rate (~1 Hz), is negligible compared to stored strain energy [33]. Nevertheless, viscoelasticity may play a critical role under pathological conditions where the deformation rate is increased, such as in atrial fibrillation, or when dealing with highly dissipative structures such as lipid pools in atherosclerotic plaques [34,35]. Despite the relevance of viscoelastic properties to pathological conditions, standardized testing protocols have yet to be developed for the exploration of its relation to disease onset and progression [36]. Notably, if viscous dissipation and inertial effects are neglected, all temporal terms in the governing equations are canceled, rendering the problem a quasi-static process (which is the most common approach applied to vascular wall mechanics).

2.1.1. Constitutive Equations

Passive Properties

Linearized elasticity falls short in describing the complex behavior of biological tissue, not only because the mechanical response is highly non-linear but also the material undergoes finite motions and deformations. In the case of non-linear behavior, it is usually convenient to employ the formulation of hyperelasticity and express the constitutive equations as the relation of a scalar stored energy density function to the deformation gradient tensor or the strain tensor (while other definitions of stretch or strain tensors are also possible and common). The scalar energy density function represents the amount of deformational energy stored per unit volume and is defined in such a way that the stress tensors can be obtained from their derivatives with respect to the strain or stretch tensor.

The passive behavior of cardiovascular tissue is characterized by an increasing resistance to deformation with strain. This behavior is represented by an increasing slope in the stress versus strain/stretch and strain energy versus strain/stretch curves, with a close to zero slope at zero strain and rapidly increasing at physiological ranges. This behavior has been attributed to the structural characteristics of the ECM components. It has been suggested that the increasing resistance to deformation with strain is owed to the progressive engagement of wavy bundles of elastin and collagen fibers to support the mechanical loads (Figure 3).

Hyperelastic isotropic material models, such as the Neo–Hookean and the Mooney–Rivlin constitutive equations, can accurately describe the behavior of amorphous bodies such as lipid pools in atherosclerotic formations, and fit some portions of the pressure–volume relation of blood vessels. The symmetry of these material models allows its representation as to the linear combination of the deformation gradient invariants weighted by the material properties. This formulation allows the determination of unique material properties for a given mechanical behavior. However, these models fail to reproduce the characteristic highly nonlinear and anisotropic behavior of cardiovascular tissue (Figures 2 and 3). This was originally addressed by the use of phenomenological equations, e.g., the Fung orthotropic exponential model being one of the most commonly used. Its success relies on its relative simplicity, widespread numerical implementation, and accuracy in the prediction of stress-strain curves [24,26]. Guccione et al. proposed modifications to Fung's orthotropic model based on myofiber structure and orientation to tailor myocardial tissue behavior [38]. These phenomenological equations usually consist of two terms contributing to the strain energy function. First, the contribution of volume changes is written as a function of the determinant (third invariant) of the deformation gradient tensor. The second term is the deviatoric contribution to the strain energy function, defined to be proportional to an exponential function of the components of the strain tensor. The proportional constant sets the scale of the material stiffness, and the function of the strain tensor components defines the material anisotropy. Part of the success of Fung-like equations in describing the stiffening of cardiovascular tissue with strain relies on the exponential functional form to quantify the effect of strain increments on deformational energy. However, the fitted constants of phenomenological equations lack physical interpretation which is desirable for studies aiming to relate material properties with pathological conditions [39,40].

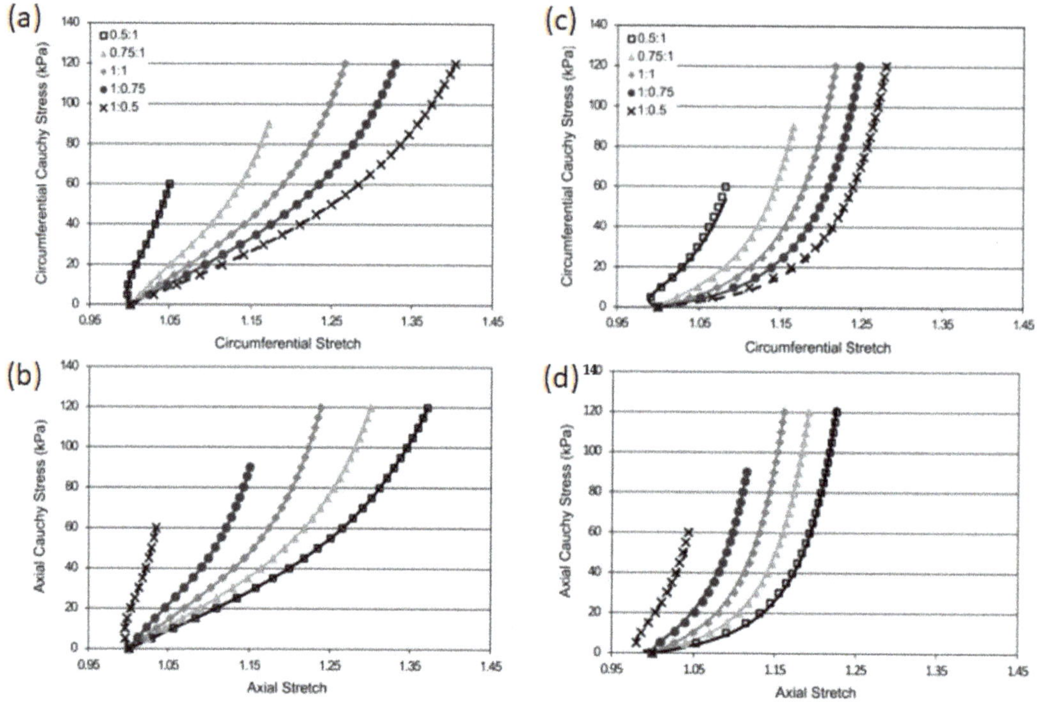

Figure 3. Representative biaxial stress-stretch behavior of healthy cardiovascular tissue. The stress and slope increase with the stretch/strain in any direction. This suggests that the cardiovascular tissue stiffens with stretch/strain which is hypothesized to be a consequence of the progressive engagement of ECM components to resist further deformation. This behavior is modeled by exponential functions of the deformation tensor components and/or invariants. Mark symbols in the figures show experimental biaxial test data of the human thoracic aorta for (**a**,**b**) young patients (20 to 35 years of age), and for (**c**,**d**) older patients (57 to 71 years of age). Solid lines represent the best-fit approximation with a four-fiber family constitutive equation. Reprinted/adapted with permission from Ref. [37], 2014, Elsevier.

In the past decades, many microstructure-inspired constitutive models have been proposed to specifically suit cardiovascular tissue behavior [39,41]. Fiber-family models have been particularly successful in reproducing the anisotropic behavior of the vascular wall while keeping physiological meaning to some of the fitting constants, the Holzapfel–Ogden model and its many variants being the most popular for cardiovascular tissue. These models assume that families of 1D fibers, each with specific mechanical behavior, orientation distribution, and volume fraction, are embedded within an isotropic continuum matrix. The isotropic component of the strain energy function is usually defined as a function proportional to the first invariant of the deformation gradient tensor. The contribution of fiber families is a weighted sum of exponential functions, the weighting factors being the fiber family material parameters. The exponential functions are defined depending on deformation tensor invariants and the relative orientation of fibers to strain, such that the contribution of a fiber family is maximized if the strain deformation occurs in the direction of the fibers [37]. Again, exponential functions are employed to mimic the stiffening effect of strain on cardiovascular tissue (Figure 3), this time being directly attributed to the ECM fiber components. Many improvements to these models have been proposed to account for different coupling effects such as inextensibility of fibers or cross-linking of the fiber ensembles [42–45].

Active Properties

Adequate modeling of active contraction is key for an accurate description of cardiovascular function in general, being particularly critical for modeling the heart. Conceptually, there are two possible approaches, the active stress models are the most common and they assume the stress tensor can be decomposed as the sum of a passive and active component, while active strain models assume a product decomposition of the deformation gradient.

Most active contraction models used in the inverse analysis of ventricular mechanics through continuum mechanics are simplifications of more complex bio-chemo-mechanical models such as the work of Hunter et al. [46]. The latter proposes a four-state variable model which includes the passive elasticity of myocardial tissue, the binding of calcium ions (Ca^{2+}) to troponin C and its release, tropomyosin movement kinetics, the myofiber length, and the kinetics of cross-bridge tension build-up under perturbation of myofilament length. In practice, the detailed information required for the evaluation of this model is out of reach, so simplified models assume the active stress acts mostly lengthways the direction of myofibers, with a magnitude that is proportional to the fiber length and the activation status. The activation status is often expressed as a time-dependent spatially heterogeneous function ranging from 0 to 1 [47]. Active strain function will impose the relative shortening lengthways of myofiber directions as a function of location and time along the cardiac cycle.

Typically, the activation state is assumed to be instantaneously homogeneous within the region of study, however, it is known that cellular activation propagates as an electrical wave and the excitation-contraction coupling poses a complex electromechanical problem [48]. This propagation has been modeled macroscopically as a reaction-diffusion problem of the electric potential through the intracellular and extracellular domains, thus known as the bidomain model. The monodomain model is a simplification that assumes the same propagation anisotropy for both domains. Some interesting research has been developed to apply inverse modeling to fit the parameters of mono and bidomain equations using patient-specific electrocardiography [49,50]. These, however, fall out of the scope of this review, and the interested reader is encouraged to study the abundant literature on the inverse problem of electrocardiography [51,52].

The application of active contraction models requires the specification of the local myofiber orientations. Patient-specific myofibers orientation can be resolved via diffusion tensor MR imaging (DT MRI) [53]. Being a relatively novel technique, these scans are rarely available from medical records of cardiovascular-disease patients, although their relevance in the biomedical field is a growing topic of discussion [54]. Therefore, myofiber orientation is either assumed a priori with simplified models or obtained through diffeomorphic transformations with the employment of precomputed cardiac atlases [55]. Bayer et al. proposed a Laplace–Dirichlet rule-based algorithm for assigning myofiber orientation to computational heart models that showed good agreement with DT MRI measurements. This algorithm consists of the resolution of the Laplace equation on the simulation domain with appropriate Dirichlet boundary conditions constrained by the following rules: the longitudinal fiber direction is parallel to the endocardial and epicardial surfaces, the longitudinal fibers rotate clockwise throughout the ventricular wall from a positive helical angle at the endocardium to a negative helical angle at the epicardium (both imposed by the user), fibers in the papillary muscles and trabeculae are assumed parallel to the long axis of these structures, the transverse fiber direction is perpendicular to longitudinal fibers, and fiber orientation in the septum is continuous with the ventricular walls [55]. Similarly, Potse et al. proposed a rule-based algorithm to define myofiber orientations assuming that longitudinal fibers are orthogonal to the local vector pointing to the shortest path between endocardium and pericardium, with a clockwise varying helical angle [56]. Rijcken et al. derived an equation for longitudinal and transverse myofiber orientation by solving an optimization problem, which maximized the ejection while maintaining fiber strain as homogeneous as possible on idealized geometries [57].

2.2. Fluid Mechanics of Blood Flow

For the study of fluids, it is more practical to implement an Eulerian formulation of the governing equations. This formulation is obtained by applying the Reynolds transport theorem to the equations for mass and momentum balance. Thus, this formulation solves the relation between flow driving forces, flow velocity, and deformation rates (Figure 2). A simplifying assumption applicable to biological systems is the incompressibility of the fluids, as most of them are either liquids or gases moving at subsonic velocities. Additionally, it is convenient to decompose the stress tensor into a spherical tensor representing the hydrostatic pressure and a deviatoric stress tensor. With this decomposition, constitutive equations can be designed to specifically relate the deviatoric stress components to the viscous dissipation of momentum.

Blood flow is generally assumed to be laminar throughout the circulatory system. The main arguments for this assumption are the pulsatile nature of the flow, the reduced dimensions of the vessels, and relatively low velocities, each contributing to the viscous effects overcoming the inertial forces and preventing turbulent random motion. However, it has been argued that transition to turbulent flows could be achieved locally in stenotic arteries. The use of a laminar model to study those cases could lead to an underestimation of wall shear stress, and stress oscillation [58,59]. Unlike most conventional engineering flows, blood flow is pulsatile and contained by compliant conduits of complex geometry. Since the 1950s, Womersley [60], McDonald [61], Taylor [62], Pedley [63], and others, developed analytical and experimental studies of pulsatile flow in mammals, identifying the most relevant parameters and features of this type of flow, thus setting the bases for modern hemodynamics.

Besides the intricacies of pulsatile flow in distensible conduits, the blood itself is a complex fluid. Blood consists of a suspension of cells in an aqueous solution of proteins and minerals called plasma. Plasma occupies approximately 55% of the blood volume, the rest being mainly occupied by red blood cells, white blood cells, and platelets. The rheological behavior of blood depends on how its constituents interact with each other and with the vessel walls, in consequence, this behavior is non-linear and highly dependent on the volumetric composition of blood, the flow conditions, and vessel dimensions. Modeling the complex interactions of blood constituents is a challenging statistical mechanics problem [64]. Some researchers have shown that cell aggregation and disaggregation are relevant to accurately describing blood rheology, especially in capillary flows where the cell size is comparable to the vessel diameter. Multiscale approaches have been successful in coupling the behavior of single cells as elastic entities with the transport equations of fluid flow, which are relevant for the study of clotting, aggregation, and platelet activation [65,66]. These approaches are computationally expensive, making them impractical for the study of large vessels.

Constitutive Equations

In the study of large and medium vessels (the ones that can be feasibly resolved with standard medical imaging), blood is often assumed to be a single-phase continuum. This approximation is reasonable given the relatively small size of cell aggregates compared to the vessel dimensions (and thickness of the boundary layer), and the relative relevance of inertia on flow motion [67]. For these cases, phenomenological constitutive equations describing the macroscopic behavior of flow are often applied. The linear Newtonian fluid is the simplest and most commonly employed model, providing reasonable results in vessels with diameters down to 200 μm [67]. Constitutive equations, such as the Casson,

Herschel–Bulkley, and Carreu–Yasuda, incorporate the shear-thinning effect on apparent viscosity by introducing a yield shear stress term [68]. To account for the effect of the volumetric share of cell suspension, recent works have included the hematocrit as an independent variable for the estimation of the effective viscosity [69].

2.3. Fluid-Structure Interactions (FSI)

Mechanics of the vascular wall and hemodynamics have been mostly studied as isolated problems; however, the function of the cardiovascular system is the result of complex interactions between blood, the actively contractile cardiac tissue, and the compliant vascular walls. The interaction of fluids and solids can conceptually be achieved by coupling the boundary conditions on the interface between the solid and the fluid, such that the field of displacements, velocities, and stresses are continuous and derivable at all points in a monolithic fully coupled approach. This, however, poses many implementation difficulties for complex 3D domains that can only be solved numerically. In addition, the typically large deformations of the cardiovascular walls cannot be handled by linearized methods used in conventional engineering applications.

The immersed boundary method, introduced by Peskin, was originally developed for the study of flow around heart valves and was rapidly adopted for many other applications [70]. In this approach, the Eulerian variables of fluid dynamics describing the surrounding flow are defined on a fixed computational grid, while the Lagrangian variables, accounting for the deformation of the tissue structures, are defined in a curvilinear computational grid that can be displaced with no conforming constraints in respect to the Eulerian grid. The moving solid boundary interacts with the fixed fluid domain by means of elastic body forces which are modulated by Dirac delta-like functions [71,72]. The fictitious domain method is a generalization of the immersed boundary method, which solves the coupling of the Lagrangian and Eulerian domains by the use of Lagrange multipliers instead of the concept of body forces [73]. This method is computationally less demanding as it does not require fitting the interface boundary at the cost of impaired accuracy near the interface.

Similarly, Figueroa et al. proposed the coupled momentum method, consisting of changing the non-slip condition on the fluid boundary to a traction condition, which is strongly coupled to the degrees of freedom of modified thin-membrane elements. This allows the formulation of the solid equations on the same Eulerian frame as in the fluid equations. In consequence, the fluid–solid interface mesh remains fixed, while the boundary nodes will have nonzero velocities [74]. Another well-established method for FSI simulation is the arbitrary Lagrangian–Eulerian (ALE) algorithm, which allows the arbitrary convective motion of the computational nodes of the discretization grid with respect to a fixed reference frame. Typically, the nodes on the fluid-solid interface are treated with a Lagrangian formulation. To deal with large or heterogeneous deformations of the interface, several implementations include the re-discretization of the computational domain to avoid the influence of ill-shaped deformed elements. The drawbacks of this method are the computational expense of re-meshing the domain, and the induced inaccuracies by transferring solutions from the degenerated mesh to the new one [75,76].

2.4. Growth and Remodeling Models by the Constrained Mixture Theory

One of the most relevant characteristics of living tissue is its capability to adapt in response to chemical and mechanical stimuli. This adaptation comes with microstructural reconfigurations, which alter the mass composition and the resulting contributions and properties of the tissue constituents. The understanding of the effect of mechanical stimulation on normal and pathological growth and remodeling of soft tissues is an active field of study that bridges biomechanics and mechanobiology.

In 1994, Rodriguez et al. proposed a general continuum formulation for the finite volumetric growth modulated by mechanical stress [77]. The theory of adaptation of living tissues was further developed by Humphrey and Rajagopal who proposed the constrained mixture theory, a mathematical framework to predict not only the growth but also the remodeling of biological tissues under transient mechanical and chemical stimulation [78,79]. The constrained mixture theory is based on the continuum theory of mixtures; that is, each component complies with a modified version of the governing principles of motion in the Eulerian formulation. The modification involves the addition of mass source/sink terms that account for the rate of synthesis or degradation of the constituent in its respective mass balance equation and the component-to-component interaction forces in the momentum balance equation. These source/sink terms respond to a series of constraints of physical and chemical nature and are dependent on the local distribution of strains, stress, and current composition. Constitutive equations must be defined for each constituent, and the overall properties of the construct can be calculated as a combination of its constituents, where simplified linearized forms weighted by their volume fraction are typically chosen [32,80].

2.5. Summary

In Table 1 we summarize the highlights of the governing principles of cardiovascular biomechanics through a continuum mechanics approach.

Table 1. Summary of governing principles of biomechanics.

Section	Highlights
2.	Inverse modeling of the cardiovascular system is usually grounded on classical continuum mechanics theory. The fundamental principles of mass and energy conservation are complemented by constitutive equations that describe the mechanical behavior of the material of interest. Constitutive material models can be either based on empirical evidence (phenomenological) or analytical expressions inspired by theory Once the model is defined through the selection of governing principles and constitutive equations, the problem is particularized by setting the domain of analysis and adequate boundary conditions.
2.1	Cardiovascular tissue is a complex multilayered structure that displays non-linear viscoelastic behavior, residual stress, and active contraction and distention. Structural mechanics of tissue is usually done with a Lagrangian formulation. The theory of finite hyperelasticity is applied to address the non-linear behavior and relatively large deformations. The adequate modeling of the passive behavior of cardiovascular tissue requires accounting for its structural anisotropy and the typical stiffening effect of strain/stretch. Active contraction and distention are the consequence of ion-based chemical signaling that triggers the contraction of actin-myosin sliding filaments, which determines the muscular tone. Active behavior is modeled by either adding an active stress or active strain components to the momentum balance. The additional active stress/strain is assumed to occur along myofiber directions and to depend on the cellular activation status. The geometrical distribution of the activation status can be determined by solving a reaction-diffusion problem. The patient-specific orientation of myofibers can be assessed by diffusion tensor MRI. However, the most common approach is to assume myofibers follow a standard orientation for which several models are available.
2.2	Blood is a suspension of cells in an aqueous solution of proteins and minerals that undergoes a pulsatile flow in vivo. Blood flow mechanics is typically studied with an Eulerian formulation. Assuming Newtonian fluid behavior and laminar flow are reasonable and typical approximations to model the blood flow in large vessels. Transition to turbulence flows may be relevant in the study of stenotic arteries. Phenomenological constitutive equations are available to model the shear-thinning effect on apparent viscosity.
2.3	The function of the cardiovascular system is the result of complex interactions between blood, the actively contractile cardiac tissue, and the compliant vascular walls. The interaction of blood flow and cardiovascular tissue requires specialized numerical formulations. There are several available formulations with different levels of complexity, one of which is the arbitrary Lagrangian–Eulerian algorithm which is complex and computationally expensive.
2.4	Living tissue has the capability to adapt in response to chemical and mechanical stimuli. The constrained mixture theory has been proposed to model the growth and remodeling of living tissue by solving sets of balance equations for each constituent of the tissue under study. The balance equations must be adequately constrained to account for the component-to-component interactions. The constrained mixture theory can introduce models to account for the reconfiguration of constituents under chemical/mechanical stimuli (remodeling).

3. Numerical Methods

The above-described system of governing and constitutive equations can only be solved analytically for a reduced group of oversimplified cases. Thus, mechanical analyses of complex biological systems require the application of numerical methods to obtain approximate solutions. It has been claimed that the development of numerical methods was key to the foundations of modern biomechanics [81,82]. Many of the early simulation analyses of the cardiovascular system and components were developed with in-house codes, but the popularization of commercial software boosted the production of computational research in biomechanics [82]. More recently, open-source specialized software for numerical biomechanics, such as SimVascular and FEBio [83,84], have risen from the collaborative effort of academic groups aiming to incorporate relevant bio-chemo-mechanical models of biological systems into simulation pipelines. Many different options exist for the numerical solution of time-dependent 3D problems. Mesh-based methods are the most popular approaches, particularly the finite volume and finite element methods which are reviewed in the following pages.

3.1. Finite Volume Method

The finite volume method (FVM) is conceptually straightforward. The domain of study is discretized in a series of non-overlapping finite volumes, and the governing equations, usually expressed in Eulerian formulation, are converted into algebraic expressions by integrating them over each discrete volume. The balance equations are applied on a node located in the center of the finite volume, while the flux terms are calculated at its faces (Figure 4a). This allows first and second-order approximations of derivatives. The surface flow for a given shared face is set identical and in opposite direction for the adjacent discrete volumes, and equal to a boundary condition at the edge of the domain. By doing so, the balance equations are held at the whole domain and within each finite volume, which is one of the most attractive features of the FVM. Additionally, since the calculation of properties happens in the center of each volume, it is relatively easy to implement boundary conditions of a higher order [85]. Numerical implementation of this method is also straightforward in the case of structured meshes, becoming more complex for unstructured meshes due to the bookkeeping necessary for the calculations of interface flux balances.

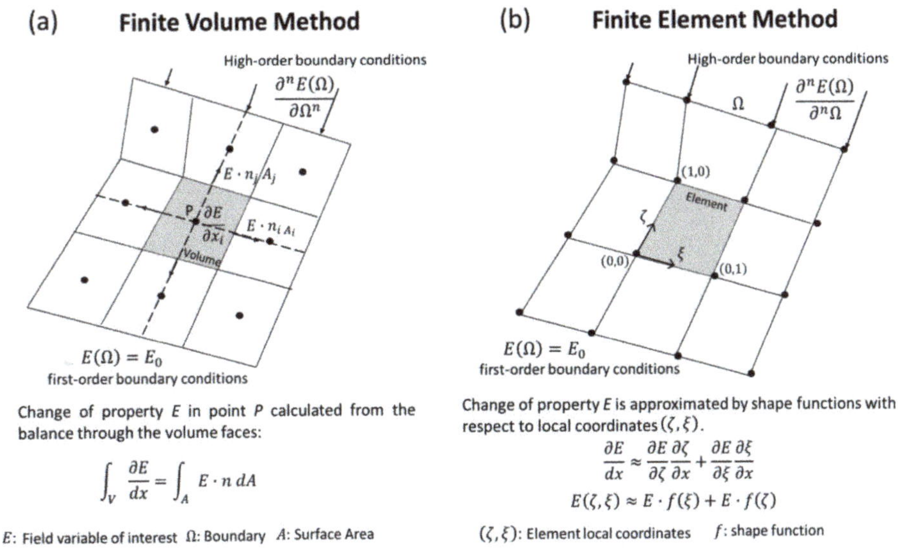

Figure 4. Diagram of finite volume (FVM) and finite element methods (FEM) approximation principles. (**a**) In FVM, the domain is discretized in finite volumes, and balance equations are solved at the center of each volume. (**b**) In FEM, the domain is discretized in finite elements, and the variables distribution is assumed to follow a prescribed shape function within each finite element.

The use of FVM for the solution of convection-diffusion problems was first introduced in the early 1960s by Tikhonov and Samarskii [86]. Since then, FVM has been particularly successful in its application to computational fluid dynamics, as many of the current commercial computational fluid dynamics (CFD) software suites are based on this method. Biomechanical applications of this method mostly focus on hemodynamic and tracheobronchial airway simulations. However, this method can be applied to other boundary value problems such as electromagnetics and structural mechanics [87,88].

3.2. Finite Element Method (FEM)

The finite element method (FEM) consists of the discretization of the domain of study on simple geometrical elements (or finite elements), where the unknown fields are discretized as linear combinations of shape functions of any order, linear and quadratic being the most common. The shape functions are typically defined at each element depending on local and normalized coordinates (Figure 4b). The local governing equations for each element are then assembled and organized in a matricial system of algebraic equations. Finally, the solution is approximated by minimizing the weighted error associated with each element. Several weighting rules have been proposed, the Galerkin method and its variations being the most widely used [89]. By converging to the solution through the minimization of an error function and not through the exact solution of balance equations, FEM is said to be formulated in a "weak" form. However, the weak form is equivalent to the exact solution in the limit of refining the domain discretization. In fact, it has been widely shown that mesh-independent FEM solutions do not show any practical difference from the output of more conservative numerical methods such as FVM [90,91].

FEM was developed in the early 1950s to perform structural analysis for the aerospace industry and was soon applied to study the biomechanics of musculoskeletal and cardiovascular tissue [81,82]. As early as 1968, FEM was used to study the non-linear viscoelastic behavior of arteriole tissue [92]. This technique has been traditionally used for the solution of solid mechanics problems; however, it has also been used to solve the governing equations of other physical phenomena, including fluid mechanics [81]. Regarding the convenience of relying on a single solver engine, many multiphysics simulation software suites have introduced FEM formulations for fluid mechanics, [93] which also facilitates the implementation of FSI simulations [94].

3.3. Summary

In Table 2 we summarize the highlights of the principles and formulations of the numerical methods typically applied on cardiovascular biomechanics research.

Table 2. Summary of numerical methods.

Section	Highlights
3.	Mechanical analyses of complex biological systems require the application of numerical methods to obtain approximate solutions. There are several numerical methods available to solve the governing principles of continuum mechanics, including popular mesh-based methods such as the finite volume method (FVM) and finite element method (FEM). Mesh-based methods discretize the domain of study on spaces of finite size and iteratively solve the governing equations on each finite space simultaneously. FVM is based on a "strong" formulation that solves exactly the balance equations on the center of each finite volume. FEM is based on a "weak" formulation that assumes the unknown variable to follow a prescribed shape function within each finite element. The method converges to the solution by minimizing the weighted error induced by the discretization and use of the shape functions. FVM and FEM offer equivalent solutions to a variety of Multiphysics problems.

4. Inverse Problems

Modeling physical phenomena can be thought of as a mapping operation, where a set of inputs (B) is transformed into a set of outputs (E) by applying the model operator (M) such that $M(B) = E$. In the realm of physics, there must be a cause–effect relation between the inputs and outputs, and forward modeling consists of designing and applying a mapping function capable of producing outputs that closely follow experimental measurements (e), meaning that the difference $E - e$ should be close to zero (Figure 5a) [95].

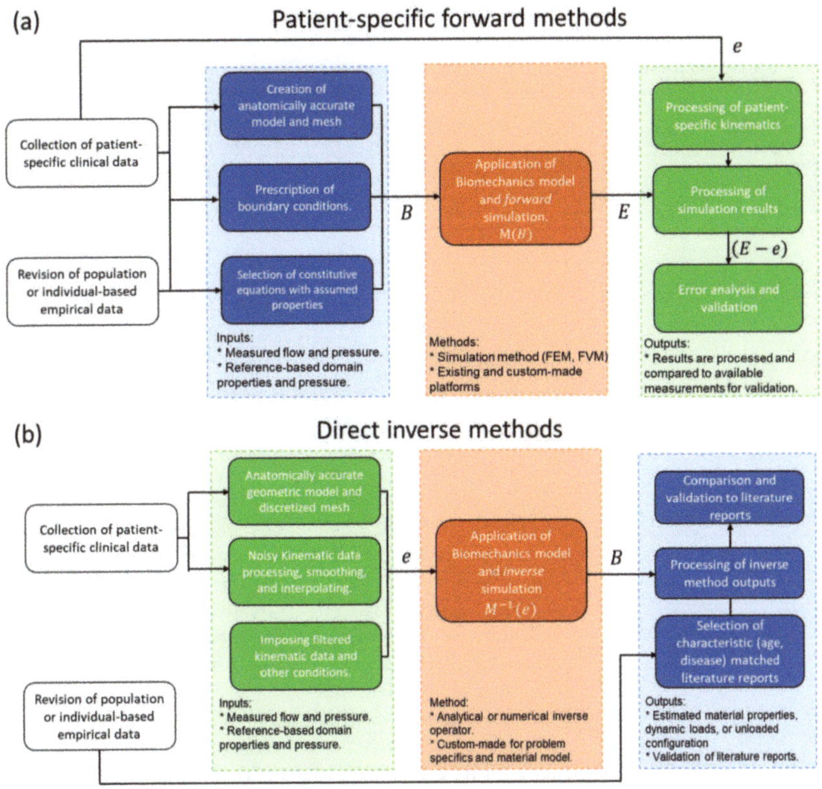

Figure 5. Data processing pipeline for patient-specific (**a**) forward problems and (**b**) direct inverse problems. Symbols, B: forward problem inputs, E: forward problem outputs, e: experimental data.

Conceptually, solving an inverse problem consists of using the measured effects to estimate the causes. That is, solving a problem of the type $B = M^{-1}(e)$, which could be straightforward if M was a bijective function, with M and M^{-1} being continuous and differentiable, and e was a continuous and smooth distribution (Figure 5b). The main difficulties with inverse problems are the possible nonlinearity of the inverse mapping function, the multiplicity of solutions, and the sparsity and noise of the measured effect data [15,95].

The development of advanced measuring techniques along with advances in computer science brought attention to the practical applications of inverse problems. A growing body of research has been built to address the afore-mentioned difficulties and to apply the inverse modeling methodologies to problems from many different engineering applications. To attend to the necessity of opening wide discussion of concepts, methodologies, and methods related to inverse formulations, specialized journals started circulating by the late 80s, e.g., the Inverse Problems and Inverse Problems in Engineering Journals (today

Inverse Problems in Science and Engineering) among many others [95]. In this section, several solution methodologies for inverse problems in mechanics are reviewed, highlighting their respective advantages and drawbacks when incorporated into cardiovascular biomechanics analyses.

4.1. Direct Inverse Methods

The direct solution of inverse problems by the deduction of the inverse mapping function (M^{-1}) is only possible for oversimplified cases; however, specialized mathematics has been developed for the direct solution of some specific problems. In the case of finite elasticity, one relevant inverse problem is the retrieval of the mechanical properties of the domain of interest from the applied loads and measured displacement field. Several methods have been proposed to solve this problem directly, e.g., the reciprocity gap method which has been used to retrieve the distribution of elastic properties and to resolve the location of cracks in solid bodies from image-derived displacement fields. This method linearizes the inverse problem by assuming that the same elasticity tensor can resolve both the measured displacement field and a slightly perturbed version of it [15]. Another alternative for inverse elasticity is the application of the virtual work principle. This requires a complete description of the deformation field as a starting point, then the virtual work identity is defined by arbitrarily selecting a virtual field function. These functions can be tailored to specific constitutive equations to convert the virtual work identity into a set of algebraic equations from which the components of the elasticity tensor can be resolved [96,97].

Another relevant problem on inverse elasticity is the resolution of the unloaded geometric reference configuration with the applied loads, material properties, and deformed configuration as inputs. This problem has many applications in manufacturing engineering and is key to the study of patient-specific biomechanics. Govindjee and Mihalic proposed a finite element implementation for the direct solution to this problem [98,99]. The proposed method exploits the duality of the equations of finite hyperelasticity when the role of the reference and deformed coordinates are interchanged. In the absence of body forces and assuming material homogeneity, the FEM implementation of the inverse problem can be formulated similarly to the conventional FEM problem, requiring slight changes in the definition of elements and shape functions. The authors highlight that the application of the method to buckling problems can lead to multiple solutions for a given input and the resulting method was highly sensitive to input variations.

Direct solutions to inverse problems are computationally efficient; however, solution methods are not generalizable and need to be tailored for each specific inverse problem and each constitutive equation, making its implementation on existing numerical solvers a non-trivial process [100]. Another limitation of direct solutions is the requirement of a smooth continuous function of the measured effects (e), which cannot be satisfied by discrete empirical measurements affected by random error. The experimental variability and noise of the input data could be incompatible with the assumed model, which could dampen the convergence to a valid solution or any solution at all [95]. An option to deal with this issue is through preprocessing of the input data with smoothening and interpolation operations.

4.2. Iterative Inverse Methods

An alternative method for the solution of inverse problems is the iterative approach. This consists of optimization algorithms that iteratively solve the forward problem while varying the input parameters (B) until an error function defined as the difference between target experimental data (e) and the forward problem output (E) is minimized [101]. The advantages of this method are its easy generalization to any kind of inverse problem, its capability to operate on top of any existing solver for the forward problem, the existence of methods to reduce solution multiplicity, and its inherent capacity to handle scattered and noisy experimental data (Figure 6) [102]. All these advantages come with a detrimental increase in computational resource requirements, given by the repetitive solution of the forward problem. To reduce computational expense, some formulations have proposed the use of surrogate simpler models for the forward problem at the initial stages of the optimization process [103,104]. Some statistical tools based on Bayesian data analysis and inference have been implemented to improve the performance of iterative inverse methods for cases where the error distribution of the measured target is known or can be safely assumed [105].

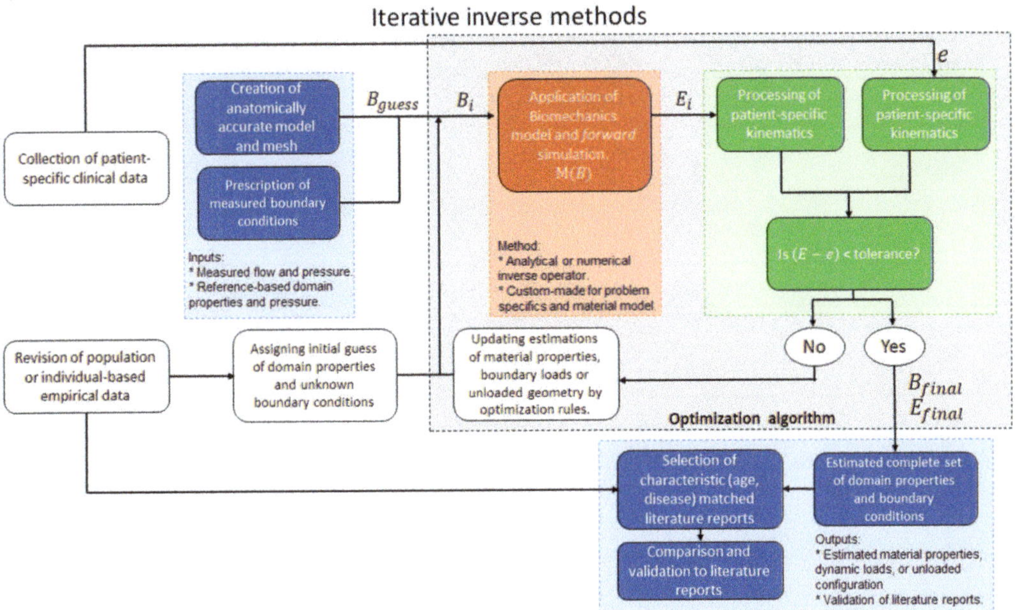

Figure 6. Data processing pipeline for patient-specific iterative inverse methods. Symbols, B: forward problem inputs, E: forward problem outputs, e: experimental data.

The core concept of iterative inverse methods is the solution of an optimization problem that drives the solution into reproducing the target data with the smallest possible error. The definition of an optimization problem requires the selection of an appropriate target function to minimize, an optimization algorithm suited for the particular characteristics of the forward problem and optimization parameters, and the implementation of parameter constraints to point and restrict the convergence of the algorithm into desirable outputs. In this section, we briefly discuss the most common optimization target functions, algorithms, and constraints used on inverse problems in biomechanics.

4.2.1. Target Function

Most inverse studies implement a single target function for optimization, although optimization of multiple targets is feasible. The target function is often defined in terms of an error between the forward problem output and experimental measurements. Since the target function is defined as an error to be minimized but not exactly reduced to zero in a point-wise fashion, the iterative inverse method can converge to reasonable solutions even if the measured data are scattered and affected by random error.

The selection of an adequate target function must comply with at least two conditions. First, the target function must be compatible with the solution space of the forward model; otherwise, convergence may never be achieved, e.g., pulse wave velocity is incompatible as a target with CFD models assuming rigid walls. Second, the target function must be representative of all the aspects of the modeled phenomena, e.g., a target function based solely on output flow estimation is not adequate for FSI models since the contribution of the elastic behavior of the wall can be miscalculated. In FSI studies, pulse wave velocity or multiple target functions dealing with pressure and flow velocity are more reasonable options [106].

Structural Tissue Mechanics

First attempts to assess patient-specific mechanical properties of blood vessels were based on the knowledge of pressure vs. volume or pressure vs. area changes. Nevertheless, this unidimensional information can only be used to fit simple models that assume material homogeneity and isotropy, which departs from the known complexity of myocardium and arterial tissue [107]. The least-squared error to a pressure-volume curve as an optimization target increases the number of comparison points and allows the fitting of non-linear material models [108]. However, pressure–volume curves can only be measured in practice through invasive catheterization and are not generally available on a patient-specific basis. Alternatively, such data could be obtained from other sources such as normalized models with self-similarity or statistically obtained atlases. For example, Klotz et al. found that normalized pressure–volume curves of the left ventricle (LV) have a consistent profile, regardless of etiology across large mammal species [109]. This normalized pressure–volume function has been extensively used for forward and inverse analyses of the LV when direct pressure measurements are unavailable [110].

Some studies use the high-resolution information from MRI or computed tomography (CT) to obtain accurate geometric models of arteries at diastole and systole. The diastolic configuration is discretized into the mesh, which is then mapped into the systolic geometry by incorporating kinematic assumptions such as negligible axial and torsional displacements [111]. Then, the target optimization function can be set to minimize the simulated to mapped displacement errors. Due to the non-uniform distributions of nodal displacements, this technique allows the estimation of heterogeneous distribution of stiffness from anisotropic material models [112]. To avoid the requirement of node-to-node correspondence, and the related displacement assumptions while using geometric (non-kinematic) information, some authors proposed the minimization of the least-square-error of the distance between the loaded/deformed simulation mesh to the surface of the segmented anatomy model at systole, or systolic shape matching [113,114].

The explicit displacement field distribution made available by ultrasound speckle tracking, MRI tagging, or DENSE MRI, allows for defining a more direct target function by minimizing the least-squared error of the nodal displacement between the simulation results and the image-based measurement. The direct comparison of simulation-to-measured displacements can be achieved by locating mesh nodes in the location of speckles/tags/voxels, or by interpolating the measured displacement field into the mesh [14]. To reduce the effect of noise, some authors prefer defining the target function in terms of a region-wise averaged strain, instead of displacement distributions [115]. By averaging the strain field over a region, the effect of the noise is dampened. However, this requires an adequate discretization of the domain on regions of similar boundary conditions and material properties, while keeping discretization regions small enough to produce a smooth distribution of strain estimates. The use of region-wise averages of strain is widely used in the analysis of heart mechanics, and there is even a standardized discretization of the left ventricle. However, defining adequate regions for smaller and thin vessels is challenging [13]. Furthermore, defining the target function solely on strain measurements rules out the effect of possible translational/rotational rigid body motions.

The use of stress fields as target data can potentially reduce solution multiplicity on the fitting of material parameters, and yield results that more accurately describe the mechanical behavior of the tissue under study. The definition of such targets, however, requires a priori knowledge of the boundary loads, and resulting stress distribution within the deformed domain of study. In practice, some controlled in vitro experiments have successfully applied inverse models with stress-based targets by having accurate measurements of forces and deformations in three orthogonal directions on samples of reduced size [18,116,117]. The definition of stress-based targets for patient-specific in vivo applications could be extremely beneficial to improve the accuracy and uniqueness of the solution. However, it would require the implantation of load sensors on and within the tissue of interest. Given that in vivo tissue samples are not isolated, as they are in controlled in vitro experiments, further assumptions on material behavior and boundary conditions are required.

Fluid Mechanics and FSI

Cardiovascular catheterization pressure measurement is considered the reference standard on patient-specific hemodynamics, as it constitutes a direct assessment of pressure and dimensions within the blood vessels or the cardiac cavities using high-accuracy transducers. When available, most inverse models of computational fluid dynamics use a least-square-error of the time-dependent pressure function as the optimization target, while image-based flow data is used as boundary conditions [103,118,119]. Models incorporating FSI can instead use the pulse wave velocity as an optimization target that accounts for both the hemodynamics and elastic properties of the vessel [120]. The carotid-femoral pulse wave velocity is considered the gold standard for systemic arterial stiffness assessment, which is calculated as the patient-specific distance between the carotid and femoral artery and the time delay between the pressure wave measured at those locations. Local estimations of pulse wave velocity can also be obtained from invasive catheterization and by flow-to-area ratios from doppler ultrasound or phase-contrast MRI [121]. However, CFD cardiovascular modeling often relies on rigid wall simplification which significantly reduces the computational cost of the forward problem. Furthermore, cardiac catheterization is an invasive procedure and may not be available, so target pressure data is either non-available or non-reproducible owing to model limitations. In these cases, either least-square-error of nodal velocity between simulation results and 4D flow MRI assessment or branch flow distributions have been used as target functions [122].

4.2.2. Optimization Algorithms

The development of algorithms for numerical optimization is a broad and active field of research. It is not the aim of this article to carry out a comprehensive review of all the available optimization techniques, but rather to list the methods most commonly used in the field of biomechanics, providing a rationale for their selection with specific problems. In the following sections, we loosely follow the classification proposed by Kochenderfer and Wheeler based on the characteristics of the target function [123]. Given the nature and complexity of inverse biomechanics problems, we only consider optimization algorithms that deal with continuous variables and multiple optimization parameters.

Updating by Differentiation of Target Function

In those cases where the target function is continuous and derivable, derivative information can be used to estimate the descent path towards the minimum. First and second-order algorithms refer to optimization methods that incorporate numerical evaluations of the local Jacobian and the Hessian matrix, respectively. To reduce the risk of convergence to local minima, stochastic sampling of these derivatives is incorporated.

First-order algorithms can only deal with relatively simple problems and are not suited for inverse biomechanics. However, they are used in other relevant applications, such as the automation of image processing and segmentation for the generation of geometric models [11]. Second-order algorithms have been used in the solution of inverse arterial and myocardial mechanics for the estimation of anisotropic material constants. The most commonly used are the Levenberg–Marquardt [34,107], Broyden–Fletcher-Goldfarb–Shanno (BFGS), limited BFGS (L-BFGS), and sequential quadratic programming [5].

Updating with No Differentiation of Target Function

Some optimization algorithms do not require derivative information of the target function to operate. These methods are not as fast as gradient-based counterparts when applied to derivable functions. However, they are advantageous in cases where the functions are not derivable, there are regions with invalid solutions or singularities, the function response is noisy, or the target presents multiple local minima. Since the target functions on simulation-based inverse problems are not analytical functions, but instead, are the simulation outputs, it is prone to some numerical problems, e.g., non-valid solutions due to forward problem divergence. In consequence, most recent inverse method developments have incorporated gradient-free algorithms. The most common algorithms can be classified into two groups direct methods and population methods.

Direct methods incorporate deterministic algorithms based on patterns or geometrical constructs for sampling the domain and carry a direct comparison of the target function value. This comparison is then used to define the location of the next sampling point. Powell and Nelder–Mead algorithms have been particularly popular in biomechanics applications and inverse analyses [124,125].

The main feature of population methods, in contrast to direct methods, is that the initial seed is not a single point in the parameter hyperspace but a pool of candidate optimum solutions. On each iteration, a new pool of candidate solutions is generated by altering the input parameter values following different recombination rules from parent candidates and stochastic variations. Then, each new candidate is evaluated and a new pool is selected to build the next generation. These algorithms have proven to be particularly useful when dealing with noisy target functions, and with multiple local minima. The popular genetic algorithms and particle swarm methods stand out due to their multiple applications, including the solution of inverse problems [117,126–128]. These methods require intensive sampling, so they are contraindicated for the solution of inverse problems when the forward problem is computationally expensive [15].

Statistics-Based Methods

Some statistical methods applied in the field of biomechanics rely on the use of Bayesian inferences, also known as inverse probability. Unlike the methods described previously, Bayesian inference-based methods provide not only an estimation of the parameters to be fitted but also a confidence interval for such values. The method requires a set of measured data along with its probability distribution (which can be often assumed normal due to random experimental error), a predictive model, and a prior probability distribution for the model parameters. The latter can be estimated from previous experiments and published data or can be simply assumed as uniform within a given range. Then, a selection of parameter combinations is used to run the prediction model and compare it to the experimental data. Finally, the Bayes theorem is used to produce a map for the probability of the model that reproduces the experimental data in a parameter hyperspace. This map is used to update the prior parameter probability distribution to iteratively repeat the process [129].

There are many different computational implementations, some of the methods used on patient-specific inverse problems are the Gaussian process regression, Kalman filters in its many variations, and linear-quadratic-Gaussian estimations. These methods differ mostly on how the sampling is carried out, how the parameter probability distribution is assumed or calculated on each step, and how the model predictions and experimental measurements are weighted to determine the converged parameter solution [130–132].

4.2.3. Constraints

A series of constraints can be implemented on the optimization algorithms to restrict solution spaces and parameter values. These constraints can be used to ensure the physiological and physical meaning of the results and to funnel down solution multiplicity. One of the most relevant constraints required for material parameter estimation on tissue mechanics is compliance with the second law of thermodynamics. One of the required conditions for this compliance is that the strain energy function must be positive convex, which restricts the relative value of material parameters [24,133].

The assumption of material incompressibility is another common constraint imposed on cardiac and arterial wall mechanics. Full incompressibility introduces singularities to the solution of numerical formulations; therefore, nearly incompressible behavior is enforced by restraining the relative values of material properties. However, experimental and in silico evidence have shown that cardiovascular tissues are compressible to some degree and that myocardial volume varies throughout the cardiac cycle [134]. The most recent in vivo measurements of myocardium compressibility in human and large-mammal animal models agree on estimating peak compressibility between 1% and 20% [134–136] Moreover, it has been shown that the accuracy of heart-mechanics models is significantly increased if this compressibility effect is considered [137]. Thus, the incompressibility constraint is a reasonable yet rough approximation that must be carefully considered in simulation analyses [134,138,139].

Microstructure-based models allow the introduction of physiologically and structural meaningful constraints to material parameters, e.g., maximum possible fiber stiffness, or maximum cellular volume fraction. Inequality type constraints can restrain material parameters within expected physiological ranges. Inequality relations between model parameters can also be introduced to address structural component differences, e.g., collagen fibers are typically stiffer than elastin fibers. In the study of FSI inverse problems, pulse wave velocity is constrained by the maximum possible speed of sound on the liquid media, and some authors have introduced constraints on the maximum volume change of the fluid-solid domain [140].

In addition, constraints can also be introduced to promote numerical stability of the solution, or to smooth the solution when the parameters to be fit are temporal or spatial distributions, e.g., the first-order Tikhonov regularization functional has been used in the estimation of heterogeneous material parameter distributions [141].

4.3. Summary

In Table 3 we summarize the highlights of direct and iterative solution methods of inverse problems.

Table 3. Summary of inverse methods.

Section	Highlights
4.	Solving an inverse problem consists of using measured effects to estimate the causes. The main difficulties with inverse problems are the possible nonlinearity of the inverse mapping function, the multiplicity of solutions, and the sparsity and noise of the measured effect data.
4.1	The direct solution of inverse problems by the deduction of the inverse mapping function is only possible for simple cases. There are specialized mathematical solutions for specific problems of finite elasticity. Some relevant problems of inverse elasticity that have direct inverse solutions are (1) the solution of material properties from boundary loads and domain displacements. (2) The solution of the unloaded configuration from the applied loads, material properties and deformed configuration. Direct solutions of inverse problems are computationally efficient. However, direct solutions are not generalizable and require continuous smooth functions of the measured input often not compatible with noise and scarce experimental data.
4.2	Inverse problems can also be solved through an iterative weak approach. This consist of iteratively solving a forward simulation problem to minimize an error function between simulation outputs and target measurements while fitting the sets of unknowns. The iterative solution methods of inverse problems are generalizable, can handle noisy and scarce experimental target data, and can operate on top of existing simulation software. However, iterative methods are computationally expensive. The selection of the target function to be minimize needs to be consistent with the nature of the problem and the characteristics of the biomechanical model. The inverse method can be implemented though a variety of optimization methods. For the solutions of biomechanical inverse problems, optimization methods with no differentiation of the target function are preferred. Population-based optimization algorithms can solve global minima of multiparametric functions with an increased toll of computational expense. Statistic-based optimization method can incorporate previously reported data which can reduce convergence time and provide probability distributions of results rather than single deterministic values. Convergence times can be improved and solution multiplicity narrowed by the implementation of solution constraints. Constraints can be based on physical laws and limits or on previous experience. Constraints can also be implemented to promote numerical stability and smoothness of the converged solution.

5. Medical Imaging-Based Kinematics

Early attempts to use medical imaging to assess the stiffness of blood vessels relied on the measurement of the luminal area change between diastolic and systolic configurations. This area change is used in several clinical risk markers, such as the β-index, that have shown a good correlation with the occurrence of certain cardiovascular pathologies such as atherosclerotic damage, hypertension, diabetes, and Marfan syndrome, as well as to tobacco exposure, obesity, aging, and other risk factors [142–144]. However, the predictive capabilities of these factors are inconsistent among different arterial locations and pathologies, most likely due to the oversimplification of the problem without any account of vascular mechanics [142].

Multiple previous studies have considered inverse problems applied to in vitro marker-tracking kinematics of surgical and cadaveric tissue samples [145,146]. In these works, direct or fluid-driven mechanical loads are applied to the tissue sample to induce controlled deformation through an in vitro experiment setup. Physical or digital markers are fixed to the samples, and their displacement is captured by high-speed, high-resolution cameras. These studies are less affected by resolution limitations and noise than in vivo studies and can be applied to structures that are difficult to capture with medical imaging such as heart valve leaflets [147]. Some notable drawbacks of in vitro testing of explanted tissue include neglecting active contractility, loss of in vivo boundary conditions, potential tissue damage during excision, experimental setup and marker placement, and degradation of the living tissue after extraction.

In vivo medical imaging has evolved to provide not only anatomical geometric information but also detailed kinematics measurements. The accuracy and availability of these techniques are limited by image resolution, signal-to-noise ratio (SNR), the occurrence of artifacts, and practical obstacles related to testing costs and health hazards [122,148–151]. In the following subsections, we review the available techniques for assessing in vivo image-based kinematics for tissue deformation (Table 4) and blood flow (Table 5), fundamental principles, typical image resolution, and some specific applications.

Table 4. Image-based technique for assessment of tissue motion.

Technology	Technique	Principle	Resolution	Applications
US	Speckle Tracking	Acoustic response to the interaction of ultrasound signals with tissue fibers.	Spatial and displacement resolution < 1 mm/pixel Real-time temporal resolution.	Identification of: septal defects, CHD, valve structure. Assessment of cardiac and aortic function.
MRI	Tissue tagging	Local perturbation of myocardium magnetization with selective radiofrequency saturation sequences	Spatial and displacement resolution ~1 mm Tag spacing ~4 mm 25 images per cardiac cycle.	Assessment of cardiac function; motion and deformation of myocardium, skeletal muscle, lung tissue and tongue.
	DENSE MRI	Applied magnetic field gradients produce a phase shift on proton spins proportional to its relative displacement.	Pixel size ~2.5 mm for myocardial motion [149], ~1.3 mm for aortic motion [150] Displacement resolution < 0.1 mm. 30 images per cardiac cycle	Assessment of myocardial and aortic motion, deformation, and function.

5.1. Ultrasound Technology (US)

Ultrasound (US) uses high frequency (2 to 15 MHz) acoustic waves to create real-time 2D in vivo images of tissues, organs, and blood pools using piezoelectric transducers. As in any wave, higher frequencies are associated with smaller wavelengths, higher penetration power, and improved image resolution [152,153]. Volume rendering from ultrasound images has led to three-dimensional, time-resolved ultrasound (4D US), and real-time imaging [154]. US is relatively inexpensive, portable, and safe, so it has become a customary tool in many clinical applications such as anesthesia, critical care, prenatal care, and pain management. Its application to cardiology, commonly known as echocardiography, was introduced in the 1950s, and currently is the more ubiquitous diagnostic tool to assess cardiovascular structure and function. With an approximate lateral resolution of 1 mm/pixel, this technique allows the estimation of heart chamber size, valve structure, identification of structural abnormalities such as seen in congenital heart defects (CHD), and determination of systolic and diastolic function (Figure 7c) [155]. Nevertheless, echocardiography presents some intrinsic limitations regarding accuracy and repeatability, particularly in patients with complex flow patterns related to congenital heart disease, aortic regurgitation, or dissection, in which case it is recommended to complement the study with other imaging techniques [156]. Intravascular ultrasound technology (IVUS) was developed following the principle that accuracy and resolution are improved as the transducer is closer to the tissue of interest. This technology involves placing a miniature ultrasound probe at the end of a catheter and then introducing the catheter into the vessel of interest in order to resolve the surrounding structures with greater detail than allowed by standard external US. This invasive technique is mostly used to study the conditions and progression of atherosclerosis in patients with coronary and carotid artery disease [157]. US technology can also provide blood flow and tissue kinematic information through the use of echo-Doppler and speckle tracking techniques.

Table 5. Image-based technique for assessment of blood flow.

Technology	Technique	Principle	Resolution	Applications
US	Echo and Vector Doppler	Measurement of frequency shift of the reflected acoustic wave.	Spatial resolution <1 mm/pixel	Identification of: septal defects, CHD, valve structure. Assessment of cardiac and arterial function. Prenatal care.
MRI	2D PC	Applied magnetic field gradients produce a phase shift on proton spins proportional to its relative velocity.	Pixel size ~1.5 mm 30 images per cardiac cycle	Assessment of cardiac, arterial, and venous flow, cardiac output, regurgitant flow, pulse wave velocity.
	4D flow		Pixel size ~2.5 mm 25 images per cardiac cycle	Same as 2D PC plus measurements of wall shear stress, vorticity and pressure drop.

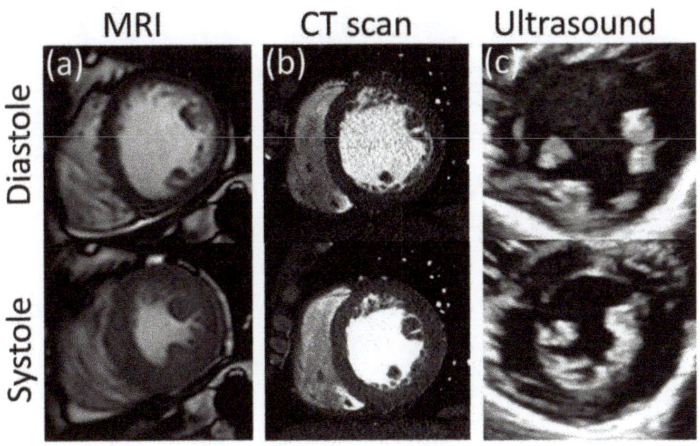

Figure 7. Resolution comparison of left ventricular myocardium at diastole and systole with clinical grade (**a**) MRI, (**b**) CT (Reprinted/adapted with permission from Ref. [158], 2019, Korean Society of Echocardiography, open access), and (**c**) 2D ultrasound.

5.1.1. Echo and Vector Doppler

Echo Doppler estimates the velocity of blood and tissue through the use of the Doppler equation. By measuring the frequency shift from the original ultrasound wave and the reflected echo, the local velocity can be determined. The main shortcoming of this technique is its dependence on the angle between the original ultrasound wave (position of the transducer) and the displacement direction, which can introduce large intra- and inter-observer variability. Dependency on the transducer angle was solved by the introduction of vector Doppler techniques, which additionally provide in-plane velocity components. This is achieved by the simultaneous measurement of two doppler signals, either from two crossing beams from different transducers, or from a single transducer with two different in-plane receivers [159]. This technique is of great use in clinical practice for the qualitative assessment of blood flow and tissue displacement [160] and used in early studies of patient-specific hemodynamics to impose inlet and outlet flow boundary conditions [161,162].

5.1.2. Speckle Tracking

Speckle tracking is a relatively novel technique developed during the early 2000s for the measurement of tissue 3D displacement and deformation. Speckles are defined as image features/spots generated by the acoustic response of tissue fibers to ultrasound signals. Single speckles are analyzed in identifiable kernels that are followed along the cardiac cycle. Postprocessing techniques allow the averaging of kernel displacements over several cardiac cycles to reduce the effects of noise [163]. The spatial and temporal resolution of speckle tracking is remarkable, providing hundreds of frames per second for pixels of <1 mm size [164]. This resolution allows not only the study of the myocardium but also the mechanics of arterial walls and aortic aneurysms [165,166]. Displacement measurements are limited by kernel size (~1 mm) and show reproducibility issues common to any US-based technology (Figure 8a) [167,168].

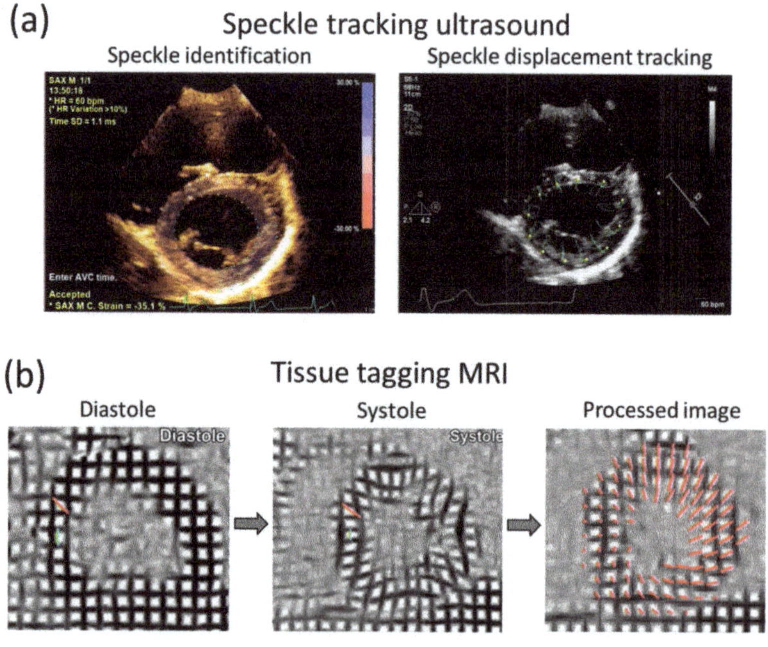

Figure 8. *Cont.*

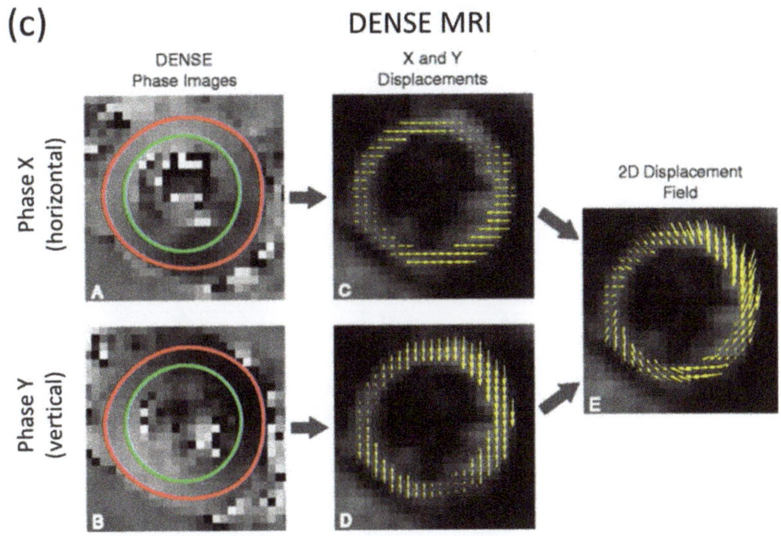

Figure 8. Examples of image-based kinematics of the left ventricular myocardium with (**a**) Speckle tracking ultrasound (Reprinted/adapted with permission from Ref. [169]. 2020, Alessandra M. Ferraro et al.; open access) (Left) Dots indicate speckle-kernel location and identification, (right) green lines indicate trajectory through the cardiac cycle. (**b**) Tissue tagging (Reprinted/adapted with permission from Ref. [170]. 2012, The Radiological Society of North America). Tissue tagging estimates kinematics by tag-to-tag tracking from diastole to systole. Transversal (green lines) and diagonal (red lines) tag-to-tag dimensions are measured at diastole (left column) and systole (middle column), their difference can be used to measure displacement and deformation (red lines in right column). (**c**) DENSE MRI (Reprinted/adapted with permission from Ref. [171]. 2015, Wehner et al.; licensee BioMed Central, open access). DENSE MRI resolves pixel-wise displacements by processing phase data for each direction. Red and green contours represent segmented luminal and adventitial boundaries, yellow arrows represent the phase-encoded displacement.

5.2. Magnetic Resonance Imaging (MRI)

Magnetic resonance imaging (MRI) offers superior quantitative utility compared to ultrasound as it offers high-resolution 2D and 3D visualization of the heart and major arteries referenced to a fixed coordinate system, providing greater accuracy for anatomic and volumetric assessment of heart chambers and wall thickness [160]. In addition, various MRI sequences have been specially designed to assess other valuable data such as fiber orientation, tissue displacement, and blood velocity [172,173].

The fundamental principle of this technology consists of the use of magnetic fields to align hydrogen protons in the body. After the magnetic field is interrupted, the protons return to a lower energy state by emitting radio signals that can be captured and utilized to create imaging data. For clinical applications, base magnetic fields with strengths ranging between 1.5 to 7 T are used to excite the protons to a base level. Then, a second time-varying radiofrequency magnetic field is used to induce changes in tissue magnetization. Tissues with different hydrogen-protons content will respond to the oscillating radiofrequency with different characteristic responses, which can be used to resolve various tissue components [174].

The first reported human magnetic resonance image dates from 1977 and was a single image that required 5 h to capture [175]. Nowadays, an MRI image can be captured in a single breath-hold with a resolution under 1 mm/pixel and temporal resolutions of 25 frames per cardiac cycle (Figure 7a). This noninvasive technique generally poses minimal risk to patients unless they have non-compatible ferromagnetic implanted devices or other internalized materials, or they have complications related to the contrast agents needed for some of the MRI modalities. Due to the confined space within an MRI scanner, the test can generate anxiety and discomfort for patients with claustrophobia. Given the expense of the equipment, maintenance, and required staffing, it does not have as wide an availability as US, particularly in smaller medical facilities.

There are several options for MRI-based assessment of kinematics of both cardiovascular soft tissues and blood flow. Four prominent examples will be discussed below, including: MRI tagging and displacement encoding with stimulated echoes (DENSE) for tissue displacement, and phase contrast and 4D flow MRI for blood flow quantification.

5.2.1. Tissue Tagging

This technique, first introduced by Zerhouni et al., in 1988, was specifically designed to quantitatively assess the transmural motion of the myocardium [176]. Image markers, or tags, are created by locally perturbing the magnetization of the tissue, either by selective radiofrequency saturation sequences or through modulation of the magnetization vector by gradient fields. Tags are created on a thin section orthogonal to the image plane at diastole, then followed by regular time-resolved imaging. Electrocardiographic gating is used to consistently apply the tagging radiofrequency at diastole. Early works reported decrement of tag resolution at systole as the magnetic saturation exponentially decays over time; nevertheless, this problem is palliated with the use of larger magnetization energy [177]. Ibrahim et al. showed that tag lines were still clearly identifiable at the end of the cardiac cycle on human hearts with a 7 T MRI scan [178]. Special radio-frequency sequences, such as spatial modulation of magnetization (SPAMM) and delays alternating with nutation for transient excitation (DANTE) [179,180], allowed the creation of 2D orthogonal tagging grids that facilitate the kinematic analysis. Tag sizes can be only as small as the pixel-size resolution (>1 mm) with typical tag spacing of about 5 mm (Figure 8b). The technique allows for 25 to 30 images per cardiac cycle, requiring about 20 s of scan time per tagging sequence [170,181]. Multiple studies on phantoms have shown this technique to be superior to US speckle-tracking in terms of accuracy and repeatability. Due to the spatial resolution limitations of this technique, it has only been successfully applied to study the kinematics of relatively thick tissues such as the myocardium, skeletal muscle, lung tissue, and the tongue [182].

5.2.2. Phase-Contrast

Phase-contrast (PC) MRI utilizes the intrinsic phase of the magnetic signal to retrieve kinematic information. When a magnetic field gradient is applied to a body, the spins of the protons develop a phase shift that is proportional to its relative velocity. When two consecutive and opposing gradients are applied, stationary protons will show no phase shift. However, moving protons will show different degrees of phase shifting as they change their position with respect to the gradient [183]. This information can then be used to encode the velocity and displacement of protons.

Since the kinematic information is encoded in the phase information, and thus is independent of image markers, this technique allows measurement at scales below pixel-size resolution [12,149]. However, the technique is sensitive to Eddy currents, concomitant gradients (Maxwell terms), and nonlinearities in the gradient field. These effects increase the signal-to-noise ratio (SNR) and produce offset errors that are both spatial- and time-dependent. SNR has been shown to increase with the magnetization energy and has been estimated to range between 20 for 1.5 T scanners to around 60 for 7 T scanners. Offset error correction requires the implementation of rectification algorithms in the postprocessing stage [184,185].

2D CINE PC-MRI and 4D Flow MRI

The application of consecutive opposite magnetic field gradients is known as a bipolar gradient. After a bipolar gradient is applied, the net phase shift of static protons is zero, so only the mobile protons will show a phase shift. From the latter, faster protons will experience a greater difference in applied gradients as they physically move longer distances than slower protons, which in turn produces greater phase shifts. The end result is that the phase shift is proportional to the proton velocity. However, because phase angles are limited (from 0 to 2 π), only a certain range of velocities can be directly quantified [183]. That is, for a given gradient, there is a maximal velocity that can be measured before aliasing occurs, called the encoding velocity (VENC). Encoding velocity is inversely proportional to the magnitude of the gradient; thus, by manipulating the strength of the gradient, it is possible to manipulate the range of velocities that can be encoded. Setting the encoding velocity is a tradeoff between the risk of aliasing and the minimum measurable velocity by the discrete scale [186].

Standard 2D Cine PC MRI, typically applied to estimate through-plane velocity, has become part of clinical practice in the treatment of cardiovascular disease, specifically, for the calculation of flow in large arteries and their main branches, cardiac output, and quantitative assessment of regurgitation and retrograde flows. This technique is also used for the qualitative assessment of flow patterns in large arteries and heart chambers. PC MRI data is usually recorded in DICOM format images with 8-bit or 16-bit pixels, that is 256 or 65,536 possible discrete levels, respectively. Phase data is typically encoded within 4095 values for the whole 0 to 2 π range, with pixel sizes around 1.5 mm [187].

However, standard 2D PC MRI can only provide the dimensional component of velocity perpendicular (through-plane) or parallel (in-plane) to the imaging plane, and thus is inadequate to estimate relevant hemodynamic metrics requiring three-dimensional flow information, such as vorticity and wall shear stress [188,189]. The logical evolution of this technique led to 4D flow MRI, which allows the volumetric and temporal resolution of three orthogonal components of velocity. This is achieved by applying consecutive bipolar gradients to three orthogonal directions on stacked planes. This requires the collection and processing of a significantly greater amount of data (three spatial dimensions and three velocity directions over several timesteps through the cardiac cycle), thus requiring special approaches to keep reasonable scanning times. Some hardware improvements include multi-receiver channels, phased-array coils, and parallel imaging technology. Other developments are related to improving the efficiency of data sampling, and averaging over several cardiac samples, namely radial undersampling, kt-GRAPPA, kt-BLAST, and kt-SENSE [190]. Additionally, the convex gradient optimization technique offers improved resolution and accuracy while maintaining the essential characteristics of velocity encoding [191]. For thoracic and abdominal applications, 4D flow scanning times range from 5 to 15 min, with voxel sizes of around 2.5 mm and temporal resolutions of 25 datasets per cardiac cycle (Figure 9). This technique has proven its value through many different in vivo patient-specific studies of normal and pathological hemodynamics in the heart [192], aorta [58,193], pulmonary artery, and complex single ventricle circulation [194–196]. Other applications include the evaluation of drug treatment effects [197] and surgical intervention outcomes [198–201].

Displacement Encoding with Stimulated Echoes (DENSE)

DENSE MRI is a modified version of PC MRI that improves phase contrast to measure slow velocity displacements while maintaining moderate gradient magnitudes, thus allowing the kinematic measurement of slow-moving tissue. This is achieved by manipulating the spin phase with stimulated echoes [149].

DENSE MRI was introduced in 1999 by Aletras et al., for the study of myocardial mechanics [149]. Since then, multiple developments have been proposed to optimize its assessment of human myocardial kinematics [202–204], minimize the effects of artifacts and breathing [205,206], and automate the unwrapping of DENSE MRI phase data [207]. Some potential clinical applications of DENSE MRI include the identification of biomarkers of early cardiac dysfunction [208,209], assessment of the response to cardiac resynchronization therapy [210,211], and identification of infarct transmurally for early postmyocardial infarction [212]. This technique has also been applied to assess the heterogeneous displacement, stretch, and circumferential strain around the aortic wall at different locations along its length [10,13,213,214]. Notably, a recent in vitro validation study of aortic DENSE MRI on wire-embedded polymer aortic phantoms revealed a final mean regional error in the quantification of the circumferential strain of <1% strain [215]. The potential for measuring in vivo shear and radial strains of the aortic wall has also been explored, though its repeatability is significantly less than the quantification of circumferential strain due to the thinness of the vascular wall [216]. Beyond the heart and aorta, other applications of DENSE MRI include the dynamics of the human brain and the cerebrovasculature [217].

The image resolution of typical cardiac DENSE MRI applications is around 2.5 mm/pixel; however, recent advances with the use of spiral k-space sampling DENSE MRI allow resolutions down to 1.3 mm/pixel to assess the kinematics of arterial walls [10,13]. Since the displacement is encoded in the MRI phase data and thus does not depend on tag-line parameters or tracking of image features, displacement is resolved at a scale below pixel size, with reported displacement uncertainties of approximately 0.09 mm [150]. Comparative studies on controlled in vitro experiments with gelatin phantoms, and in vivo strain measurements on human myocardium showed that DENSE MRI provides better accuracy and reproducibility than tissue tagging (Figure 8c) [181,183].

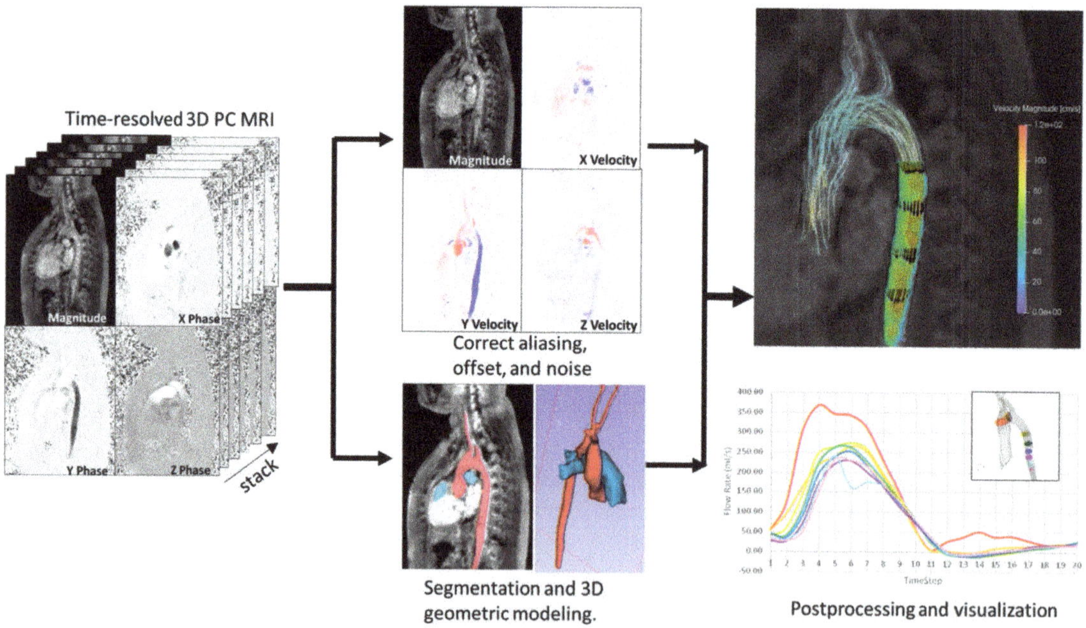

Figure 9. Processing pipeline of 4D flow MRI scans. (**Left**) Velocity-sensitive phase images are generated by 3D velocity-encoding subtracted from reference images. (**Middle**) Velocity estimations are corrected for errors due to noise, aliasing, and eddy currents. A 3D segment is created to define the region of interest. (**Right**) Velocity data are postprocessed to produce hemodynamic factors and useful plots and visualization.

5.2.3. Other Relevant MRI-Based Scanning Modalities

Magnetic resonance imaging can also provide relevant information for inverse modeling other than anatomic and kinematic information. In this section, we briefly review MRI sequences that allow the resolution of tissue structure. The image-based resolution of tissue structure and compositional heterogeneity can be used to tailor constitutive models to the volumetric share and orientation of fibrous structures. With regards to patient-specific modeling of cardiovascular disease, these techniques could help identify the location, extension, and severity of lesions, and thus more accurately divide the patient-specific anatomic models into unique regions with particular sets of mechanical properties. By supplying such patient-specific material heterogeneity as a prescribed input, the accuracy, convergence time, and solution multiplicity of inverse models of cardiovascular disease could be significantly improved.

Spin-to-lattice, and spin-to-spin relaxation times, also known as T1 and T2, respectively, are common MRI parameters typically used for highlighting the difference between fat and water. By definition, T1 is a shorter relaxation time than T2, so T1-weighted images highlight fat structures with large proton density, whereas T2 weighted images highlight both fat and water-rich structures. These sequences are commonly used in clinical practice to resolve scar tissue, blood pools, and edemas. Rapid T1 and T2 mapping combines both measurements to resolve an estimation of the extracellular volume fraction that has shown to be a robust marker for several cardiomyopathies, with a strong correlation to histological measurements [218].

Diffusion tensor MRI (DT MRI) is an imaging sequence that uses similar principles to PC MRI. With this technique, special magnetic gradients are designed to cancel out the signal from static water molecules while preserving the magnitude and orientation of moving molecules. Within tissues, water molecules diffuse by Brownian thermal motion, and in fibrous structures, this diffusion occurs preferentially in the fiber orientation. This technique has been mostly applied for the imaging of the white matter and axon orientation in the brain, and more recently to resolve myofiber orientation in the heart [53].

Other MRI-based techniques, such as gadolinium-enhanced MRI and perfusion tests, have been developed to specifically image cardiovascular scars, thereby allowing the quantification of lesion severity. Gadolinium is a contrast agent used to increase the SNR of MRI. The cellular membranes of healthy cardiomyocytes are almost impermeable to gadolinium contrast agents. As a result, following intravascular injection, gadolinium perfuses throughout the myocardium via the branches of the coronary arteries while being excluded from the intracellular space of viable cardiomyocytes due to the impermeability of cell membranes. For this reason, gadolinium can be used to measure the extracellular volume fraction of healthy myocardia from T1 mapping sequences. When myocardial cell membranes are ruptured, as is seen in infarction, a larger portion of gadolinium is accumulated. The contrast can now occupy the no-longer enclosed intracellular space, allowing the assessment of the location and severity of cell-rupturing injuries [219].

The MRI perfusion stress test can assess the severity of coronary artery insufficiency. The quality of blood perfusion into the cardiac wall is resolved through the use of contrast agents at rest and under stress conditions. Increased cardiac stress state can be induced by either exercise or the use of pharmacological stressors. Pharmacologically induced stress is preferred over exercise-induced stress as it renders more reproducible results and is easier to implement in clinical practice [220]. Typical pharmacological stressors include vasodilators (adenosine or regadenoson) or chronotropic inotropic agents (dobutamine). This technique exposes the patient to hazards associated with the use of contrast agents and pharmacological stressors and is generally reserved for patients with confirmed coronary artery disease [221].

5.3. Computerized Tomography (CT)

CT consists of a mobile X-ray source that rotates around a focal point to produce scans from different angles. The result is a high-resolution stack of 2D images that can be time-resolved. The use of intravascular contrast agents is common for studies of the vascular system to improve the visibility of the blood vessels. CT scans can provide better resolution than all the other techniques described above with pixel sizes of about 0.5 mm (Figure 7b). There is no special feature to assess kinematics from CT scans, although its superior temporal and spatial resolution has been used to measure the dynamic change in cross-sectional area and shape of blood vessels during the cardiac cycle, from which homogenized values of circumferential strain for a given cross-section can be estimated. From there, kinematics can be inferred from tracking a given anatomical feature or making reasonable assumptions about rigid body rotation and torsion [11,222].

The use of ionizing radiation makes this technique potentially hazardous; thus risk-benefit of a CT study should be seriously considered. This limits its use in serial follow-up, particularly in pediatric patients, to avoid repetitive exposure to radiation [223]. However, it avoids the risk of unknown or contraindicated implanted metallic object/devices associated with MRI and is typically capable of much shorter scan times than MRI, making it ideal for trauma or other acute emergencies.

5.4. Summary

In Table 6, we summarize the highlights of medical imaging techniques that provide kinematic data, and other useful information for inverse modeling.

Table 6. Summary of medical imaging-based kinematics.

Section	Highlights
5.	Early assessments of in vivo stiffness of blood vessels relied on measurements of luminal area changes. However, the predictive capabilities of these factors are inconsistent among different arterial locations and pathologies. In vivo medical imaging has evolved to provide not only anatomical geometric information but also detailed kinematic measurements. The accuracy and availability of these techniques are limited by image resolution, signal-to-noise ratio (SNR), the occurrence of artifacts, and practical obstacles related to testing costs and health hazards.
5.1	Ultrasound (US) uses high-frequency (2 to 15 MHz) acoustic waves to create real-time in vivo images of tissues, organs, and blood pools using piezoelectric transducers with lateral resolution of 1 mm/pixel. US is relatively inexpensive, portable, and safe, so it has become a customary tool in many clinical applications. However, the accuracy and reproducibility of US-derived measurements are limited in comparison to MRI-based measurements. Blood flow velocity can be assessed with echo and vector doppler technology. Tissue displacement can be measured using speckle tracking technology, which consists of image tracking the acoustic response of tissue fibers to ultrasound signals.
5.2	Magnetic resonance imaging (MRI) offers superior quantitative utility compared to ultrasound as it can offer higher resolution and accuracy of measurements of anatomical features. MRI generally poses minimal hazard to patients unless they have implanted medical devices/objects or suffer from claustrophobia. However, the technique requires specialized equipment and trained staff, which limits availability compared to US. MRI-based techniques for assessment of tissue kinematics include tissue tagging and DENSE MRI. Tissue tagging is based on image tracking of magnetically induced markers, while DENSE MRI encodes the tissue displacement on the phase of the MR signal. Tissue tagging and DENSE MRI have been used to assess the kinematics of the myocardium. However, the superior resolution and accuracy of DENSE MRI allow the assessment of aortic kinematics. Phase-contrast (PC) MRI is a technique that allows the time-resolved quantification of blood flow velocity in or through a 2D plane by encoding the velocity in the phase of the MRI signal. PC MRI has been generalized to 3D spaces at the expense of decreased spatial and temporal resolution. The resulting technique is called 4D flow MRI. PC MRI and 4D flow MRI have been applied to the study of healthy and pathological hemodynamics of the heart and large arteries and are currently implemented in clinical practice for the assessment of aortic and pulmonary diseases. MRI can provide other complementary information relevant for inverse modeling analyses of the cardiovascular system. Diffusion-tensor MRI can resolve the orientation of tissue fibers based on the principle that the Brownian displacement of water molecules occurs preferentially in the direction of fibers. Gadolinium-enhanced (GE) MRI can be used to resolve the size and severity of cardiac lesions. Since healthy cell membranes are impermeable to gadolinium, this contrast agent occupies a larger volume in injured tissue where cell membrane integrity is compromised. Perfusion stress tests use contrast agents and MRI imaging to assess the severity of coronary artery insufficiency. This is performed by comparing the perfusion of contrast agents in the myocardium at rest and at a stress state (high heart rate).
5.3	Computerized tomography (CT) provides the best resolution among all the medical imaging techniques with pixel sizes around 0.5 mm. The high-resolution CT images can be used to assess cardiovascular kinematics through image tracking of anatomical features. However, this requires the introduction of assumptions of displacement modes. CT scans are based on X-ray technology with inherent ionizing radiation hazards.

6. Applications to Cardiovascular Medicine

One of the most relevant outputs of modeling in cardiovascular mechanics is the estimation of wall stress distributions, either in the vascular walls of major arteries or the myocardium of the heart. Mechanical stresses and strains, and their spatial and temporal evolutions, are measures of physiological significance because they may be potential indicators of myocardial and arterial function, may serve as risk stratification factors for tissue failure and rupture, and provide specific measures of the mechanical stimuli modulating biological adaptation. Applications of patient-specific models of cardiovascular mechanics include supporting diagnosis and risk stratification, providing visualizations and insights of deformation and loads on tissue structures, population-based analyses, and supporting and/or challenging mechanistic hypotheses of normal and pathological growth and remodeling.

For patient-specific forward problems in cardiovascular simulation, an anatomical geometric model is typically retrieved from medical imaging and discretized into a computational mesh over a domain of interest. Pulse pressure of blood is typically used as a load boundary condition acting on the luminal or endocardial surface, and any available image-based kinematic information is either used to prescribe velocity boundary conditions (for CFD analyses) or to validate the output of the simulations [224,225]. Forward simulations require the assumption of many parameters that cannot be directly measured or were not collected (e.g., myocardial and arterial wall composition and mechanical properties, blood properties, or focalized blood pressure measurements [226]). Due to these limitations, it has been argued that forward-simulation results should not be taken as absolute quantitative results, but instead, interpreted qualitatively and comparatively in terms of patterns, distributions, and trends of derived stresses and other relevant metrics [227,228]. This caution should especially be emphasized in pathological cases, where the normal physiological function is impaired and assumptions applicable to normal and healthy tissues do not hold [229]. Conversely, patient-specific data can be input into inverse methods to solve for the parameters that are unavailable or cannot be directly measured, potentially reducing the number of required assumptions and improving the ability of the model to fit the observed data. In the study of the cardiovascular system, inverse problems can provide patient-specific estimations of tissue properties, composition, local pressure gradients and stress distributions from image-derived wall deformation, and blood flow dynamics.

The development of patient-specific inverse analyses of cardiovascular mechanics has advanced considerably recently thanks to continuous technological improvements in imaging hardware and software, decreasing cost, increased imaging availability, improvements in image-based kinematics acquisition, and postprocessing, simulation engineering, and significant increases in computational power (Figure 1). Notably, the modern era of computationally robust image-based cardiovascular inverse modeling began with the study of animal models by the end of the 20th century. A pioneering work on in vivo image-based inverse modeling of cardiovascular tissue was published in 1995 by Moulton et al. This research on a canine animal model used a single slice MRI with radio-tagging to retrieve the anatomy and displacement of the short axis plane of the heart [107]. A non-linear error-gradient-based optimization algorithm minimized the least-square error of FEM simulated and MRI-derived strains, by iterating over the constants of the Fung material model considered without any muscular activation component. The boundary conditions were the trans-ventricular pressure measured from catheterization and the restriction of two degrees of freedom of a single computational node. An improved approach was proposed by Walker et al., who applied the inverse method to study the mechanics and properties of infarcted sheep hearts [230] and the effect of surgical intervention [231]. Therein, the authors employed MRI-based 3D models of the left ventricle and MRI tissue tagging to estimate the diastole-to-systole strain field. The latter was used as a target for fitting the material parameters through an iterative inverse formulation. The active contraction was simulated by a time-dependent homogeneous active stress model, and catheter measurements of ventricular pressure were used as boundary conditions. These studies found that fiber and cross-fiber stress are significantly larger at the infarct border zone relative to non-infarct regions. Additionally, the inverse model was employed to evaluate the benefits of diverse treatments and suggested that aneurysm plication decreases the myofiber stretch without compromising stroke volume, which the authors highlighted as one of the benefits delivered by such intervention.

These early works present all the elements of more recent medical image-based inverse analyses: an image-based kinematic target, an optimization algorithm, and a parametric function to be optimized to estimate in vivo case-specific information that cannot be directly assessed without an invasive procedure. These studies were limited by the available computational power at the time. Walker et al. reported a total of 16 h for each iteration of their forward cardiac model using a Silicon Graphics Octane II workstation with a capacity of about 250 MHz, which was a cutting-edge multiprocessor workstation at the time. Currently, the processing capacity of a desktop workstation is at least ten-fold greater (i.e., 3 to 4 GHz). Furthermore, many parallelization and cloud-computing options are now available to augment the speed of simulations. The technology is now mature enough for the evaluation of patient-specific inverse analyses on complex biomechanical models of clinical relevance.

Though there are many instances of image-based inverse analyses on animal models and explanted tissues [137], in this review we aim to highlight the potential clinical applications of inverse methods. Thus, in this section, we present a detailed review of in vivo patient-specific inverse problems applied to elements of the human cardiovascular system along with a few pioneering and groundbreaking studies on animals. Figure 10 summarizes the anatomical references and location of focalized pathologies studied by the inverse-modeling applications reviewed herein.

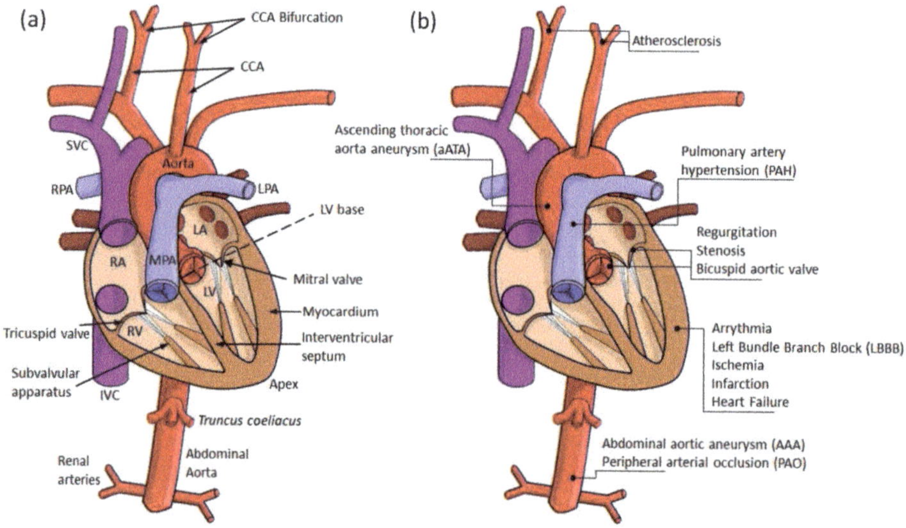

Figure 10. (**a**) Schematic representation of the human heart with anatomical references. (**b**) Location of human pathologies studied with inverse models. Acronyms: CCA, common carotid artery; IVC, inferior vena cava; LA, left atrium; LPA, left pulmonary artery; LV, left ventricle; MPA, main pulmonary artery; SVC, superior vena cava; RA, right atrium; RPA, right pulmonary artery; RV, right ventricle.

6.1. The Unloaded Reference Configuration in Cardiovascular Mechanics

Blood vessels, in particular those of the arterial tree, function under physiological pressure load at all times and are axially pre-stretched; thus, none of the patient-specific configurations resolved by in vivo imaging is truly a stress-free or zero-strain configuration [232]. It is well established that image-based estimations of material properties and stress distributions are sensitive to the selection of the reference configuration. Furthermore, image-based in vivo estimations of material properties assuming the diastolic configuration as a zero-strain stress-free reference lead to significant disagreements with experimental measurements made on excised tissue [233]. That means an adequate selection of the reference configuration is key for the accurate solution of inverse problems of cardiovascular tissue mechanics.

The solution of an unloaded configuration from the deformed geometry, mechanical loads, and material properties is a classical inverse problem with existing direct and iterative solutions [234]. In patient-specific analyses, however, the material properties are also unknown. Thus, the solution to this problem requires the specification of at least two deformed and loaded states as input data [47]. In the case of myocardium, it is often assumed that the transition from unloaded to diastolic configurations is purely passive [235].

In this first subsection, we review previous contributions related to finding patient-specific unloaded and stress-free configurations without the estimation of mechanical properties. Since the methods described can be applied to any pressure vessel, we include developments regardless of the specific tissue application. Research works that incorporate the unloaded or stress-free configuration on the patient-specific estimation of tissue properties and loads are reviewed on the following sub-sections separated by the corresponding tissue of interest.

The direct inverse FEM formulation by Govindjee and Mihalic (c.f. Section 4.1) for the direct solution of the unloaded configuration was first applied to cardiovascular tissue by Lu et al. [111]. The inverse elastostatic approach was used to find the unloaded configuration of an abdominal aortic aneurysm (AAA), assumed to be loaded at a luminal pressure of 100 mmHg, and to behave as an isotropic hyperelastic material with population-averaged material constants. The authors concluded that the selection of diastole as the zero-stress reference leads to the overestimation of stress at systole. A similar approach was applied by Peirlinck et al., who incorporated the inverse elastostatic formulation as a module for the Abaqus FEM solver [100]. The method was applied to an iliac artery ideal model, an image-based porcine biventricular model, a human AAA, and a patient-specific 4-chamber heart model (Figure 11). The method was tested with hyperelastic and fiber-reinforced anisotropic material models. Material constants and pressure loads were imposed based on established reference values from the literature. The authors highlight the convenient modular implementation, computational efficiency, and solution uniqueness as the main advantages of their proposed method.

Several iterative methods have been specifically proposed to solve the zero-pressure configuration for blood vessels. One of the first contributions was proposed by Raghavan et al., who developed an optimization framework for an arbitrary parameter k such that the coordinates of the unknown zero-pressure reference geometry $(\boldsymbol{x}_0)$ can be approximated by the difference of the in vivo deformed configuration $(\boldsymbol{x}_i)$ minus k times the displacement produced by the pressure load on that configuration $(\boldsymbol{U})$, i.e., $\boldsymbol{x}_0 = \boldsymbol{x}_i - k\boldsymbol{U}$. The main conceptual limitation of this method is the assumption that the backward deformation field is linearly related to the forward deformation field through the factor k. This method was then applied to estimate the unloaded configuration of a patient-specific AAA [236].

The backward displacement method was introduced by Rajagopal et al., in 2007, for breast biomechanics and by Bols et al., in 2013, for cardiovascular tissue [234,237]. This method solves the unloaded configuration using the fixed-point interactions proposed by Sellier et al. [100]. It consists in approaching the zero-pressure geometry by iteratively updating the reference configuration, calculated by subtracting the nodal displacement vector between the updated deformed configuration and the target in vivo configuration until a required error tolerance is reached. Rivero et al. successfully applied a similar pullback algorithm to 12 patient-patient specific models of AAA built from CT scans which were assumed to be at uniform diastolic pressure on the image-based deformed geometry. They tested isotropic and anisotropic material models, assuming material homogeneity with reference material constants from the literature [238]. Rausch et al. proposed an augmented Sellier's method based on Aitken's delta-squared process, by introducing an augmentation parameter to accelerate the convergence rate and increase the chances of convergence. The method was applied to find the unloaded geometries of a thrombus and heart valve leaflets from animal models with geometries and properties collected from previous studies [239]. More recently, Das et al. proposed the shrink-and-fit algorithm, that assumes the unloaded configuration is a shrink analogous to the loaded reference geometry. On each iteration step, the coordinates of each node are mapped into a smaller geometry affected by a circumferential and axial shrink factor, the new geometry is loaded by the reference pressure until the least squared error of the nodal coordinates of the reference and inflated model is minimized. The method was then applied to resolve the unloaded configuration of an ideal and a patient-specific artery model assuming Mooney-Rivlin hyperelastic behavior and employing the Nelder–Mead optimization algorithm [240].

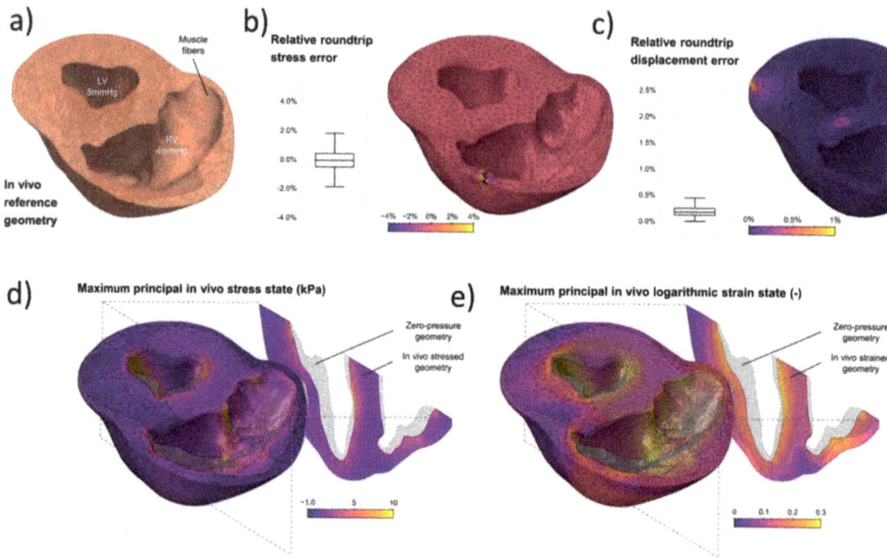

Figure 11. Results and accuracy of the direct inverse elastostatic problem implemented on a FEM solver and applied to a porcine biventricular model. (**a**) Reference loaded configuration reconstructed from MRI scans. (**b**) Relative stress error of the in vivo loaded configuration and solution of the forward inflation problem from the estimated unloaded configuration (roundtrip solution). (**c**) Relative displacement error the in vivo loaded configuration and solution of the forward inflation problem from the estimated unloaded configuration (roundtrip solution). (**d**) Colormap representation of the maximum principal stress distribution of the loaded configuration on top of the estimated unloaded configuration in gray shade. (**e**) Colormap representation of the maximum principal strain distribution on top of the estimated unloaded configuration in gray shade. (Reprinted/adapted with permission from Ref. [100]. 2018, Elsevier).

A different iterative approach is to solve the strain and stress distribution that balances the applied loads acting on the image-derived anatomic configurations without the resolution of the unloaded geometry [232]. Some methods that fit this category are the backward incremental algorithm and the modified updated Lagrangian formulation. In the backward incremental algorithm, small increments of the pressure load are applied to the reference geometry, the resulting stress state is mapped to the reference geometry as the initial condition for the next pressure increment until static equilibrium is reached. This method was applied by de Putter et al., using patient-specific AAA geometries and pressure loads to determine the stress distribution at diastole while assuming isotropic Neo-Hookean material behavior with population-averaged constants [241]. Similarly, the modified updated Lagrangian formulation applies consecutive small loads increments on the image-based reference configuration to build up an incremental multiplicative update of an independent deformation gradient. This method was used by Gee et al., to study the diastolic stress distribution of three patient-specific AAA geometries derived from CT scans with population-averaged pressure loads and material constants for an isotropic Neo–Hookean constitutive equation. In this work, the outcomes of the modified updated Lagrangian formulation are compared to direct solutions with inverse FEM, concluding that both methods yield similar diastolic stress distributions although the latter seemed more prone to solution multiplicity and buckling [242,243]. However, iterative methods may require suboptimal convergence times, and on some occasions, convergence could fail altogether [100].

It is important to highlight that even at an unloaded configuration, cardiovascular tissue is not truly stress-free. This fact has been widely proven by opening angle experiments at different arteries and layers of the heart wall. The residual stress responsible for this recoil effect exists without any distending pressure, being the possible result of non-uniform growth and remodeling over the patient's entire lifespan. The latter implies that residual stress cannot be resolved solely from load-deformation data. Indeed, the most common technique for the estimation of residual strain relies on the quantification of the opening angle after a stress-relieving cut. Some specialized studies have collected opening angle data from multiple locations of the cardiovascular system through experimental tests on human cadaveric tissue. These experiments have shown that the opening angle, and thus the preexisting residual stress, depends on specific tissue location and individual factors such as age and health conditions. Consequently, generic or averaged opening-angle derived residual stress can hardly be used for patient-specific analyses, especially, in pathological cases. The constrained mixture theory provides a consistent framework for the estimation of residual stress through the modeling of growth and remodeling and could be the key, along with the image-based resolution of tissue composition, for a truly patient-specific estimation of a stress-free reference configuration [32,244].

6.2. *The Heart*

The relatively large thickness of cardiac tissue allowed the resolution of image-based kinematics even at the early stages of this technology. For this reason, along with the key role of the heart as the driving element of circulation, the heart was the first physiological system subject to patient-specific inverse analyses. Sermesant et al. and Aguado-Sierra et al. proposed comprehensive patient-specific models for cardiac function including the resolution of the unloaded configuration, bioelectrical activity, passive and active tissue properties, and hemodynamics [245,246]. These authors evaluated the possibility of solving such inverse problems with data acquired with medical imaging and electrocardiography and concluded that such comprehensive models easily became overparametrized, and computationally expensive to be solved by the available resources at the time. In consequence, most inverse models focus on only one or a few of their constituents instead of the whole heart. In the following subsections, we classify the research approaches based on the variables chosen to be solved by the inverse method.

6.2.1. Properties of the Healthy and Infarcted Ventricular Wall

The ventricular wall is a complex multilayered composite responsible for delivering the driving force to pump blood throughout the cardiovascular system. The myocardium is the functional layer of the ventricular wall, containing the myofibers responsible for the active contraction of the muscle and the structural collagen fibers that contribute to its bulk mechanical properties. An accurate understanding of myocardial mechanics is key for the diagnosis and treatment of diverse cardiac pathologies, and potentially, predicts and stratifies the risk of heart failure after infarct. Therefore, many studies have focused on the estimation of mechanical properties of healthy myocardium, and more interestingly, estimating the effects of ischemia, and quantifying the properties of infarcted cardiac tissue to yield a truly patient-specific risk assessment of cardiac failure. Most developments relied on FEM for the solution of a forward problem (summarized in Table 7).

Table 7. Literature review of iterative inverse models for the analysis of human heart tissue mechanics.

Study	Clinical Data		Forward Problem		Inverse Problem	
Rumindo et al., 2020 [247]	Population	12 H	Reference	End of diastole	Target function	Least squared error to Klotz P-V
	Pathology	None	Passive model	Hom. Guccione		
	Data	Cine MRI	Active model	1 eq. active stress	Opt. algorithm	Nelder Mead.
	Anatomy	LV with RBFO by Rijcken et al.	Boundary	ICP, TF epicardium Constrained base		
Zhang et al., 2020 [17]	Population	1H 5D	Reference	Early diastole	Target function	Volume change error and segment-wise strain error
	Pathology	FMR-CAD	Passive model	Het. Guccione		
	Data	Cine MRI, TT, Stress perfusion MRI, GE MRI, 4D US, Hand cuff pressure	Active model	2 eq. active stress		Non-specified
	Anatomy	BV in 17 AHA regions with RBFO by Bayer et al.	Boundary	ICP, TF epicardium, Constrained base	Opt. algorithm	
Balaban et al., 2018 [141]	Population	1D	Reference	End of diastole	Target function	Deformation gradient error.
	Pathology	LBBB and CI	Passive model	Het. Holzapfel-Ogden		
	Data	4D US, USST, GE MRI, ICP	Active model	None		Sequential quadratic programming with a first-order Tikhonov functional constraint
	Anatomy	LV in 17 AHA regions with RBFO by Bayer et al.	Boundary	ICP, Constrained apex, EF at base.	Opt. algorithm	
Wang et al., 2018 [248]	Population	5H 19D	Reference	Diastasis	Target function	Least-squared error to P-V curve.
	Pathology	HFrEF, HFpEF	Passive model	Hom. Guccione		
	Data	Cine MRI, ICP	Active model	None	Opt. algorithm	Non-specified
	Anatomy	LV with RBFO by Nielsen et al.	Boundary	* IPC * Constrained base		
Finsberg et al., 2019 [249]	Population	6H 12D	Reference	Unloaded	Target function	Coordinate error for passive properties. P-V curve and strain error for active properties.
	Pathology	PAH	Passive model	Hom. Holzapfel-Ogden		
	Data	Cine MRI, ICP	Active model	1 eq. active strain	Opt. algorithm	Sequential quadratic programming algorithm
	Anatomy	BV with RBFO by Bayer et al.	Boundary	ICP, EF at base, EF pericardium		
Palit et al., 2018 [108]	Population	5H	Reference	Early Diastole	Target function	Least squared error to Klotz P-V
	Pathology	None	Passive model	Hom. Holzapfel-Ogden		
	Data	Cine MRI	Active model	None	Opt. algorithm	Genetic Algorithm
	Anatomy	BV with RBFO by Bayer et al.	Boundary	* Assumed ICP * Constrained base		
Finsberg et al., 2018 [235]	Population	7H 7D	Reference	Unloaded	Target function	Coordinate error for passive properties. P-V curve and strain error for active properties.
	Pathology	LBBB	Passive model	Hom. Holzapfel-Ogden		
	Data	4D US, USST, ICP	Active model	1 eq. active stress, 1 eq. active strain	Opt. algorithm	Sequential quadratic programming algorithm
	Anatomy	LV with RBFO by Bayer et al.	Boundary	ICP, EF at base, EF pericardium		
Asner et al., 2015, 2017 [250,251]	Population	3H 3P	Reference	End of diastole	Target function	P-V curve and nodal displacement error
	Pathology	Dilated cardiomyopathy	Passive model	Hom. Holzapfel-Ogden		
	Data	Cine MRI, TT, PC MRI	Active model	1 eq. active stress		Shamanskii–Newton Raphson algorithm
	Anatomy	LV with fiber orientation from canine histology	Boundary	Weak formulation for volume and displacement	Opt. algorithm	
Nasopoulou et al., 2017 [252]	Population	1H 7D	Reference	Lower ventricular pressure	Target function	Energy balance error and displacement error
	Pathology	Arrythmia	Passive model	Hom. Guccione		
	Data	Cine MRI, ICP	Active model	None		Non-specified
	Anatomy	LV with fiber orientation from canine histology	Boundary	ICP, displacement at apex and base, TF epicardium	Opt. algorithm	

Table 7. *Cont.*

Study	Clinical Data		Forward Problem		Inverse Problem	
Gao et al., 2017 [253]	Population	27H 11D	Reference	End of diastole	Target function	Volume error and strain error.
	Pathology	Acute myocardial infraction	Passive model	Het. Holzapfel-Ogden		
	Data	Cine MRI, GE MRI, Hand-cuff pressure	Active model	2eq. active stress	Opt. algorithm	Gaussian processes and automatic relevance determination algorithm
	Anatomy	LV in 17 AHA regions with RBFO by Potse et al.	Boundary	ICP, TF epicardium		
Genet et al., 2014 [110]	Population	5H	Reference	Early diastole	Target function	Least-squared-error to normalized Klotz P-V curve
	Pathology	None	Passive model	Hom. Guccione		
	Data	Cine MRI, TT	Active model	Hom. 1eq. active stress	Opt. algorithm	Derivative-free quadratic approximation algorithm
	Anatomy	LV with fiber orientation from canine histology	Boundary	Volume change, TF epicardium, Constrained base.		
Marchesseau et al., 2013 [254]	Population	8H 3D	Reference	End of diastole	Target function	Volume change error
	Pathology	HFrEF	Passive model	Region heterogeneous Mooney-Rivlin		
	Data	Cine MRI, ICP, Electrophysiology	Active model	2 eq. active stress	Opt. algorithm	Kalman filter
	Anatomy	BV divided in 17 regions with RBFO by Bayer et al.	Boundary	ICP, TF epicardium, Constrained base		
Xi et al., 2013, 2011a, 2011b [47,255,256]	Population	1H 2D	Reference	Unloaded	Target function	Nodal coordinate error
	Pathology	HFrEF	Passive model	Hom. Guccione		
	Data	Cine MRI, TT, ICP	Active model	1 eq. active stress	Opt. algorithm	Parameter sweeping
	Anatomy	LV with fiber orientation from canine histology	Boundary	ICP, TF epicardium, Displacement at apex and base.		

Abbreviations and acronyms: Clinical data: AHA, American Heart Association; BV, biventricular; D, diseased; GE, gadolinium enriched; H, healthy; ICP, intracardiac pressure; LV, left ventricle; MRI, magnetic resonance imaging; TT, tissue tagging; RBFO, ruled-based fiber orientation; US, ultrasound; USST, ultrasound speckle tracking. Pathologies: CI, cardiac infraction; FMR CAD: functional mitral regurgitation associated to coronary artery disease; HFpEF, heart failure with preserved ejection fraction; HFrEF, heart failure with reduced ejection fraction; LBBB, left bundle branch block. Forward problem: EF, elastic foundation; eq., equation; Het., heterogeneous; Hom., homogeneous; ICP, intracardiac pressure; TF, traction free.

Homogeneous Models

The assumption of material homogeneity is a common and convenient simplification for forward and inverse models. It limits the number of parameters to be fit while still reproducing the overall mechanics of the organ with reasonable accuracy. Even though the myocardium is highly complex and spatially heterogeneous, homogeneous models may be deemed to be adequate for the study of healthy hearts, or when the aim of the analysis is not centered on the study of focalized lesions. In the study of tissues with steep localized changes in structure and properties, as in infarcted myocardium, the material homogeneous models cannot reproduce the localized strain and stiffness distributions on the infarct itself and the infarct borderzone, providing only averaged estimations of local deformation and material properties. However, these averaged properties can still be used as a measure of lesion severity by comparative studies of healthy cases.

One of the first simplified inverse models of the left ventricle (LV) was introduced by Hassabalah et al., to study the compressibility of the myocardium [257]. An idealized truncated ellipsoid matching MRI-derived averaged dimensions of a human LV was used as a computational domain. Fiber orientations were assumed helical with the linear transmural distribution established from DT-MRI data from Helm et al. [258]. The myocardium was modeled as a homogeneous fiber-reinforced Ogden hyperelastic material. The active tension of myofibers was assumed to be proportional to the pressure load, the latter being prescribed as a boundary condition at the endocardial surface. A uniform elastic foundation was applied to the pericardium to simulate the interaction of the heart with the surrounding organs, and all displacements were fixed on a lateral node. All material parameters were fixed except for the bulk modulus, which was optimized to fit a measured pressure-volume curve. The authors conceptually divided the cardiac cycle into the following consecutive stages: atrial systole, isovolumetric contraction, rapid ejection, isovolumetric relaxation, rapid filling, and reduced filling. This study suggested that the volume of the myocardium changed slightly during the cardiac cycle. According to this, myocardium behaves as an incompressible tissue only during rapid and reduced ejection and isovolumetric relaxation stages, while showing some degree of compressibility in the atrial systole, isovolumetric contraction, and filling stages. These observations are in agreement with in vivo compressibility measurements in large mammals [134].

A more sophisticated inverse analysis was presented by Xi et al., in two consecutive papers published in 2011, introducing patient-specific geometries [255,256]. MRI-based models of the LV at end-diastole were assumed as the zero-strain reference and discretized with Hermit-cubic finite elements. MRI tissue tagging was used to assess the diastole-to-systole displacement distribution and then interpolated into the nodes of the FEM mesh. The Fung–Guccione constitutive equation was selected to model the uniform passive properties of the myocardium. Myofiber orientations were assigned with a rule-based algorithm from Bayer et al. [55] based on canine serial histology from Usyk et al. [259]. Boundary conditions consisted of catheter measured ventricular pressure increments, zero traction at the epicardium, and apical and base displacement from tissue tagging. The least squared nodal displacement error was minimized using a reduced-order Kalman filter. The method was applied to one healthy heart and two patients with diastolic heart failure with impaired ejection fraction. The authors found a large difference in material parameters between healthy and heart failure patients, although the authors recognized the results were likely not unique for the given dataset. In addition, they found that passive behavior alone could not fully describe the deformation state at early diastole. They addressed this issue by introducing a time-dependent homogeneous active tension model and the backward displacement method to estimate the unloaded configuration [47]. Two different minimization problems are solved iteratively: first, the estimation of passive properties by consecutive simulation of deflation from early diastole to unloaded configuration followed by inflation to end-diastole; second the estimation of active properties by inflation from late diastole to systole (Figure 12b).

The target minimization function for both iterative loops was defined as the error in nodal coordinates between simulations and interpolated tissue-tagging measurements for their corresponding end of process configuration. According to their results, the residual activation state from early to end-diastole was larger for patients with heart failure, which may indicate that diastolic relaxation is impaired after cardiac failure due to the compensation mechanism to maintain cardiac function.

Genet et al. applied methods to define patient-specific anatomic models and define active and passive behavior similar to Xi et al. [47], although with no available patient-specific pressure data and the assumption of end-diastole as the stress-free reference [110]. Boundary conditions constrained all displacement on the basal plane of the ventricle and the dynamic boundary condition was a prescribed volume change, instead of the pressure increment. In the absence of pressure data, Genet et al. assumed a well-established normalized LV pressure-volume curve of Klotz et al. [109] as the optimization target. MRI tissue tagging measurements were used to validate the converged results, which showed good agreement with image-derived circumferential and axial strain. Regarding fiber stress distribution, the authors found that the end of diastole myofiber stress peaked near the subendocardial wall. They also highlighted that the transmural variation of the end-of-systole myofiber stress was nonmonotonic and was maximal at the mid-wall of the ventricle.

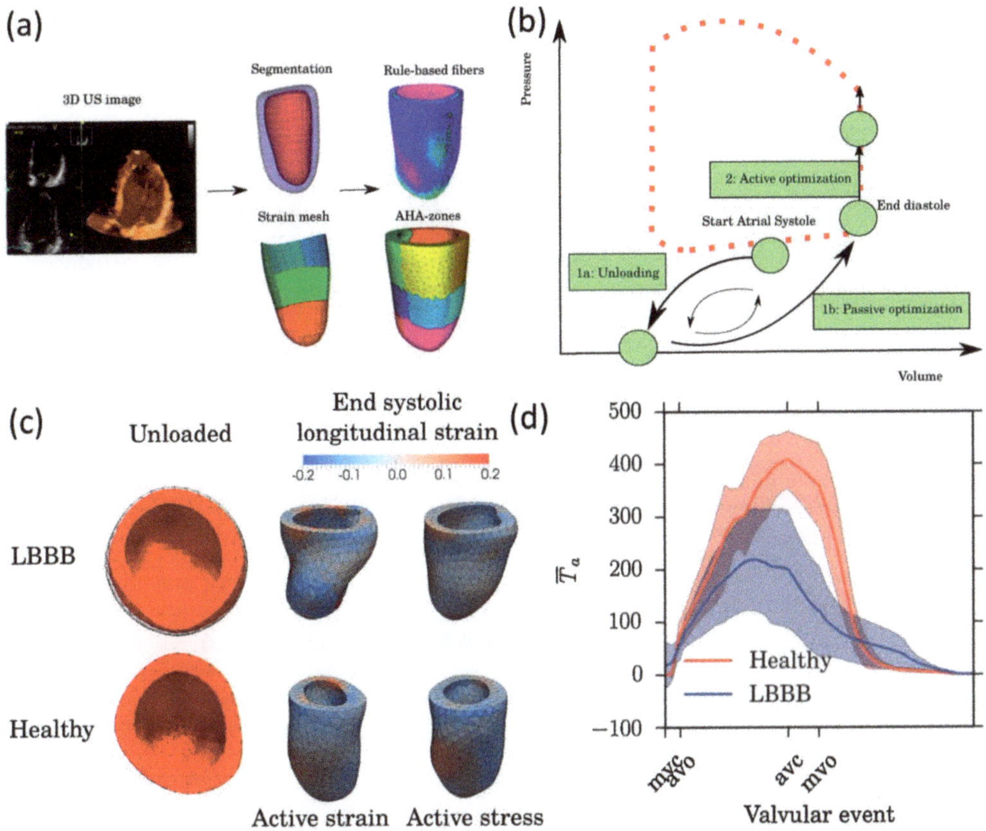

Figure 12. Inverse analysis of left ventricular mechanics. (Reprinted/adapted with permission from Ref. [235]. 2018, Elsevier). (**a**) Preprocessing pipeline, from left to right: medical image-based segmentation and kinematics, anatomic model generation, discretization, partition into 17 AHA standard regions. (**b**) Optimization loops, between unloaded and diastolic configuration for passive properties and between diastole and systole for active contraction parameters. (**c**) Resulting unloaded configuration and strain distributions for a healthy volunteer and an LBBB patient, using active stress and active strain approaches. (**d**) Comparison of activation parameters over time for a healthy individual and an LBBB patient showing the effect of impaired bioelectrical function. Acronyms: mvo, mitral valve opening; mvc, mitral valve closure; avo, aortic valve opening; avc, aortic valve closure.

Solution multiplicity has been one of the main concerns about inverse methods, which motivated Nasopoulou et al. to explore how the definition of the optimization target functions can be designed to improve material property identifiability and solution uniqueness [252]. Sets of cine MRI and catheter pressure measurements were gathered from 7 cardiac resynchronization therapy (CRT) patients and one healthy volunteer. The configuration corresponding to the lower ventricular pressure was assumed to be stress-free. Patient-specific LV models were built at the reference configuration and a warping algorithm was used to estimate the displacement of the ventricular wall from cine MRI. The myocardium was assumed homogeneous and purely passive with a Fung–Guccione constitutive equation. Myofibers orientation was assumed to follow a linear transmural distribution following the findings of Streeter et al., on canine left ventricles [260]. Uniform pressure on the endocardium and image-derived displacements on the basal plane were imposed as boundary conditions. Two target functions were defined, one based on the displacement error, and the other defined as a normalized error of the pressure-energy input and stored strain energy. Two optimization processes were implemented consecutively to minimize the two error functions, which constrained the number of possible solutions. The authors concluded that a single purely geometric target function is unable to constrain the parameter space, while the application of the energy-based target function isolates one of the material parameters, that in conjunction with a geometry-based target provides a unique estimation of parameter sets.

Most of the inverse modeling approaches dealing with the heart, either constrain or prescribe measurement-derived magnitudes of displacements on the basal plane and/or the apex. Asner et al. highlighted the necessity of imposing more physiologically meaningful boundary conditions for the adequate assessment of cardiac mechanics. These authors proposed a method to impose consistent boundary conditions for ventricular mechanics based on non-invasive tests alone [250,251]. The proposed method was applied to synthetic datasets for validation generated in silico with idealized geometries and known material properties, motion, and loads. Then, the method was applied to patient-specific datasets from three healthy volunteers and three moderately dilated cardiomyopathy patients. Cine MRI, tissue tagging, and PC MRI were collected and used as imaging data either to set up the forward problem or as target data for minimization. End of diastole configuration was used to build anatomic models of the LV which were assumed to be at the zero-strain reference. Tissue tagging-derived displacements were interpolated to the FEM mesh to obtain a smooth displacement field and PC MRI was used to estimate the stroke volume. The myocardium wall was assumed to follow a reduced-order Holzapfel-Ogden constitutive equation with a time-dependent homogeneous active stress model. Myofiber orientations were assumed to follow a linear transmural distribution based on the work of Streeter et al., on canine ventricles [260]. Ventricular pressure–volume relation was assumed to follow the normalized Klotz LV pressure-volume curve, and the PC MRI-derived diastole-to-systole flow ratio was correlated to the pressure pulse amplitude. The authors proposed a data-based method for imposing boundary conditions through the use of Lagrange multipliers and the minimization of energy potentials. The endocardial boundary condition was defined in terms of volume change, while the basal plane and the epicardial node boundary conditions were defined in terms of a virtual force proportional to their correspondent displacements. A Shamanskii–Newton Raphson procedure was used to resolve the material properties and boundary condition multipliers. Parameter fitting was solved in two steps: first, the passive material parameters were solved by minimizing a displacement-based error function from tissue tagging data between early- and end-of-diastole; second, the active components were fitted by minimizing an error function defined as a weighted average of nodal displacement error and pressure-volume curve error. The authors highlight the potential bias introduced by hard displacement restrictions as boundary conditions, which prevents the reproduction of naturally occurring torsional modes of deformation. They also highlight that the direct imposition of noise MRI-derived displacement as a boundary condition can introduce computational issues associated with continuity and solution smoothness, concluding that the proposed weak formulation of boundary conditions is advantageous.

Palit et al. performed an inverse analysis of biventricular models with a microstructural material model [108]. Steady-state free precession (SSFP) cine MRI was used to build anatomical models, and to calculate the diastole-to-systole volume change of both ventricles from five healthy adult volunteers. A purely passive Holzapfel–Ogden material model was imposed with fiber orientation following the Laplace–Dirichlet rule-based algorithm by Bayer et al. [55]. Early diastole was assumed as the stress-free reference configuration. A normalized Klotz pressure–volume curve was set as the optimization target for a genetic algorithm. The authors introduced empiric constraints on the constitutive equation for the maximum absolute and relative values of shear-related terms to reduce the sampling space. In addition, they carried out a sensitivity analysis of the results on the assumed parameters.

They concluded that variations within the normal ranges of interventricular pressure and fiber orientation did not produce significant changes in material parameters estimations.

Wang et al. explored potential differences in LV stiffness among heart failure patients with preserved and reduced ejection fractions [248]. Cine MRI and catheter pressure measurements were collected from 8, 11, and 5 individuals with reduced, preserved, and normal ejection fractions, respectively. Anatomical models of the LV were customized for all cine MRI images with an interactive guide-point modeling tool and volumes were matched to pressure measurements. The diastasis state (immediately after rapid filling) was assumed as the reference stress-free configuration. The Fung–Guccione model was used to describe myocardium mechanical behavior neglecting the active component. Myocardium fiber orientation was defined following the rule-based algorithm proposed by Nielsen et al., based on the fibrous structure of canine hearts [261]. Boundary conditions consisted of uniform pressure at the endocardium and constraints to the displacement of the basal plane. Image-based endocardial and pericardial surfaces were projected into finite element model predictions for each time step and the mean-squared error was minimized by fitting the passive material properties. Results showed no significant differences in ventricular stiffness between groups, although patients with reduced ejection fraction presented elevated diastolic stress levels.

Rumindo et al. explored the variability of in vivo estimations of passive and active properties of the LV in healthy individuals through inverse methods [247]. This retrospective study gathered cardiac MRI datasets from 21 volunteers with normal cardiac function. End systolic and diastolic volumes were calculated from MRI segmentation, and end-systolic configurations were assumed the stress-free reference and used for geometric modeling and meshing. The Fung–Guccione material equation and a time-dependent homogeneous active stress model were used to describe mechanical behavior. Myofiber orientation was assigned following the equations proposed by Rijcken et al., to optimize cardiac ejection [57]. Models were uniformly pressurized in the endocardium while constraining all displacements on the basal plane and assuming the epicardium to be traction-free. The Nelder–Mead algorithm was used to fit the material parameters by minimizing the least-squared error of the pressure–volume relation to the normalized Klotz curve. Population-based statistics were calculated showing that results were consistent within this population of similar characteristics. The authors highlighted the variety of reported Fung–Guccione material parameters from among different studies and discussed the relevance of the selection of a reference configuration for the estimation of passive properties.

Heterogeneous Models

Modeling of material heterogeneity can provide better fits to kinematic data, resolve property changes, and identify the location and severity of myocardial lesions. However, this comes with an increased modeling effort and computational expense. By assuming material heterogeneity on inverse methods, the number of parameters to be fitted increases, posing a burden on the optimization algorithm and complicating the solution of the forward problem. A common approach is to approximate spatial variations of myocardium properties and microstructure with region-wise heterogeneities. The simulation domain of the myocardium is divided into segments, each one with its own set of homogeneous material properties. The American Heart Association (AHA) proposed the division of the left ventricle into 17 standardized LV segments which have been adopted extensively in the study of myocardium mechanics (Figure 12a) [262].

One of the earliest inverse analyses of biventricular models with region-wise material heterogeneity was proposed by Marchesseau et al. [254]. The study gathered cine MRI datasets from 8 healthy volunteers and 3 heart failure patients with impaired ejection fractions. Cine MRI was used to estimate the volume change of both ventricles, to identify the location of the epicardial surface on several time-steps over the cardiac cycle, and to estimate displacements with a diffeomorphic free form deformation algorithm. End-diastole was used as the reference configuration and to build a deformable FEM mesh.

Electromechanical behavior was modeled with a Bestel–Clément–Sorine model, which consists of a Mooney–Rivlin hyperelastic material matrix reinforced with fibers with passive elastic and time-dependent active components. Fiber orientation was assumed to follow the Laplace–Dirichlet rule-based algorithm by Bayer et al. [55]. The active component was assumed to have a viscous dissipation component and was modeled by a two-differential equation system solving for the time-dependent active stress and sarcomere stiffness as a function of an activation state variable. Parameter fitting is carried out by applying Kalman filters in two steps: first, a general fit is achieved with the overall pressure-volume curve, followed by a parameter refinement for each sector using sector-specific displacements and change of LV section volume. The model was able to locate the infarcted regions by assigning them lower contractility, while healthy patients converged to more homogeneous property distributions and normal active function.

In 2017, Gao et al. performed an inverse analysis on 27 healthy subjects and 11 patients with acute myocardial infarction [253]. Gadolinium-enhanced and cine MRI were applied to identify the location of infarcted regions and to calculate the volume change of the LV. Anatomic models were built at end-diastole, which was assumed as the zero-strain reference. The LV systolic blood pressure was approximated by the sphygmomanometer systolic measurements. The anatomic models were divided into the 17 standard AHA regions, and circumferential strains were calculated for each region through a b-spline deformable registration algorithm. Non-infarcted tissue was modeled as a Holzapfel–Ogden material with a sophisticated differential-algebraic model for active stress. Myofiber orientation was defined by the minimum-distance rule-based algorithm Potse et al. [56]. Infarcted tissue was assumed 50-fold stiffer than regular tissue with no active contraction. A Bayesian approach with Gaussian processes and automatic relevance determination algorithm was used to fit material properties and active contraction parameters by minimizing a weighted function of the volume error and region-wise circumferential strain error. Results showed that active tension was larger in infarcted hearts, which agrees with the early observations of Xi et al. and Marchensseau et al., for which the authors hypothesized the existence of a compensation mechanism for infarcted hearts to preserve stroke volume.

In 2018, Finsberg et al. compared the LV contraction between healthy adult volunteers and patients with blocked or delayed electrical activation impulses, a condition called left bundle branch block (LBBB) [235]. The study was carried out on a population of 7 individuals per group. Four-dimensional (4D) echocardiography was used to build patient-specific anatomic models and FEM meshes. Ventricular volume was measured at 10 different instances within the cardiac cycle. Ultrasound speckle tracking was used to estimate the piece-wise strain field, consisting of circumferential, radial, and longitudinal strain at each of the 17 standard regions. Direct pressure measurements were obtained through catheterization for the LBBB patients. The myocardium was assumed to follow a uniform Holzapfel–Ogden material model, while two models for active contraction (active strain and active stress) were tested. Myofiber orientation was assigned following the Laplace–Dirichlet rule-based algorithm by Bayer et al. [55]. Rigid-body translation and rotation were constrained by an elastic foundation boundary condition on the basal plane imposed as a collection of linear springs with uniform elastic constants. Two iterative inverse models were solved consecutively in each case: first, the passive isotropic material properties and unloaded configuration were estimated with a backward displacement algorithm using the geometric and pressure information at early and late diastole. Second, the active and anisotropic material properties were obtained by minimizing an error function defined as a weighted average of ventricular volume and strain error (Figure 12b). Minimizations were carried out with a sequential quadratic programming algorithm, and maximum value constraints were imposed on active model parameters. Results suggested that the myocardium wall was more compliant for the healthy group (Figure 12c) and that active contraction was significantly lower for the LBBB, which is consistent with an impaired propagation of the activation pulse (Figure 12d). Both the active stress and active strain models showed equivalent results. A similar methodology was later applied to

12 patients with pulmonary hypertension and 6 healthy human controls, using cine MRI and hyperelastic warping to estimate regional strains [249]. This study found that larger right ventricular contractility affected the right-to-left ventricle volume ratio, the latter being a clinical risk factor for pulmonary hypertension. The authors suggest that this mechanistic relation between ventricular contractility and interventricular volume ratio could provide further insights into pulmonary artery hypertension risk stratification.

Zhang et al. studied the local effect of ischemia with the segment-wise heterogeneity approach [17]. Five patients with functional mitral regurgitation associated with coronary artery disease and treated with percutaneous revascularization were retrospectively recruited. The population was complemented by one healthy volunteer. The treatment protocol included cardiac MRI and transthoracic echocardiography before and 3 months after revascularization. Gadolinium-enhanced MRI allowed the identification of infarcted scar tissue and an MRI stress perfusion test was used to assess the location and severity of ischemia. With this image-derived information, a normalized scale for infract and ischemia severity was assigned to each region. MRI-derived patient-specific 3D biventricular models at early diastole were used to define the geometrical model and assigned to be the zero-stress reference. MRI tissue tagging was used to estimate average strains in all 17 standard regions. Left and right ventricular pressure were estimated from sphygmomanometry and concomitant transthoracic echocardiography, respectively. The Fung–Guccione material model was used to describe passive behavior and a time-dependent heterogeneous-by-region active stress model was implemented. Myofiber orientation was prescribed following the Laplace–Dirichlet rule-based algorithm by Bayer et al. [55]. Measured left ventricular pressure was applied to the endocardium of the stress-free early diastolic model while constraining axial displacements on the basal plane. Boundary loads consisted of right ventricular pressure at the septum and a traction-free condition at the epicardium. Passive and active material parameters were defined in terms of a scale of the infarction and ischemia severity. This ischemia effect factor modulated different responses with regions identified with zero severity behaving like healthy tissue and becoming stiffer and less actively contractile with larger lesion severity. Material parameters and the ischemia effect factor were fitted for each one of the 17 regions by minimizing a weighted function of the mean square error of diastole-to-systole volume change and the region-wise average strain. Results agree with previous studies on predicting the stiffening of regions corresponding to infarcted tissue and border zone. Additionally, the model allowed the estimation of the ischemia effect on tissue stiffening and the recovery of compliance after revascularization treatment.

One of the main limitations of the above studies is the assumption of either spatial material homogeneity or segment-wise heterogeneity, however, material properties are likely to vary continuously throughout the myocardium. To address this, Balaban et al. proposed an iterative inverse method to resolve the heterogeneous distribution of mechanical properties on an LV model from a 64-year-old heart with systolic heart failure, LBBB, coronary artery disease, and chronic infarction in the inferior section of the LV [141]. 4D echocardiography was used to obtain the anatomic model and FEM mesh at early atrial systole. Speckle tracking was used to estimate systolic strain averaged over the 17 standard regions, and pressure was measured by catheterization. Gadolinium-enhanced MRI was used to identify the location of infarcted fibrotic tissue. The Holzapfel-Ogden material model was implemented allowing spatial variations of the scalar material parameters with a piece-wise linear representation with fiber orientation following the rule-based algorithm proposed by Bayer et al. [55]. Active tension was neglected, and end-diastole configuration was assumed stress-free. Rigid body motion was constrained by impeding axial displacement at the basal plane and apex, and by an in-plane elastic foundation at the base plane imposed as a collection of linear springs with uniform elastic constant. A sequential quadratic programming algorithm was applied to estimate the almost 3000 spatially distributed material parameters. To favor convergence to smooth distributions, optimization was constrained by a first-order Tikhonov functional. Results show that estimated strains were lower, and the material stiffer, in regions corresponding to infarcted tissue and its immediate surroundings identified by gadolinium-enhanced MRI.

6.3. Valves and Leaflets

Each one of the chambers of the heart is equipped with a discharge valve to ensure unidirectional blood flow, acting mostly passively to changes in transvalvular pressures. The atrioventricular valves are the mitral and tricuspid, for the left and right sides of the heart respectively. These valves typically define the basal plane and separate the atria from the ventricles (Figure 10a). They are structurally supported by the papillary muscles and chordae tendineae to hold the valves closed during systole and avoid ventricle-to-atria backward flow. The pulmonary and aortic valves regulate blood flow from the ventricles to their homonym arteries and are not supported by any subvalvular apparatus. The main element of heart valves are fibrous structures called leaflets or cusps, that flap to allow or impede blood flow. In normal conditions, only the mitral valve has two leaflets while the other valves have three [263].

Heart valve disease is mostly related to regurgitation, stenosis, and atresia. The former consists of backflow due to deficient closing, stenosis is the hardening and thickening of the leaflets, preventing the valve to open properly and result in increased load in the heart, while the latter is a congenital disease where the heart valve is partially or completely absent. Heart valve malfunction can lead to several complications such as heart failure, blood clotting, stroke, and death. Heart valve disease is most common on the left side, as the aortic and mitral valves are loaded with larger pressures, and in consequence, they have received more attention from the medical and scientific community. However, attention to right heart valves has significantly grown in the last two decades along with the awareness of pulmonary artery diseases [264].

There is a considerable body of research on the forward modeling of heart valve function accounting for structural and FSI mechanics, usually validated against in vitro experiments [265]. However, leaflets are typically thin structures (<1.5 mm) showing complex displacement patterns, which renders them extremely challenging to resolve through in vivo imaging techniques. Owing to this, most inverse analyses of valve mechanics are based on in vitro experiments on excised or synthetic valves, where the leaflet displacement is resolved with the use of physical markers [266–273], or with high-resolution cameras [147,274].

In vivo inverse modeling of ovine heart valves function has been achieved by the use of fluoroscopic markers implanted on the surface of mitral valve leaflets [275,276], a technique that cannot be pursued in human studies. More recently, Lee et al. applied ultrasound technology to assess the anatomy and displacement of the mitral valve of ovine animal models to explore the use of inverse modeling, and in vivo mechanical properties and stress distribution were successfully estimated [271,277].

Aggarwal et al. estimated the residual strain on human aortic valves by combining in vivo imaging with measurements on explanted tissues [278]. The authors collected in vivo transesophageal 3D echocardiographic images of the aortic valve from five open-heart transplant patients at three configurations: fully open, just-coapted, and fully loaded. Each aortic valve leaflet was excised during surgery and then imaged in a flattened configuration ex vivo. Strains were calculated between the ex vivo stress-free configuration and the three in vivo configurations from echocardiography segmentation by the application of a spline parametrization algorithm. Results suggest that leaflets are significantly pre-strained with respect to the excised reference even at the just-coapted configuration where the transvalvular pressure load is negligible. Results also showed that leaflet deformation is larger in the radial direction if compared to the circumferential direction, the latter being structurally stiffer due to the alignment of collagen fibers.

The work of Aly et al. stands out as one of the few in vivo works on human heart valves for the generation of transient anatomical models [279]. In this work, 4D ultrasound was collected from 28 patients, half with normal mitral valve anatomy and function, the other half with ischemic mitral valve regurgitation. An automatic inverse algorithm uses the manual identification of five key landmarks on the leaflet anatomy as input. Then, Kalman filter optimization is used to build anatomical models at different instants of the cardiac cycle. According to the authors, this algorithm could be used as the base for more comprehensive inverse modeling to assess leaflet material properties.

6.4. Arterial Wall

Changes in mechanical properties of arterial walls have been associated with the onset of multiple cardiovascular pathologies (e.g., atherosclerosis, dissection, stenosis) and remains an important predictor of cardiovascular morbidity and mortality in clinical practice. This motivated the development of early techniques for the non-invasive assessment of arterial stiffness through the evaluation of luminal area change and pulse wave velocity. These techniques, although useful, can only provide a gross estimation of material properties as they introduce many assumptions and simplifications related to homogeneity, perivascular support, and linearized behavior.

The image-based resolution of vascular tissue kinematics is technically challenging; the main reason being the relative thinness of vascular walls. For example, the ascending aorta has a typical thickness of about 2.5 mm, which decreases to about 1.5 mm at the abdominal aorta, and the pulmonary artery is only about 0.2 mm thick. These length scales are comparable to the highest resolutions available on imaging techniques, for which luminal area changes (either with or without contrast agents) remained the main input for early inverse analyses of arteries. However, recent developments in ultrasound speckle tracking and DENSE MRI techniques make available arterial wall displacement measurements on a meaningful number of pixels. Most approaches rely on FEM for the solution of the forward problem (summarized in Table 8).

Table 8. Literature review of iterative inverse models for the analysis of human arterial wall mechanics.

Study	Clinical Data		Forward Problem		Inverse Problem	
Bracamonte et al., 2022, 2021, 2020 [14,150,280]	Population	27H	Reference	Diastole	Target function	Least-squared displacement error
	Pathology	None	Passive model	Hom. Fung orthotropic		
	Data	Cine and DENSE MRI	Active model	None	Opt. algorithm	Constrained Powell
	Anatomy	IAA, DTA, DAA	Boundary	LP, Het. EF at adventitia		
Pourmodheji et al., 2021 [5]	Population	2D	Reference	Diastole	Target function	P-V curve error
	Pathology	PAH and CHD	Passive model	Constrained 4-fiber family		
	Data	IVP, Cine MRI, PC MRI	Active model	None	Opt. algorithm	L-BFGS
	Anatomy	Pulmonary Artery	Boundary	LP, TF adventitia		
Giuseppe et al., 2021 [281] Farzaneh et al., 2019 [112]	Population	30D	Reference	Diastole	Target function	Systolic shape.
	Pathology	aATA, w BAV and TAV	Passive model	Het. Linear elastic		
	Data	CT scans	Active model	None	Opt. algorithm	Direct solution
	Anatomy	Thoracic aorta	Boundary	LP and shape change		
Disseldorp et al., 2019, 2016 [282,283]	Population	30H 65D, 40D	Reference	Unloaded	Target function	Displacement error
	Pathology	AAA	Passive model	Hom. Neo-Hookean		
	Data	4D US, ST, CT scan, Hand cuff pressure	Active model	None	Opt. algorithm	Nelder-Mead
	Anatomy	IAA	Boundary	LP, AP		
Maso Talou et al., 2018 [284]	Population	4D	Reference	Unloaded	Target function	Displacement error
	Pathology	Atherosclerosis	Passive model	Het. Neo-Hookean		
	Data	IVUS	Active model	None	Opt. algorithm	Kalman filter
	Anatomy	Carotid artery bifurcation	Boundary	LP, EF at adventitia		

Table 8. *Cont.*

Study	Clinical Data		Forward Problem		Inverse Problem	
Liu et al., 2018 [113]	Population	4D	Reference	Diastole	Target function	Systolic shape error
	Pathology	aATA	Passive model	Hom. Holzapfel-Ogden		
	Data	CT scans	Active model	None	Opt. algorithm	multi-resolution direct search method
	Anatomy	Ascending Aorta	Boundary	LP, AP		
Wittek et al., 2016 [125]	Population	5H 1D	Reference	Axially unloaded	Target function	Displacement error
	Pathology	PAO	Passive model	Hom. Holzapfel-Ogden		
	Data	4D US, ST, Hand cuff pressure	Active model	None	Opt. algorithm	Nelder-Mead with stochastic Montecarlo sampling
	Anatomy	IAA	Boundary	LP, AP		
Wang et al., 2017 [285] Liu et al., 2012 [286]	Population	8D	Reference	Unloaded	Target function	Area change error
	Pathology	Atherosclerosis	Passive model	Mooney-Rivlin		
	Data	Cine MRI, MC MRI, Hand cuff pressure	Active model	None	Opt. algorithm	L-BFGS-B
	Anatomy	Carotid artery bifurcation	Boundary	LP, TF adventitia		
Krishnan et al., 2015 [225]	Population	4D	Reference	Unloaded	Target function	Least-squared strain error
	Pathology	aATA	Passive model	Hom. Holzapfel-Ogden		
	Data	CT scan, DENSE MRI	Active model	None	Opt. algorithm	Non-specified
	Anatomy	Ascending Aorta	Boundary	LP, TF adventitia		
Karatolios et al., 2013 [164]	Population	6H 2D	Reference	Axially unloaded	Target function	Displacement error
	Pathology	AAA	Passive model	Hom. Holzapfel-Ogden		
	Data	4D US, ST, Hand cuff pressure	Active model	None	Opt. algorithm	Nelder-Mead
	Anatomy	Abdominal aorta.	Boundary	LP, AP		
Wittek et al., 2013 [115]	Population	5H	Reference	Axially unloaded	Target function	Displacement error
	Pathology	None	Passive model	Hom. Holzapfel-Ogden		
	Data	4D US, ST, Hand cuff pressure	Active model	None	Opt. algorithm	Nelder-Mead
	Anatomy	IAA	Boundary	LP, AP		
Franquet et al., 2013 [114]	Population	2H	Reference	Diastole	Target function	Systolic shape error
	Pathology	None	Passive model	Hom. Linear isotropic		
	Data	Cine MRI, AT pressure	Active model	None	Opt. algorithm	Levenberg–Marquardt
	Anatomy	CCA	Boundary	LP, EF at adventitia		
Masson et al., 2010 [287]	Population	2H	Reference	Cut-open stress-free	Target function	Pressure waveform error
	Pathology	None	Passive model	Hom. Holzapfel-Ogden		
	Data	2D US, AT pressure	Active model	1 eq. active stress	Opt. algorithm	Levenberg–Marquardt
	Anatomy	CCA (idealized)	Boundary	Area change, Non-linear EF at adventitia.		
Taviani et al., 2008 [288]	Population	3H	Reference	Diastole	Target function	Area change error
	Pathology	None	Passive model	Hom. Linear isotropic		
	Data	Cine MRI, AT pressure	Active model	None	Opt. algorithm	Non-specified
	Anatomy	CCA	Boundary	LP, TF adventitia		

Abbreviations and acronyms: Clinical data: AT, applanation tonometry; CCA, common carotid artery; CT, computerized tomography; D, diseased; DAA, distal aortic arch; DTA, descending thoracic artery; H, healthy; IVUS, intravascular ultrasound; MC MRI, Multi-contrast magnetic resonance imaging; PAO, peripheral arterial occlusion; US, ultrasound; USST, ultrasound speckle tracking. Pathologies: AAA, abdominal aortic aneurysm; aATA, ascending aorta thoracic aneurysm; BAV, bicuspid aortic valve; CHD, congenital heart defect; TAV, tricuspid aortic valve. Forward problem: AP, adventitial pressure; EF, elastic foundation; eq., equation; Het., heterogeneous; Hom., homogeneous; LP, luminal pressure; TF, traction-free.

One of the earliest works on inverse arterial mechanics was introduced by Taviani et al., in 2008 [288]. Cine MRI was used to assess the cross-sectional geometry and distension of the common carotid artery of three healthy volunteers, while applanation tonometry was utilized to gather pressure wave data. The wall was assumed to behave as a nearly incompressible linear-elastic isotropic material with the diastolic configuration as the unloaded stress-free reference. The luminal surface was loaded with the measured pressure increment, while the adventitial surface was assumed traction-free. An optimization algorithm iterated over the elastic modulus while minimizing the normalized distance between the simulated and measured lumen. The method was successfully validated with a silicon rubber phantom and provided consistent results among all healthy adult volunteers. This inverse model of the common carotid artery was improved by Franquet et al., who incorporated the effect of perivascular support by attaching the adventitial surface to a homogeneous compressible-elastic boundary with fixed properties and a third embedded body representing the superior vena cava [114]. A Levenberg–Marquardt optimization algorithm was used to minimize a shape-based error function that accounted for pixel-wise signal intensity to define the location of the lumen. Additionally, the authors studied the effect of variability on the luminal area and wall thickness estimations used to define the reference configuration. The method was again validated against a silicon rubber phantom and applied to two adult healthy volunteers showing good agreement with estimations of elastic moduli reported in classical literature.

To incorporate the effect of residual and pre-stresses on the loaded diastolic configuration, and to fit a more complex material model, Masson et al., proposed a semi-analytical approach [289]. Clinical data from two adult volunteers 33 and 64 years of age consisted of 2D ultrasound on the common carotid artery, which was used for the resolution of luminal area change and the thickness of the intima-media layers. Additionally, planar tonometry was used to estimate the pressure wave. The carotid artery was assumed to be a pre-stressed bi-layered idealized straight cylinder. The passive material properties were assumed to follow an incompressible four-fiber family elastic constitutive equation, and active tension was assumed to act on the circumferential direction according to a single-equation active stress model. Perivascular support was modeled as a uniform adventitial pressure that exponentially increases with area increments. The forward problem was formulated as the solution of the luminal pressure corresponding to area changes assuming purely radial displacements. A Levenberg–Marquardt optimization algorithm was used to minimize the least-squared error of the predicted and measured pressure waveform. The optimization algorithm fitted 14 parameters including pre-stress parameters (opening angle and axial pre-stretch), material parameters for the two material layers, and active tension constants. The method successfully reproduced the pressure waveform while adjusting the material parameters. The authors reported that prestretch and active stress constants were similar among both patients, but passive material parameters reflected stiffer material for the older subject.

One of the first uses of image-based kinematics to estimate the anisotropic mechanical properties of a realistic large artery model was introduced by Wittek et al., in 2013 [115]. 4D ultrasound records with speckle tracking of the abdominal aorta were retrospectively collected from five healthy adult volunteers in segments proximal to the *truncus coeliacus*. Diastolic and systolic pressures were measured at the brachial artery with a sphygmomanometer. Diastolic 3D models of about 50 mm in length were segmented from ultrasound images assuming a fixed wall thickness of 1.6 mm. This configuration was assumed to be axially pre-strained by a quantity estimated by an empirical correlation. The arterial wall was assumed to behave as a modified Holzapfel–Ogden material. Perivascular support was modeled as a uniform adventitial pressure of 20 mmHg. A Nelder–Mead optimization algorithm was applied to iterate over the parameters of the material model to minimize the error of Biot's strain tensor between the benchmark measurement-derived model and the simulation. Each iteration consisted of the solution of three sequential problems: first, the inverse solution of the unloaded configuration for the given diastolic pressure and axial prestrain through a backward displacement algorithm; second, the stretch from diastolic to systolic configuration by the imposition of measurement-derived displacements to produce the benchmark model; and finally, the inflation from diastolic to systolic geometry through incrementing luminal pressure for the simulation. The resulting material parameters were used to produce stress–strain plots, which showed reasonable agreement with experimental biaxial test data from excised tissue. This method was further refined in 2016 by improving the error function and optimization algorithm. The error function was based solely on image-based estimations of strain instead of the benchmark model output. The deterministic Nelder–Mead algorithm was complemented with a stochastic Monte Carlo algorithm for the iterative generation of parameters to avoid convergence to local minima [125]. The improved method was applied to three clinical ultrasound datasets from a healthy adult volunteer, a patient with peripheral arterial occlusion, and an AAA patient. Results predicted stiffer material behavior of the arterial wall for diseased individuals when compared to results on healthy volunteers.

Pourmodheji et al. collected cine and PC MRI images, and intracardiac catheterization pressures from a pediatric patient with pulmonary hypertension and a cardiac transplant control subject. A 3D model of the main pulmonary artery with its proximal left and right branches was created at the diastolic configuration. The material model was assumed as a homogeneous constrained mixture of elastin fibers, four families of collagen fibers, and an incompressible continuum of smooth muscle cells. The constrained mixture theory was applied to prescribe pre-stretches to each constituent to balance the diastolic pressure load. An L-BFGS optimization algorithm was applied to iterate over the material parameters to minimize the cumulative error to the measured pressure–area curve at the main pulmonary artery. The model suggests that pulmonary hypertension-induced remodeling led to the stiffening of elastin fibers and wall thickening.

All of the above models assume that arteries are uniformly loaded at the luminal and adventitial surfaces. In the lumen, arterial tissue is subjected to blood pressure; however, loads and reactions on the adventitial surface are typically complex. Without appropriate adventitial boundary conditions, the deformation of a pressurized blood vessel at systole results in a homogeneously deformed configuration following the principle of minimal strain energy [14,115]. However, different image-based in vivo analyses have shown that large vessels may undergo heterogeneous deformations from diastole to systole, an effect that is not reproduced on standard in vitro pressurization setups or in silico experiments without appropriate adventitial boundary conditions [13,115,290].

These observations supported the hypothesis that the interaction of healthy blood vessels with diverse perivascular structures may induce the in vivo deformational heterogeneity [14,166]. To address this, Bracamonte et al. proposed a heterogeneous elastic foundation approach, consisting of the attachment of static linear springs of heterogeneous stiffness to the adventitial surface of arterial models. The distribution of stiffness of the elastic boundary was discretized to piece-wise constant regions and fitted through an iterative inverse algorithm to reproduce the heterogeneous deformation of the vessel [14]. For this study, retrospective cine and 2D DENSE MRI data were collected at the infrarenal abdominal aorta from nine healthy adult volunteers of diverse ages. DENSE MRI data were processed to obtain the spatial distribution of the diastole-to-systole displacement and then interpolated onto a FEM mesh built from the segmented diastolic configuration. The material was assumed to follow the Fung material model at a plane-strain state with the diastolic configuration as the unloaded stress-free reference. The Powell optimization algorithm was employed to iterate over the material parameters and elastic boundary stiffness distribution to minimize the least-squared error of the nodal displacement. Estimated material parameters reproduced the stiffening effect of aging. The elastic boundary stiffness distribution was independent of discretization and consistent among patients. Notably, it showed good agreement with the location of known anatomical features of the perivascular space, such that the vicinity to the vertebrae corresponded to the stiffest boundary, whereas the region adjacent to the peritoneal cavity resulted in the most compliant boundary.

The authors found that this approach properly captured the mechanics of the infrarenal aorta but failed to reproduce displacement measurements of the descending thoracic aorta, where the aortic wall shows both distention due to pressurization and bulk motion (Figure 13a,b). This bulk motion was hypothesized to be driven by the interactions with the adjacent beating heart. These interactions were modeled by incorporating a moving elastic foundation boundary approach [280]. This was implemented by attaching linear springs of homogeneous effective stiffness to the adventitial surface of the 2D aortic model, which was then allowed to displace radially (either inwards or outwards) to best reproduce the target bulk motion and heterogeneous wall deformation upon luminal pressurization (Figure 13c). The method was applied to a collection of retrospective cine and 2D DENSE MRI data at the infrarenal abdominal aorta, descending thoracic aorta, and descending aortic arch from 27 healthy adult volunteers of diverse, ranging from 19 to 65 years of age. A similar optimization algorithm was applied, although in this new model, the fitted elastic boundary parameters were the material model constants and spring displacement distribution, which translated directly to adventitial load distribution (Figure 13c). A parametric study was performed to study the effect of the moving elastic boundary parameters on the resulting estimations of distributed adventitial loads, which revealed that averaged adventitial load and adventitial load distributions were seemingly independent of elastic boundary parameters within the range that yielded physiologically meaningful results [280]. The proposed method converged to elastic regions that were located around relevant anatomical features (Figure 13d), and peak loads were found at locations where the heart pushes the aorta against the vertebrae (Figure 13e). Results suggest that adventitial load increases with age (Figure 13f), and that the thoracic aorta carries a larger adventitial surface load than the abdominal aorta, most likely due to the interactions with the beating heart (Figure 13g) [150].

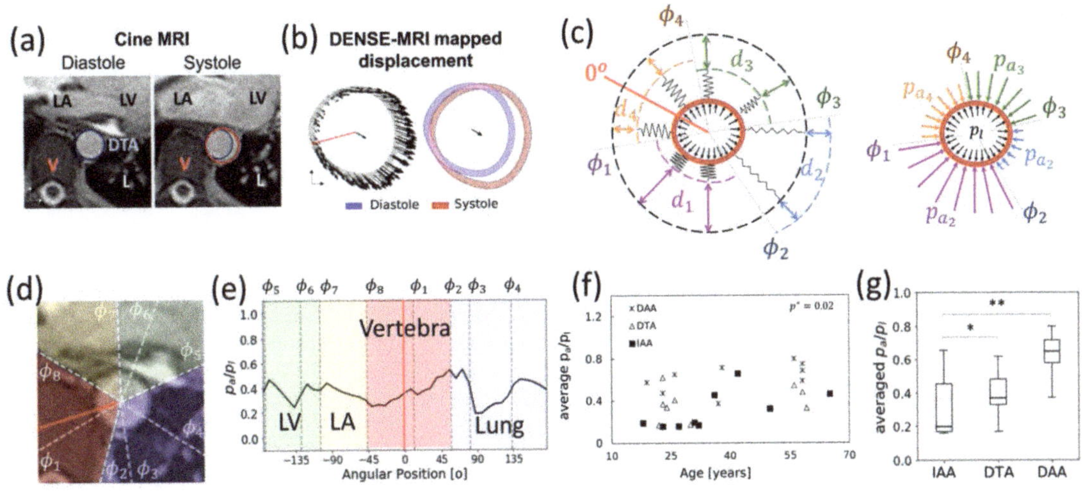

Figure 13. Inverse analysis of perivascular interactions at the descending thoracic aorta (DTA), based on results from Bracamonte et al., 2021 [150]. (**a**) Surroundings of the DTA. (**b**) DENSE MRI-derived displacement quiver representation and mapping into deformed (systolic) configuration. (**c**) Moving elastic foundation implementation and equivalent adventitial load distribution. (**d**) Patient-specific elastic boundary, and (**e**) adventitial load distribution. (**f**) Adventitial load increments with age. (**g**) Average adventitial load at different aortic locations (* p-value < 0.05; ** p-value < 0.01). Symbols: ϕ elastic boundary region angular delimiter, d moving elastic boundary displacement, p_a adventitial force per unit area, p_l luminal pressure increment, LA left atrium, LV left ventricle, V vertebra, L lung, IAA Infrarenal abdominal aorta, DTA descending thoracic aorta, DAA distal aortic arch. Red line (0°) is the angular reference selected as the closest location of the vertebra to the aortic wall.

6.4.1. Aneurysms

Aneurysms are enlarged blood vessels caused by the remodeling of its wall. When local wall stress exceeds wall strength, rupture occurs, which carries significant morbidity and mortality. Brain and aortic aneurysms are the common manifestations of this disease. Aortic aneurysms have an incidence of 5 to 10 cases per 100,000 and are responsible for approximately 15,000 deaths per year just in the United States [291,292]. Maximum aneurysm diameter and expansion rate are currently the main criteria for diagnostics and risk assessment [293]. Notably, though rupture risk increases with a maximum diameter on average for the entire population, diameter alone struggles to predict rupture for any given individual. Thus, further research is ongoing to develop more reliable metrics for predicting rupture based on biomechanics [13].

For example, Karatolios et al. applied the inverse modeling approach of Wittek et al., (2013) to study the strain distribution in two abdominal aortic aneurysms (AAA) of two adults and the abdominal aorta of six healthy controls [164]. Results suggested that peak strains in AAAs are time-delayed (in late systole) with respect to their occurrence in healthy aortas. This work was followed by an extensive retrospective study published by van Disseldorp et al., in 2016, which gathered information from 40 AAAs patients that underwent CT scans and 4D Ultrasound with speckle tracking [282]. Patient-specific 3D models of the abdominal aneurysm were generated from the CT scans with a fixed wall thickness of 2 mm for all cases. The arterial wall was assumed to behave as a neo-Hookean material. The shear modulus was estimated iteratively to minimize the diastole-to-systole nodal displacement between forward FEM simulations and speckle-tracking derived measurements (Figure 14a). The error function was designed ad hoc so that regions with more precise and reliable measurements carried more weight when calculating error. For each iteration, the pressurization from diastole-to-systole was preceded by the estimation of the diastolic stress applying the backward increment algorithm. Systolic pressure was assumed to be 140 mmHg for all cases and the reference geometry built from CT scans was assumed to be at a mean arterial pressure of 105 mmHg. Interestingly, results from the study suggested that aneurysms with larger diameters tend to be stiffer. An extension of this work was published in 2019 by van Disseldorp et al., with a comparative study of material properties from 30 healthy volunteers and 65 AAA patients using 4D Ultrasound datasets [283]. Healthy cases were grouped by age, whereas AAA patients were grouped by aneurysm diameter. Segmentation and parameter estimation followed the same methodology as previously; however, patient-specific diastolic and systolic pressures were measured from a sphygmomanometer and used as boundary conditions for the backward increment method and forward simulations. The analysis showed a significant difference in stiffness between age-matched healthy volunteers and AAA patients even at the early stages of the disease (Figure 14b). The study suggests that most of the stiffening occurred at the onset of the disease with slight further increases as the aneurysm grows (Figure 14c). Additionally, a significant correlation between peak stress and aneurysm size was found, which is consistent with the general correlation of aneurysm size with wall rupture.

Krishnan et al., performed inverse model analyses on ascending thoracic aorta aneurysms (aTAA) [286]. These authors collected CT angiography and DENSE MRI sequences from four patients. Three-dimensional models of aortic aneurysms were built from the CT scans at systolic configuration. The Ogden isotropic hyperelastic constitutive equation was selected as the material model. They applied an iterative updating algorithm to find the set of material parameters that minimized the least-square error of simulated strains against DENSE MRI-derived estimations. The iterative algorithm consisted of three steps: first, a deflation step to 0 mm Hg (assume to be the zero-stress reference), followed by the inflation to the assumed 120 mm Hg at systole, and finally, the deflation to diastolic pressure of 80 mm Hg. This study revealed that the estimated peak principal stress is circumferential and about 25% greater than the average stress in aTAAs and located in the inner and outer curvature of the arch towards the pulmonary artery.

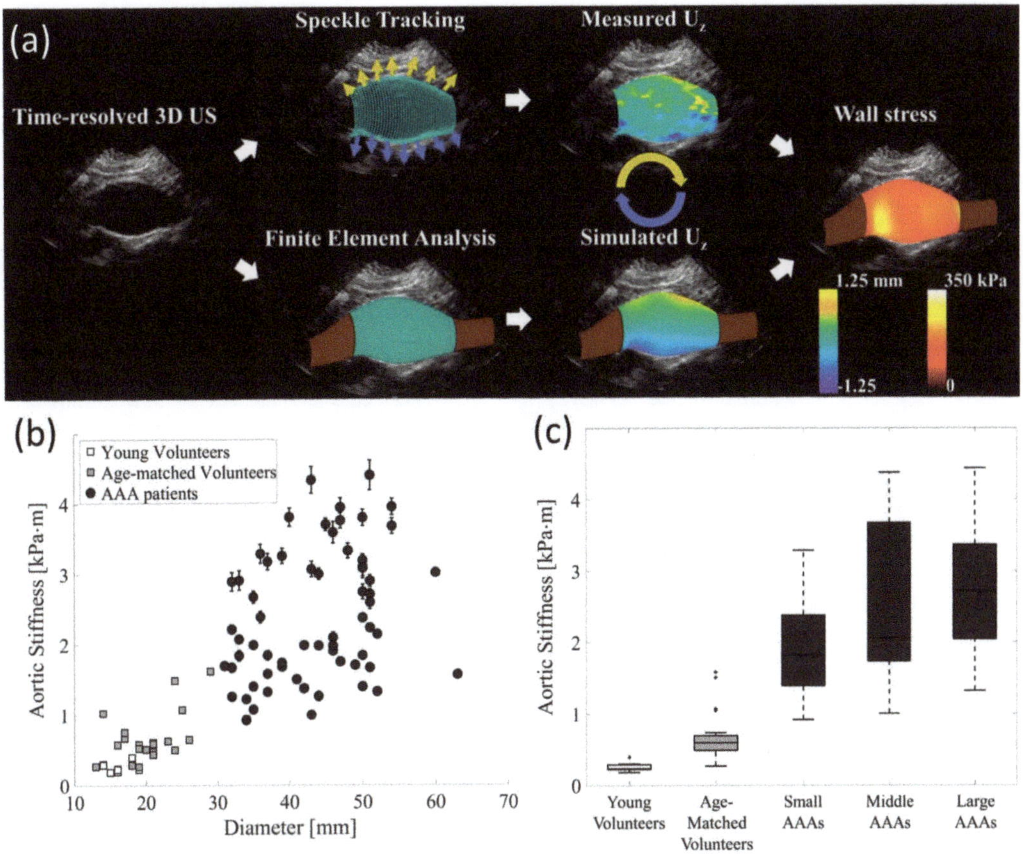

Figure 14. Inverse analysis of abdominal aneurysm mechanics (Reprinted/adapted with permission from Ref. [283]. 2018, Oxford University Press). (**a**) Patient-specific data processing algorithm and typical outputs. (**b**) Aortic stiffness versus maximum aortic diameter for healthy volunteers (gray squares) and AAA patients (black circles). (**c**) Population-based statistics of aortic stiffness in a box and whisker plot with dots representing outliers. Results suggest that most wall stiffening occurs at early stages of the disease when the aneurysm diameter is still relatively small.

Liu et al. explored new methods to reduce the computational cost of inverse analyses while studying the mechanical properties of aTAAs. First, they investigated a method based on the computation of wall stress by solving a simplified statically determinate problem to obtain an "almost true" stress field [104]. They collected retrospective CT angiography from 4 patients with aTAA who went through surgical repair with tissue excision used for ex vivo biaxial testing. The geometry was built at the systolic configuration and assumed to be loaded at 120 mmHg. The material was modeled with a Holzapfel-Ogden constitutive equation. The backward displacement algorithm was used on each iteration to calculate the unloaded configuration assumed to be stress-free. An iterative inverse method was applied to obtain an estimation of material parameters using a constrained gradient-free trust-region optimization algorithm. Each iteration consisted of two steps: first, computing an almost true stress field from the in vivo geometries and loading conditions by using the Laplace law for statically determinate stiff thin-wall vessels; and second, calculating the stress distribution with the updated material parameters. The target function for the optimization algorithm was defined as the least-squared error of the simulated to almost true stress.

Constraints consisted of upper and lower limits for material parameters extracted from the literature. Estimated material properties showed good agreement with results from patient-specific mechanical tests from excised tissue while decreasing the computational cost relative to regular iterative inverse approaches. Subsequently, Liu et al. used the same database and material model to explore the effectiveness of the multi-resolution direct search method as the optimization algorithm [113]. This algorithm works by decomposing the search for the optimal material parameters with a multi-scale representation of the parameter hyperspace. The target function to be minimized was defined in terms of the distance between surface nodes and the location of the segmented surface at systole. The converged material properties successfully reproduced the strain energy curves from biaxial testing while considerably reducing the computational cost of the inverse approach.

All these studies assumed material homogeneity of the aortic wall, which is a major limitation for the study of aneurysms. In vitro mechanical tests and histology analyses have been performed on aneurysms from human cadavers revealing both structural and mechanical heterogeneity [294,295]. Farzaneh et al. studied material heterogeneity on three aTAA patients from which CT scans were collected [112]. Medical images were used to build 3D models of the aneurysms at diastole and systole, and these models were used to estimate the local strain state. Each element on the wall surface was assumed to be part of an ellipsoid sharing the center to the cross-section of the vessel and was assumed to behave as a linearly elastic material. The stiffness was directly calculated element-wise from local balance equations. Their results suggested that diseased tissue was stiffer in the bulging part of the aneurysm and generally stiffer than the adjacent non-aneurystic tissue. Giuseppe et al. further applied this methodology to a cohort of 30 aTAA patients, 12 with bicuspid aortic valves, and the remaining with normal tricuspid valves [281]. Wall stiffness distribution was heterogeneous for each individual, however, regional differences appeared to be marginal within the cohort due to interindividual variability. Notably, this study found no significant differences in stiffness nor its distribution between the bicuspid and tricuspid valve groups, suggesting that no distinction should be made in the surgical management of aneurysms between these groups.

6.4.2. Atherosclerotic Plaques

Atherosclerosis is a chronic inflammatory disease that manifests as the hardening and occlusion of arteries due to the build-up of plaque on the lumen of the arterial wall. Atherosclerotic plaque is a mixture of fatty substances, cholesterol, calcium, and cellular waste, usually enclosed in a fibrous cap. Atherosclerotic lesions are generated at specific regions of the arterial tree, mostly in the vicinity of branch points, the outer wall of bifurcations, and the inner wall of curves [157]. Among many possible associated complications, plaque can break and detach, generating thrombosis, acute myocardial infarction, and stroke. Thus, the in vivo evaluation of the mechanical properties of atherosclerotic plaques and their mechanical environment could support the assessment of risk associated with plaque rupture. One of the earliest inverse analyses of atherosclerotic plaques was proposed by Liu et al., in 2012 [286]. This study was performed on 12 patients with carotid artery atherosclerosis. For each patient, a set of cine MRI, 3D multi-contrast MRI, and sphygmomanometry were collected. Two-dimensional models of the diseased sections were built from MRI images at diastole, including lipid pools resolved by multi-contrast MRI. The arterial wall and plaque were assumed uniform and to behave as a Mooney–Rivlin hyperelastic material, while the lipid pools were assumed to be isotropic linear elastic. An L-BFGS-B optimization method was applied to fit the material properties of the wall plaque until the error between the simulated and measured diastole-to-systole area change was minimized. Each iteration included the estimation of the unloaded configuration by the shrink-and-fit algorithm, and a forward FEM problem for the inflation from the unloaded configuration to the systolic configuration applying uniform luminal pressurization. The authors found the estimations of material stiffness show reasonable agreement with reported data from experimental studies. An analysis of stress distribution indicated that, for all cases, peak stress was located at the thin cap covering the lipid core. This study was further refined by Wang et al., in 2017 [285], with similar imaging and functional data acquired for 8 patients with carotid atherosclerosis with follow-up tests after 18 months. The material models, optimization algorithm, target function, and iteration steps were the same as previously; however, a total of eight slices were analyzed from each carotid artery and modeled as a 3D thin layer so that axial prestretch could be included in the estimation of the unloaded configuration. Results revealed high patient-to-patient variability on plaque stiffness, which was significantly larger in the hypertensive cases. The authors also found that estimations of material properties of the plaque can significantly change over time, with stiffness increments being the most common scenario. Huang et al. further explored these results with FSI simulations based on patient-specific estimations of atherosclerotic tissues with patient-specific measurements of pressure gradients by applanation tonometry and confirmed that flow and pressure-induced stresses peak at the fibrous cap that covers the lipid core, which could offer support to explain the main mechanisms of plaque rupture [296].

The main limitation of previous studies is that the current resolution of non-invasive imaging techniques is insufficient to resolve the displacement of atheroma plaques in small vessels such as the carotid artery. To overcome this, Maso Talou et al. utilized intravascular ultrasound technology [284]. This work analyzed data from 4 atherosclerotic lesions which were modeled as 3D thin cross-sectional slices. Each model was single-layered and divided into six circumferential sections, each portion being assumed materially homogeneous and following the Neo–Hookean hyperelastic material model. Perivascular tethering was modeled as a homogeneous elastic media of fixed stiffness. Kalman filters were used to estimate material parameters for each section while minimizing the diastolic-to-systolic displacements. Each iteration included the estimation of diastolic stress distribution by a backward increment method assuming a pressure load of 80 mmHg and population average-based axial stretch. From this preloaded state, a forward inflation problem to systolic pressure was then solved. Parallelization techniques were employed to reduce computer processing times achieving convergence between 12 h and three days. Sensitivity of the results to numerical and model parameters was carried out, finding that perivascular elastic properties have a significant effect on material parameter predictions. The estimated material parameters agreed with the magnitudes reported from available experimental data.

6.5. Hemodynamics

In general, computational modeling of hemodynamics is more resource-consuming than tissue mechanics, as simulations need to account for transient effects and deal with the difficulties introduced by the non-linearities of convection and momentum dissipation. This makes the application of inverse modeling to hemodynamics a challenging task.

The use of simplified 0D (lumped) and 1D models can significantly reduce the computational cost. These simplified models have been used on a patient-specific basis and implemented onto inverse modeling approaches to provide useful systemic information about flow distribution, vascular resistance, and the systemic effect of drug treatments [297,298]. However, these approaches cannot exploit the detailed features offered by modern image-based kinematics as they only deal with 2D integrated or averaged metrics. Furthermore, despite all assumptions and simplifications, inverse approaches to lumped and 1D models are still prone to solution multiplicity [119]. With our focus on inverse modeling based on image-based kinematics, these approaches employing 1D simplified models fall outside the scope of this review.

To deal with the computational expense of the forward problem on inverse hemodynamics, Lassila et al. proposed a method for parametrizing the Navier–Stokes equations and patient-specific geometries to reduce the basis of the partial differential equations. The parameterized model is iteratively solved until the algorithm is close to the final solution. At this point, the inverse method then switches to the solution of the full-forward problem using FVM. This method was tested using deterministic and Bayesian optimization algorithms showing promising results on the solution of test cases involving rigid-wall and FSI simulations [103]. Herein, we review some of the existing research on inverse hemodynamics separating approaches that assume rigid-wall flow boundaries from those using FSI approaches.

6.5.1. Rigid Wall Models

Romarowski et al. applied an iterative inverse method for the hemodynamic study of three descending thoracic aortic aneurysms. CT scans were used to build the 3D models that included the ascending and descending aorta [118]. PC MRI sequences were collected at the ascending aorta (above the aortic bulb), the suprarenal abdominal aorta, and all three branches of the aortic arch. Diastolic and systolic pressures were collected from sphygmomanometry. The authors observed that balances with the inlet and outlet flow rates measured with PC MRI did not comply with the conservation of mass principle. The forward problem was defined by applying the PC MRI-derived velocity distribution in the ascending aorta as an inlet boundary condition. At all four outlets, a surrogate three element Windkessel model of unknown parameters was imposed as a boundary condition, while blood was assumed to be an incompressible Newtonian fluid. The forward problem was solved by a FEM solver. An optimization algorithm was applied to minimize the least-squared error of the measured blood flow at the outlets to simulation estimates, by fitting the surrogate model parameters. The authors highlight that this weak approach allows distributing the error related to measurement noise while enforcing mass conservation. Similarly, Gaidzik et al. used PC MRI data from a healthy volunteer to find the pressure gradient distribution in the circle of Willis, an important cerebral arterial system [299]. In this work, Kalman filters are iterated over pressure boundary conditions to adjust the simulated flows to PC MRI measurements with an FVM solver for the forward problem. Noise-to-signal ratios were used to incorporate the measurement uncertainty into the data analysis. The authors highlight that the outputs of the inverse methods yield smaller uncertainties than CFD or 4D flow MRI data analysis alone.

Rispoli et al. proposed a modification to the implementation of FVM for fluid dynamics problems, to introduce the minimization of simulated nodal velocity components to 4D flow data measurements in the linearized SIMPLER algorithm [300]. The method required the smoothing and interpolation of coarse 4D flow MRI data to the FVM mesh. 4D flow MRI-derived velocities were directly used as inlet and outlet boundary conditions. The minimization problem and FVM solution were solved simultaneously using a version of the iterative Runge–Kutta algorithm. This method allowed the simultaneous solution of the simulation and inverse problems, thus reducing the computational expense. As a proof of concept, the method was applied to anatomy and 4D flow MRI scans of a healthy human carotid artery. The method was incorporated into a custom-made solver that required special discretization into a structured mesh in the Cartesian space. Töger et al. further developed this approach by incorporating the nodal velocity error minimization approach into a discontinuous Galerkin FEM formulation, allowing the solution of unstructured meshes [122]. The method was validated to in vitro measurements with laser particle imaging velocimetry in a pulsating flow loop with an abrupt change of cross-sectional area to induce complex flow patterns. Then, the method was applied to a healthy-human proximal cerebral artery. CT angiography was used to build the 3D anatomic model, 4D flow data were collected at a resolution of 0.7 mm voxel size with a 7 T scan, and PC MRI scans were collected at inlet and outlet planes with a resolution of 0.5 mm/px. Moreover, 4D flow data were spatially and temporally smoothed and interpolated into the FEM mesh, while PC MRI data were integrated to enforce inlet and outlet transient plug flow as boundary conditions. The method showed errors below 1% on velocity distribution for in vitro validation tests, and the proof of concept on in vivo datasets demonstrated the potential of the proposed methodology for future human studies.

6.5.2. Fluid-Structure Interaction (FSI) Models

Fluid–structure interaction simulation is itself a complex, resource-consuming process, and its incorporation with inverse models is challenging. Some of the early work by Moireau, Chapelle, D'Elia, Perego, among others, set the bases for inverse modeling of FSI by calibrating models to in vitro experiments and synthetic datasets [301–303].

In 2014, Bertoglio et al. proposed the use of Kalman Filters to estimate the material properties of several regions of the aorta from inverse FSI [304]. Available clinical data included SSFP MRI, intravascular pressure measurements at the ascending, thoracic, and abdominal aorta, and PC MRI measurements at four planes along the aorta. The aorta was divided into four sections each one assumed to follow the Mooney–Rivlin material model. The arbitrary Lagrangian–Eulerian algorithm was implemented to couple fluid and wall mechanics. Kalman filter optimization was used to minimize an error function based on all available clinical measurements weighted by the associated uncertainty while fitting the regional material parameters. Results reproduced the expected stiffness distribution, with stiffer distal descending aorta.

Zambrano et al. proposed an iterative inverse method for the study of the pulmonary artery [305]. Intravascular pressure measurements, PC MRI at the main branches of the pulmonary artery, and cine MRI were collected from a pulmonary hypertensive adult patient and a healthy volunteer with no reported cardiovascular disease. A 3D model from the main pulmonary artery (MPA) down to the 4th branch generation was built from MRI images at the end-diastole configuration, which was considered stress-free. MRI-derived diameter changes were calculated at the main pulmonary artery and coupled to pressure measurements.

The arterial wall was assumed homogeneous and isotropic linear elastic throughout the entire domain. The fluid–structure interactions were modeled with the coupled momentum method. Boundary conditions consisted of PC MRI-derived inlet flow and three-parameter Windkessel models in the outlets. The elastic modulus of the wall and Windkessel boundary parameters were calibrated by iterating in two nested loops. In the inner loop, the Windkessel parameters were adjusted until the error to the measured pressure waveform is minimized, while the outer loop adjusted the elastic modulus until the error to the measured pressure-area curve is minimized (Figure 15a). On each iteration, the forward problem was solved until solution periodicity was confirmed. The proposed model was able to reproduce the expected increase in arterial stiffness and vascular flow resistance in the hypertensive patient. In a follow-up study, the methodology was applied to a cohort of six individuals with pulmonary artery hypertension and five healthy volunteers [306]. A statistical analysis of the results revealed that the hypertensive group showed significantly larger wall stiffness, regurgitant flow, and distal vascular resistance, with significantly smaller time-averaged wall shear stress (Figure 15b). Interestingly, a linear correlation between the estimated wall elastic modulus and the magnitude of retrograde flow volume was found, which further supports the hypothesized relation between irregular flow patterns and the pathological remodeling of vascular tissue.

6.6. Summary

In Table 9 we summarize the highlights of inverse analyses for cardiovascular mechanics applications and notable results.

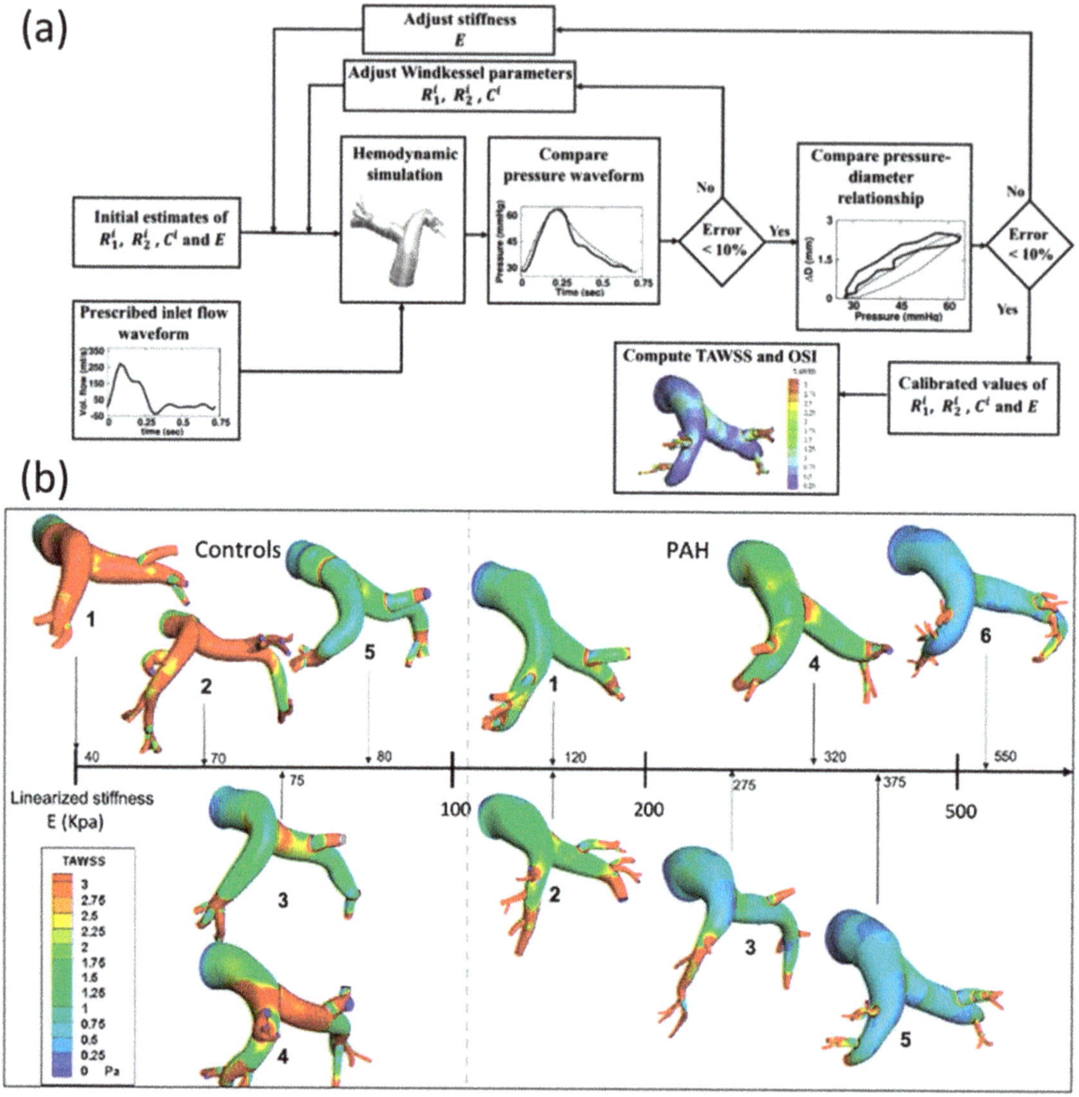

Figure 15. Iterative inverse method and results for the study of pulmonary artery hypertension. (**a**) Double optimization loop for the inverse resolution of distal vasculature Winkesel model parameters and wall elastic modulus from phase contrast MRI data. (Reprinted/adapted with permission from Ref. [305]. 2018, Elsevier) (**b**) Biomechanical parameters of pulmonary artery from healthy individuals and pulmonary artery hypertension patients. Models are organized from left to right according to the wall elastic modulus (stiffness) scale, colormaps show the time-averaged wall shear stress (TAWSS) distribution (Reprinted/adapted with permission from Ref. [306]. 2021, Zambrano et al.; open access).

Table 9. Summary of medical imaging-based kinematics.

Section	Highlights
6.	The development of patient-specific inverse analyses of cardiovascular mechanics has advanced considerably recently thanks to continuous technological improvements in imaging hardware and software, decreasing cost, increased imaging availability, improvements in image-based kinematics acquisition and postprocessing, simulation engineering, and significant increases in computational power.
6.1	Blood vessels, in particular those of the arterial tree, function under physiological pressure load at all times and are axially pre-stretched; thus, none of the patient-specific configurations resolved by in vivo imaging is truly a stress-free or zero-strain configuration. The unloaded configuration of cardiovascular tissue is not truly stress-free. The residual stress is hypothesized to be the product of heterogeneous growth and remodeling of tissue. For patient-specific analyses, the material properties and zero-stress configuration are unknown. Thus, the solution to this problem requires the specification of at least two deformed and loaded states as input data. Direct methods for the solution of inverse elastostatic problems to determine the unloaded configuration of the heart and arteries have been incorporated into FEM solvers for hyperelastic and fiber-family material models. Several iterative methods for the solution of the unloaded configurations have been proposed. All these methods have in common that a single point or a collection of points on the surface are fixed, while forward inflation problems from unloaded configuration iterations to the known loaded configurations are solved until a convergence criterion is satisfied. Unloaded configuration iterations are estimated either by shrinking the known loaded configuration or by taking "backward" inflation steps. An alternative iterative approach is to solve the strain and stress distribution that balances the applied loads acting on the image-derived anatomic configurations without the resolution of the unloaded geometry.
6.2	The inverse modeling of the heart as a whole is currently unfeasible due to the complexity of the system and computational limitations. An accurate understanding of myocardial mechanics is key for the diagnosis and treatment of diverse cardiac pathologies, and potentially, to predict and stratify the risk of heart failure after infarct. The assumption of material homogeneity is a common and convenient simplification for forward and inverse models. Homogeneous models may be deemed to be adequate for the study of healthy hearts, or when the aim of the analysis is not centered on the study of focalized lesions. Homogeneous models can quantify the stiffening effect of infarct lesions and predict the natural compensation of the active component of the heart to maintain cardiac function after infarction. Modeling of material heterogeneity of the heart can provide better fits to kinematic data, can resolve property changes, and identify the location and severity of myocardial lesions. This comes with an increment of model complexity and computational expense. A common approach is to approximate spatial variations of myocardial properties and microstructure with region-wise heterogeneities. AHA standard division of the left ventricle is often used to define region-wise heterogeneity. Heterogenous models of the myocardium can identify the material properties of the infarcted zone, the border zone, and the unaffected tissue. Heterogeneous models can accurately predict how impaired activation of the myocardium affects the cardiac function in patients with left bundle branch block (LBBB). Inverse analyses with heterogeneous models have been used to predict the effect of ischemia on cardiac function, and its recovery after revascularization treatment.
6.3	Heart valves and leaflets are thin structures with complex motion that are difficult to resolve through in vivo imaging techniques. Owing to this, most studies on these structures are carried out in vitro. Recent developments in US imaging of heart valves are the first steps toward the in vivo inverse modeling of these structures.
6.4	Changes in mechanical properties of arterial walls have been associated with the onset of multiple cardiovascular pathologies and remain an important predictor of cardiovascular morbidity and mortality in clinical practice. The image-based resolution of vascular tissue kinematics is technically challenging due to the relative thinness of vascular walls. Inverse analyses of healthy arteries have been used to assess the stiffening effect of aging and to explore the effect of perivascular interaction on aortic mechanics. Aneurysms are a potentially fatal condition that consist of the enlargement of blood vessels caused by the remodeling of its wall. Aneurysmal rupture risk increases with maximum diameter on average for the entire population, although diameter alone struggles to predict rupture for any given individual. Inverse modeling has been used to obtain heterogeneous maps of mechanical stress and strain in thoracic and abdominal aneurysms and to assess the effect of disease progression on tissue stiffening. Atherosclerosis is a chronic inflammatory disease that manifests as the hardening and occlusion of arteries due to the build-up of plaque on the lumen of the arterial wall. The in vivo evaluation of the mechanical properties of atherosclerotic plaques and their mechanical environment through inverse modeling could support the assessment of risk associated with plaque rupture.
6.5	Computational modeling of hemodynamics is more resource consuming than tissue mechanics. Statistical analyses have shown that outputs of the inverse methods yield smaller uncertainties than CFD or 4D flow MRI data analysis alone. Inverse modeling of the fluid–structure interaction of the blood flow in the pulmonary arteries has been used to identify relevant markers of pulmonary artery hypertension. Among these markers are wall stiffness, wall shear stress and oscillation, pulse wave velocity, and regurgitant flow.

7. Closing Remarks

Inverse modeling is an analysis tool that can provide detailed information about domain properties and loading conditions using kinematic measurements as inputs. When applied to collected data from controlled in vitro experiments it can provide dynamic information with high levels of accuracy and reliability. In biomedical research, inverse modeling has been coupled with microscope-based imaging techniques to yield relevant information on the response of cardiovascular and engineered tissue to mechanical stimuli at the cellular level. These contributions hold relevant scientific value in the fields of mechanobiology and tissue engineering, however, the extrapolation of these results to patient-specific cases is limited.

There is great interest in the development of reliable patient-specific non-invasive medical tools to assess the onset and progression of cardiovascular disease. This has led to significant advances in non-invasive medical imaging, including improvements in resolution, scan time, operational costs, availability, and the ability to quantify detailed regional kinematic information. Inverse biomechanical analyses can exploit this available clinical data to provide patient-specific estimations of dynamic parameters that otherwise require invasive (and potentially risky) procedures, such as vascular catheterization, or cannot be measured at all. Inverse modeling fits dynamical unknowns to kinematic data, which would be simply assumed with fixed values on classical forward modeling approaches. However, inverse modeling cannot entirely substitute measurements of absolute pressure (required to define the loading boundary conditions); instead, this technique can be used to estimate other relevant biomarkers defined in terms of pressure or load differences, such as vascular flow resistance. As highlighted in this review of the clinical applications of these methodologies, inverse analyses can estimate stiffness for healthy and diseased cardiac and vascular tissues, identify and delineate pathological lesions, resolve tissue composition, and quantify mechanical loads and stresses during in vivo function. Inverse modeling can also provide physiological rationales for empirically derived risk factors, such as aneurysmal diameter and ventricular volume, as well as yield new sets of physiologically meaningful risk markers. In addition, inverse modeling can deliver insights into how biological tissues respond and adapt to pathology and/or therapies through comparative studies, such as regional changes in active contraction within infarcted hearts or tissue growth and remodeling in aneurysmal arteries.

Despite all these advantages, the incorporation of patient-specific inverse-modeling in clinical practice still faces several challenges, including the presence of multiple solutions, uncertainty regarding patient-specific stress-free reference configurations, computational costs, and the lack of required clinical and imaging data. The multiplicity of solutions is a common challenge to any inverse problem, and the solution set can be reduced by constraining the optimization parameters within ranges of expected values, incorporating regularization functionals, sampling stochastic parameters, designing special optimization target functions, and, for the specific case of Bayesian approaches, providing probability distributions of parameters from previous experiences.

A step towards resolving patient-specific stress-free references for tissue mechanics is the inverse solution of unloaded configurations through direct and iterative methods. However, it is generally accepted that unloaded blood vessels are not truly stress-free due to the existence of residual stress/strains which are influenced by the heterogeneous growth and continuous remodeling of the tissue, including the prestretch of key extracellular matrix components such as collagen. This issue could potentially be addressed by the implementation of a constrained mixture theory and the in vivo resolution of tissue microstructure via medical imaging.

The computational cost of iterative inverse methods is often addressed by simplifications of the forward problem, the use of surrogate models for early optimization stages, utilization of more efficient iterative optimization methods, and the use of parallel computing. Furthermore, the ongoing increase of computational power may allow the solution of complex problems that escapes the reach of current technology.

Similarly, it is reasonable to expect that medical imaging technology will continue to evolve, making them more readily available in healthcare practice. The development of data-driven techniques for the support of clinical decision making and treatment planning could also motivate the implementation of image-based kinematics in routine health care.

Inverse modeling is just one of many patient-specific techniques that have been proposed as a useful support for clinical practice. Machine learning has been increasingly explored in the last two decades for incorporation into the new field of precision medicine [307]. This technique consists of training decision-making algorithms with annotated large datasets, which when combined with the application of statistical principles, can return valuable evidence-based information from raw clinical data [308]. The main advantage of machine learning techniques is that once the algorithm has been trained, results can be obtained in short times with low associated computational cost. However, the outcomes are highly dependent on the quality of the annotated dataset used for training, as they are not the result of a physiology-based simulation but on statistical probabilities calculated from collected evidence. Thus, this approach can potentially fail if unique or unexpected conditions are presented.

An additional advantage of simulation-based techniques is their predictive capabilities. Founded on physical and physiological principles, patient-specific inverse problems can be coupled to mechanobiology-inspired growth and remodeling models to potentially predict the progression of diseases and/or the effect of treatments [106]. In conclusion, image-based inverse modeling is a promising quantitative tool to generate and analyze clinically relevant physiological data through a non-invasive approach with the ultimate goal of providing improved patient-specific diagnostic and prognostic assessments of diverse cardiovascular diseases in order to improve outcomes, reduce costs, and increase the quality of life.

Author Contributions: Conceptualization, J.H.B. and J.S.S.; methodology, J.H.B., S.K.S. and J.S.S.; data collection, J.H.B., U.T.T. and J.S.W.; data analysis and interpretation, J.H.B., U.T.T., J.S.W. and J.S.S.; writing—original draft preparation J.H.B.; writing—review and editing J.H.B., S.K.S., J.S.W., U.T.T. and J.S.S.; visualization J.H.B. and U.T.T.; supervision J.S.S.; project administration, J.S.S.; funding acquisition, J.H.B., S.K.S. and U.T.T. All authors have read and agreed to the published version of the manuscript.

Funding: This work was supported by the American Heart Association and the D.C. Women's Board (award number 834649/2021), the American Heart Association (award number 20PRE35210730/2020), and the National Institute of Health (grant number NIH NHLBI K23HL135352).

Acknowledgments: We thank Jenna Gaspar, Lindsay Jordan, Nickolas Roberts, Raymond Hang and the VCU College of Engineering DURI Dean's Undergraduate Research Initiative Program.

Conflicts of Interest: The authors declare no conflict of interest.

References

1. Kakisis, J.D.; Liapis, C.D.; Sumpio, B.E. Effects of Cyclic Strain on Vascular Cells. *Endothelium* **2004**, *11*, 17–28. [CrossRef] [PubMed]
2. Chang, H.-I.; Wang, Y. Cell response to surface and architecture of tissue engineering scaffolds. In *Regenerative Medicine and Tissue Engineering—Cells and Biomaterials*; Eberli, D., Ed.; InTech Open Access Publisher: Rijeka, Croatia, 2011. [CrossRef]
3. Butler, D.L.; Goldstein, S.A.; Guilak, F. Functional Tissue Engineering: The Role of Biomechanics. *J. Biomech. Eng.* **2000**, *122*, 570–575. [CrossRef] [PubMed]
4. Tang, B.T.; Pickard, S.S.; Chan, F.P.; Tsao, P.S.; Taylor, C.A.; Feinstein, J.A. Wall Shear Stress is Decreased in the Pulmonary Arteries of Patients with Pulmonary Arterial Hypertension: An Image-Based, Computational Fluid Dynamics Study. *Pulm. Circ.* **2012**, *2*, 470–476. [CrossRef] [PubMed]
5. Pourmodheji, R.; Jiang, Z.; Tossas-Betancourt, C.; Figueroa, C.A.; Baek, S.; Lee, L.-C. Inverse modeling framework for characterizing patient-specific microstructural changes in the pulmonary arteries. *J. Mech. Behav. Biomed. Mater.* **2021**, *119*, 104448. [CrossRef] [PubMed]
6. Watton, P.N.; Hill, N.A. Evolving mechanical properties of a model of abdominal aortic aneurysm. *Biomech. Model. Mechanobiol.* **2009**, *8*, 25–42. [CrossRef] [PubMed]
7. Neal, M.L.; Kerckhoffs, R. Current progress in patient-specific modeling. *Brief. Bioinform.* **2010**, *11*, 111–126. [CrossRef] [PubMed]

8. Marsden, A.; Esmaily-Moghadam, M. Multiscale Modeling of Cardiovascular Flows for Clinical Decision Support. *Appl. Mech. Rev.* **2015**, *67*, 030804. [CrossRef]
9. Itatani, K.; Miyazaki, S.; Furusawa, T.; Numata, S.; Yamazaki, S.; Morimoto, K.; Makino, R.; Morichi, H.; Nishino, T.; Yaku, H. New imaging tools in cardiovascular medicine: Computational fluid dynamics and 4D flow MRI. *Gen. Thorac. Cardiovasc. Surg.* **2017**, *65*, 611–621. [CrossRef]
10. Wilson, J.S.; Zhong, X.; Hair, J.B.; Taylor, W.R.; Oshinski, J.N. In Vivo Quantification of Regional Circumferential Green Strain in the Thoracic and Abdominal Aorta by Two-Dimensional Spiral Cine DENSE MRI. *J. Biomech. Eng.* **2019**, *141*, 0609011–06090111. [CrossRef]
11. Kong, F.; Shadden, S.C. Automating Model Generation for Image-Based Cardiac Flow Simulation. *J. Biomech. Eng.* **2020**, *142*, 1110111–11101113. [CrossRef]
12. Markl, M.; Schnell, S.; Barker, A. 4D Flow Imaging: Current Status to Future Clinical Applications. *Curr. Cardiol. Rep.* **2014**, *16*, 481. [CrossRef]
13. Wilson, J.S.; Taylor, W.R.; Oshinski, J. Assessment of the regional distribution of normalized circumferential strain in the thoracic and abdominal aorta using DENSE cardiovascular magnetic resonance. *J. Cardiovasc. Magn. Reson.* **2019**, *21*, 59. [CrossRef]
14. Bracamonte, J.H.; Wilson, J.S.; Soares, J.S. Assessing Patient-Specific Mechanical Properties of Aortic Wall and Peri-Aortic Structures from In Vivo DENSE Magnetic Resonance Imaging Using an Inverse Finite Element Method and Elastic Foundation Boundary Conditions. *J. Biomech. Eng.* **2020**, *142*, 1210111–12101113. [CrossRef]
15. Bonnet, M.; Constantinescu, A. Inverse problems in elasticity. *Inverse Probl.* **2005**, *21*, R1–R50. [CrossRef]
16. Wang, W.; Wang, D.; Falisse, A.; Severijns, P.; Overbergh, T.; Moke, L.; Scheys, L.; De Groote, F.; Jonkers, I. A Dynamic Optimization Approach for Solving Spine Kinematics While Calibrating Subject-Specific Mechanical Properties. *Ann. Biomed. Eng.* **2021**, *49*, 2311–2322. [CrossRef]
17. Zhang, Y.; Wang, V.Y.; Morgan, A.E.; Kim, J.; Ge, L.; Guccione, J.M.; Weinsaft, J.W.; Ratcliffe, M.B. A Novel MRI-Based Finite Element Modeling Method for Calculation of Myocardial Ischemia Effect in Patients with Functional Mitral Regurgitation. *Front. Physiol.* **2020**, *11*, 158. [CrossRef]
18. Avazmohammadi, R.; Li, D.S.; Leahy, T.; Shih, E.; Soares, J.S.; Gorman, J.H.; Gorman, R.C.; Sacks, M.S. An integrated inverse model-experimental approach to determine soft tissue three-dimensional constitutive parameters: Application to post-infarcted myocardium. *Biomech. Model. Mechanobiol.* **2018**, *17*, 31–53. [CrossRef]
19. Butler, J.P.; Tolić-Nørrelykke, I.M.; Fabry, B.; Fredberg, J.J. Traction fields, moments, and strain energy that cells exert on their surroundings. *Am. J. Physiol. Physiol.* **2002**, *282*, C595–C605. [CrossRef]
20. Tambe, D.T.; Croutelle, U.; Trepat, X.; Park, C.Y.; Kim, J.H.; Millet, E.; Butler, J.P.; Fredberg, J.J. Monolayer Stress Microscopy: Limitations, Artifacts, and Accuracy of Recovered Intercellular Stresses. *PLoS ONE* **2013**, *8*, e55172. [CrossRef]
21. Islam, M.M.; Steward, R.L. Probing Endothelial Cell Mechanics through Connexin 43 Disruption. *Exp. Mech.* **2019**, *59*, 327–336. [CrossRef]
22. Banerjee, I.; Carrion, K.; Serrano, R.; Dyo, J.; Sasik, R.; Lund, S.; Willems, E.; Aceves, S.; Meili, R.; Mercola, M.; et al. Cyclic stretch of embryonic cardiomyocytes increases proliferation, growth, and expression while repressing Tgf-β signaling. *J. Mol. Cell. Cardiol.* **2015**, *79*, 133–144. [CrossRef]
23. Pasqualini, F.S.; Agarwal, A.; O'Connor, B.; Liu, Q.; Sheehy, S.P.; Parker, K.K. Traction force microscopy of engineered cardiac tissues. *PLoS ONE* **2018**, *13*, e0194706. [CrossRef]
24. Ateshian, G.A.; Costa, K.D. A frame-invariant formulation of Fung elasticity. *J. Biomech. Eng.* **2009**, *42*, 781–785. [CrossRef]
25. Fung, Y.C. Elasticity of soft tissues in simple elongation. *Am. J. Physiol. Content* **1967**, *213*, 1532–1544. [CrossRef] [PubMed]
26. Chuong, C.J.; Fung, Y.C. Three-Dimensional Stress Distribution in Arteries. *J. Biomech. Eng.* **1983**, *105*, 268–274. [CrossRef] [PubMed]
27. Pfaller, M.R.; Hörmann, J.M.; Weigl, M.; Nagler, A.; Chabiniok, R.; Bertoglio, C.; Wall, W.A. The importance of the pericardium for cardiac biomechanics: From physiology to computational modeling. *Biomech. Model. Mechanobiol.* **2019**, *18*, 503–529. [CrossRef] [PubMed]
28. Liu, Y.; Dang, C.; Garcia, M.; Gregersen, H.; Kassab, G.S. Surrounding tissues affect the passive mechanics of the vessel wall: Theory and experiment. *Am. J. Physiol. Circ. Physiol.* **2007**, *293*, 3290–3300. [CrossRef]
29. Humphrey, J.D. Mechanics of the Arterial Wall: Review and Directions. *Crit. Rev. Biomed. Eng.* **1995**, *23*, 1–162. [CrossRef]
30. Sokolis, D.P.; Savva, G.D.; Papadodima, S.A.; Kourkoulis, S.K. Regional distribution of circumferential residual strains in the human aorta according to age and gender. *J. Mech. Behav. Biomed. Mater.* **2017**, *67*, 87–100. [CrossRef]
31. Sokolis, D.P.; Bompas, A.; Papadodima, S.A.; Kourkoulis, S.K. Variation of Axial Residual Strains along the Course and Circumference of Human Aorta Considering Age and Gender. *J. Biomech. Eng.* **2020**, *142*, 0210031–02100313. [CrossRef]
32. Cardamone, L.; Valentín, A.; Eberth, J.F.; Humphrey, J.D. Origin of axial prestretch and residual stress in arteries. *Biomech. Model. Mechanobiol.* **2009**, *8*, 431–446. [CrossRef]
33. Holzapfel, G.; Gasser, T.; Stadler, M. A structural model for the viscoelastic behavior of arterial walls: Continuum formulation and finite element analysis. *Eur. J. Mech. A Solids* **2002**, *21*, 441–463. [CrossRef]
34. Humphrey, J.D.; Na, S. Elastodynamics and Arterial Wall Stress. *Ann. Biomed. Eng.* **2002**, *30*, 509–523. [CrossRef]
35. Akyildiz, A.C.; Speelman, L.; Gijsen, F.J.H. Mechanical properties of human atherosclerotic intima tissue. *J. Biomech.* **2014**, *47*, 773–783. [CrossRef]

36. Wang, Z.; Golob, M.J.; Chesler, N.C. Viscoelastic Properties of Cardiovascular Tissues. *Viscoelast. Viscoplast. Mater.* **2016**, *2*, 64. [CrossRef]
37. Roccabianca, S.; Figueroa, C.A.; Tellides, G.; Humphrey, J.D. Quantification of regional differences in aortic stiffness in the aging human. *J. Mech. Behav. Biomed. Mater.* **2014**, *29*, 618–634. [CrossRef]
38. Guccione, J.M.; McCulloch, A.D.; Waldman, L.K. Passive Material Properties of Intact Ventricular Myocardium Determined from a Cylindrical Model. *J. Biomech. Eng.* **1991**, *113*, 42–55. [CrossRef]
39. Polzer, S.; Gasser, T.; Novak, K.; Man, V.; Tichy, M.; Skacel, P.; Bursa, J. Structure-based constitutive model can accurately predict planar biaxial properties of aortic wall tissue. *Acta Biomater.* **2015**, *14*, 133–145. [CrossRef]
40. Spronck, B.; Humphrey, J.D. Arterial Stiffness: Different Metrics, Different Meanings. *J. Biomech. Eng.* **2019**, *141*, 0910041–09100412. [CrossRef]
41. Schroeder, F.; Polzer, S.; Slažanský, M.; Man, V.; Skácel, P. Predictive capabilities of various constitutive models for arterial tissue. *J. Mech. Behav. Biomed. Mater.* **2018**, *78*, 369–380. [CrossRef]
42. Holzapfel, G.A.; Ogden, R.W. An arterial constitutive model accounting for collagen content and cross-linking. *J. Mech. Phys. Solids* **2020**, *136*, 103682. [CrossRef]
43. Avazmohammadi, R.; Hill, M.; Simon, M.; Zhang, W.; Sacks, M.S. A novel constitutive model for passive right ventricular myocardium: Evidence for myofiber–collagen fiber mechanical coupling. *Biomech. Model. Mechanobiol.* **2017**, *16*, 561–581. [CrossRef]
44. Sacks, M.S.; Zhang, W.; Wognum, S. A novel fibre-ensemble level constitutive model for exogenous cross-linked collagenous tissues. *Interface Focus* **2016**, *6*, 20150090. [CrossRef]
45. Horgan, C.; Ogden, R.; Saccomandi, G. A theory of stress softening of elastomers based on finite chain extensibility. *Proc. R. Soc. A Math. Phys. Eng. Sci.* **2004**, *460*, 1737–1754. [CrossRef]
46. Hunter, P.J.; McCulloch, A.D.; Ter Keurs, H.E.D.J. Modelling the mechanical properties of cardiac muscle. *Prog. Biophys. Mol. Biol.* **1998**, *69*, 289–331. [CrossRef]
47. Xi, J.; Lamata, P.; Niederer, S.; Land, S.; Shi, W.; Zhuang, X.; Ourselin, S.; Duckett, S.G.; Shetty, A.K.; Rinaldi, C.A.; et al. The estimation of patient-specific cardiac diastolic functions from clinical measurements. *Med. Image Anal.* **2013**, *17*, 133–146. [CrossRef]
48. Bers, D.M. Cardiac excitation–contraction coupling. *Nature* **2002**, *415*, 198–205. [CrossRef]
49. Pullan, A.J.; Cheng, L.K.; Nash, M.P.; Ghodrati, A.; MacLeod, R.; Brooks, D.H. The Inverse Problem of Electrocardiography. In *Comprehensive Electrocardiology*; Springer: London, UK, 2010; pp. 299–344. [CrossRef]
50. Bhagirath, P.; Van Der Graaf, A.; De Hooge, J.; De Groot, N.; Götte, M. Integrated whole-heart computational workflow for inverse potential mapping and personalized simulations. *J. Transl. Med.* **2016**, *14*, 147. [CrossRef]
51. Van Oosterom, A. The inverse problem of bioelectricity: An evaluation. *Med. Biol. Eng. Comput.* **2012**, *50*, 891–902. [CrossRef]
52. Marchesseau, S.; Sermesant, M.; Billet, F.; Delingette, H.; Ayache, N. Personalization of Electromechanical Models of the Cardiac Ventricular Function by Heterogeneous Clinical Data Assimilation. *Multi-Modality Card. Imaging Process. Anal.* **2015**, *17*, 293–330. [CrossRef]
53. Kung, G.L.; Vaseghi, M.; Gahm, J.K.; Shevtsov, J.; Garfinkel, A.; Shivkumar, K.; Ennis, D.B. Microstructural Infarct Border Zone Remodeling in the Post-infarct Swine Heart Measured by Diffusion Tensor MRI. *Front. Physiol.* **2018**, *9*, 826. [CrossRef] [PubMed]
54. Khalique, Z.; Ferreira, P.; Scott, A.D.; Nielles-Vallespin, S.; Firmin, D.N.; Pennell, D.J. Diffusion Tensor Cardiovascular Magnetic Resonance Imaging. *JACC Cardiovasc. Imaging* **2020**, *13*, 1235–1255. [CrossRef]
55. Bayer, J.D.; Blake, R.C.; Plank, G.; Trayanova, N.A. A Novel Rule-Based Algorithm for Assigning Myocardial Fiber Orientation to Computational Heart Models. *Ann. Biomed. Eng.* **2012**, *40*, 2243–2254. [CrossRef] [PubMed]
56. Potse, M.; Dube, B.; Richer, J.; Vinet, A.; Gulrajani, R.M. A Comparison of Monodomain and Bidomain Reaction-Diffusion Models for Action Potential Propagation in the Human Heart. *IEEE Trans. Biomed. Eng.* **2006**, *53*, 2425–2435. [CrossRef]
57. Rijcken, J.; Bovendeerd, P.H.M.; Schoofs, A.J.G.; Van Campen, D.H.; Arts, T. Optimization of Cardiac Fiber Orientation for Homogeneous Fiber Strain During Ejection. *Ann. Biomed. Eng.* **1999**, *27*, 289–297. [CrossRef]
58. Barker, A.J.; van Ooij, P.; Bandi, K.; Garcia, J.; Albaghdadi, M.; McCarthy, P.; Bonow, R.O.; Carr, J.; Collins, J.; Malaisrie, S.C.; et al. Viscous energy loss in the presence of abnormal aortic flow. *Magn. Reson. Med.* **2014**, *72*, 620–628. [CrossRef]
59. Mahalingam, A.; Gawandalkar, U.U.; Kini, G.; Buradi, A.; Araki, T.; Ikeda, N.; Nicolaides, A.; Laird, J.R.; Saba, L.; Suri, J.S. Numerical analysis of the effect of turbulence transition on the hemodynamic parameters in human coronary arteries. *Cardiovasc. Diagn. Ther.* **2016**, *6*, 208–220. [CrossRef]
60. Womersley, J.R. Method for the calculation of velocity, rate of flow and viscous drag in arteries when the pressure gradient is known. *J. Physiol.* **1955**, *127*, 553–563. [CrossRef]
61. McDonald, D.A. The relation of pulsatile pressure to flow in arteries. *J. Physiol.* **1955**, *127*, 533–552. [CrossRef]
62. Taylor, M.G. The Discrepancy between Steady- and Oscillatory-Flow Calibration of Flowmeters of the 'Bristle'and 'Pendulum' Types: A Theoretical Study. *Phys. Med. Biol.* **1958**, *2*, 324–337. [CrossRef]
63. Pedley, T.J.; Schroter, R.C.; Sudlow, M.F. Pressure flow relations in branched tubes. *J. Physiol.* **1969**, *204*, 114. Available online: https://europepmc.org/article/med/5824621 (accessed on 29 November 2021).
64. Williams, M.C.; Rosenblatt, J.S.; Soane, D.S. Theory of Blood Rheology Based on a Statistical Mechanics Treatment of Rouleaux, and Comparisons with Data. *Int. J. Polym. Mater. Polym. Biomater.* **1993**, *21*, 57–63. [CrossRef]

65. Soares, J.S.; Gao, C.; Alemu, Y.; Slepian, M.; Bluestein, D. Simulation of platelets suspension flowing through a stenosis model using a dissipative particle dynamics approach. *Ann. Biomed. Eng.* **2013**, *41*, 2318–2333. [CrossRef]
66. Fedosov, D.A.; Noguchi, H.; Gompper, G. Multiscale modeling of blood flow: From single cells to blood rheology. *Biomech. Model. Mechanobiol.* **2013**, *132*, 239–258. [CrossRef]
67. Lei, H.; Fedosov, D.A.; Caswell, B.; Karniadakis, G.E. Blood flow in small tubes: Quantifying the transition to the non-continuum regime. *J. Fluid Mech.* **2013**, *722*, 214–239. [CrossRef]
68. Kim, S.; Namgung, B.; Ong, P.K.; Cho, Y.I.; Chun, K.J.; Lim, D. Determination of rheological properties of whole blood with a scanning capillary-tube rheometer using constitutive models. *J. Mech. Sci. Technol.* **2009**, *23*, 1718–1726. [CrossRef]
69. Ameenuddin, M.; Anand, M.; Massoudi, M. Effects of shear-dependent viscosity and hematocrit on blood flow. *Appl. Math. Comput.* **2019**, *356*, 299–311. [CrossRef]
70. Peskin, C.S. The immersed boundary method. *Acta Numer.* **2002**, *11*, 479–517. [CrossRef]
71. Flamini, V.; DeAnda, A.; Griffith, B.E. Immersed boundary-finite element model of fluid–structure interaction in the aortic root. *Theor. Comput. Fluid Dyn.* **2015**, *30*, 139–164. [CrossRef]
72. Yang, X.; Zhang, X.; Li, Z.; He, G.W. A smoothing technique for discrete delta functions with application to immersed boundary method in moving boundary simulations. *J. Comput. Phys.* **2009**, *228*, 7821–7836. [CrossRef]
73. van Loon, R.; Anderson, P.D.; van de Vosse, F.; Sherwin, S. Comparison of various fluid–structure interaction methods for deformable bodies. *Comput. Struct.* **2007**, *85*, 833–843. [CrossRef]
74. Figueroa, C.A.; Vignon-Clementel, I.E.; Jansen, K.E.; Hughes, T.J.R.; Taylor, C.A. A coupled momentum method for modeling blood flow in three-dimensional deformable arteries. *Comput. Methods Appl. Mech. Eng.* **2006**, *195*, 5685–5706. [CrossRef]
75. Moireau, P.; Xiao, N.; Astorino, M.; Figueroa, C.A.; Chapelle, D.; Taylor, C.A.; Gerbeau, J.F. External tissue support and fluid–structure simulation in blood flows. *Biomech. Model. Mechanobiol.* **2012**, *11*, 1–18. [CrossRef] [PubMed]
76. Souli, M.; Ouahsine, A.; Lewin, L. ALE formulation for fluid–structure interaction problems. *Comput. Methods Appl. Mech. Eng.* **2000**, *190*, 659–675. [CrossRef]
77. Rodriguez, E.K.; Hoger, A.; McCulloch, A.D. Stress-dependent finite growth in soft elastic tissues. *J. Biomech.* **1994**, *27*, 455–467. [CrossRef]
78. Humphrey, J.D. Constrained Mixture Models of Soft Tissue Growth and Remodeling—Twenty Years after. *J. Elast.* **2021**, *145*, 49–75. [CrossRef]
79. Humphrey, J.D.; Rajagopal, K. A Constrained Mixture Model for Growth and Remodeling of Soft Tissues. *Math. Model. Methods Appl. Sci.* **2002**, *12*, 407–430. [CrossRef]
80. Cyron, C.J.; Aydin, R.C.; Humphrey, J.D. A homogenized constrained mixture (and mechanical analog) model for growth and remodeling of soft tissue. *Biomech. Model. Mechanobiol.* **2016**, *15*, 1389–1403. [CrossRef]
81. Taylor, C.A.; Hughes, T.J.R.; Zarins, C.K. Finite element modeling of blood flow in arteries. *Comput. Methods Appl. Mech. Eng.* **1998**, *158*, 155–196. [CrossRef]
82. Simon, B.R.; Kaufmann, M.V.; McAfee, M.A.; Baldwin, A.L. Finite Element Models for Arterial Wall Mechanics. *J. Biomech. Eng.* **1993**, *115*, 489–496. [CrossRef]
83. Updegrove, A.; Wilson, N.M.; Merkow, J.; Lan, H.; Marsden, A.L.; Shadden, S.C. SimVascular: An Open Source Pipeline for Cardiovascular Simulation. *Ann. Biomed. Eng.* **2016**, *45*, 525–541. [CrossRef]
84. Maas, S.A.; Ellis, B.J.; Ateshian, G.A.; Weiss, J.A. FEBio: Finite Elements for Biomechanics. *J. Biomech. Eng.* **2012**, *134*, 011005. [CrossRef]
85. Moukalled, F.D.F.; Mangani, L. *The Finite Volume Method in Computational Fluid Dynamics*; Springer: New York, NY, USA, 2015. [CrossRef]
86. Tikhonov, A.N.; Samarskii, A.A. Homogeneous difference schemes on non-uniform nets. *USSR Comput. Math. Math. Phys.* **1963**, *2*, 927–953. [CrossRef]
87. Cardiff, P.; Demirdžić, I. Thirty Years of the Finite Volume Method for Solid Mechanics. *Arch. Comput. Methods Eng.* **2021**, *28*, 3721–3780. [CrossRef]
88. Li, Y.; Berthiau, G.; Feliachi, M.; Cheriet, A. 3D Finite Volume Modeling of ENDE Using Electromagnetic T-Formulation. *J. Sens.* **2012**, *2012*, 1–6. [CrossRef]
89. Mackerle, J. Finite element modelling and simulations in cardiovascular mechanics and cardiology: A bibliography 1993–2004. *Comput. Methods Biomech. Biomed. Eng.* **2005**, *8*, 59–81. [CrossRef]
90. Fallah, N.A.; Bailey, C.; Cross, M.; Taylor, G. Comparison of finite element and finite volume methods application in geometrically nonlinear stress analysis. *Appl. Math. Model.* **2000**, *24*, 439–455. [CrossRef]
91. Jeong, W.; Seong, J. Comparison of effects on technical variances of computational fluid dynamics (CFD) software based on finite element and finite volume methods. *Int. J. Mech. Sci.* **2014**, *78*, 19–26. [CrossRef]
92. Wiederhielm, C.A.; Kobayashi, A.S.; Stromberg, D.D.; Woo, S.L.Y. Structural response of relaxed and constricted arterioles. *J. Biomech.* **1968**, *1*, 259–270. [CrossRef]
93. Ateshian, G.A.; Shim, J.J.; Maas, S.A.; Weiss, J.A. Finite Element Framework for Computational Fluid Dynamics in FEBio. *J. Biomech. Eng.* **2018**, *140*, 0210011–02100117. [CrossRef]
94. Shim, J.J.; Maas, S.A.; Weiss, J.A.; Ateshian, G.A. A Formulation for Fluid Structure-Interactions in FEBio Using Mixture Theory. *J. Biomech. Eng.* **2019**, *141*, 0510101–05101015. [CrossRef]

95. Sabatier, P.C. Inverse Problems—An introduction. *Inverse Probl.* **1985**, *1*, 302. [CrossRef]
96. Grédiac, M.; Toussaint, E.; Pierron, F. Special virtual fields for the direct determination of material parameters with the virtual fields method. 2—Application to in-plane properties. *Int. J. Solids Struct.* **2002**, *39*, 2707–2730. [CrossRef]
97. Grediac, M.; Toussaint, E.; Pierron, F. Special virtual fields for the direct determination of material parameters with the virtual fields method. 1—Principle and definition. *Int. J. Solids Struct.* **2002**, *39*, 2691–2705. [CrossRef]
98. Govindjee, S.; Mihalic, P.A. Computational methods for inverse finite elastostatics. *Comput. Methods Appl. Mech. Eng.* **1996**, *136*, 47–57. [CrossRef]
99. Govindjee, S.; Mihalic, P.A. Computational methods for inverse deformations in quasi-incompressible finite elasticity. *Int. J. Numer. Methods Eng.* **1998**, *43*, 821–838. [CrossRef]
100. Peirlinck, M.; De Beule, M.; Segers, P.; Rebelo, N. A modular inverse elastostatics approach to resolve the pressure-induced stress state for in vivo imaging based cardiovascular modeling. *J. Mech. Behav. Biomed. Mater.* **2018**, *85*, 124–133. [CrossRef]
101. Tessler, A.; Spangler, J.L. A least-squares variational method for full-field reconstruction of elastic deformations in shear-deformable plates and shells. *Comput. Methods Appl. Mech. Eng.* **2005**, *194*, 327–339. [CrossRef]
102. Beck, J.V.; A Woodbury, K.A. Inverse problems and parameter estimation: Integration of measurements and analysis. *Meas. Sci. Technol.* **1998**, *9*, 839–847. [CrossRef]
103. Lassila, T.; Manzoni, A.; Quarteroni, A.; Rozza, G. A reduced computational and geometrical framework for inverse problems in hemodynamics. *Int. J. Numer. Methods Biomed. Eng.* **2013**, *29*, 741–776. [CrossRef]
104. Liu, M.; Liang, L.; Sun, W. A new inverse method for estimation of in vivo mechanical properties of the aortic wall. *J. Mech. Behav. Biomed. Mater.* **2017**, *72*, 148–158. [CrossRef] [PubMed]
105. Cotter, S.L.; Dashti, M.; Robinson, J.C.; Stuart, A.M. Bayesian inverse problems for functions and applications to fluid mechanics. *Inverse Probl.* **2009**, *25*, 115008. [CrossRef]
106. Marsden, A.L. Optimization in Cardiovascular Modeling. *Annu. Rev. Fluid Mech.* **2014**, *46*, 519–546. [CrossRef]
107. Moulton, M.J.; Creswell, L.L.; Actis, R.L.; Myers, K.W.; Vannier, M.W.; Szabo, B.A.; Pasque, M.K. An inverse approach to determining myocardial material properties. *J. Biomech.* **1995**, *28*, 935–948. [CrossRef]
108. Palit, A.; Bhudia, S.; Arvanitis, T.; Turley, G.; Williams, M. In vivo estimation of passive biomechanical properties of human myocardium. *Med. Biol. Eng. Comput.* **2018**, *56*, 1615–1631. [CrossRef]
109. Klotz, S.; Hay, I.; Dickstein, M.L.; Yi, G.-H.; Wang, J.; Maurer, M.S.; Kass, D.A.; Burkhoff, D. Single-beat estimation of end-diastolic pressure-volume relationship: A novel method with potential for noninvasive application. *Am. J. Physiol. Circ. Physiol.* **2006**, *291*, 403–412. [CrossRef]
110. Genet, M.; Lee, L.C.; Nguyen, R.; Haraldsson, H.; Acevedo-Bolton, G.; Zhang, Z.; Ge, L.; Ordovas, K.; Kozerke, S.; Guccione, J.M. Distribution of normal human left ventricular myofiber stress at end diastole and end systole: A target for in silico design of heart failure treatments. *J. Appl. Physiol.* **2014**, *117*, 142–152. [CrossRef]
111. Lu, J.; Zhou, X.; Raghavan, M.L. Inverse elastostatic stress analysis in pre-deformed biological structures: Demonstration using abdominal aortic aneurysms. *J. Biomech.* **2007**, *40*, 693–696. [CrossRef]
112. Farzaneh, S.; Trabelsi, O.; Avril, S. Inverse identification of local stiffness across ascending thoracic aortic aneurysms. *Biomech. Model. Mechanobiol.* **2019**, *18*, 137–153. [CrossRef]
113. Liu, M.; Liang, L.; Sun, W. Estimation of in vivo mechanical properties of the aortic wall: A multi-resolution direct search approach. *J. Mech. Behav. Biomed. Mater.* **2018**, *77*, 649–659. [CrossRef]
114. Franquet, A.; Avril, S.; Le Riche, R.; Badel, P.; Schneider, F.C.; Li, Z.Y.; Boissier, C.; Favre, J.P. A New Method for the In Vivo Identification of Mechanical Properties in Arteries from Cine MRI Images: Theoretical Framework and Validation. *IEEE Trans. Med. Imaging* **2013**, *32*, 1448–1461. [CrossRef] [PubMed]
115. Wittek, A.; Karatolios, K.; Bihari, P.; Schmitz-Rixen, T.; Moosdorf, R.; Vogt, S.; Blase, C. In vivo determination of elastic properties of the human aorta based on 4D ultrasound data. *J. Mech. Behav. Biomed. Mater.* **2013**, *27*, 167–183. [CrossRef] [PubMed]
116. Candito, A.; Palacio-Torralba, J.; Jiménez-Aguilar, E.; Good, D.W.; McNeill, A.; Reuben, R.L.; Chen, Y. Identification of tumor nodule in soft tissue: An inverse finite-element framework based on mechanical characterization. *Int. J. Numer. Methods Biomed. Eng.* **2020**, *36*, e3369. [CrossRef] [PubMed]
117. Chawla, A.; Mukherjee, S.; Karthikeyan, B. Characterization of human passive muscles for impact loads using genetic algorithm and inverse finite element methods. *Biomech. Model. Mechanobiol.* **2008**, *8*, 67–76. [CrossRef] [PubMed]
118. Romarowski, R.M.; Lefieux, A.; Morganti, S.; Veneziani, A.; Auricchio, F. Patient-specific CFD modelling in the thoracic aorta with PC-MRI–based boundary conditions: A least-square three-element Windkessel approach. *Int. J. Numer. Methods Biomed. Eng.* **2018**, *34*, e3134. [CrossRef] [PubMed]
119. Quick, C.M.; Young, W.L.; Noordergraaf, A. Infinite number of solutions to the hemodynamic inverse problem. *Am. J. Physiol. Circ. Physiol.* **2001**, *280*, H1472–H1479. [CrossRef] [PubMed]
120. Pewowaruk, R.; Roldán-Alzate, A. 4D Flow MRI Estimation of Boundary Conditions for Patient Specific Cardiovascular Simulation. *Ann. Biomed. Eng.* **2019**, *47*, 1786–1798. [CrossRef] [PubMed]
121. Peng, H.-H.; Chung, H.-W.; Yu, H.-Y.; Tseng, W.-Y.I. Estimation of pulse wave velocity in main pulmonary artery with phase contrast MRI: Preliminary investigation. *J. Magn. Reson. Imaging* **2006**, *24*, 1303–1310. [CrossRef]
122. Töger, J.; Zahr, M.J.; Aristokleous, N.; Bloch, K.M.; Carlsson, M.; Persson, P.-O. Blood flow imaging by optimal matching of computational fluid dynamics to 4D-flow data. *Magn. Reson. Med.* **2020**, *84*, 2231–2245. [CrossRef]

123. Kochenderfer, M.J.; Wheeler, T.A. *Algorithms for Optimization*; The MIT Press: Cambridge, MA, USA, 2019.
124. Peña, J.A.; Corral, V.; Martínez, M.A.; Peña, E. Over length quantification of the multiaxial mechanical properties of the ascending, descending and abdominal aorta using Digital Image Correlation. *J. Mech. Behav. Biomed. Mater.* **2018**, *77*, 434–445. [CrossRef]
125. Wittek, A.; Derwich, W.; Karatolios, K.; Fritzen, C.P.; Vogt, S.; Schmitz-Rixen, T.; Blase, C. A finite element updating approach for identification of the anisotropic hyperelastic properties of normal and diseased aortic walls from 4D ultrasound strain imaging. *J. Mech. Behav. Biomed. Mater.* **2016**, *58*, 122–138. [CrossRef]
126. Khalil, A.S.; Bouma, B.E.; Mofrad, M.R.K. A Combined FEM/Genetic Algorithm for Vascular Soft Tissue Elasticity Estimation. *Cardiovasc. Eng.* **2006**, *6*, 93–102. [CrossRef]
127. Aboelkassem, Y.; Savic, D. Particle swarm optimizer for arterial blood flow models. *Comput. Methods Progr. Biomed.* **2021**, *201*, 105933. [CrossRef]
128. Carniel, T.A.; de Castro, P.B.; Santos, A.L.G.; Roesler, C.R.d.M.; Breitenbach, E.R.; Salmoria, G.V.; Morozo, M.A.; Colaço, P.A.; Fiori, M.A.; Fancello, E.A. Mechanical characterization of hydrolysis effects on the stiffness of bioabsorbable polymeric filaments: An experimental and modeling approach based on a simple constitutive damage model. *Polym. Polym. Compos.* **2021**, *29*, S262–S273. [CrossRef]
129. Shahriari, B.; Swersky, K.; Wang, Z.; Adams, R.P.; De Freitas, N. Taking the Human Out of the Loop: A Review of Bayesian Optimization. *Proc. IEEE* **2016**, *104*, 148–175. [CrossRef]
130. Kim, Y.; Bang, H. Introduction to Kalman Filter and Its Applications. In *Introduction and Implementations of the Kalman Filter*; Govaers, F., Ed.; IntechOpen: London, UK, 2018. [CrossRef]
131. Alexanderian, A. Optimal experimental design for infinite-dimensional Bayesian inverse problems governed by PDEs: A review. *Inverse Probl.* **2021**, *37*, 043001. [CrossRef]
132. Ferruzzi, J.; Di Achille, P.; Tellides, G.; Humphrey, J.D. Combining in vivo and in vitro biomechanical data reveals key roles of perivascular tethering in central artery function. *PLoS ONE* **2018**, *13*, e0201379. [CrossRef]
133. Sun, W.; Sacks, M.S. Finite element implementation of a generalized Fung-elastic constitutive model for planar soft tissues. *Biomech. Model. Mechanobiol.* **2005**, *4*, 190–199. [CrossRef]
134. Avazmohammadi, R.; Soares, J.S.; Li, D.S.; Eperjesi, T.; Pilla, J.; Gorman, R.C.; Sacks, M.S. On the in vivo systolic compressibility of left ventricular free wall myocardium in the normal and infarcted heart. *J. Biomech.* **2020**, *107*, 109767. [CrossRef]
135. Rodriguez, I.; Ennis, D.B.; Wen, H. Noninvasive measurement of myocardial tissue volume change during systolic contraction and diastolic relaxation in the canine left ventricle. *Magn. Reson. Med.* **2006**, *55*, 484–490. [CrossRef]
136. Zhong, X.; Spottiswoode, B.S.; Meyer, C.H.; Kramer, C.M.; Epstein, F.H. Imaging three-dimensional myocardial mechanics using navigator-gated volumetric spiral cine DENSE MRI. *Magn. Reson. Med.* **2010**, *64*, 1089–1097. [CrossRef] [PubMed]
137. Liu, H.; Soares, J.S.; Walmsley, J.; Li, D.S.; Raut, S.; Avazmohammadi, R.; Iaizzo, P.; Palmer, M.; Gorman, J.H.; Gorman, R.C.; et al. The impact of myocardial compressibility on organ-level simulations of the normal and infarcted heart. *Sci. Rep.* **2021**, *11*, 1–15. [CrossRef] [PubMed]
138. Girerd, X.J.; Acar, C.; Mourad, J.J.; Boutouyrie, P.; Safar, M.E.; Laurent, S. Incompressibility of the human arterial wall: An in vitro ultrasound study. *J. Hypertens. Suppl.* **1992**, *10*, S111–S114. [CrossRef] [PubMed]
139. Nolan, D.; McGarry, J.P. On the Compressibility of Arterial Tissue. *Ann. Biomed. Eng.* **2015**, *44*, 993–1007. [CrossRef]
140. Liu, J.; Yang, W.; Lan, I.S.; Marsden, A.L. Fluid-structure interaction modeling of blood flow in the pulmonary arteries using the unified continuum and variational multiscale formulation. *Mech. Res. Commun.* **2020**, *107*, 103556. [CrossRef]
141. Balaban, G.; Finsberg, H.; Funke, S.; Håland, T.F.; Hopp, E.; Sundnes, J.; Wall, S.; Rognes, M.E. In vivo estimation of elastic heterogeneity in an infarcted human heart. *Biomech. Model. Mechanobiol.* **2018**, *17*, 1317–1329. [CrossRef]
142. Wohlfahrt, P.; Krajčoviechová, A.; Seidlerová, J.; Mayer, O.; Bruthans, J.; Filipovský, J.; Laurent, S.; Cífková, R. Arterial stiffness parameters: How do they differ? *Atherosclerosis* **2013**, *231*, 359–364. [CrossRef]
143. Vitarelli, A.; Conde, Y.; Cimino, E.; D'Angeli, I.; D'Orazio, S.; Stellato, S.; Padella, V.; Caranci, F. Aortic Wall Mechanics in the Marfan Syndrome Assessed by Transesophageal Tissue Doppler Echocardiography. *Am. J. Cardiol.* **2006**, *97*, 571–577. [CrossRef]
144. Mack, W.J.; Islam, T.; Lee, Z.; Selzer, R.H.; Hodis, H.N. Environmental tobacco smoke and carotid arterial stiffness. *Prev. Med.* **2003**, *37*, 148–154. [CrossRef]
145. Azadani, A.N.; Chitsaz, S.; Matthews, P.B.; Jaussaud, N.; Leung, J.; Tsinman, T.; Ge, L.; Tseng, E.E. Comparison of Mechanical Properties of Human Ascending Aorta and Aortic Sinuses. *Ann. Thorac. Surg.* **2012**, *93*, 87–94. [CrossRef]
146. Genovese, K.; Humphrey, J.D. Multimodal optical measurementin vitroof surface deformations and wall thickness of the pressurized aortic arch. *J. Biomed. Opt.* **2015**, *20*, 046005. [CrossRef]
147. Abbasi, M.; Barakat, M.S.; Vahidkhah, K.; Azadani, A.N. Characterization of three-dimensional anisotropic heart valve tissue mechanical properties using inverse finite element analysis. *J. Mech. Behav. Biomed. Mater.* **2016**, *62*, 33–44. [CrossRef]
148. Hess, A.T.; Bissell, M.M.; Ntusi, N.A.B.; Lewis, A.J.M.; Tunnicliffe, E.M.; Greiser, A.; Stalder, A.F.; Francis, J.M.; Myerson, S.G.; Neubauer, S.; et al. Aortic 4D flow: Quantification of signal-to-noise ratio as a function of field strength and contrast enhancement for 1.5 T, 3 T, and 7 T. *Magn. Reson. Med.* **2015**, *73*, 1864–1871. [CrossRef] [PubMed]
149. Aletras, A.H.; Ding, S.; Balaban, R.S.; Wen, H. DENSE: Displacement Encoding with Stimulated Echoes in Cardiac Functional MRI. *J. Magn. Reson.* **1999**, *137*, 247–252. [CrossRef]
150. Bracamonte, J.H.; Wilson, J.S.; Soares, J.S. Quantification of the heterogeneous effect of static and dynamic perivascular struc-tures on patient-specific local aortic wall mechanics using inverse finite element modeling and DENSE MRI. *J. Biomech.* **2009**, *11*, 123.

151. Busch, J.; Giese, D.; Kozerke, S. Image-based background phase error correction in 4D flow MRI revisited. *J. Magn. Reson. Imaging* **2017**, *46*, 1516–1525. [CrossRef]
152. Ng, A.; Swanevelder, J. Resolution in ultrasound imaging. *Contin. Educ. Anaesth. Crit. Care Pain* **2011**, *11*, 186–192. [CrossRef]
153. Sassaroli, E.; Crake, C.; Scorza, A.; Kim, D.; Park, M. Image quality evaluation of ultrasound imaging systems: Advanced B-modes. *J. Appl. Clin. Med. Phys.* **2019**, *20*, 115–124. [CrossRef] [PubMed]
154. Linte, C.A.; Moore, J.; Wiles, A.D.; Wedlake, C.; Peters, T.M. Virtual reality-enhanced ultrasound guidance: A novel technique for intracardiac interventions. *Comput. Aided Surg.* **2008**, *13*, 82–94. [CrossRef] [PubMed]
155. Olson, I.; Brabender, J.; Thorson, K.; Lopez, L. 3D Echocardiography. In *Multimodality Imaging Innovations in Adult Congenital Heart Disease*; Springer Nature: Berlin, Germany, 2021; pp. 3–25.
156. Malik, S.B.; Chen, N.; Parker, R.A.; Hsu, J.Y. Transthoracic Echocardiography: Pitfalls and Limitations as Delineated at Cardiac CT and MR Imaging. *RadioGraphics* **2017**, *37*, 383–406. [CrossRef] [PubMed]
157. Ohayon, J.; Finet, G.; Le Floc'H, S.; Cloutier, G.; Gharib, A.; Heroux, J.; Pettigrew, R.I. Biomechanics of Atherosclerotic Coronary Plaque: Site, Stability and In Vivo Elasticity Modeling. *Ann. Biomed. Eng.* **2014**, *42*, 269–279. [CrossRef]
158. Ko, S.M.; Hwang, S.H.; Lee, H.-J. Role of Cardiac Computed Tomography in the Diagnosis of Left Ventricular Myocardial Diseases. *J. Cardiovasc. Imaging* **2019**, *27*, 73–92. [CrossRef]
159. Dunmire, B.; Beach, K.W.; Labs, K.H.; Plett, M.; Strandness, D.E. Cross-beam vector Doppler ultrasound for angle-independent velocity measurements. *Ultrasound Med. Biol.* **2000**, *26*, 1213–1235. [CrossRef]
160. Truong, U.T.; Kutty, S.; Broberg, C.S.; Sahn, D.J. Multimodality Imaging in Congenital Heart Disease: An Update. *Curr. Cardiovasc. Imaging Rep.* **2012**, *5*, 481–490. [CrossRef]
161. Zhao, S.Z.; Xu, X.Y.; Hughes, A.D.; Thom, S.A.; Stanton, A.V.; Ariff, B.; Long, Q. Blood flow and vessel mechanics in a physiologically realistic model of a human carotid arterial bifurcation. *J. Biomech.* **2000**, *33*, 975–984. [CrossRef]
162. Bracamonte-Baran, W.; Bracamonte-Baran, J.; Baritto-Loreto, M.; D'Alessandro-Martínez, A. Computational Fluid Dynamics applied to the study of blood flow in the human aortic arch and its main branches. *Ing. Investig. Tecnol.* **2016**, *17*, 45–60. [CrossRef]
163. Mondillo, S.; Galderisi, M.; Mele, D.; Cameli, M.; Lomoriello, V.S.; Zacà, V.; Ballo, P.; D'Andrea, A.; Muraru, D.; Losi, M.; et al. Speckle-Tracking Echocardiography. *J. Ultrasound Med.* **2011**, *30*, 71–83. [CrossRef]
164. Karatolios, K.; Wittek, A.; Nwe, T.H.; Bihari, P.; Shelke, A.; Josef, D.; Schmitz-Rixen, T.; Geks, J.; Maisch, B.; Blase, C.; et al. Method for Aortic Wall Strain Measurement with Three-Dimensional Ultrasound Speckle Tracking and Fitted Finite Element Analysis. *Ann. Thorac. Surg.* **2013**, *96*, 1664–1671. [CrossRef]
165. Bihari, P.; Shelke, A.; Nwe, T.; Mularczyk, M.; Nelson, K.; Schmandra, T.; Knez, P.; Schmitz-Rixen, T. Strain Measurement of Abdominal Aortic Aneurysm with Real-time 3D Ultrasound Speckle Tracking. *Eur. J. Vasc. Endovasc. Surg.* **2013**, *45*, 315–323. [CrossRef]
166. Petterson, N.J.; van Disseldorp, M.R.H.M.; van Sambeek, F.R.; van de Vosse, F.N.; Lopata, R.G.P. Including surrounding tissue improves ultrasound-based 3D mechanical characterization of abdominal aortic aneurysms. *J. Biomech.* **2019**, *85*, 126–133. [CrossRef]
167. Cho, G.-Y.; Chan, J.; Leano, R.; Strudwick, M.; Marwick, T.H. Comparison of Two-Dimensional Speckle and Tissue Velocity Based Strain and Validation with Harmonic Phase Magnetic Resonance Imaging. *Am. J. Cardiol.* **2006**, *97*, 1661–1666. [CrossRef]
168. Pedrizzetti, G.; Claus, P.; Kilner, P.J.; Nagel, E. Principles of cardiovascular magnetic resonance feature tracking and echocardiographic speckle tracking for informed clinical use. *J. Cardiovasc. Magn. Reson.* **2016**, *18*, 1–12. [CrossRef]
169. Ferraro, A.M.; Adar, A.; Ghelani, S.J.; Sleeper, L.A.; Levy, P.T.; Rathod, R.H.; Marx, G.R.; Harrild, D.M. Speckle tracking echocardiographically-based analysis of ventricular strain in children: An intervendor comparison. *Cardiovasc. Ultrasound* **2020**, *18*, 1–12. [CrossRef]
170. Jeung, M.-Y.; Germain, P.; Croisille, P.; El Ghannudi, S.; Roy, C.; Gangi, A. Myocardial Tagging with MR Imaging: Overview of Normal and Pathologic Findings. *RadioGraphics* **2012**, *32*, 1381–1398. [CrossRef]
171. Wehner, G.J.; Suever, J.D.; Haggerty, C.M.; Jing, L.; Powell, D.K.; Hamlet, S.M.; Grabau, J.D.; Mojsejenko, W.D.; Zhong, X.; Epstein, F.H.; et al. Validation of in vivo 2D displacements from spiral cine DENSE at 3 T. *J. Cardiovasc. Magn. Reson.* **2015**, *17*, 5. [CrossRef]
172. Campbell-Washburn, A.E.; Tavallaei, M.A.; Pop, M.; Grant, E.K.; Chubb, H.; Rhode, K.; Wright, G.A. Real-time MRI guidance of cardiac interventions. *J. Magn. Reson. Imaging* **2017**, *46*, 935–950. [CrossRef]
173. Pop, M.; Ghugre, N.R.; Ramanan, V.; Morikawa, L.; Stanisz, G.; Dick, A.J.; Wright, G.A. Quantification of fibrosis in infarcted swine hearts byex vivolate gadolinium-enhancement and diffusion-weighted MRI methods. *Phys. Med. Biol.* **2013**, *58*, 5009–5028. [CrossRef]
174. Dall'Armellina, E.; Bissell, M.M.; Broadbent, D.A.; Plein, S. MRI T1 Mapping: Myocardial Fibrosis. In *Multimodality Imaging Innovations in Adult Congenital Heart Disease*; Springer Nature: Berlin, Germany, 2021; pp. 49–61.
175. Edelman, R.R. The History of MR Imaging as Seen through the Pages of Radiology. *Radiology* **2014**, *273*, S181–S200. [CrossRef] [PubMed]
176. Zerhouni, E.A.; Parish, D.M.; Rogers, W.J.; Yang, A.; Shapiro, E.P. Human heart: Tagging with MR imaging—A method for noninvasive assessment of myocardial motion. *Radiology* **1988**, *169*, 59–63. [CrossRef] [PubMed]
177. Kramer, C.M. Role of Cardiac MR Imaging in Cardiomyopathies. *J. Nucl. Med.* **2015**, *56*, 39S–45S. [CrossRef] [PubMed]
178. Ibrahim, E.-S.H.; Arpinar, V.E.; Muftuler, L.T.; Stojanovska, J.; Nencka, A.S.; Koch, K.M. Cardiac functional magnetic resonance imaging at 7 T: Image quality optimization and ultra-high field capabilities. *World J. Radiol.* **2020**, *12*, 231–246. [CrossRef]

179. Li, L.; Miller, K.L.; Jezzard, P. DANTE-prepared pulse trains: A novel approach to motion-sensitized and motion-suppressed quantitative magnetic resonance imaging. *Magn. Reson. Med.* **2012**, *68*, 1423–1438. [CrossRef]
180. Qian, Z.; Metaxas, D.; Axel, L. Non-tracking-based 2d strain estimation in tagged MRI. In Proceedings of the 2008 5th IEEE Internaional Symposium of Biomedicinal Imaging from Nano to Macro, Paris, France, 14–17 May 2008; pp. 528–531. [CrossRef]
181. Kar, J.; Knutsen, A.K.; Cupps, B.P.; Pasque, M.K. A Validation of Two-Dimensional In Vivo Regional Strain Computed from Displacement Encoding with Stimulated Echoes (DENSE), in Reference to Tagged Magnetic Resonance Imaging and Studies in Repeatability. *Ann. Biomed. Eng.* **2014**, *42*, 541–554. [CrossRef]
182. Axel, L. Biomechanical Dynamics of the Heart with MRI. *Annu. Rev. Biomed. Eng.* **2002**, *4*, 321–347. [CrossRef]
183. Wymer, D.T.; Patel, K.P.; Burke, W.F.; Bhatia, V.K. Phase-Contrast MRI: Physics, Techniques, and Clinical Applications. *Radio-Graphics* **2020**, *40*, 122–140. [CrossRef]
184. Lorenz, R.; Bock, J.; Snyder, J.; Korvink, J.G.; Jung, B.A.; Markl, M. Influence of eddy current, Maxwell and gradient field corrections on 3D flow visualization of 3D CINE PC-MRI data. *Magn. Reson. Med.* **2014**, *72*, 33–40. [CrossRef]
185. Gatehouse, P.D.; Rolf, M.P.; Graves, M.J.; Hofman, M.B.; Totman, J.; Werner, B.; Quest, R.A.; Liu, Y.; von Spiczak, J.; Dieringer, M.; et al. Flow measurement by cardiovascular magnetic resonance: A multi-centre multi-vendor study of background phase offset errors that can compromise the accuracy of derived regurgitant or shunt flow measurements. *J. Cardiovasc. Magn. Reson.* **2010**, *12*, 1–8. [CrossRef]
186. Stalder, A.F.; Russe, M.F.; Frydrychowicz, A.; Bock, J.; Hennig, J.; Markl, M. Quantitative 2D and 3D phase contrast MRI: Optimized analysis of blood flow and vessel wall parameters. *Magn. Reson. Med.* **2008**, *60*, 1218–1231. [CrossRef]
187. Minderhoud, S.C.S.; Van Der Velde, N.; Wentzel, J.J.; Van Der Geest, R.J.; Attrach, M.; Wielopolski, P.A.; Budde, R.P.J.; Helbing, W.A.; Roos-Hesselink, J.W.; Hirsch, A. The clinical impact of phase offset errors and different correction methods in cardiovascular magnetic resonance phase contrast imaging: A multi-scanner study. *J. Cardiovasc. Magn. Reson.* **2020**, *22*, 1–13. [CrossRef]
188. Sieren, M.M.; Berlin, C.; Oechtering, T.H.; Hunold, P.; Drömann, D.; Barkhausen, J.; Frydrychowicz, A. Comparison of 4D Flow MRI to 2D Flow MRI in the pulmonary arteries in healthy volunteers and patients with pulmonary hypertension. *PLoS ONE* **2019**, *14*, e0224121. [CrossRef]
189. Bollache, E.; van Ooij, P.; Powell, A.; Carr, J.; Markl, M.; Barker, A.J. Comparison of 4D flow and 2D velocity-encoded phase contrast MRI sequences for the evaluation of aortic hemodynamics. *Int. J. Cardiovasc. Imaging* **2016**, *32*, 1529–1541. [CrossRef] [PubMed]
190. Stankovic, Z.; Allen, B.D.; Garcia, J.; Jarvis, K.B.; Markl, M. 4D flow imaging with MRI. *Cardiovasc. Diagn. Ther.* **2014**, *4*, 173–192. [CrossRef] [PubMed]
191. Middione, M.J.; Wu, H.H.; Ennis, D.B. Convex gradient optimization for increased spatiotemporal resolution and improved accuracy in phase contrast MRI. *Magn. Reson. Med.* **2013**, *72*, 1552–1564. [CrossRef] [PubMed]
192. Lantz, J.; Gupta, V.; Henriksson, L.; Karlsson, M.; Persson, A.; Carlhäll, C.-J.; Ebbers, T. Intracardiac Flow at 4D CT: Comparison with 4D Flow MRI. *Radiology* **2018**, *289*, 51–58. [CrossRef] [PubMed]
193. Markl, M.; Wallis, W.; Brendecke, S.; Simon, J.; Frydrychowicz, A.; Harloff, A. Estimation of global aortic pulse wave velocity by flow-sensitive 4D MRI. *Magn. Reson. Med.* **2010**, *63*, 1575–1582. [CrossRef] [PubMed]
194. Barker, A.J.; Roldán-Alzate, A.; Entezari, P.; Shah, S.J.; Chesler, N.; Wieben, O.; Markl, M.; Francois, C.J. Four-dimensional flow assessment of pulmonary artery flow and wall shear stress in adult pulmonary arterial hypertension: Results from two institutions. *Magn. Reson. Med.* **2015**, *73*, 1904–1913. [CrossRef] [PubMed]
195. Schäfer, M.; Ivy, D.D.; Abman, S.H.; Stenmark, K.; Browne, L.P.; Barker, A.J.; Mitchell, M.B.; Morgan, G.J.; Wilson, N.; Shah, A.; et al. Differences in pulmonary arterial flow hemodynamics between children and adults with pulmonary arterial hypertension as assessed by 4D-flow CMR studies. *Am. J. Physiol. Circ. Physiol.* **2019**, *316*, H1091–H1104. [CrossRef]
196. Friesen, R.M.; Schäfer, M.; Ivy, D.D.; Abman, S.H.; Stenmark, K.; Browne, L.P.; Barker, A.J.; Hunter, K.S.; Truong, U. Proximal pulmonary vascular stiffness as a prognostic factor in children with pulmonary arterial hypertension. *Eur. Heart J. Cardiovasc. Imaging* **2019**, *20*, 209–217. [CrossRef]
197. Schäfer, M.; Frank, B.S.; Ivy, D.D.; Abman, S.H.; Stenmark, K.R.; Mitchell, M.B.; Browne, L.P.; Barker, A.J.; Hunter, K.S.; Kheyfets, V.; et al. Short-Term Effects of Inhaled Nitric Oxide on Right Ventricular Flow Hemodynamics by 4-Dimensional–Flow Magnetic Resonance Imaging in Children with Pulmonary Arterial Hypertension. *J. Am. Heart Assoc.* **2021**, *10*, 20548. [CrossRef]
198. Jarvis, K.; Schnell, S.; Barker, A.J.; Rose, M.; Robinson, J.D.; Rigsby, C.K.; Markl, M. Caval to pulmonary 3D flow distribution in patients with Fontan circulation and impact of potential 4D flow MRI error sources. *Magn. Reson. Med.* **2019**, *81*, 1205–1218. [CrossRef]
199. McLennan, D.; Schäfer, M.; Mitchell, M.B.; Morgan, G.J.; Ivy, D.; Barker, A.J.; Jacobsen, R. Usefulness of 4D-Flow MRI in Mapping Flow Distribution Through Failing Fontan Circulation Prior to Cardiac Intervention. *Pediatr. Cardiol.* **2019**, *40*, 1093–1096. [CrossRef]
200. Cheng, A.L.; Wee, C.P.; Pahlevan, N.M.; Wood, J.C. A 4D flow MRI evaluation of the impact of shear-dependent fluid viscosity on in vitro Fontan circulation flow. *Am. J. Physiol. Circ. Physiol.* **2019**, *317*, H1243–H1253. [CrossRef]
201. Schäfer, M.; Barker, A.J.; Jaggers, J.; Morgan, G.J.; Stone, M.L.; Truong, U.; Browne, L.P.; Malone, L.; Ivy, D.D.; Mitchell, M.B. Abnormal aortic flow conduction is associated with increased viscous energy loss in patients with repaired tetralogy of Fallot. *Eur. J. Cardio-Thorac. Surg.* **2020**, *57*, 588–595. [CrossRef]

202. Kim, D.; Gilson, W.D.; Kramer, C.M.; Epstein, F.H. Myocardial Tissue Tracking with Two-Dimensional Cine Displacement-Encoded MR Imaging: Development and Initial Evaluation. *Radiology* **2004**, *230*, 862–871. [CrossRef]
203. Aletras, A.H.; Wen, H. Mixed echo train acquisition displacement encoding with stimulated echoes: An optimized DENSE method for in vivo functional imaging of the human heart. *Magn. Reson. Med.* **2001**, *46*, 523–534. [CrossRef]
204. Perotti, L.E.; Magrath, P.; Verzhbinsky, I.A.; Aliotta, E.; Moulin, K.; Ennis, D.B. Microstructurally Anchored Cardiac Kinematics by Combining In Vivo DENSE MRI and cDTI. *Lect. Notes Comput. Sci.* **2017**, *10263*, 381–391. [CrossRef]
205. Abdi, M.; Feng, X.; Sun, C.; Bilchick, K.C.; Meyer, C.H.; Epstein, F.H. Suppression of artifact-generating echoes in cine DENSE using deep learning. *Magn. Reson. Med.* **2021**, *86*, 2095–2104. [CrossRef]
206. Cai, X.; Frederick, H.; Epstein, F.H. Free-breathing cine DENSE MRI using phase cycling with matchmaking and stimulated-echo image-based navigators. *Magn. Reson. Med.* **2018**, *80*, 1907–1921. [CrossRef]
207. Ghadimi, S.; Auger, D.A.; Feng, X.; Sun, C.; Meyer, C.H.; Bilchick, K.C.; Cao, J.J.; Scott, A.D.; Oshinski, J.N.; Ennis, D.B.; et al. Fully-automated global and segmental strain analysis of DENSE cardiovascular magnetic resonance using deep learning for segmentation and phase unwrapping. *J. Cardiovasc. Magn. Reson.* **2021**, *23*, 1–13. [CrossRef]
208. Magrath, P.; Maforo, N.; Renella, P.; Nelson, S.F.; Halnon, N.; Ennis, D.B. Cardiac MRI biomarkers for Duchenne muscular dystrophy. *Biomarkers Med.* **2018**, *12*, 1271–1289. [CrossRef]
209. Naresh, N.K.; Misener, S.; Zhang, Z.; Yang, C.; Ruh, A.; Bertolino, N.; Epstein, F.H.; Collins, J.D.; Markl, M.; Procissi, D.; et al. Cardiac MRI Myocardial Functional and Tissue Characterization Detects Early Cardiac Dysfunction in a Mouse Model of Chemotherapy-Induced Cardiotoxicity. *NMR Biomed.* **2020**, *33*, e4327. [CrossRef] [PubMed]
210. Gao, X.; Abdi, M.; Auger, D.A.; Sun, C.; Hanson, C.A.; Robinson, A.A.; Schumann, C.; Oomen, P.J.; Ratcliffe, S.; Malhotra, R.; et al. Cardiac Magnetic Resonance Assessment of Response to Cardiac Resynchronization Therapy and Programming Strategies. *JACC Cardiovasc. Imaging* **2021**, *14*, 2369–2383. [CrossRef] [PubMed]
211. Bilchick, K.C.; Auger, D.A.; Abdishektaei, M.; Mathew, R.; Sohn, M.-W.; Cai, X.; Sun, C.; Narayan, A.; Malhotra, R.; Darby, A.; et al. CMR DENSE and the Seattle Heart Failure Model Inform Survival and Arrhythmia Risk after CRT. *JACC Cardiovasc. Imaging* **2019**, *13*, 924–936. [CrossRef] [PubMed]
212. Mangion, K.; Loughrey, C.M.; Auger, D.A.; McComb, C.; Lee, M.M.; Corcoran, D.; McEntegart, M.; Davie, A.; Good, R.; Lindsay, M.; et al. Displacement Encoding with Stimulated Echoes Enables the Identification of Infarct Transmurality Early Postmyocardial Infarction. *J. Magn. Reson. Imaging* **2020**, *52*, 1722–1731. [CrossRef]
213. Iffrig, E.; Wilson, J.S.; Zhong, X.; Oshinski, J.N. Demonstration of circumferential heterogeneity in displacement and strain in the abdominal aortic wall by spiral cine DENSE MRI. *J. Magn. Reson. Imaging* **2019**, *49*, 731–743. [CrossRef]
214. Haraldsson, H.; Hope, M.; Acevedo-Bolton, G.; Tseng, E.; Zhong, X.; Epstein, F.H.; Ge, L.; Saloner, D. Feasibility of asymmetric stretch assessment in the ascending aortic wall with DENSE cardiovascular magnetic resonance. *J. Cardiovasc. Magn. Reson.* **2014**, *16*, 1–8. [CrossRef]
215. Wilson, J.S.; Islam, M.; Oshinski, J.N. In Vitro Validation of Regional Circumferential Strain Assessment in a Phantom Aortic Model Using Cine Displacement Encoding with Stimulated Echoes MRI. *J. Magn. Reson. Imaging* **2021**, *12*, 27972. [CrossRef]
216. Jones, P.A.; Wilson, J.S. The Potential for Quantifying Regional Distributions of Radial and Shear Strain in the Thoracic and Abdominal Aortic Wall Using Spiral Cine DENSE Magnetic Resonance Imaging. *J. Biomech. Eng.* **2021**, *143*. [CrossRef]
217. Abderezaei, J.; Martinez, J.; Terem, I.; Fabris, G.; Pionteck, A.; Yang, Y.; Holdsworth, S.J.; Nael, K.; Kurt, M. Amplified Flow Imaging (aFlow): A Novel MRI-Based Tool to Unravel the Coupled Dynamics between the Human Brain and Cerebrovasculature. *IEEE Trans. Med. Imaging* **2020**, *39*, 4113–4123. [CrossRef]
218. Kim, P.K.; Hong, Y.; Im, D.J.; Suh, Y.J.; Park, C.H.; Kim, J.Y.; Chang, S.; Lee, H.J.; Hur, J.; Kim, Y.J.; et al. Myocardial T1 and T2 Mapping: Techniques and Clinical Applications. *Korean J. Radiol.* **2017**, *18*, 113–131. [CrossRef]
219. Arai, A.E. Gadolinium Can Depict Area at Risk and Myocardial Infarction: A Double-Edged Sword? *JACC Cardiovasc. Imaging* **2011**, *4*, 619–621. [CrossRef]
220. Sharedalal, P.; Gerard, P.; Jain, D. Pharmacological stress myocardial perfusion imaging after an inadequate exercise stress test. *J. Nucl. Cardiol.* **2021**, *13*, 6613. [CrossRef]
221. Shehata, M.L.; Basha, T.A.; Hayeri, M.R.; Hartung, D.; Teytelboym, O.M.; Vogel-Claussen, J. MR Myocardial Perfusion Imaging: Insights on Techniques, Analysis, Interpretation, and Findings. *RadioGraphics* **2014**, *34*, 1636–1657. [CrossRef]
222. Kwon, S.T.; Burek, W.; Dupay, A.C.; Farsad, M.; Baek, S.; Park, E.-A.; Lee, W. Interaction of expanding abdominal aortic aneurysm with surrounding tissue: Retrospective CT image studies. *S. Pac. J. Nat. Appl. Sci.* **2015**, *1*, e150.
223. Kofler, J.M.; Cody, D.D.; Morin, R.L. CT Protocol Review and Optimization. *J. Am. Coll. Radiol.* **2014**, *11*, 267–270. [CrossRef]
224. Bordones, A.D.; Leroux, M.; Kheyfets, V.O.; Wu, Y.-A.; Chen, C.-Y.; Finol, E.A. Computational Fluid Dynamics Modeling of the Human Pulmonary Arteries with Experimental Validation. *Ann. Biomed. Eng.* **2018**, *46*, 1309–1324. [CrossRef]
225. Krishnan, K.; Ge, L.; Haraldsson, H.; Hope, M.D.; Saloner, D.A.; Guccione, J.M.; Tseng, E.E. Ascending thoracic aortic aneurysm wall stress analysis using patient-specific finite element modeling of in vivo magnetic resonance imaging. *Interact. Cardiovasc. Thorac. Surg.* **2015**, *21*, 471–480. [CrossRef]
226. Steinman, D.A. Image-based computational fluid dynamics modeling in realistic arterial geometries. *Ann. Biomed. Eng.* **2002**, *30*, 483–497. [CrossRef]

227. Kheyfets, V.O.; Rios, L.; Smith, T.; Schroeder, T.; Mueller, J.; Murali, S.; Lasorda, D.; Zikos, A.; Spotti, J.; Reilly, J.J.; et al. Patient-specific computational modeling of blood flow in the pulmonary arterial circulation. *Comput. Methods Progr. Biomed.* **2015**, *120*, 88–101. [CrossRef]
228. Van Bakel, T.M.J.; Lau, K.D.; Hirsch-Romano, J.; Trimarchi, S.; Dorfman, A.L.; Figueroa, C.A. Patient-Specific Modeling of Hemodynamics: Supporting Surgical Planning in a Fontan Circulation Correction. *J. Cardiovasc. Transl. Res.* **2018**, *11*, 145–155. [CrossRef]
229. Steinman, D.A. Image-based Computational Fluid Dynamics: A New Paradigm for Monitoring Hemodynamics and Atherosclerosis. *Curr. Drug Target Cardiovasc. Hematol. Disord.* **2004**, *4*, 183–197. [CrossRef]
230. Walker, J.C.; Ratcliffe, M.B.; Zhang, P.; Wallace, A.W.; Fata, B.; Hsu, E.W.; Saloner, D.; Guccione, J.M. MRI-based finite-element analysis of left ventricular aneurysm. *Am. J. Physiol. Circ. Physiol.* **2005**, *289*, H692–H700. [CrossRef]
231. Walker, J.C.; Ratcliffe, M.B.; Zhang, P.; Wallace, A.W.; Hsu, E.W.; Saloner, D.A.; Guccione, J.M. Magnetic resonance imaging-based finite element stress analysis after linear repair of left ventricular aneurysm. *J. Thorac. Cardiovasc. Surg.* **2008**, *135*, 1094–1102. [CrossRef]
232. Maas, S.A.; Erdemir, A.; Halloran, J.P.; Weiss, J.A. A general framework for application of prestrain to computational models of biological materials. *J. Mech. Behav. Biomed. Mater.* **2016**, *61*, 499–510. [CrossRef]
233. Nikou, A.; Dorsey, S.M.; McGarvey, J.R.; Gorman, J.H.; Burdick, J.A.; Pilla, J.J.; Gorman, R.C.; Wenk, J.F. Effects of using the unloaded configuration in predicting the in vivo diastolic properties of the heart. *Comput. Methods Biomech. Biomed. Eng.* **2016**, *19*, 1714–1720. [CrossRef]
234. Bols, J.; Degroote, J.; Trachet, B.; Verhegghe, B.; Segers, P.; Vierendeels, J. A computational method to assess the in vivo stresses and unloaded configuration of patient-specific blood vessels. *J. Comput. Appl. Math.* **2013**, *246*, 10–17. [CrossRef]
235. Finsberg, H.; Balaban, G.; Ross, S.; Håland, T.F.; Odland, H.H.; Sundnes, J.; Wall, S. Estimating cardiac contraction through high resolution data assimilation of a personalized mechanical model. *J. Comput. Sci.* **2018**, *24*, 85–90. [CrossRef]
236. Raghavan, M.L.; Trivedi, S.; Nagaraj, A.; McPherson, D.D.; Chandran, K.B. Three-dimensional finite element analysis of residual stress in arteries. *Ann. Biomed. Eng.* **2004**, *32*, 257–263. [CrossRef]
237. Rajagopal, V.; Chung, J.-H.; Bullivant, D.; Nielsen, P.M.F.; Nash, M.P. Determining the finite elasticity reference state from a loaded configuration. *Int. J. Numer. Methods Eng.* **2007**, *72*, 1434–1451. [CrossRef]
238. Riveros, F.; Chandra, S.; Finol, E.; Gasser, T.C.; Rodriguez, J.F. A Pull-Back Algorithm to Determine the Unloaded Vascular Geometry in Anisotropic Hyperelastic AAA Passive Mechanics. *Ann. Biomed. Eng.* **2013**, *41*, 694–708. [CrossRef]
239. Rausch, M.K.; Kuhl, E. On the effect of prestrain and residual stress in thin biological membranes. *J. Mech. Phys. Solids* **2013**, *61*, 1955–1969. [CrossRef] [PubMed]
240. Das, A.; Paul, A.; Taylor, M.D.; Banerjee, R.K. Pulsatile arterial wall-blood flow interaction with wall pre-stress computed using an inverse algorithm. *Biomed. Eng. Online* **2015**, *14*, S18. [CrossRef] [PubMed]
241. De Putter, S.; Wolters, B.; Rutten, M.; Breeuwer, M.; Gerritsen, F.; van de Vosse, F. Patient-specific initial wall stress in abdominal aortic aneurysms with a backward incremental method. *J. Biomech.* **2007**, *40*, 1081–1090. [CrossRef] [PubMed]
242. Gee, M.W.; Förster, C.; Wall, W.A. A computational strategy for prestressing patient-specific biomechanical problems under finite deformation. *Int. J. Numer. Methods Biomed. Eng.* **2009**, *26*, 52–72. [CrossRef]
243. Gee, M.W.; Reeps, C.; Eckstein, H.H.; Wall, W.A. Prestressing in finite deformation abdominal aortic aneurysm simulation. *J. Biomech.* **2009**, *42*, 1732–1739. [CrossRef]
244. Genet, M.; Rausch, M.; Lee, L.C.; Choy, S.; Zhao, X.; Kassab, G.S.; Kozerke, S.; Guccione, J.M.; Kuhl, E. Heterogeneous growth-induced prestrain in the heart. *J. Biomech.* **2015**, *48*, 2080–2089. [CrossRef]
245. Aguado-Sierra, J.; Krishnamurthy, A.; Villongco, C.; Chuang, J.; Howard, E.; Gonzales, M.J.; Omens, J.; Krummen, D.E.; Narayan, S.; Kerckhoffs, R.C.P.; et al. Patient-specific modeling of dyssynchronous heart failure: A case study. *Prog. Biophys. Mol. Biol.* **2011**, *107*, 147–155. [CrossRef]
246. Sermesant, M.; Moireau, P.; Camara, O.; Sainte-Marie, J.; Andriantsimiavona, R.; Cimrman, R.; Hill, D.L.G.; Chapelle, D.; Razavi, R. Cardiac function estimation from MRI using a heart model and data assimilation: Advances and difficulties. *Med. Image Anal.* **2006**, *10*, 642–656. [CrossRef]
247. Rumindo, G.K.; Ohayon, J.; Croisille, P.; Clarysse, P. In vivo estimation of normal left ventricular stiffness and contractility based on routine cine MR acquisition. *Med. Eng. Phys.* **2020**, *85*, 16–26. [CrossRef]
248. Wang, Z.J.; Wang, V.Y.; Bradley, C.P.; Nash, M.P.; Young, A.A.; Cao, J.J. Left Ventricular Diastolic Myocardial Stiffness and End-Diastolic Myofibre Stress in Human Heart Failure Using Personalised Biomechanical Analysis. *J. Cardiovasc. Transl. Res.* **2018**, *11*, 346–356. [CrossRef]
249. Finsberg, H.; Xi, C.; Zhao, X.; Le Tan, J.; Genet, M.; Sundnes, J.S.; Lee, L.C.; Zhong, L.; Wall, S.T. Computational quantification of patient-specific changes in ventricular dynamics associated with pulmonary hypertension. *Am. J. Physiol. Circ. Physiol.* **2019**, *317*, H1363–H1375. [CrossRef]
250. Asner, L.; Hadjicharalambous, M.; Chabiniok, R.; Peresutti, D.; Sammut, E.; Wong, J.; Carr-White, G.; Chowienczyk, P.; Lee, J.; King, A.; et al. Estimation of passive and active properties in the human heart using 3D tagged MRI. *Biomech. Model. Mechanobiol.* **2015**, *15*, 1121–1139. [CrossRef]

251. Asner, L.; Hadjicharalambous, M.; Chabiniok, R.; Peressutti, D.; Sammut, E.; Wong, J.; Carr-White, G.; Razavi, R.; King, A.P.; Smith, N.; et al. Patient-specific modeling for left ventricular mechanics using data-driven boundary energies. *Comput. Methods Appl. Mech. Eng.* **2016**, *314*, 269–295. [CrossRef]
252. Nasopoulou, A.; Shetty, A.; Lee, J.; Nordsletten, D.; Rinaldi, C.; Lamata, P.; Niederer, S. Improved identifiability of myocardial material parameters by an energy-based cost function. *Biomech. Model. Mechanobiol.* **2017**, *16*, 971–988. [CrossRef]
253. Gao, H.; Aderhold, A.; Mangion, K.; Luo, X.; Husmeier, D.; Berry, C. Changes and classification in myocardial contractile function in the left ventricle following acute myocardial infarction. *J. R. Soc. Interface* **2017**, *14*, 20170203. [CrossRef]
254. Marchesseau, S.; Delingette, H.; Sermesant, M.; Cabrera-Lozoya, R.; Tobon-Gomez, C.; Moireau, P.; Ventura, R.F.; Lekadir, K.; Hernandez, A.; Garreau, M.; et al. Personalization of a cardiac electromechanical model using reduced order unscented Kalman filtering from regional volumes. *Med. Image Anal.* **2013**, *17*, 816–829. [CrossRef]
255. Xi, J.; Lamata, P.; Lee, J.; Moireau, P.; Chapelle, D.; Smith, N. Myocardial transversely isotropic material parameter estimation from in-silico measurements based on a reduced-order unscented Kalman filter. *J. Mech. Behav. Biomed. Mater.* **2011**, *4*, 1090–1102. [CrossRef]
256. Xi, J.; Lamata, P.; Shi, W.; Niederer, S.; Land, S.; Rueckert, D.; Duckett, S.G.; Shetty, A.K.; Rinaldi, C.A.; Razavi, R.; et al. An Automatic Data Assimilation Framework for Patient-Specific Myocardial Mechanical Parameter Estimation. In *Functional Imaging and Modeling of the Heart*; Springer: Berlin/Heidelberg, Germany, 2011; pp. 392–400. [CrossRef]
257. Hassaballah, A.I.; Hassan, M.A.; Mardi, A.N.; Hamdi, M. An Inverse Finite Element Method for Determining the Tissue Compressibility of Human Left Ventricular Wall during the Cardiac Cycle. *PLoS ONE* **2013**, *8*, e82703. [CrossRef]
258. Helm, P.; Beg, M.F.; Miller, M.I.; Winslow, R.L. Measuring and Mapping Cardiac Fiber and Laminar Architecture Using Diffusion Tensor MR Imaging. *Ann. N. Y. Acad. Sci.* **2005**, *1047*, 296–307. [CrossRef]
259. Usyk, T.P.; Mazhari, R.; McCulloch, A.D. Effect of Laminar Orthotropic Myofiber Architecture on Regional Stress and Strain in the Canine Left Ventricle. *J. Elast.* **2000**, *61*, 143–164. [CrossRef]
260. Streeter, D.D.; Spotnitz, H.M.; Patel, D.P.; Ross, J.; Sonnenblick, E.H. Fiber Orientation in the Canine Left Ventricle during Diastole and Systole. *Circ. Res.* **1969**, *24*, 339–347. [CrossRef]
261. Nielsen, P.M.; Le Grice, I.J.; Smaill, B.H.; Hunter, P.J. Mathematical model of geometry and fibrous structure of the heart. *Am. J. Physiol. Circ. Physiol.* **1991**, *260*, H1365–H1378. [CrossRef]
262. Cerqueira, M.D.; Weissman, N.J.; Dilsizian, V.; Jacobs, A.K.; Kaul, S.; Laskey, W.K.; Pennell, D.J.; Rumberger, J.A.; Ryan, T.; Verani, M.S. Standardized myocardial segmentation and nomenclature for tomographic imaging of the heart: A statement for healthcare professionals from the Cardiac Imaging Committee of the Council on Clinical Cardiology of the American Heart Association. *Circulation* **2002**, *105*, 539–542. [CrossRef]
263. Misfeld, M.; Sievers, H.-H. Heart valve macro- and microstructure. *Philos. Trans. R. Soc. B Biol. Sci.* **2007**, *362*, 1421–1436. [CrossRef] [PubMed]
264. Lee, C.-H.; Laurence, D.W.; Ross, C.J.; Kramer, K.E.; Babu, A.R.; Johnson, E.L.; Hsu, M.-C.; Aggarwal, A.; Mir, A.; Burkhart, H.M.; et al. Mechanics of the Tricuspid Valve—From Clinical Diagnosis/Treatment, In-Vivo and In-Vitro Investigations, to Patient-Specific Biomechanical Modeling. *Bioengineering* **2019**, *6*, 47. [CrossRef]
265. Votta, E.; Le, T.B.; Stevanella, M.; Fusini, L.; Caiani, E.G.; Redaelli, A.; Sotiropoulos, F. Toward patient-specific simulations of cardiac valves: State-of-the-art and future directions. *J. Biomech.* **2013**, *46*, 217–228. [CrossRef]
266. Rausch, M.K.; Famaey, N.; Shultz, T.O.; Bothe, W.; Miller, D.C.; Kuhl, E. Mechanics of the mitral valve. *Biomech. Model. Mechanobiol.* **2012**, *12*, 1053–1071. [CrossRef]
267. Aggarwal, A.; Sacks, M.S. An inverse modeling approach for semilunar heart valve leaflet mechanics: Exploitation of tissue structure. *Biomech. Model. Mechanobiol.* **2016**, *15*, 909–932. [CrossRef]
268. Einstein, D.R.; Freed, A.D.; Stander, N.; Fata, B.; Vesely, I. Inverse Parameter Fitting of Biological Tissues: A Response Surface Approach. *Ann. Biomed. Eng.* **2005**, *33*, 1819–1830. [CrossRef] [PubMed]
269. Bark, D.L.; Dasi, L.P. The Impact of Fluid Inertia on In Vivo Estimation of Mitral Valve Leaflet Constitutive Properties and Mechanics. *Ann. Biomed. Eng.* **2015**, *44*, 1425–1435. [CrossRef]
270. Goth, W.; Potter, S.; Allen, A.C.B.; Zoldan, J.; Sacks, M.S.; Tunnell, J.W. Non-Destructive Reflectance Mapping of Collagen Fiber Alignment in Heart Valve Leaflets. *Ann. Biomed. Eng.* **2019**, *47*, 1250–1264. [CrossRef] [PubMed]
271. Lee, C.-H.; Amini, R.; Gorman, R.; Gorman, J.; Sacks, M. An inverse modeling approach for stress estimation in mitral valve anterior leaflet valvuloplasty for in-vivo valvular biomaterial assessment. *J. Biomech.* **2014**, *47*, 2055–2063. [CrossRef] [PubMed]
272. Hwang, J.-W.; Kim, S.M.; Park, S.-J.; Cho, E.J.; Kim, E.K.; Chang, S.-A.; Lee, S.-C.; Choe, Y.H.; Park, S.W. Assessment of reverse remodeling predicted by myocardial deformation on tissue tracking in patients with severe aortic stenosis: A cardiovascular magnetic resonance imaging study. *J. Cardiovasc. Magn. Reson.* **2017**, *19*, 80. [CrossRef] [PubMed]
273. Aggarwal, A.; Aguilar, V.S.; Lee, C.-H.; Ferrari, G.; Gorman, J.H.; Gorman, R.C.; Sacks, M.S. Patient-Specific Modeling of Heart Valves: From Image to Simulation. In *Functional Imaging and Modeling of the Heart*; Springer: Berlin/Heidelberg, Germany, 2013; pp. 141–149. [CrossRef]
274. Abbasi, M.; Azadani, A.N. Leaflet stress and strain distributions following incomplete transcatheter aortic valve expansion. *J. Biomech.* **2015**, *48*, 3663–3671. [CrossRef]

275. Krishnamurthy, G.; Ennis, D.B.; Itoh, A.; Bothe, W.; Swanson, J.C.; Karlsson, M.; Kuhl, E.; Miller, D.C.; Ingels, N.B. Material properties of the ovine mitral valve anterior leaflet in vivo from inverse finite element analysis. *Am. J. Physiol. Circ. Physiol.* **2008**, *295*, H1141–H1149. [CrossRef]
276. Itoh, A.; Krishnamurthy, G.; Swanson, J.C.; Ennis, D.B.; Bothe, W.; Kuhl, E.; Karlsson, M.; Davis, L.R.; Miller, D.C.; Ingels, N.B. Active stiffening of mitral valve leaflets in the beating heart. *Am. J. Physiol. Circ. Physiol.* **2009**, *296*, H1766–H1773. [CrossRef]
277. Lee, C.-H.; Zhang, W.; Feaver, K.; Gorman, R.C.; Gorman, J.H.; Sacks, M.S. On the in vivo function of the mitral heart valve leaflet: Insights into tissue–interstitial cell biomechanical coupling. *Biomech. Model. Mechanobiol.* **2017**, *16*, 1613–1632. [CrossRef]
278. Aggarwal, A.; Pouch, A.M.; Lai, E.; Lesicko, J.; Yushkevich, P.A.; Gorman, J.H.; Gorman, R.C.; Sacks, M.S. In-vivo heterogeneous functional and residual strains in human aortic valve leaflets. *J. Biomech.* **2016**, *49*, 2481–2490. [CrossRef]
279. Aly, A.H.; Lai, E.K.; Yushkevich, N.; Stoffers, R.H.; Gorman, J.H.; Cheung, A.T.; Gorman, R.C.; Yushkevich, P.A.; Pouch, A.M. In Vivo Image-Based 4D Modeling of Competent and Regurgitant Mitral Valve Dynamics. *Exp. Mech.* **2020**, *61*, 159–169. [CrossRef]
280. Bracamonte, J.; Wilson, J.S.; Soares, J.S. Modeling Patient-Specific Periaortic Interactions with Static and Dynamic Structures Using a Moving Heterogeneous Elastic Foundation Boundary Condition. In *Lecture Notes in Computer Science*; Springer: Cham, Switzerland, 2021; pp. 315–327. [CrossRef]
281. Di Giuseppe, M.; Farzaneh, S.; Zingales, M.; Pasta, S.; Avril, S. Patient-specific computational evaluation of stiffness distribution in ascending thoracic aortic aneurysm. *J. Biomech.* **2021**, *119*, 110321. [CrossRef]
282. Van Disseldorp, E.M.J.; Petterson, N.J.; Rutten, M.C.M.; van de Vosse, F.N.; van Sambeek, M.R.H.M.; Lopata, R. Patient Specific Wall Stress Analysis and Mechanical Characterization of Abdominal Aortic Aneurysms Using 4D Ultrasound. *Eur. J. Vasc. Endovasc. Surg.* **2016**, *52*, 635–642. [CrossRef]
283. Van Disseldorp, E.M.J.; Petterson, N.J.; Van De Vosse, F.N.; Van Sambeek, M.R.H.M.; Lopata, R.G.P. Quantification of aortic stiffness and wall stress in healthy volunteers and abdominal aortic aneurysm patients using time-resolved 3D ultrasound: A comparison study. *Eur. Heart J.—Cardiovasc. Imaging* **2019**, *20*, 185–191. [CrossRef]
284. Talou, G.D.M.; Blanco, P.J.; Ares, G.D.; Bezerra, C.G.; Lemos, P.A.; Feijóo, R.A. Mechanical Characterization of the Vessel Wall by Data Assimilation of Intravascular Ultrasound Studies. *Front. Physiol.* **2018**, *9*, 292. [CrossRef]
285. Wang, Q.; Canton, G.; Guo, J.; Guo, X.; Hatsukami, T.S.; Billiar, K.L.; Yuan, C.; Wu, Z.; Tang, D. MRI-based patient-specific human carotid atherosclerotic vessel material property variations in patients, vessel location and long-term follow up. *PLoS ONE* **2017**, *12*, e0180829. [CrossRef]
286. Liu, H.; Canton, G.; Yuan, C.; Yang, C.; Billiar, K.; Teng, Z.; Hoffman, A.H.; Tang, D. Using In Vivo Cine and 3D Multi-Contrast MRI to Determine Human Atherosclerotic Carotid Artery Material Properties and Circumferential Shrinkage Rate and Their Impact on Stress/Strain Predictions. *J. Biomech. Eng.* **2012**, *134*, 011008–0110089. [CrossRef]
287. Masson, I.; Boutouyrie, P.; Laurent, S.; Humphrey, J.D.; Zidi, M. Characterization of arterial wall mechanical behavior and stresses from human clinical data. *J. Biomech.* **2008**, *41*, 2618–2627. [CrossRef]
288. Taviani, V.; Sutcliffe, M.P.F.; Wong, P.; Li, Z.-Y.; Young, V.; Graves, M.J.; Gillard, J.H. In vivo non-invasive high resolution MR-based method for the determination of the elastic modulus of arterial vessels. *IEEE Trans. Biomed. Eng.* **2008**, *55*, 5569–5572. [CrossRef]
289. Masson, I.; Fassot, C.; Zidi, M. Finite dynamic deformations of a hyperelastic, anisotropic, incompressible and prestressed tube. Applications to in vivo arteries. *Eur. J. Mech. A Solids* **2010**, *29*, 523–529. [CrossRef]
290. Labrosse, M.R.; Gerson, E.R.; Veinot, J.P.; Beller, C.J. Mechanical characterization of human aortas from pressurization testing and a paradigm shift for circumferential residual stress. *J. Mech. Behav. Biomed. Mater.* **2013**, *17*, 44–55. [CrossRef]
291. Abdulameer, H.; Al Taii, H.; Al-Kindi, S.G.; Milner, R. Epidemiology of fatal ruptured aortic aneurysms in the United States (1999–2016). *J. Vasc. Surg.* **2019**, *69*, 378–384. [CrossRef]
292. American Heart Association. *Heart Disease and Stroke Statistics-2019 At-a-Glance Heart Disease, Stroke and Other Cardio-Vascular Diseases*; American Heart Association: Dallas, TX, USA, 2019; Available online: https://professional.heart.org/es/science-news/-/media/22cf5db5b1a24b38a435fcecb42d588b.ashx (accessed on 27 November 2019).
293. Saeyeldin, A.A.; Velasquez, C.A.; Mahmood, S.U.B.; Brownstein, A.J.; Zafar, M.; Ziganshin, B.A.; Elefteriades, J.A. Thoracic aortic aneurysm: Unlocking the "silent killer" secrets. *Gen. Thorac. Cardiovasc. Surg.* **2017**, *67*, 1–11. [CrossRef] [PubMed]
294. Jana, S.; Hu, M.; Shen, M.; Kassiri, Z. Extracellular matrix, regional heterogeneity of the aorta, and aortic aneurysm. *Exp. Mol. Med.* **2019**, *51*, 1–15. [CrossRef] [PubMed]
295. Thubrikar, M.J.; Labrosse, M.; Robic, F. Mechanical properties of abdominal aortic aneurysm wall. *J. Med. Eng. Technol.* **2001**, *25*, 133–142. [CrossRef] [PubMed]
296. Huang, X.; Yang, C.; Yuan, C.; Liu, F.; Canton, G.; Zheng, J.; Woodard, P.K.; Sicard, G.A.; Tang, D. Patient-Specific Artery Shrinkage and 3D Zero-Stress State in Multi-Component 3D FSI Models for Carotid Atherosclerotic Plaques Based on In Vivo MRI Data. *Mol Cell Biomech.* **2009**, *6*, 121–134.
297. Quick, C.M.; Berger, D.S.; Stewart, R.H.; Laine, G.A.; Hartley, C.; Noordergraaf, A. Resolving the Hemodynamic Inverse Problem. *IEEE Trans. Biomed. Eng.* **2006**, *53*, 361–368. [CrossRef]
298. Pant, S.; Corsini, C.; Baker, C.; Hsia, T.-Y.; Pennati, G.; Vignon-Clementel, I.E. Inverse problems in reduced order models of cardiovascular haemodynamics: Aspects of data assimilation and heart rate variability. *J. R. Soc. Interface* **2017**, *14*, 20160513. [CrossRef]

299. Gaidzik, F.; Pathiraja, S.; Saalfeld, S.; Stucht, D.; Speck, O.; Thévenin, D.; Janiga, G. Hemodynamic Data Assimilation in a Subject-specific Circle of Willis Geometry. *Clin. Neuroradiol.* **2020**, *31*, 643–651. [CrossRef]
300. Rispoli, V.C.; Nielsen, J.F.; Nayak, K.S.; Carvalho, J.L.A. Computational fluid dynamics simulations of blood flow regularized by 3D phase contrast MRI. *Biomed. Eng. Online* **2015**, *14*, 1–23. [CrossRef]
301. Moireau, P.; Chapelle, D. Reduced-order Unscented Kalman Filtering with application to parameter identification in large-dimensional systems. *ESAIM Control. Optim. Calc. Var.* **2011**, *17*, 380–405. [CrossRef]
302. D'Elia, M.; Mirabella, L.; Passerini, T.; Perego, M.; Piccinelli, M.; Vergara, C.; Veneziani, A. Applications of variational data assimilation in computational hemodynamics. *Model. Simul. Appl.* **2012**, 363–394. [CrossRef]
303. Perego, M.; Veneziani, A.; Vergara, C. A Variational Approach for Estimating the Compliance of the Cardiovascular Tissue: An Inverse Fluid-Structure Interaction Problem. *SIAM J. Sci. Comput.* **2011**, *33*, 1181–1211. [CrossRef]
304. Bertoglio, C.; Barber, D.; Gaddum, N.; Valverde, I.; Rutten, M.; Beerbaum, P.; Moireau, P.; Hose, R.; Gerbeau, J.-F. Identification of artery wall stiffness: In vitro validation and in vivo results of a data assimilation procedure applied to a 3D fluid–structure interaction model. *J. Biomech.* **2014**, *47*, 1027–1034. [CrossRef]
305. Zambrano, B.A.; McLean, N.A.; Zhao, X.; Tan, J.-L.; Zhong, L.; Figueroa, C.A.; Lee, L.C.; Baek, S. Image-based computational assessment of vascular wall mechanics and hemodynamics in pulmonary arterial hypertension patients. *J. Biomech.* **2018**, *68*, 84–92. [CrossRef]
306. Zambrano, B.A.; McLean, N.; Zhao, X.; Tan, J.-L.; Zhong, L.; Figueroa, C.A.; Lee, L.C.; Baek, S. Patient-Specific Computational Analysis of Hemodynamics and Wall Mechanics and Their Interactions in Pulmonary Arterial Hypertension. *Front. Bioeng. Biotechnol.* **2020**, *8*, 1536. [CrossRef]
307. Peirlinck, M.; Costabal, F.S.; Yao, J.; Guccione, J.M.; Tripathy, S.; Wang, Y.; Ozturk, D.; Segars, P.; Morrison, T.M.; Levine, S.; et al. Precision medicine in human heart modeling. *Biomech. Model. Mechanobiol.* **2021**, *20*, 803–831. [CrossRef]
308. He, X.; Avril, S.; Lu, J. Prediction of local strength of ascending thoracic aortic aneurysms. *J. Mech. Behav. Biomed. Mater.* **2021**, *115*, 104284. [CrossRef]

applied sciences

MDPI

Article

Cardiac Diffusion Tensor Biomarkers of Chronic Infarction Based on In Vivo Data

Tanjib Rahman [1], Kévin Moulin [2,3] and Luigi E. Perotti [1,*]

1 Department of Mechanical and Aerospace Engineering, University of Central Florida, Orlando, FL 32816, USA; tanjib@knights.ucf.edu
2 CREATIS Laboratory, University of Lyon, UJM-Saint-Etienne, INSA, CNRS UMR 5520, INSERM, 69100 Villeurbanne, France; kevin.moulin@creatis.insa-lyon.fr
3 Department of Radiology, University Hospital Saint-Etienne, 42270 Saint-Priest-en-Jarez, France
* Correspondence: luigi.perotti@ucf.edu

Abstract: In vivo cardiac diffusion tensor imaging (cDTI) data were acquired in swine subjects six to ten weeks post-myocardial infarction (MI) to identify microstructural-based biomarkers of MI. Diffusion tensor invariants, diffusion tensor eigenvalues, and radial diffusivity (RD) are evaluated in the infarct, border, and remote myocardium, and compared with extracellular volume fraction (ECV) and native T1 values. Additionally, to aid the interpretation of the experimental results, the diffusion of water molecules was numerically simulated as a function of ECV. Finally, findings based on in vivo measures were confirmed using higher-resolution and higher signal-to-noise data acquired ex vivo in the same subjects. Mean diffusivity, diffusion tensor eigenvalues, and RD increased in the infarct and border regions compared to remote myocardium, while fractional anisotropy decreased. Secondary (e_2) and tertiary (e_3) eigenvalues increased more significantly than the primary eigenvalue in the infarct and border regions. These findings were confirmed by the diffusion simulations. Although ECV presented the largest increase in infarct and border regions, e_2, e_3, and RD increased the most among non-contrast-based biomarkers. RD is of special interest as it summarizes the changes occurring in the radial direction and may be more robust than e_2 or e_3 alone.

Keywords: diffusion tensor imaging; in vivo cDTI; chronic infarction; cardiac microstructure; radial diffusivity; swine infarction model

Citation: Rahman, T.; Moulin, K.; Perotti, L.E. Cardiac Diffusion Tensor Biomarkers of Chronic Infarction Based on In Vivo Data. *Appl. Sci.* **2022**, *12*, 3512. https://doi.org/10.3390/app12073512

Academic Editors: Mihaela Pop and Cristian A. Linte

Received: 24 February 2022
Accepted: 25 March 2022
Published: 30 March 2022

1. Introduction

Cardiac function in health and disease depends on cardiac microstructure, which governs the preferential directions of contraction/relaxation and the mechanical/electrical properties [1,2] of the myocardium. Cardiac microstructure can be inferred from cardiac diffusion tensor imaging (cDTI), a magnetic resonance imaging (MRI) technique that allows the mapping of tissue microstructure in vivo without the use of contrast agents. Indeed, cDTI measures the intracellular and extracellular anisotropic diffusion of water molecules, from which the preferential orientation of cardiomyocytes and sheetlets is estimated [3,4].

As cDTI probes cardiac microstructure, it also provides information on microstructural changes occurring as a result of remodeling due to cardiac diseases, for example due to chronic myocardial infarction [5,6]. Remodeling due to scar formation post-myocardial infarction may lead to increased wall stress, reduced ejection fraction, and wall thinning, affecting the overall cardiac function [7]. Understanding the changes in cardiac microstructure due to myocardial infarction (MI) could provide valuable insight into the post-MI remodeling process.

MRI sequences such as T1-mapping and late gadolinium enhancement (LGE) are commonly used to identify the location and extent of the infarcted myocardium. These imaging sequences require the use of gadolinium-based contrast agents, which can have adverse effects on patients with pre-existing renal conditions [8].

Diffusion tensor invariants, such as mean diffusivity (MD) and fractional anisotropy (FA), can be used to detect and quantify the extent of MI with the additional benefit of providing insight into cardiac microstructure. In chronic infarcted tissue, previous studies [9–11] have reported an increase in MD and a decrease in FA, indicating overall microstructural changes post-MI. However, these overall microstructural changes have not carried over to specific changes in the preferential direction of the cardiomyocytes and along the sheetlets/cross-myofiber directions. Information regarding changes in these microstructural directions can be extracted from the diffusion tensor eigenvalues. Indeed, the primary eigenvalue (e_1) corresponds to the diffusivity along the preferential direction of the cardiomyocytes [12], while the secondary (e_2) and tertiary (e_3) eigenvalues correspond to the diffusivity along the sheetlet and cross-myofiber directions [13].

The main goal of this study is to refine the analysis of cardiac microstructural changes by studying the individual diffusion tensor eigenvalues and by computing radial diffusivity, a marker of both sheetlet and cross-myofiber diffusivity, using in vivo cDTI data (radial diffusivity is defined as the average of the secondary and tertiary eigenvalues). These quantities are then compared against diffusion tensor invariants (e.g., MD, FA), native-T1, and extracellular volume fraction (ECV) to establish their sensitivity in identifying the infarcted tissue and its microstructural changes.

In the infarcted region, the extracellular volume fraction increases due to the death of cardiomyocytes. We hypothesize that this increased extracellular space will mainly occur in the radial direction of the remaining cardiomyocytes and replacement fibrosis, and will be reflected by a higher increase in e_2, e_3, and radial diffusivity compared to the increase in e_1 and MD. This hypothesis is also motivated by our previous work [14] based on ex vivo, high-resolution, and high signal-to-noise ratio (SNR) data. Moreover, to further understand the results computed from experimental data, we simulate numerically the diffusion of water molecules. The computed diffusion quantities are then compared with experimental results as a function of ECV. We conclude our study by discussing the most effective diffusion quantities to detect and characterize the remote, border, and infarct regions and their microstructural interpretation.

2. Methods

2.1. Animal Model and Infarct Induction

Animal care during infarct induction, imaging, and all experimental procedures followed protocol #2015-124 approved by the Institutional Animal Care and Use Committee of the University of California, Los Angeles.

These experiments were part of a larger study, whose scope is to characterize myocardial structure and function [15], and to identify the material behavior of the passive myocardium [16]. As part of this study, an infarct model was created using female Yorkshire swine subjects. In vivo and/or ex vivo MRI data necessary for the current study was successfully acquired in seven (N = 7) subjects. All subjects (N = 7) were included in the ex vivo analysis. Two subjects were excluded from the in vivo analysis since: (1) one subject died during MRI acquisition before in vivo cDTI data could be acquired; and (2) one subject presented a right ventricle infarct that was not visible on in vivo cDTI data due to the poor in vivo cDTI quality in the right ventricle. The remaining five (N = 5) subjects were considered for in vivo analyses.

Before the beginning of the experimental procedures, the animal subjects had time to acclimate for at least one week. At the time of the MRI exam, the subjects' body weight was 59.5 kg $\pm$ 6.6 kg (mean $\pm$ standard deviation). Myocardial infarction was induced under general anesthesia using microspheres. Ketamine (12.5 mg/kg) and midazolam (1 mg/kg) were injected intramuscularly to induce anesthesia. After induction, carprofen (4 mg/kg) and buprenorphine (0.02 mg/kg) were administered intramuscularly to provide pre-emptive analgesia. During MRI, anesthesia was maintained using isoflurane (1.5–2%) and Lactated Ringer's solution was administered (2–5 mL/kg/h).

After accessing the femoral artery using a Seldinger technique, a balloon wedge pressure catheter (7 French) was inserted and guided to the aortic sinus using a metal guidewire under X-ray fluoroscopy. Subsequently, a micro guidewire (0.014 inches) was used to select a branch of the left circumflex (LCx) or left anterior descending (LAD) artery and a balloon catheter (1.5 mm diameter) was inserted and inflated prior of injecting a volume of microspheres (90 μm Polystyrene microspheres) equal to 2.5–3.0 mL. After one minute to avoid microsphere backflow, the balloon catheter was deflated and extracted. In vivo imaging was conducted six to ten weeks after infarct induction to allow for the formation of scar tissue. Additional details regarding the experimental procedure may be found in [17].

A 3T MRI scanner (Prisma, Siemens Healthineers, Erlangen, Germany) was used for in vivo and ex vivo imaging. In vivo imaging protocols included: late gadolinium enhancement (phase-sensitive inversion recovery sequence, TE/TR = 1.6 ms/876 ms; flip angle = 20°; spatial resolution = $1.33 \times 1.33 \times 8.0$ mm^3, $N_{avg} = 1$), T1-mapping (modified look-locker inversion recovery sequence with motion correction, TE/TR = 1.04 ms/280.69 ms; flip angle = 30°; spatial resolution = $1.77 \times 1.77 \times 8.0$ mm^3, $N_{avg} = 1$), and cDTI (M1M2-nulled motion-compensated waveform sequence [18], TE/TR = 59 ms/5000 ms; flip angle = 90°; spatial resolution = $2.0 \times 2.0 \times 8.0$ mm^3; b-values = 0 s/mm^2 and 350 s/mm^2; $N_{dir} = 12$; $N_{avg} = 30$).

At the end of the in vivo MRI exam and before euthanasia, subjects were injected with a double dose of gadolinium-based contrast agent (0.6 mL/kg gadopentetate dimeglumine and 10 mL of saline solution). Euthanasia solution (0.1 mL/lb, Euthasol®, Virbac, Carros, France) was administered after 10 min to allow the circulation of the contrast agent. Hearts were then extracted, rinsed, and prepared for ex vivo imaging. To preserve ventricular geometry, the hearts were inserted into 3D-printed molds based on images acquired at mid-diastasis [19]. Subsequently, the hearts and molds were submersed in Fomblin perfluoropolyether (PFPE), and ex vivo imaging was conducted with a knee coil. On an average, ex vivo imaging began 2.5 h after euthanasia. A simplified flowchart summarizing the experimental procedure is shown in Figure 1.

Ex vivo imaging protocols included: T1-weighted GRE (TE/TR = 3.15 ms/12 ms; flip angle = 25°; acquisition matrix = 160×160; spatial resolution = $1.0 \times 1.0 \times 1.0$ mm^3, $N_{avg} = 6$), T2-weighted SE (TE/TR = 89 ms/15,460 ms; flip angle = 180°; acquisition matrix = 192×190; spatial resolution = $1.0 \times 1.0 \times 1.0$ mm^3, $N_{avg} = 8$), and cDTI (readout segmented sequence [20] with a twice-refocused spin-echo encoding [21], TE/TR = 62 ms/15,560 ms; acquisition matrix = 150×150; spatial resolution = $1.0 \times 1.0 \times 1.0$ mm^3; b-values = 0 s/mm^2 and 1000 s/mm^2; $N_{dir} = 30$; $N_{avg} = 5$).

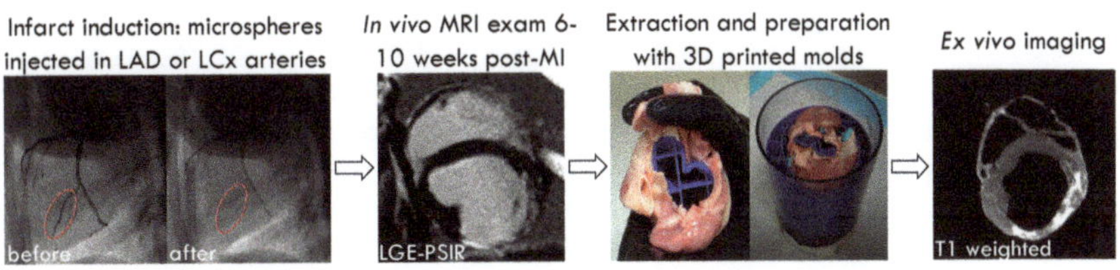

Figure 1. Experimental procedure: infarct induction and MR imaging

2.2. Regional Subdivision and Registration

To quantitatively compare diffusion tensor quantities, Native T1, and ECV, the myocardium was subdivided into remote, border, and infarct regions. The regional subdivision process was based on late gadolinium enhanced (LGE) MR images.

Following the regional subdivision technique described by Schelbert et al. [22], regions of interest were drawn on the remote regions of each slice. Mean signal intensity (μ_{remote}) and standard deviation (σ_{remote}) were computed for each subject using regions of interest across all slices. Voxels with signal intensity (SI) less than the sum of mean signal intensity and two standard deviations were labeled as the remote zone (i.e., SI $\leq \mu_{\text{remote}} + 2\sigma_{\text{remote}}$). To determine mean signal intensity (μ_{infarct}) and standard deviation (σ_{infarct}) of the infarct zone, small regions of interest (ROIs) were outlined on the most hyper-enhanced region of the remaining unlabeled myocardium of each slice. The mean signal intensities of remote and infarct zones were averaged to compute an intermediate value. Voxels with signal intensity higher than the sum of mean signal intensity and two standard deviations of the remote zone, but below the computed intermediate value (i.e., $\mu_{\text{remote}} + 2\sigma_{\text{remote}} < \text{SI} < \frac{1}{2}(\mu_{\text{remote}} + \mu_{\text{infarct}})$) were labeled as border zone. The remaining voxels (i.e., voxels with SI $\geq \frac{1}{2}(\mu_{\text{infarct}} + \mu_{\text{remote}})$) were labeled as infarct zone. Based on this approach, label maps marking remote, border, and infarct regions were created for each slice.

LGE data were acquired during diastole while cDTI, due to its motion-compensated approach, was acquired during systole. Moreover, the LGE and cDTI sequences had different resolutions and fields of view. Hence, both rigid and non-rigid registrations were necessary to superimpose the LGE-based remote, border, and infarct zones to the cDTI data. For this purpose, quaternions [23] computed from LGE and cDTI images were used to rigidly register LGE nodes (each node corresponded to a voxel in the image space) to the cDTI nodes in the cDTI image space. This step was carried out to ensure that the LGE and cDTI nodes were in the same image space (Figure 2a).

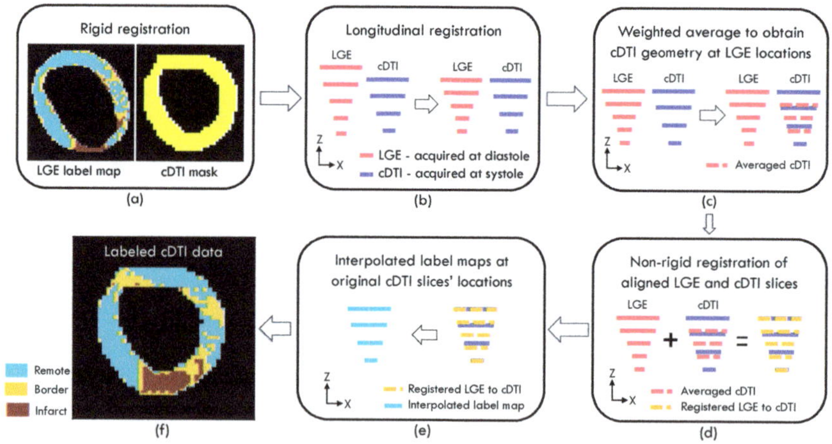

Figure 2. Diagram illustrating the registration process that was used to register LGE-based label maps to cDTI slices. (**a**) Quaternions were used to rigidly register LGE-based label maps to cDTI binary masks to ensure that both LGE and cDTI were in the same image space. (**b**) Since LGE was acquired during diastole and cDTI was acquired during systole, the longitudinal spacing between the LGE slices was reduced to reflect the contracted LV state during systole. This shortening was achieved by matching the basal and apical LGE and cDTI slices. (**c**) Although the basal and apical LGE and cDTI slices were aligned, the rest of the slices were not aligned. Hence, the cDTI slice masks at the z-axis location of the LGE slices were obtained via weighted average. (**d**) To obtain label maps at the cDTI location, aligned LGE-based label maps and weighted averaged cDTI slices were non-rigidly registered. (**e**) The cDTI label maps obtained in step (**d**) were at the z-axis location of the LGE slices. Hence, 3D interpolation was carried out to obtain label map values at the z-axis location of the original cDTI slices. (**f**) Based on the 3D interpolation, label maps for the remote (blue), border (yellow), and infarct (red) regions were obtained at the original cDTI slices' locations.

To reflect the left ventricle (LV) longitudinal shortening during systole, spacing along the z-axis between LGE-based label map slices was reduced uniformly on a subject-specific basis. This was accomplished by matching the z coordinates of the most basal and apical LGE slices with the corresponding most basal and apical cDTI slices (Figure 2b).

At this stage, even though the basal and apical slices were at the same longitudinal locations, the rest of the slices were not aligned along the z-axis. Hence, a weighted average was carried out to obtain cDTI slices at the z-axis location of the LGE slices. First, for each subject and for each cDTI slice, the endocardium and epicardium borders were detected using the Canny edge detection method [24]. Then, at each LGE z-axis location, cDTI slices were calculated by averaging the cDTI endocardium and epicardium immediately above and below that z-axis location. The distances along the z-axis between the LGE-based label map slice and the cDTI slices immediately above and below were used as weights for this process (Figure 2c).

After obtaining all the cDTI slices at the LGE z-axis locations using the method described above, the LGE-based label maps were non-rigidly registered [25] to the cDTI slices at the LGE longitudinal locations. (Figure 2d).

Finally, the resultant LGE-based label maps were applied to the original cDTI slices via 3D interpolation (Figure 2e).

At the end of this registration process, each voxel in the original cDTI slices was labeled as either remote, border, or infarct as a function of the LGE-based maps (Figure 2f). Diffusion tensor invariants, eigenvalues (e_1, e_2, and e_3), and radial diffusivity (RD) were associated with the remote, border, and infarct zones based on these maps.

2.3. Data Analysis

Diffusion tensor invariants and eigenvalues were computed after reconstructing the diffusion tensors from the acquired cDTI data. Voxelwise diffusion tensors were calculated using the freely available DiffusionRecon code [26] provided on GitHub and used in several previous studies (e.g., [4]).

ECV values at each voxel in the LV myocardium were computed from the subject Hematocrit, and from pre- and post-T1 mapping data according to [27].

To compare diffusion tensor invariants, eigenvalues, and ECV across remote, border, and infarct zones, all voxels in the LV myocardium were subdivided according to the registered LGE-based label maps. The ECV slices and cDTI slices were first rigidly and then non-rigidly registered according to the process described in Section 2.2; in this case, however, ECV slices were used instead of LGE label maps. On average, 2% (maximum 5.61%, minimum 0.61%) diffusion data were rejected due to having a mean diffusivity voxel value above the free diffusion of water (3×10^{-3} mm^2/s).

Data are visualized using diffusion tensor quantities, ECV, and native T1 maps overlaid on five short-axis slices for one representative subject, and for all subjects using raincloud plots [28] and box plots across the remote, border, and infarct regions.

Remote, border, and infarct data for all subjects were initially examined for normality using the Anderson–Darling normality test. Since the data did not pass the normality test, the pairwise non-parametric Kruskal–Wallis test with Bonferroni post hoc adjustment was used to assess differences between remote, border, and infarct data across all subjects. The same pairwise test was also run for subject-wise remote, border, and infarct data. $p < 0.01$ was considered to be significant.

2.4. Numerical Modeling

Numerical simulations were carried out to investigate the relationship between ECV and diffusion tensor quantities. Tensor invariants and eigenvalues were computed from the simulated diffusion of particles, and ECV was computed based on the generated synthetic cell structures. Individual cardiomyocytes were represented by cylinders with 9 to 20 μm diameter and 100 μm length. Cells were then connected along the axial direction to form a cellular tree with branching added between trees. Cells were added in a $0.5 \times 0.5 \times 0.5$ mm^3 voxel until a target ECV was reached. Therefore, different ECVs corresponded to different cell densities and cell-to-cell distances. The displacements due to the diffusion of 20,000 water molecules were simulated in each voxel using a random walk approach [29] with a time step Δt of 10 μs for a total duration of 51 ms, which corresponds to the duration of the diffusion encoding in vivo [30]. The native diffusion coefficient D_0 of the water molecules was 3×10^{-3} mm^2/s and 2.2×10^{-3} mm^2/s in the extra- and intracellular compartments, respectively [31]. The intracellular and extracellular compartments were kept impermeable.

The diffusion of water molecules resulted in an intra-voxel displacement distribution that was projected in 12 directions to mimic a diffusion tensor acquisition. Subsequently, the diffusion tensor corresponding to the simulated signal was reconstructed and its mean diffusivity, fractional anisotropy, eigenvalues, and radial diffusivity were calculated.

Five cell structures were generated per each target ECV. Nine target ECVs ranging from 20% to 100% were simulated, resulting in 45 different cell structures. For a given ECV and cell structure, each simulation was repeated five times, resulting in 225 diffusion simulations across ECVs and cell structures. To best approximate the simulated diffusion signal to the one acquired in the MRI experiment used in this work, only the extracellular water displacements were considered.

The code to perform the diffusion simulations described above is freely available on GitHub at [32].

3. Results

Figure 3 illustrates five short-axis representative slices of late gadolinium enhancement PSIR (LGE-PSIR), native T1, and the corresponding ECV maps along with b = 0 s/mm^2 and b = 350 s/mm^2 images of an infarcted swine heart.

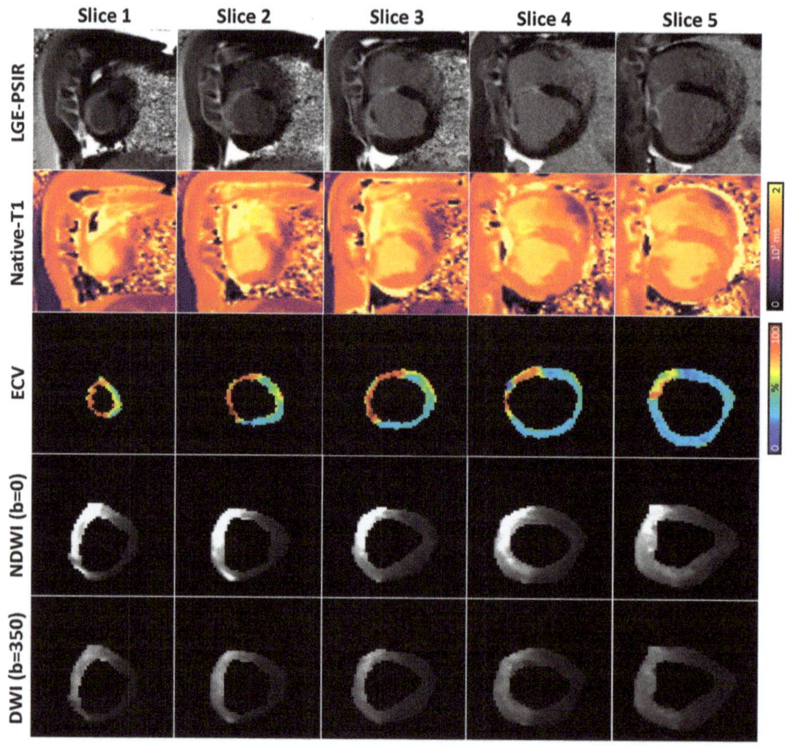

Figure 3. LGE-PSIR, native T1, ECV maps, b = 0 s/mm^2, and b = 350 s/mm^2 images for five representative short-axis slices of the same infarcted swine subject. LGE-PSIR is a contrast-based imaging technique, and native T1 and diffusion images (NDWI and DWI) are non-contrast-based techniques.

LGE-PSIR images, along with their corresponding cDTI label maps and diffusion tensor quantities maps for the same five representative short-axis slices, are reported in Figure 4. As described in the Method section, the label maps were used to subdivide the corresponding FA, MD, e_1, e_2, e_3, and RD maps in remote, border, and infarct regions. The border and infarct zones are distinguishable from the remote myocardium due to their markedly hyper-enhanced nature in the LGE-PSIR images. In the following tensor invariants and eigenvalue maps, the infarct and border zones exhibit a higher MD, e_1, e_2, e_3, RD and a lower FA compared to the remote myocardium.

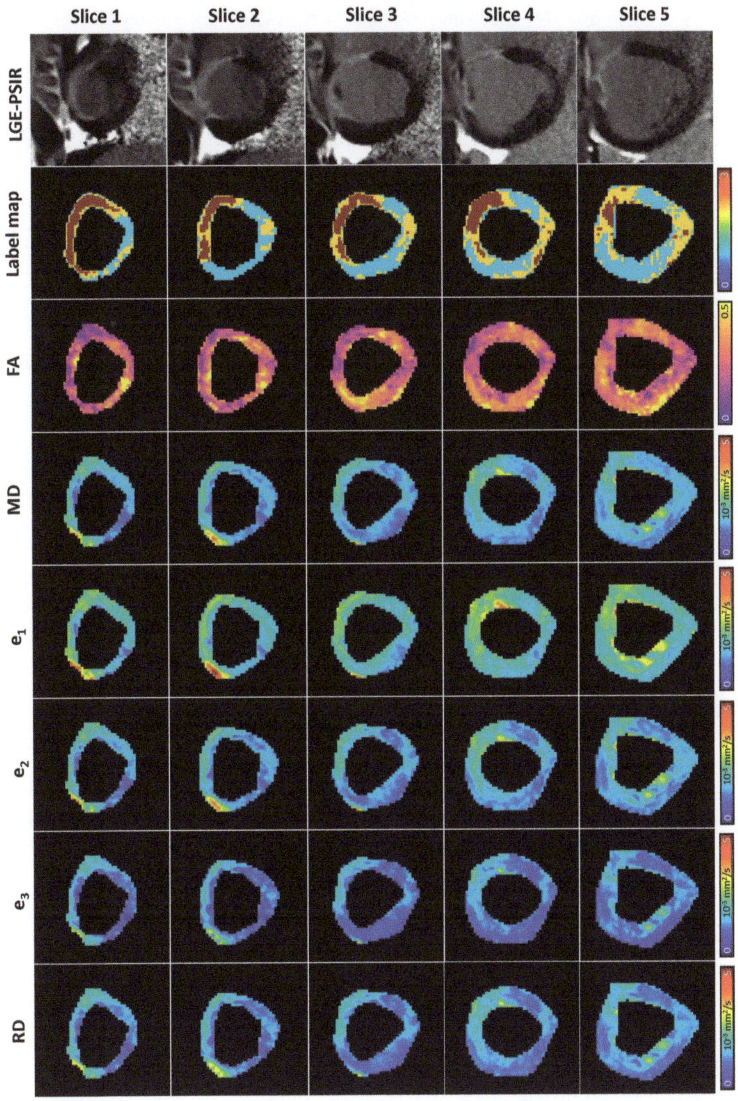

Figure 4. LGE-PSIR and corresponding label maps registered to cDTI along with fractional anisotropy (FA), mean diffusivity (MD), primary eigenvalue (e_1), secondary eigenvalue (e_2), tertiary eigenvalue (e_3), and radial diffusivity (RD) maps for five representative short-axis slices in an infarcted subject.

In Figure 5, raincloud and box plots illustrate the distributions and quantitative differences of tensor invariants, eigenvalues, radial diffusivity, ECV, and native T1 across infarct, border, and remote regions. The 1st quartile (25th percentile) and the 3rd quartile (75th percentile) are marked by the lower and upper edges of the boxplot, respectively. The bottom and top whiskers mark the smallest and largest values within 1.5 times the interquartile range measured from the 25th and 75th percentiles, respectively. A summary of the mean, median, 1st, and 3rd quartiles over all subjects for all measured quantities are listed in Table 1.

Table 1. Overall mean, median, and Q1–Q3 computed across all subjects for native T1, ECV, FA, MD, diffusion tensor eigenvalues, and RD in the infarct, border, and remote regions.

	Native T1 (ms)			**ECV**		
	Infarct	**Border**	**Remote**	**Infarct**	**Border**	**Remote**
Mean	1546	1434	1344	0.49	0.39	0.32
Median	1554	1389	1282	0.47	0.38	0.31
Q1, Q3	1363, 1700	1274, 1553	1210, 1442	0.36, 0.62	0.31, 0.46	0.27, 0.37
	FA			**MD (1×10^{-3} mm^2/s)**		
	Infarct	**Border**	**Remote**	**Infarct**	**Border**	**Remote**
Mean	0.25	0.28	0.29	1.81	1.69	1.46
Median	0.23	0.27	0.29	1.84	1.68	1.46
Q1, Q3	0.18, 0.30	0.21, 0.33	0.23, 0.35	1.52, 2.12	1.45, 1.92	1.32, 1.60
	e_1 (1×10^{-3} mm^2/s)			**e_2 (1×10^{-3} mm^2/s)**		
	Infarct	**Border**	**Remote**	**Infarct**	**Border**	**Remote**
Mean	2.27	2.17	1.91	1.80	1.65	1.40
Median	2.28	2.14	1.94	1.83	1.65	1.38
Q1, Q3	1.98, 2.55	1.90, 2.39	1.74, 2.10	1.48, 2.11	1.37, 1.92	1.22, 1.58
	e_3 (1×10^{-3} mm^2/s)			**RD (1×10^{-3} mm^2/s)**		
	Infarct	**Border**	**Remote**	**Infarct**	**Border**	**Remote**
Mean	1.38	1.25	1.06	1.59	1.45	1.23
Median	1.42	1.24	1.06	1.62	1.45	1.22
Q1, Q3	1.09, 1.71	1.01, 1.48	0.91, 1.23	1.29, 1.90	1.20, 1.69	1.08, 1.39

Among native T1, diffusion tensor invariants, eigenvalues, and RD, the largest percent differences are observed for e_2 (e_2 increases by 19.2% and 31.9% from remote to border and infarct zones, respectively), e_3 (e_3 increases by 16.1% and 33.4% from remote to border and infarct zones, respectively), and RD (RD increases by 18% and 32.3% from remote to border and infarct zones, respectively). Overall percentage changes in median values for native T1, ECV, diffusion tensor invariants, eigenvalues, and RD between infarct, border, and remote regions are detailed in Table 2.

The computed p values were less than 0.01 for the statistical analyses carried out by grouping together the data across all subjects (Table 2), therefore showing that the observations in remote, border, and infarct regions did not originate from the same distribution. However, when the same statistical analyses were performed subject-wise, the resultant p values were greater than 0.01 in a few cases, especially in the native T1 distributions between infarct and border regions. All p values resulting from the pairwise non-parametric Kruskal–Wallis test for each subject are listed in Table 3.

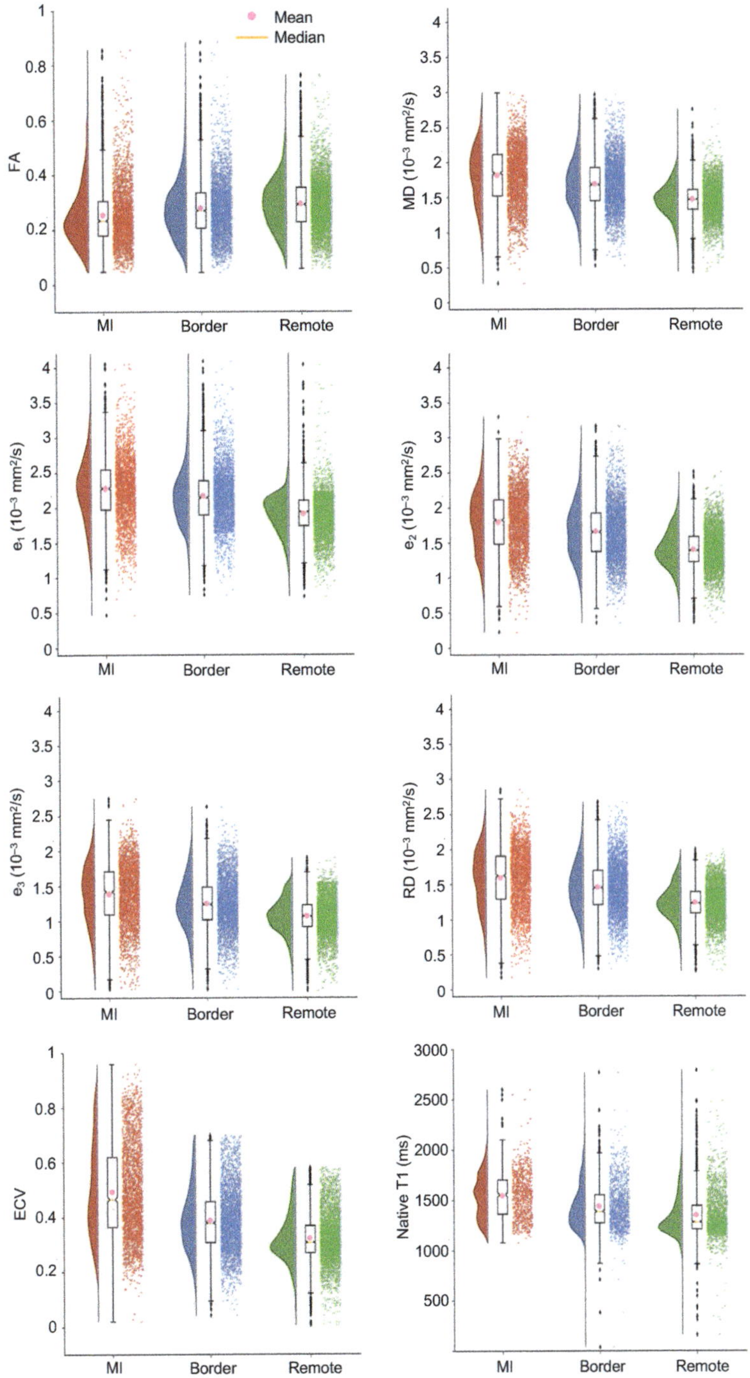

Figure 5. Raincloud plots [28] overlaid to the corresponding box plots for FA, MD, eigenvalues e_1, e_2, and e_3, radial diffusivity (RD), extracellular volume fraction (ECV), and native T1 in the infarct, border, and remote myocardial regions.

Table 2. Overall percentage change in median values and resulting p-values from non-parametric pairwise Kruskal–Wallis test for native T1, ECV, FA, MD, diffusion tensor eigenvalues, and RD between border–remote and infarct–remote regions across all subjects.

Native T1		**ECV**	
Border–Remote 8.4%, $p < 0.01$	Infarct–Remote 21.2%, $p < 0.01$	Border–Remote 22.6%, $p < 0.01$	Infarct–Remote 51.6%, $p < 0.01$
FA		**MD**	
Border–Remote −7.6%, $p < 0.01$	Infarct–Remote −19.8%, $p < 0.01$	Border–Remote 14.7%, $p < 0.01$	Infarct–Remote 26.1%, $p < 0.01$
e_1		**e_2**	
Border–Remote 10.4%, $p < 0.01$	Infarct–Remote 17.5%, $p < 0.01$	Border–Remote 19.2%, $p < 0.01$	Infarct–Remote 31.9%, $p < 0.01$
e_3		**RD**	
Border–Remote 16.1%, $p < 0.01$	Infarct–Remote 33.4%, $p < 0.01$	Border–Remote 18.0%, $p < 0.01$	Infarct–Remote 32.3%, $p < 0.01$

Table 3. Median value percentage change and p-values resulting from non-parametric pairwise Kruskal–Wallis test between infarct–border, border–remote, and infarct–remote regions for each subject separately and for all measured quantities. Values reported in red correspond to p-values greater than or equal to 0.01 or isolated cases where percentage changes are opposite to the overall observed trends.

Sub	**FA**			**MD**		
	Infarct vs. Border	**Border vs. Remote**	**Infarct vs. Remote**	**Infarct vs. Border**	**Border vs. Remote**	**Infarct vs. Remote**
S1	−18.4%, $p < 0.01$	−24.0%, $p < 0.01$	−38.0%, $p < 0.01$	14.6%, $p < 0.01$	21.3%, $p < 0.01$	39.0%, $p < 0.01$
S2	−6.5%, $p = 0.25$	−2.7%, $p < 0.01$	−9.0%, $p < 0.01$	7.0%, $p = 0.02$	5.4%, $p < 0.01$	12.8%, $p < 0.01$
S3	−11.8%, $p < 0.01$	−11.1%, $p < 0.01$	−21.6%, $p < 0.01$	15.0%, $p < 0.01$	4.6%, $p < 0.01$	20.3%, $p < 0.01$
S4	−9.5%, $p < 0.01$	−5.7%, $p = 0.1$	−14.6%, $p = 0.02$	7.3%, $p < 0.01$	2.9%, $p < 0.01$	10.3%, $p < 0.01$
S5	−10.2%, $p < 0.01$	8.3%, $p < 0.01$	−2.7%, $p < 0.01$	7.1%, $p < 0.01$	6.0%, $p < 0.01$	13.5%, $p < 0.01$
Sub	**e_1**			**e_2**		
	Infarct vs. Border	**Border vs. Remote**	**Infarct vs. Remote**	**Infarct vs. Border**	**Border vs. Remote**	**Infarct vs. Remote**
S1	9.2%, $p < 0.01$	14.1%, $p < 0.01$	24.6%, $p < 0.01$	15.3%, $p < 0.01$	25.8%, $p < 0.01$	45.0%, $p < 0.01$
S2	3.7%, $p = 0.22$	3.2%, $p < 0.01$	7.0%, $p < 0.01$	13.7%, $p < 0.01$	6.9%, $p < 0.01$	21.5%, $p < 0.01$
S3	7.8%, $p < 0.01$	3.5%, $p < 0.01$	11.6%, $p < 0.01$	16.2%, $p < 0.01$	8.5%, $p < 0.01$	26.1%, $p < 0.01$
S4	7.1%, $p < 0.01$	−0.8%, $p < 0.01$	6.2%, $p < 0.01$	8.4%, $p < 0.01$	4.5%, $p < 0.01$	13.2%, $p < 0.01$
S5	4.1%, $p < 0.01$	7.8%, $p < 0.01$	12.2%, $p < 0.01$	6.7%, $p < 0.01$	6.7%, $p < 0.01$	13.8%, $p < 0.01$
Sub	**e_3**			**RD**		
	Infarct vs. Border	**Border vs. Remote**	**Infarct vs. Remote**	**Infarct vs. Border**	**Border vs. Remote**	**Infarct vs. Remote**
S1	19.5%, $p < 0.01$	35.6%, $p < 0.01$	62.1%, $p < 0.01$	17.2%, $p < 0.01$	30.3%, $p < 0.01$	52.8%, $p < 0.01$
S2	5.1%, $p = 0.19$	8.3%, $p < 0.01$	13.8%, $p < 0.01$	10.0%, $p = 0.01$	8.7%, $p < 0.01$	19.6%, $p < 0.01$
S3	16.3%, $p < 0.01$	6.5%, $p < 0.01$	23.9%, $p < 0.01$	16.9%, $p < 0.01$	6.6%, $p < 0.01$	24.6%, $p < 0.01$
S4	10.4%, $p < 0.01$	5.2%, $p < 0.01$	16.2%, $p < 0.01$	7.7%, $p < 0.01$	4.2%, $p < 0.01$	12.2%, $p < 0.01$
S5	10.6%, $p < 0.01$	2.9%, $p < 0.01$	13.8%, $p < 0.01$	7.5%, $p < 0.01$	4.8%, $p < 0.01$	12.7%, $p < 0.01$
Sub	**Native T1**			**ECV**		
	Infarct vs. Border	**Border vs. Remote**	**Infarct vs. Remote**	**Infarct vs. Border**	**Border vs. Remote**	**Infarct vs. Remote**
S1	16.7%, $p = 0.06$	4.0%, $p = 0.88$	21.3%, $p < 0.01$	23.8%, $p < 0.01$	2.1%, $p < 0.01$	26.4%, $p < 0.01$
S2	18.7%, $p = 0.66$	14.1%, $p < 0.01$	35.4%, $p < 0.01$	61.7%, $p < 0.01$	29.2%, $p < 0.01$	108.9%, $p < 0.01$
S3	7.1%, $p = 0.11$	14.2%, $p < 0.01$	22.3%, $p < 0.01$	12.1%, $p < 0.01$	36.8%, $p < 0.01$	53.3%, $p < 0.01$
S4	2.0%, $p = 0.3$	4.5%, $p < 0.01$	6.6%, $p = 0.018$	10.1%, $p < 0.01$	11.3%, $p < 0.01$	22.6%, $p < 0.01$
S5	9.3%, $p < 0.01$	1.4%, $p = 0.1$	10.8%, $p < 0.01$	13.9%, $p < 0.01$	8.3%, $p < 0.01$	23.3%, $p < 0.01$

The results of the diffusion simulations with 20,000 water molecules and a timestep Δt of 10 µs are illustrated in Figure 6. The change in MD, FA, primary eigenvalue (e_1), and RD simulated as a function of ECV show that MD, e_1, and RD increase with respect to ECV, while FA decreases. Furthermore, as ECV increases, the simulated percentage increase in RD is larger than the increase in e_1.

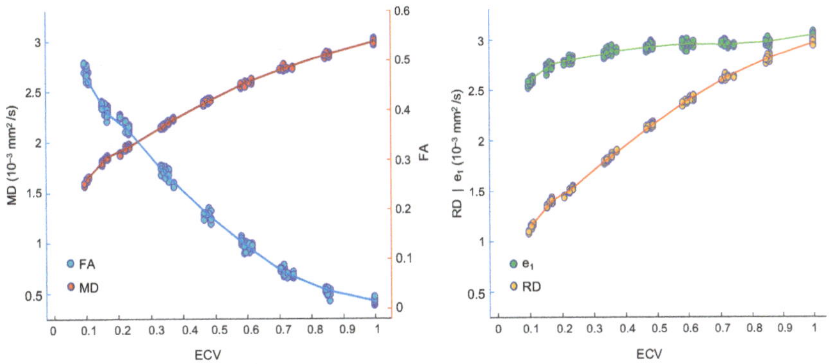

Figure 6. MD, FA, e_1, and RD as a function of simulated ECV for 45 cellular structures and 5 diffusion distributions (225 cases in total). These results are obtained from diffusion simulations with 20,000 water molecules and a timestep $\Delta t = 10$ µs. Cellular structures with similar ECV (±0.05) are clustered together and median values of each cluster are connected to better visualize the overall trend of MD, FA, e_1, and RD with increasing ECV.

Figure 7 compares in vivo and ex vivo FA, MD, e_1, e_2, e_3, and RD values. Mappings are reported for a representative slice and boxplots were used to quantitatively compare the diffusion quantities computed using in vivo and ex vivo data across all subjects. The in vivo versus ex vivo boxplot comparisons are based on the subject-wise median values for FA, MD, e_1, e_2, e_3, and RD.

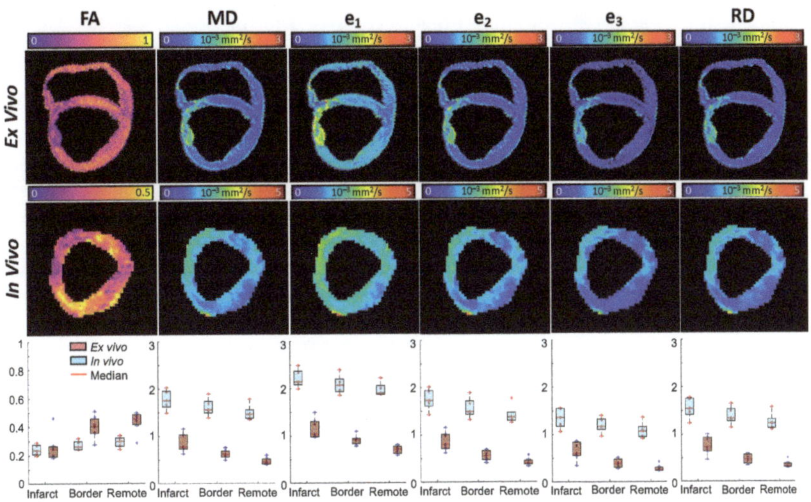

Figure 7. FA, MD, e_1, e_2, e_3, and RD mappings for a representative slice from data acquired ex vivo (top) and in vivo (mid). Ex vivo and in vivo diffusion quantities across all subjects are compared using boxplots (bottom). These boxplots are constructed based on the subject-wise medians of the diffusion quantities.

4. Discussion

In this work, MRI data acquired in vivo in swine subjects with chronic MI were used to quantitatively analyze and compare ECV, native T1, diffusion tensor invariants, radial diffusivity, and eigenvalues across the remote, border, and infarcted myocardium. Among all analyzed quantities, RD, e_2, e_3, and ECV showed the highest percentage increase in the border and infarct regions with respect to the remote myocardium.

ECV has been identified as a potential biomarker to characterize affected myocardial tissue due to various cardiac diseases such as myocardial infarction [27] and hypertension [33]. After administration, Gadolinium-based contrast agents diffuse into the extracellular space. This causes the T1-relaxation time of the myocardium to depend on local gadolinium concentration [34]. The increased extracellular space in infarcted myocardium allows retention of a higher amount of gadolinium-based contrast. Hence, myocardial regions such as the border and infarct zones, where the extracellular space is higher due to the presence of replacement fibrosis, are expected to exhibit higher ECV compared to the remote myocardium [27]. The increased median ECV of the infarct (0.47) and border (0.38) regions compared to the remote (0.31) region observed in this work agrees with this mechanism.

Although widely used, ECV maps require the use of contrast agents, which are not indicated for patients suffering from pre-existing renal conditions. Among the existing non-contrast-based imaging techniques, native T1 has been considered to be an alternative biomarker for myocardial infarction detection [35], as well as for detecting myocardial edema and diffuse fibrosis [36–38]. In the current study, native T1 values in the infarct and border regions were 21.2% and 8.4% higher than the remote myocardium, respectively.

As with the case for native T1, diffusion tensor invariants and eigenvalues can also be computed without the use of contrast agents, and convey additional information regarding microstructural changes (e.g., myocardium anisotropy and inter-cellular spacing) occurring in the infarcted tissue. Among these quantities, FA decreased whereas MD, eigenvalues, and RD increased in the border and infarct regions when compared to the remote myocardium. Across all subjects, the percent increase or decrease of these parameters was higher in the infarct region compared to the border region. These trends concur with previous findings [9–11].

Chen et al. [39] validated histologically that after chronic myocardial infarction, the infarct region is largely comprised of replacement fibrosis, and the border region is a mixture of viable myocardium and replacement fibrosis. The presence of replacement fibrosis is reflected by an increased ECV. Due to the increase in extracellular space [40], the diffusion of water molecules increases. This is reflected by the increased MD values in the infarct and border regions computed in this study from experimental in vivo cDTI data. However, the observed increase in diffusion is not isotropic; instead, the increase in e_2 and e_3 is more significant than the increase in e_1. This unequal increase in diffusion agrees with previous studies, suggesting that replacement fibrosis maintains the preferential direction of the replaced cardiomyocytes [41], and larger extracellular space is now present [40]. This corresponds to a larger diffusion increase in the e_2 and e_3 directions with respect to the increase in the e_1 direction. The larger increase in the e_2 and e_3 directions concurs with a decrease in FA and agrees with histological findings [40].

As expected, due to a smaller amount of replacement fibrosis in the border zone, the increase in e_2, e_3, MD, and ECV and decrease in FA are less significant in the border region with respect to the infarct region.

To present diffusion values in the radial direction in a concise manner, radial diffusivity (RD) was used. RD is the average of e_2 and e_3 and represents the changes occurring in the radial direction due to the increase or decrease of ECV. In this study, we showed that RD increases significantly from the remote to the border (18% increase) and infarct (32.3% increase) regions.

The results computed from experimental data concur with the findings of particle diffusion simulations. From the diffusion simulations, as ECV increases, e_1, e_2, e_3, and

MD increase, and FA decreases. These trends agree with previous studies [42] and remain consistent regardless of the number of simulated water molecules (from four to thirty thousand water molecules per representative volume), length of adopted time step (from 10^{-5} s to 10^{-7} s), and cell structures (10 to 45 cell structures have been simulated). Additionally, as ECV increases, simulated e_2 and e_3 values increase at a higher rate with respect to e_1.

Among T1-mapping, ECV and diffusion quantities, ECV demonstrated the highest change in the infarct (51.6% increase) and border (22.6% increase) regions compared to the remote myocardium. However, among the non-contrast-based methods, e_2, e_3, and RD exhibited the largest change. Moreover, the increases in e_2 (31.9% and 19.2% increases from remote to infarct and border regions, respectively), e_3 (33.4% and 16.1% increases from remote to infarct and border regions, respectively), and RD (32.3% and 18% increases from remote to infarct and border regions, respectively) were higher than native T1 (21.2% and 8.4% increases from remote to infarct and border regions, respectively).

In our previous study [14], ex vivo cDTI swine data acquired six to ten weeks post-infarction was used to compare diffusion tensor invariants, diffusion eigenvalues, and radial diffusivity across the remote, border, and infarct regions. In terms of the increase or decrease of tensor invariants, eigenvalues, and radial diffusivity in the infarct and border regions, similar trends were observed as in the present study (cf. Figure 7). However, while the diffusion eigenvalues and mean diffusivity computed using in vivo cDTI were observed to be higher than the values obtained using ex vivo data across all regions, their percentage changes were lower than the percentage changes computed using ex vivo data.

The median e_1, e_2, e_3, and RD values computed using ex vivo data (values computed using in vivo data in this study are in parenthesis) were, respectively: 1.16 (2.28) $\times 10^{-3}$ mm^2/s for e_1, 0.88 (1.83) $\times 10^{-3}$ mm^2/s for e_2, 0.72 (1.42) $\times 10^{-3}$ mm^2/s for e_3, and 0.79 (1.62) $\times 10^{-3}$ mm^2/s for RD, in the infarct region; 0.91 (2.14) $\times 10^{-3}$ mm^2/s for e_1, 0.55 (1.65) $\times 10^{-3}$ mm^2/s for e_2, 0.39 (1.24) $\times 10^{-3}$ mm^2/s for e_3, and 0.47 (1.45) $\times 10^{-3}$ mm^2/s for RD, in the border region; 0.70 (1.94) $\times 10^{-3}$ mm^2/s for e_1, 0.41 (1.38) $\times 10^{-3}$ mm^2/s for e_2, 0.28 (1.06) $\times 10^{-3}$ mm^2/s for e_3, and 0.35 (1.22) $\times 10^{-3}$ mm^2/s for RD, in the remote region.

The observed discrepancy between results obtained using in vivo and ex vivo data could be due to several factors, among which the in vivo motion artifact and perfusion. Furthermore, the MRI sequence parameters were not identical in vivo and ex vivo. The ex vivo data were acquired at a higher resolution of $1.0 \times 1.0 \times 1.0$ mm^3, higher SNR of $\approx$41 and higher b-value of 1000 s/mm^2 while the in vivo data were acquired at a resolution of $2.0 \times 2.0 \times 2.0$ mm^3 with a SNR of $\approx$16 and b-value of 350 s/mm^2. Higher resolution and SNR allow for a more accurate segmentation and subdivision of myocardium into remote, border, and infarct regions, therefore emphasizing the differences between regions.

Overall, the diffusion tensor invariants reported in this study are in good agreement with values reported in previous studies. Das et al. [43] conducted a study on 30 patients with myocardial infarction. The mean MD and FA values computed in vivo by Das et al. compared to the values computed in current study (reported in parenthesis), were: 1.83×10^{-3} (1.81×10^{-3}) mm^2/s for MD and 0.22 (0.25) for FA in the infarct region; 1.53×10^{-3} (1.69×10^{-3}) mm^2/s for MD and 0.33 (0.28) for FA in the border zone, and 1.45×10^{-3} (1.46×10^{-3}) mm^2/s for MD and 0.35 (0.29) for FA in the remote myocardium.

The computed mean ECV in Das et al. [43] was 0.60, 0.29, and 0.29 in the infarct, border, and remote regions, respectively, compared to 0.49, 0.39, and 0.32 in the current study. The discrepancies in the reported ECV values could be due to the different methods used to subdivide the myocardium into remote, border, and infarct regions. Moreover, Das et al. used in vivo human data, whereas in vivo swine data were used in the current study. The much larger cohort size (N = 30 vs. N = 5) compared to the current study could also be a reason for the observed discrepancies.

Stoeck et al. [9] conducted a study on five swine subjects with myocardial infarction. At nine weeks post-MI, reported mean MD and FA values in the infarct and remote zone were (values from the current study are reported in parenthesis): 1.66×10^{-3}

(1.81×10^{-3}) mm^2/s for MD and 0.28 (0.25) for FA in the infarct region; 1.34×10^{-3} (1.46×10^{-3}) mm^2/s for MD and 0.38 (0.29) for FA in the remote myocardium. Mean ECV values reported by Stoeck et al. were 0.83 and 0.31 in the infarct and remote regions, and in the current study, the mean ECV was 0.49 and 0.32, respectively. This discrepancy could be due to the uncertainty associated with native T1 scans and the computed ECV [44]. Differences in imaging modalities and data processing could also play a role.

This study also presents several limitations. First, in vivo cDTI data for only five subjects was considered. Although key results agree with other studies in the literature and our own ex vivo study and simulations, a larger cohort size in future studies will allow a strengthening of the current findings. As infarct size varied significantly across the imaged subjects, a larger cohort with varying infarct size will allow a better understanding of the dependence of diffusion tensor quantities on infarct size. Second, myocardial segmentation to determine the ground truth infarct, border, and remote regions based on LGE-PSIR data was carried out following the thresholding method described by Schelbert et al. [22]. Although previously adopted, uncertainties remain regarding the choice of the threshold values, which could result in small variations of the determined myocardial regions. Histological validation of the segmented remote, border, and infarct regions was not possible due to lack of histological data. Third, the difference in resolution, field of view (FOV), and cardiac phase between in vivo cDTI and LGE required interpolation and a non-rigid registration step. This might have led to the mislabeling of a small number of voxels in the boundary of the adjacent myocardial regions. Further work is required to quantify the uncertainty associated with non-rigid registration [45] along with the uncertainty due to the noise in the MRI images and observer error. Finally, regarding the diffusion simulation, a simplified geometric model was used to generate cellular structures, and simulations were carried out without considering cell permeability and intracellular diffusion. Given the higher sensitivity to extracellular diffusion of the adopted MRI sequence, we expect that the computed trends will remain representative even if a more realistic model is adopted. However, further work is needed to account for cell permeability, intracellular diffusion, multi-compartment cellular structures, and more realistic cell structures.

5. Conclusions

In vivo cDTI data that were acquired in swine subjects six to ten weeks post-myocardial infarction showed a decrease in fractional anisotropy and an increase in mean diffusivity and diffusion tensor eigenvalues in the infarct and border regions with respect to the remote myocardium. Across all subjects, the second (e_2) and third (e_3) diffusion tensor eigenvalues together with radial diffusivity (RD) showed the largest increase in the infarct and border regions compared to remote myocardium. As RD averages the changes in e_2 and e_3, it can be a potential biomarker to identify infarct regions without the use of contrast agents, and can provide additional information about microstructural changes occurring post-MI.

Supplementary Materials: The following supporting information can be downloaded at: https://www.mdpi.com/article/10.3390/app12073512/s1.

Author Contributions: L.E.P. and K.M. conceived and conducted the experiments; T.R. and K.M. designed and wrote the analysis software; T.R., K.M. and L.E.P. analyzed the data and results. All authors have read and agreed to the published version of the manuscript.

Funding: This research was supported by NIH/NHLBI K25-HL135408 grant and the University of Central Florida startup funds to L.E.P. The content of this article is solely the responsibility of the authors and does not necessarily represent the official views of the Funders.

Institutional Review Board Statement: Animal care and handling during the experimental procedures followed protocol #2015-124 approved by the University of California, Los Angeles Institutional Animal Care and Use Committee. The protocol was last approved on 30 August 2018.

Informed Consent Statement: Not applicable.

Data Availability Statement: The code to reconstruct the voxelwise diffusion tensors from cDTI data is freely available on GitHub at https://github.com/KMoulin/DiffusionRecon (Last accessed on 24 March 2022). The code to perform the diffusion simulations is also available on GitHub at https://github.com/KMoulin/DiffusionSimulation (Last accessed on 24 March 2022). In vivo cDTI, native T1, post-contrast T1, and LGE data for the analyzed swine subjects are provided as supplementary material.

Conflicts of Interest: The authors declare no conflict of interest.

References

1. Perotti, L.E.; Magrath, P.; Verzhbinsky, I.A.; Aliotta, E.; Moulin, K.; Ennis, D.B. Microstructurally anchored cardiac kinematics by combining in vivo DENSE MRI and cDTI. In Proceedings of the International Conference on Functional Imaging and Modeling of the Heart, Toronto, ON, Canada, 11–13 June 2017; pp. 381–391.
2. Hooks, D.A.; Tomlinson, K.A.; Marsden, S.G.; LeGrice, I.J.; Smaill, B.H.; Pullan, A.J.; Hunter, P.J. Cardiac microstructure: Implications for electrical propagation and defibrillation in the heart. *Circ. Res.* **2002**, *91*, 331–338. [CrossRef] [PubMed]
3. Ferreira, P.F.; Kilner, P.J.; McGill, L.A.; Nielles-Vallespin, S.; Scott, A.D.; Ho, S.Y.; McCarthy, K.P.; Haba, M.M.; Ismail, T.F.; Gatehouse, P.D.; et al. In vivo cardiovascular magnetic resonance diffusion tensor imaging shows evidence of abnormal myocardial laminar orientations and mobility in hypertrophic cardiomyopathy. *J. Cardiovasc. Magn. Reson.* **2014**, *16*, 1–16. [CrossRef] [PubMed]
4. Moulin, K.; Verzhbinsky, I.A.; Maforo, N.G.; Perotti, L.E.; Ennis, D.B. Probing cardiomyocyte mobility with multi-phase cardiac diffusion tensor MRI. *PLoS ONE* **2020**, *15*, e0241996. [CrossRef] [PubMed]
5. Celle, T.D.; Cleutjens, J.P.; Blankesteijn, W.M.; Debets, J.J.; Smits, J.F.; Janssen, B.J. Long-term structural and functional consequences of cardiac ischaemia–reperfusion injury in vivo in mice. *Exp. Physiol.* **2004**, *89*, 605–615. [CrossRef] [PubMed]
6. Blankesteijn, W.; Creemers, E.; Lutgens, E.; Cleutjens, J.; Daemen, M.; Smits, J. Dynamics of cardiac wound healing following myocardial infarction: Observations in genetically altered mice. *Acta Physiol. Scand.* **2001**, *173*, 75–82. [CrossRef]
7. Grieve, D.J.; Byrne, J.A.; Cave, A.C.; Shah, A.M. Role of oxidative stress in cardiac remodelling after myocardial infarction. *Hear. Lung Circ.* **2004**, *13*, 132–138. [CrossRef]
8. Mehran, R.; Nikolsky, E. Contrast-induced nephropathy: Definition, epidemiology, and patients at risk. *Kidney Int.* **2006**, *69*, S11–S15. [CrossRef]
9. Stoeck, C.T.; von Deuster, C.; Fuetterer, M.; Polacin, M.; Waschkies, C.F.; van Gorkum, R.J.; Kron, M.; Fleischmann, T.; Cesarovic, N.; Weisskopf, M.; et al. Cardiovascular magnetic resonance imaging of functional and microstructural changes of the heart in a longitudinal pig model of acute to chronic myocardial infarction. *J. Cardiovasc. Magn. Reson.* **2021**, *23*, 1–14. [CrossRef]
10. Kung, G.L.; Vaseghi, M.; Gahm, J.K.; Shevtsov, J.; Garfinkel, A.; Shivkumar, K.; Ennis, D.B. Microstructural infarct border zone remodeling in the post-infarct swine heart measured by diffusion tensor MRI. *Front. Physiol.* **2018**, *9*, 826. [CrossRef]
11. Wu, M.T.; Su, M.Y.M.; Huang, Y.L.; Chiou, K.R.; Yang, P.; Pan, H.B.; Reese, T.G.; Wedeen, V.J.; Tseng, W.Y.I. Sequential changes of myocardial microstructure in patients postmyocardial infarction by diffusion-tensor cardiac MR: Correlation with left ventricular structure and function. *Circ. Cardiovasc. Imaging* **2009**, *2*, 32–40. [CrossRef]
12. Scollan, D.F.; Holmes, A.; Winslow, R.; Forder, J. Histological validation of myocardial microstructure obtained from diffusion tensor magnetic resonance imaging. *Am. J. Physiol.-Heart Circ. Physiol.* **1998**, *275*, H2308–H2318. [CrossRef]
13. Scollan, D.; Holmes, A.; Zhang, J.; Winslow, R. Reconstruction of cardiac ventricular geometry and fiber orientation using magnetic resonance imaging. *Ann. Biomed. Eng.* **2000**, *28*, 934–944. [CrossRef]
14. Rahman, T.; Moulin, K.; Ennis, D.B.; Perotti, L.E. Diffusion Biomarkers in Chronic Myocardial Infarction. In Proceedings of the International Conference on Functional Imaging and Modeling of the Heartr, Stanford, CA, USA, 21–25 June 2021; pp. 137–147.
15. Verzhbinsky, I.A.; Perotti, L.E.; Moulin, K.; Cork, T.E.; Loecher, M.; Ennis, D.B. Estimating aggregate cardiomyocyte strain using *Vivo* Diffus. Displac. Encoded MRI. *IEEE Trans. Med. Imaging* **2019**, *39*, 656–667. [CrossRef]
16. Perotti, L.E.; Ponnaluri, A.V.; Krishnamoorthi, S.; Balzani, D.; Ennis, D.B.; Klug, W.S. Method for the unique identification of hyperelastic material properties using full-field measures. Application to the passive myocardium material response. *Int. J. Numer. Methods Biomed. Eng.* **2017**, *33*, e2866. [CrossRef]
17. Li, X.; Perotti, L.E.; Martinez, J.A.; Duarte-Vogel, S.M.; Ennis, D.B.; Wu, H.H. Real-time 3T MRI-guided cardiovascular catheterization in a porcine model using a glass-fiber epoxy-based guidewire. *PLoS ONE* **2020**, *15*, e0229711. [CrossRef]
18. Stoeck, C.T.; Von Deuster, C.; Genet, M.; Atkinson, D.; Kozerke, S. Second-order motion-compensated spin echo diffusion tensor imaging of the human heart. *Magn. Reson. Med.* **2016**, *75*, 1669–1676. [CrossRef]
19. Cork, T.E.; Perotti, L.E.; Verzhbinsky, I.A.; Loecher, M.; Ennis, D.B. High-Resolution Ex Vivo Microstructural MRI After Restoring Ventricular Geometry via 3D Printing. In Proceedings of the International Conference on Functional Imaging and Modeling of the Heart, Bordeaux, France, 6–8 June 2019; pp. 177–186.
20. Porter, D.A.; Heidemann, R.M. High resolution diffusion-weighted imaging using readout-segmented echo-planar imaging, parallel imaging and a two-dimensional navigator-based reacquisition. *Magn. Reson. Med. Off. J. Int. Soc. Magn. Reson. Med.* **2009**, *62*, 468–475. [CrossRef]

21. Reese, T.G.; Heid, O.; Weisskoff, R.; Wedeen, V. Reduction of eddy-current-induced distortion in diffusion MRI using a twice-refocused spin echo. *Magn. Reson. Med. Off. J. Int. Soc. Magn. Reson. Med.* **2003**, *49*, 177–182. [CrossRef]
22. Schelbert, E.B.; Hsu, L.Y.; Anderson, S.A.; Mohanty, B.D.; Karim, S.M.; Kellman, P.; Aletras, A.H.; Arai, A.E. Late gadolinium-enhancement cardiac magnetic resonance identifies postinfarction myocardial fibrosis and the border zone at the near cellular level in ex vivo rat heart. *Circ. Cardiovasc. Imaging* **2010**, *3*, 743–752. [CrossRef]
23. Li, X.; Morgan, P.S.; Ashburner, J.; Smith, J.; Rorden, C. The first step for neuroimaging data analysis: DICOM to NIfTI conversion. *J. Neurosci. Methods* **2016**, *264*, 47–56. [CrossRef]
24. Canny, J. A computational approach to edge detection. *IEEE Trans. Pattern Anal. Mach. Intell.* **1986**, *PAMI-8*, 679–698. [CrossRef]
25. Thirion, J.P. Image matching as a diffusion process: An analogy with Maxwell's demons. *Med. Image Anal.* **1998**, *2*, 243–260. [CrossRef]
26. Moulin, K. DiffusionRecon. Available online: https://github.com/KMoulin/DiffusionRecon (accessed on 24 January 2022).
27. Haaf, P.; Garg, P.; Messroghli, D.R.; Broadbent, D.A.; Greenwood, J.P.; Plein, S. Cardiac T1 mapping and extracellular volume (ECV) in clinical practice: A comprehensive review. *J. Cardiovasc. Magn. Reson.* **2017**, *18*, 89. [CrossRef]
28. Allen, M.; Poggiali, D.; Whitaker, K.; Marshall, T.R.; Kievit, R.A. Raincloud plots: A multi-platform tool for robust data visualization. *Wellcome Open Res.* **2019**, *4*, 63. [CrossRef]
29. Balls, G.T.; Frank, L.R. A simulation environment for diffusion weighted MR experiments in complex media. *Magn. Reson. Med. Off. J. Int. Soc. Magn. Reson. Med.* **2009**, *62*, 771–778. [CrossRef]
30. Berry, D.B.; Regner, B.; Galinsky, V.; Ward, S.R.; Frank, L.R. Relationships between tissue microstructure and the diffusion tensor in simulated skeletal muscle. *Magn. Reson. Med.* **2018**, *80*, 317–329. [CrossRef]
31. Moulin, K.; Aliotta, E.; Ennis, D.B. Effect of flow-encoding strength on intravoxel incoherent motion in the liver. *Magn. Reson. Med.* **2019**, *81*, 1521–1533. [CrossRef]
32. Moulin, K. DiffusionSimulation. Available online: https://github.com/KMoulin/DiffusionSimulation (accessed on 14 February 2022).
33. Messroghli, D.R.; Nordmeyer, S.; Dietrich, T.; Dirsch, O.; Kaschina, E.; Savvatis, K.; O h Ici, D.; Klein, C.; Berger, F.; Kuehne, T. Assessment of diffuse myocardial fibrosis in rats using small-animal Look-Locker inversion recovery T1 mapping. *Circ. Cardiovasc. Imaging* **2011**, *4*, 636–640. [CrossRef]
34. Messroghli, D.R.; Walters, K.; Plein, S.; Sparrow, P.; Friedrich, M.G.; Ridgway, J.P.; Sivananthan, M.U. Myocardial T1 mapping: Application to patients with acute and chronic myocardial infarction. *Magn. Reson. Med.* **2007**, *58*, 34–40. [CrossRef]
35. Kali, A.; Choi, E.Y.; Sharif, B.; Kim, Y.J.; Bi, X.; Spottiswoode, B.; Cokic, I.; Yang, H.J.; Tighiouart, M.; Conte, A.H.; et al. Native T1 mapping by 3-T CMR imaging for characterization of chronic myocardial infarctions. *JACC Cardiovasc. Imaging* **2015**, *8*, 1019–1030. [CrossRef]
36. Arcari, L.; Engel, J.; Freiwald, T.; Zhou, H.; Zainal, H.; Gawor, M.; Buettner, S.; Geiger, H.; Hauser, I.; Nagel, E.; et al. Cardiac biomarkers in chronic kidney disease are independently associated with myocardial edema and diffuse fibrosis by cardiovascular magnetic resonance. *J. Cardiovasc. Magn. Reson.* **2021**, *23*, 1–14. [CrossRef] [PubMed]
37. Camastra, G.; Arcari, L.; Ciolina, F.; Danti, M.; Cacciotti, L. Cardiac magnetic resonance imaging of transient myocardial dysfunction in a patient treated with checkpoint-targeted immunotherapy. *Eur. J. Cancer* **2021**, *144*, 389–391. [CrossRef] [PubMed]
38. Kolentinis, M.; Le, M.; Nagel, E.; Puntmann, V.O. Contemporary cardiac MRI in chronic coronary artery disease. *Eur. Cardiol. Rev.* **2020**, *15*, e50. [CrossRef] [PubMed]
39. Chen, J.; Song, S.K.; Liu, W.; McLean, M.; Allen, J.S.; Tan, J.; Wickline, S.A.; Yu, X. Remodeling of cardiac fiber structure after infarction in rats quantified with diffusion tensor MRI. *Am. J. Physiol.-Heart Circ. Physiol.* **2003**, *285*, H946–H954. [CrossRef]
40. Pop, M.; Ghugre, N.R.; Ramanan, V.; Morikawa, L.; Stanisz, G.; Dick, A.J.; Wright, G.A. Quantification of fibrosis in infarcted swine hearts by ex vivo late gadolinium-enhancement and diffusion-weighted MRI methods. *Phys. Med. Biol.* **2013**, *58*, 5009. [CrossRef]
41. Pashakhanloo, F.; Herzka, D.A.; Mori, S.; Zviman, M.; Halperin, H.; Gai, N.; Bluemke, D.A.; Trayanova, N.A.; McVeigh, E.R. Submillimeter diffusion tensor imaging and late gadolinium enhancement cardiovascular magnetic resonance of chronic myocardial infarction. *J. Cardiovasc. Magn. Reson.* **2017**, *19*, 9. [CrossRef]
42. Rose, J.N.; Nielles-Vallespin, S.; Ferreira, P.F.; Firmin, D.N.; Scott, A.D.; Doorly, D.J. Novel insights into in-vivo diffusion tensor cardiovascular magnetic resonance using computational modelling and a histology-based virtual microstructure. *Magn. Reson. Med.* **2019**, *81*, 2759–2773. [CrossRef]
43. Das, A.; Kelly, C.; Teh, I.; Stoeck, C.T.; Kozerke, S.; Chowdhary, A.; Brown, L.A.; Saunderson, C.E.; Craven, T.P.; Chew, P.G.; et al. Acute Microstructural Changes after ST-Segment Elevation Myocardial Infarction Assessed with Diffusion Tensor Imaging. *Radiology* **2021**, *299*, 86–96. [CrossRef]
44. Gottbrecht, M.; Kramer, C.M.; Salerno, M. Native T1 and extracellular volume measurements by cardiac MRI in healthy adults: A meta-analysis. *Radiology* **2019**, *290*, 317–326. [CrossRef]
45. Risholm, P.; Pieper, S.; Samset, E.; Wells, W.M. Summarizing and visualizing uncertainty in non-rigid registration. In Proceedings of the International Conference on Medical Image Computing and Computer-Assisted Intervention, Beijing, China, 20–24 September 2010; pp. 554–561.

applied sciences

MDPI

Article

Subclinical Atherosclerosis Progression in Obese Children with Relevant Cardiometabolic Risk Factors Can Be Assessed through Carotid Intima Media Thickness

Monica-Simina Mihuta [1], Corina Paul [2,*], Adrian Ciulpan [2], Farah Dacca [3], Iulian Puiu Velea [2], Ioana Mozos [4] and Dana Stoian [5,6]

1 PhD School Department, Victor Babes University of Medicine and Pharmacy, 300041 Timisoara, Romania; simina.mihuta@umft.ro
2 Department of Pediatrics, Victor Babes University of Medicine and Pharmacy, 300041 Timisoara, Romania; ciulpan.adrian@umft.ro (A.C.); ivelea@umft.ro (I.P.V.)
3 Independent Researcher, 300041 Timișoara, Romania; farah.dacca1402@gmail.com
4 Center for Translational Research and Systems Medicine, Department of Functional Sciences—Pathophysiology, Victor Babeş University of Medicine and Pharmacy, 300041 Timişoara, Romania; ioanamozos@umft.ro
5 2nd Department of Internal Medicine, Victor Babes University of Medicine and Pharmacy, 300041 Timisoara, Romania; stoian.dana@umft.ro
6 Center of Molecular Research in Nephrology and Vascular Disease, Faculty of Medicine, Victor Babes University of Medicine and Pharmacy, 300041 Timisoara, Romania
* Correspondence: paul.corina@umft.ro

Citation: Mihuta, M.-S.; Paul, C.; Ciulpan, A.; Dacca, F.; Velea, I.P.; Mozos, I.; Stoian, D. Subclinical Atherosclerosis Progression in Obese Children with Relevant Cardiometabolic Risk Factors Can Be Assessed through Carotid Intima Media Thickness. *Appl. Sci.* **2021**, *11*, 10721. https://doi.org/10.3390/app112210721

Academic Editors: Mihaela Pop and Cristian A. Linte

Received: 30 September 2021
Accepted: 11 November 2021
Published: 13 November 2021

Featured Application: The proposal to use intima-media thickness as a non-invasive and cost-efficient biomarker of subclinical atherosclerosis in obese children.

Abstract: Given the growing obesity rates among children, a more complete evaluation of their potential cardiometabolic risk is needed. Carotid intima-media thickness (CIMT), a marker of endothelial distress and a predictor of atherosclerotic progression in adulthood, may complete the day-to-day evaluation of children at risk. Multiple risk factors act as additional precipitant causes of atherosclerosis. We analyzed 60 patients aged 6–17 years old by measuring their CIMT using the Aixplorer MACH 30 echography machine automatic measurement software. All subjects were clinically and anamnestically assessed to identify risk factors. CIMT values are significantly higher in older children and boys. Over 20 kg weight gain during pregnancy and other at-risk disorders ($p = 0.047$), family history of cardiovascular risk ($p = 0.049$), hypertension ($p = 0.012$), and smoking ($p = 0.015$) are linked to increased CIMT. Our study also supports international data on artificial postnatal nutrition, high/low birth weight, and sedentary lifestyle being linked to increased CIMT. Significant correlations were detected between CIMT and the entire lipid panel. Weight excess and abdominal adiposity in children is clearly linked to increased CIMT. Moreover, waist circumference and TG/HDL-c are significant predictors of CIMT. Although each parameter of the lipid panel is correlated to CIMT, fasting glucose is not.

Keywords: cardiometabolic risk; carotid intima-media thickness; childhood obesity; subclinical atherosclerosis

1. Introduction

Weight excess has been one of the largest public health problems of modern society for decades and, in the obesogenic context of the COVID-19 pandemic, obesity has been affecting children at rates faster than we have ever encountered before [1]. In many countries, not only are the lockdown conditions, which increased sedentary behavior, to blame, but also the worsening of financial status, which has led to reduced access to qualitative foods and more stress within families [2].

The importance of a comprehensive evaluation of obesity in children is of paramount importance for early detection of cardiovascular and metabolic complications. Atherosclerosis and high blood pressure, frequent complications of obesity, are slow but steady ongoing processes in obese children, manifesting in adolescence and early adulthood, depending on the severity of the weight excess [3].

Increased carotid intima-media thickness (CIMT) is one of the first observable signs of subclinical atherosclerosis, making this ultrasonographic technique probably the most accurate non-invasive method of detecting early stages of atherosclerosis. For a long time, in children, these early stages have been ordinarily estimated by monitoring the lipid and glucose metabolism alterations (lipid profile, fasting glucose levels, oral glucose tolerance tests, serum insulin, blood pressure values, etc.). Lately, cardiovascular risk has been proven to be well and reliably assessed through imaging technologies in pediatric patients as well [4]. Beauloye et al. showed that blood parameters that are classically assessed in obese children are significantly correlated to CIMT [5]. Therefore, studies estimate that the use of CIMT, a non-invasive, non-painful, non-radiating, cost-effective, and easily reproducible method, can become one of the pillars of risk evaluation for pediatric patients at risk.

Nevertheless, the value of CIMT in children as a predictor of risk is still a subject of research. Although cut-off points for CIMT are still a subject of discussion, there is significant evidence that its values reach higher points in children with obesity. According to Farello et al., both in metabolically healthy and unhealthy obese children, CIMT reaches higher values. The same study showed that early detection of high CIMT in children with metabolic syndrome is a predictor of cardiovascular disease in young adults [6].

Hence, one of the goals of the present study, other than assessing CIMT in relation to obesity, is to evaluate some of the other risks associated with higher CIMT in children, and their effect on the vascular thickness when associated with obesity. Most of the risk factors we have analyzed are already set in stone as aggravating factors of cardiometabolic disease in adults and many are on the way to being settled for the pediatric population as well.

Advancing age, masculine sex, and high BMI are three factors known to have a direct impact on CIMT values in both children and adults [7–9]. However, risks associated with pregnancy, mother's health, perinatal aspects regarding weight or nutrition, family history of cardiometabolic disease, exposure to smoking, high values for blood pressure, and a lack of physical activity on daily basis may also be risks that can impact vascular integrity or worsen the atherosclerotic progression.

Previous studies place postnatal nutrition in a stronghold regarding breastfeeding's universal protective characteristics, on the condition that it not be too prolonged [10,11]. There is evidence that artificial milk is associated with obesity, higher CIMT values, arterial hypertension, and insulin resistance [12–15].

Data on birth weight as a risk factor for increased CIMT show that low birth for gestational age is the most prominent risk, and the more severe the weight deficit is, the more powerful the risk [16]. On the other side of the spectrum, higher than normal birth weight is also a risk factor for high CIMT in young adult life [17].

Another important risk factor is maternal health during pregnancy. It is a factor that can affect CIMT values indirectly by inducing a pathology that places the newborn at risk for high CIMT. For instance, obesity in children is correlated to motherhood obesity and gestational diabetes [11,18]. High blood pressure in a pregnant woman tracks to her child, especially in their adult life. In fact, a mother's entire cardiovascular profile is proven to be linked to her offspring's [19]. In utero exposure to autoimmune Hashimoto thyroiditis is associated with an increased risk of the offspring developing a thyroid disorder, including Hashimoto thyroiditis, in childhood or adolescence [20]. With regard to CIMT, Hashimoto thyroiditis in children, especially in adolescent girls, is an aggravating factor of atherosclerosis, regardless of thyroid function, due to the maintenance of a status of chronic inflammation that affects endothelial integrity [21]. Moreover, smoking during pregnancy causes low birth for gestational age and hence, high CIMT values in the child's young adulthood [16].

Family history obtained by targeted anamnesis and medical records is another important part of a full evaluation. From genetic predispositions to behavioral components of certain pathologies, they all echo the lifetime development of the offspring. A family history of premature cardiovascular disease reflects higher values of CIMT and the association of other factors like oxidant status, insulin resistance, and dyslipidemia [22].

Smoking cigarettes increases the values of CIMT directly and indirectly, so much so that the vascular age of a regular smoker is about 6–7 years older than of non-smokers. Smoking aggravates the effects of metabolic syndrome and age on CIMT, despite the fact that smokers are usually leaner and have a good glucose-insulin homeostasis [23]. There are also differences between how smoking affects the genders: Men are more affected by smoking in terms of arterial stiffness, probably due to the lack of estrogen-given protection; however, a lower exposure to cigarette smoke increases arterial stiffness in women [24]. Moreover, there is irrepressible evidence that passive (second-hand) smoking in children has the same dire adverse effects on vascular health by increasing the risk of atherosclerosis [25,26]. Furthermore, we mention a significantly higher risk of becoming a smoker among adolescents who have both parents as smokers [27].

Sedentary behavior is not just one of the main causes for weight gain, but is also involved in glucose metabolism as a promoter of insulin resistance, and in lipid metabolism as a maintainer of dyslipidemia. It also decreases the cardiovascular adaptability for effort. Previous studies have shown that leading a sedentary lifestyle as children increases cardiometabolic risk and carotid plaque development in young adulthood [28].

Alterations of glucose and lipid metabolisms are consequences of excess fat tissue, unhealthy lifestyle, and genetic risk factors. High levels of LDL cholesterol, total cholesterol, non-HDL cholesterol, triglycerides, low values of HDL cholesterol, and high ratios between total cholesterol and HDL-c and triglycerides and HDL-c, respectively, all induce, either separately or combined, defective lipid vascular clearance and excess lipoprotein storage in the sub-endothelial space. Pathological values of any of the aforementioned parameters can be associated with increased CIMT [29]. Insulin resistance, diagnosed by clinical and laboratory findings, is positively correlated to CIMT values and is a pathological precursor of type 2 diabetes in both children and adults [30].

Non-invasive evaluation of cardiovascular and metabolic risk in individuals with pathologies that present vascular implications can be based on the assessment of arterial stiffness and atherosclerosis progression and the evaluation of inflammatory markers [31].

Therefore, given the growing obesity rates among children and even among our patients, we have designed an observational study for our overweight and obese patients, with the scope of assessing the importance of CIMT in a more comprehensive clinical evaluation and cardiometabolic risk assessment. To achieve this scope, we have identified certain risk factors in our patients and analyzed their impact on CIMT values in obese and overweight children as opposed to normal-weight ones. The particularity of our study is that apart from analyzing CIMT in the context of weight excess in children, it shows to what extent certain risk factors correlate to CIMT increase in children. We believe this approach broadens our understanding of how easily identifiable risk factors influence CIMT, a fact that can be useful to clinicians who rely on CIMT in their day-to-day activity.

2. Materials and Methods

The observational study was performed in our US endocrinology unit from January 2021 until May 2021 on 60 children. The study was approved by the Ethics Committee of Scientific Research (CECS) of the University of Medicine and Pharmacy Victor Babes Timisoara and respects the ethical guidelines of the Helsinki Declaration.

The study was centered on the impact of excess adipose tissue on carotid intima-media thickness and how CIMT correlates to identifiable genetic and epigenetic risk factors.

Three study groups were defined, depending on the severity of weight excess: obese and overweight, and normal-weight patients as controls.

2.1. Inclusion Criteria

- Obese group—patients with a BMI score $\geq$ 95th percentile for age and sex, overweight group—BMI ranging from the 85th percentile to 95th percentile, and the control group—BMI ranging from the 5th percentile to the 85th.
- Both sexes were included, and ages ranged from 6 to 17.

2.2. Exclusion Criteria

- Secondary obesity causes: Cushing syndrome, thyroid disfunctions with hypothyroidism, insulin-dependent diabetes mellitus, polycystic ovarian syndrome, hypothalamic injury/disorders, genetic syndromes like Prader–Willi syndrome [32], grelin–leptin dysfunction [33], and use of medication that can induce weight gain (glucocorticosteroids, sulphonylureas, tricyclic antidepressants, antipsychotics) [34].

Prior to any examination, informed consent forms were administered to the patient's parent/legal guardian and a verbal agreement was given by the child, after exhaustive explanations regarding the study and further examinations.

The main analysis consisted of the measurement of CIMT by carotid ultrasonography and the comparison of the values found between groups, depending on certain risk factors.

In addition to the ultrasonography measurement, we performed a clinical examination (weight, height, waist circumference, and blood pressure measurements) and a targeted anamnesis to detect the presence of certain risk factors: postnatal nutrition (breastfed/formula-fed), birth weight (<2500 g/>3500 g/normal weight), pregnancy-associated risk factors (no pathology/>20 kg surplus/gestational diabetes/gestational hypertension/autoimmune thyroiditis/smoking during pregnancy), family history (no pathologies/obesity/dyslipidemia/type 2 diabetes/coronary disease/stroke/autoimmune thyroiditis), smoking during pregnancy (yes/no), smoking by the patient (yes/no), and physical activity (normal/sedentary *). Moreover, the following blood parameters were selected from each patient's previous 6 months of medical history: fasting glucose (mg/dL), HDL cholesterol (mg/dL), LDL cholesterol (mg/dL), total cholesterol (mg/dL), and triglycerides (mg/dL).

* We considered sedentary a subject who performed no sport and/or less than 1 h of physical activity/day.

2.3. Ultrasonography Technique

The Aixplorer MACH 30 echography machine (SuperSonic Imagine, Aix-en-Provence, France) was utilized to perform the carotid ultrasonography. We used 2 ultrasound probes: SuperLinear SL 10-2 (2–10 MHz) and SL 18-5 (5–18 MHz). CIMT values were determined automatically by Aixplorer MACH 30 software (SuperSonic Imagine, Aix-en-Provence, France). We performed 3 measurements on both the right and the left common carotid artery and used the mean of all measurements in our analysis.

Examination starts by choosing the appropriate setting: vascular probe and carotid evaluation (B-mode setting). The patient lies in a supine position with their neck tilted backwards in an extended position. The exploration of the right and left carotids starts by transversal scanning, starting from the clavicle and guiding the probe upwards to locate the carotid bulb (bifurcation of the common carotid artery into the internal and external carotid arteries). At this point, the probe is rotated 180°, fixating the region of interest to distinguish the carotid lumen and carotid walls clearly, with the carotid bulb visible on the left of the screen and a clear intima-media in the 1–2 cm caudally from the carotid bulb. After the image is frozen in the most accurate instance, the Aixplorer MACH 30 software automatically measures the space between the intimal–luminal and the medial–adventitial interfaces on the posterior (far) wall of the left and right carotids [35–37].

Among other advantages, the novelty of this measurement is that it eliminates human-eye mistakes [38,39].

Examples of CIMT measurements (Figure 1):

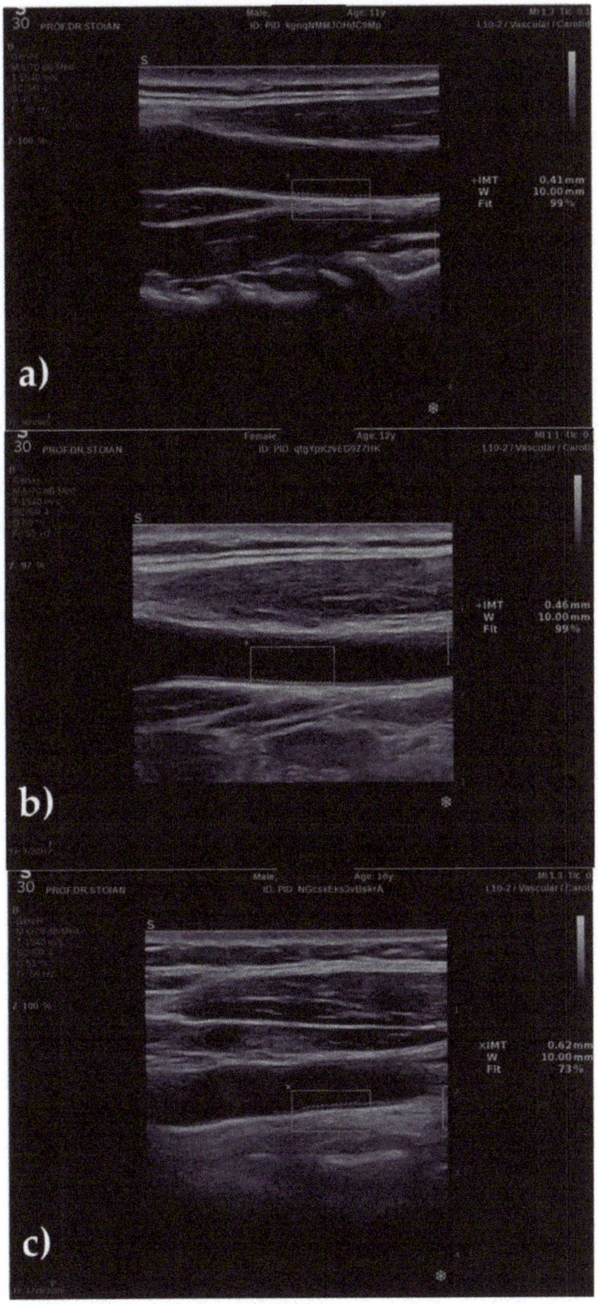

Figure 1. Examples of CIMT measurements. (**a**) an 11-year-old normal-weight boy. (**b**) a 12-year-old overweight girl. (**c**) a 16-year-old obese boy.

2.4. Statistical Analysis

Data were collected and statistically analyzed using Microsoft Excel and SPSS statistical software version 17 (SPSS Inc., Chicago, IL, USA). We used statistical tests to assess differences between study groups and the prevalence of risk factors, as well as their impact on the objects of our research.

Normality of variable distribution was checked in SPSS prior to statistical analysis (Shapiro–Wilk test). For normally distributed variables we used means, Student's T test and Pearson's correlation, whereas for non-normally distributed variables we used medians, the Mann–Whitney test, and Spearman's correlation. Subjects were divided into three main groups—obese, overweight, and normal weight (control)—and subgroups (by age, sex, and risk factors). Categories were analyzed individually and/or in pairs, mainly focusing on CIMT. Statistical significance was considered $p = 0.05$. For cases of multiple analysis on the same data (2 by 2 tests on 3 groups), we performed ANOVA (single factor) tests and post-hoc tests (Bonferroni corrected) in Microsoft Excel and adjusted the p-values according to the Bonferroni corrected α in order to keep the significance threshold at 0.05. Multivariable regression analysis was performed in SPSS with the stepwise method, with CIMT as the dependent variable.

3. Results

The study included 60 children, aged 6 to 18, of both sexes. They were divided by BMI scores into three groups: obese group (13 boys and 7 girls), overweight group (10 boys and 5 girls), and normal-weight group (12 boys and 13 girls). Extremely significant differences were detected between the CIMT values of the three groups: obese vs. normal weight, $p << 0.001$; obese vs. overweight, $p = 0.037$; and overweight vs. normal weight, $p = 0.001$ (Bonferroni-corrected p-values). That means that CIMT mean values grew as BMI and weight excess severity grew. The overall CIMT values for the three groups combined were normally distributed (Shapiro–Wilk test significance = 0.081). However, the obese and the normal-weight groups presented non-normal distribution for CIMT values.

3.1. CIMT with Regard to Age

Children from each weight group were further subdivided into three age subgroups: pre-pubertal (<12 years old), pubertal (12–15 years old), and post-pubertal (≥16 years old). The p-values were Bonferroni corrected for all comparisons in Table 1, with a significance threshold of 0.05.

We did not detect statistically significant differences between the age subgroups in the case of obese children (Table 1), with all three groups having similar values for CIMT (p-values for obese children are not presented in their Bonferroni-corrected form because they were too high, corrected $\alpha = 0.016$). In the overweight group (Table 1), we detected higher values of CIMT in pubertal compared to pre-pubertal patients and even higher values of CIMT for post-pubertal patients compared to pre-pubertal ones ($p = 0.024$). However, only the latter comparison was statistically significant.

In the normal-weight controls, pre-pubertal and pubertal children scored significantly lower CIMT averages, but children 16 years old and over had an CIMT average closer to the obese and overweight lots (Table 1).

CIMT values in obese, overweight, and normal-weight post-pubescent children always scored the highest values. For children under 12 years of age, we found significantly higher values for CIMT in obese children compared to overweight children ($p = 0.015$) and compared to the control group ($p << 0.001$) (Bonferroni-corrected p values). The same finding was detected in overweight children under 12 compared to normal-weight children under 12, but the p-value was not statistically significant.

Table 1. CIMT differences with regard to age.

Obese Group	Group Number	% of Entire Group	Mean CIMT (mm)
<12 years	I	55%	0.48
12–15 years	II	25%	0.49
≥16 years	III	20%	0.50
Comparison	I vs. II	I vs. III	II vs. III
p-value	0.71	0.59	0.88
Overweight Group	**Group Number**	**% of Entire Group**	**Mean CIMT (mm)**
<12 years	I	34%	0.43
12–15 years	II	53%	0.48
≥16 years	III	13%	0.54
Comparison	I vs. II	I vs. III	II vs. III
p-value	0.11	0.024	0.21
Normal-Weight Group	**Group Number**	**% of Entire Group**	**Mean CIMT (mm)**
<12 years	I	48%	0.35
12–15 years	II	40%	0.4
≥16 years	III	12%	0.51
Comparison	I vs. II	I vs. III	II vs. III
p-value	0.042	0.0006	0.012

3.2. CIMT with Regard to Gender

In the overweight and normal-weight lots, boys demonstrated higher CIMT averages than girls; nevertheless, the variation was statistically significant only in the overweight section, $p = 0.043$. In the group of obese children, the CIMT averages of boys and girls were almost identical (Table 2).

Table 2. CIMT with regard to gender.

Group	Sex	% of Entire Group	Mean CIMT (mm)	*p*-Value
Obese	Girls	35%	0.5135	0.99
	Boys	65%	0.5134	
Overweight	Girls	33%	0.43	0.043
	Boys	67%	0.49	
Normal	Girls	52%	0.38	0.55
	Boys	48%	0.4	

3.3. Assessment of Risk Factors

We further present the comparison between CIMT values in the context of the presence of certain risk factors within the analyzed groups: overweight vs. normal weight and obese vs. normal weight. We acknowledge the fact that when divided into subgroups according to certain risk factors, the lots become too small to reflect true statistical significance. Nevertheless, our intention is to show the trend of the analyzed data and how it matches with similar published data on the subject.

3.3.1. Obese Patients

Obese patients showed CIMT values close to 0.50 mm, with high percentages for formula nutrition, abnormal birth weight, pregnancy-related risks, pathological family history and sedentary lifestyle (Table 3).

Table 3. Mean CIMT in the presence of different risk factors in obese children.

Risk Factor		% of Total Obese Children	Mean CIMT (mm)	p-Value
Postnatal food	Formula	65%	0.53	0.99
	Breastmilk	35%	0.53	
Birth weight	<2500 g	25%	0.54	0.62
	>3500 g	25%	0.53	
	normal	50%	0.52	
Biological mother's health	Risk factors present	70%	0.55	0.047
	No risk factors	30%	0.48	
Family history	Risk factors present	75%	0.54	0.11
	No risk factors	25%	0.48	
Smoking	Smoker	15%	0.61	0.015
	Non-smoker	85%	0.51	
High blood pressure	Yes	20%	0.58	0.1
	No	80%	0.51	
Lifestyle	Sedentary	65%	0.55	0.89
	Normal	35%	0.49	
CIMT median for entire obese group (mm)				0.50

Postnatal nutrition. 65% of our obese patients received formula nutrition as nurslings. The CIMT mean of formula-fed children did not differ from the CIMT mean of the entire obese group, and it was almost equal to the CIMT mean of the breastfed group. The differences between the two subgroups were statistically insignificant ($p = 0.989$).

Birth weight. Half of our obese patients had abnormal birth weight (BW), either higher (>3500 g) or lower (<2500 g) than normal. In children with low BW, CIMT mean ($\overline{X} = 0.546$ mm) was higher than the mean of the entire obese sample, whereas in children with high BW, the CIMT mean was closer to the mean of the obese sample. No statistical significance was detected when comparing subgroups ($p = 0.62$). Obese children with normal BW had a slightly lower CIMT mean than those with abnormal BW.

Biological mother's health during pregnancy. Up to 70% of the obese patients were born to mothers who presented different pathologies during their pregnancy (Figure 2). Obese children who were born to unhealthy mothers had statistically significantly higher CIMT values than the entire obese sample and obese children who were born to healthy mothers ($p = 0.047$). Over 20 kg weight gain during pregnancy was the most encountered risk factor (30% of cases), followed closely by the presence of autoimmune thyroiditis (25%). No mother admitted to smoking during pregnancy in this group.

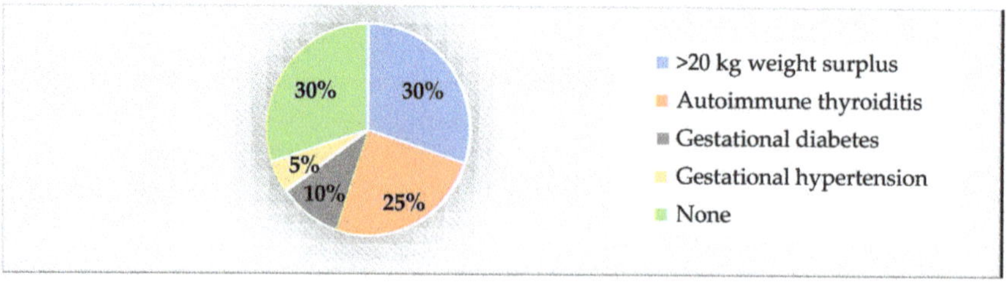

Figure 2. Mother's health during pregnancy, in obese children.

- Family history. Family history was an important context for the obese group: 75% of obese children had at least one significant cardiometabolic risk factor within their close family history (Figure 3). Obesity was present in 40% of the cases, followed by autoimmune thyroiditis and type 2 diabetes. CIMT reached statistically significant higher values in children with positive family history than in those without ($p = 0.049$).

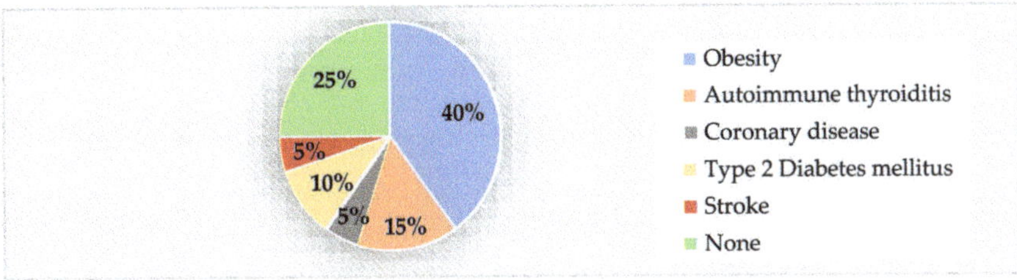

Figure 3. Family history of risk factors, in obese children.

- High blood pressure. High blood pressure was detected in 20% of the obese patients. Although CIMT scored higher values in this group than in the one with healthy blood pressure levels, the difference was not statistically significant ($p = 0.102$).
- Smoking. Smoking proved to be a significant risk factor. Only 15% of the adolescents pertaining to the obese group declared that they were smokers. Their CIMT mean was highest than any of the analyzed subgroups and the differences were statistically significant ($p = 0.015$). Moreover, smoking was strongly and positively correlated with higher values of CIMT ($r = 0.53$).
- Lifestyle. In the obese group, 65% of children declared a sedentary lifestyle (practicing no sport and having less than an hour/day of physical activity). CIMT values were higher than in children with healthy physical activity, but the differences were not statistically significant.

3.3.2. Overweight Patients

Overall CIMT mean values were lower for overweight children than for obese children (Table 4).

Table 4. Mean CIMT in the presence of different risk factors in overweight children.

Risk Factor		% of Total Overweight Children	Mean CIMT (mm)	*p*-Value
Postnatal food	Formula	47%	0.5	0.023
	Breast milk	53%	0.44	
Birth weight	<2500 g	20%	0.45	0.48
	>3500 g	33%	0.5	
	Normal	47%	0.46	
Biological mother's health	Risk factors present	53%	0.49	0.042
	No risk factors	47%	0.44	
Family history	Risk factors present	60%	0.49	0.049
	No risk factors	40%	0.44	
Smoking	Smoker	13%	0.53	0.06
	Non-smoker	87%	0.46	
High blood pressure	Yes	13%	0.55	0.012
	No	87%	0.46	
Lifestyle	Sedentary	40%	0.49	0.32
	Normal	60%	0.46	
CIMT $\overline{X}$ for entire overweight group (mm)				0.47

- Postnatal nutrition. In the overweight group, the nutrition factor was divided almost equally between children, with 47% having received formula as nurslings. However, the mean CIMT was significantly higher in children fed with formula ($\overline{X}$ = 0.5 mm) than in breastfed children ($\overline{X}$ = 0.44 mm), p = 0.023.
- Birth weight. Exactly one third of the overweight patients were born with a birth weight higher than 3500 g and scored higher CIMT values ($\overline{X}$ = 0.5 mm) than the low BW and normal BW groups ($\overline{X}$ = 0.45 mm and $\overline{X}$ = 0.46 mm, respectively). No statistical differences were detected between subgroups.
- Biological mother's health during pregnancy. A total of 53% of overweight patients were born to mothers who had problematic pregnancies. These children showed higher CIMT values than those who were born to healthy mothers (Table 4) and the differences between them were statistically significant (p = 0.042). The most encountered risk factor was weight gain of over 20 kg during pregnancy (20% of cases), followed by autoimmune thyroiditis and gestational diabetes in equal percentages (13%). A total of 7% of mothers admitted to smoking during pregnancy in this group (Figure 4).

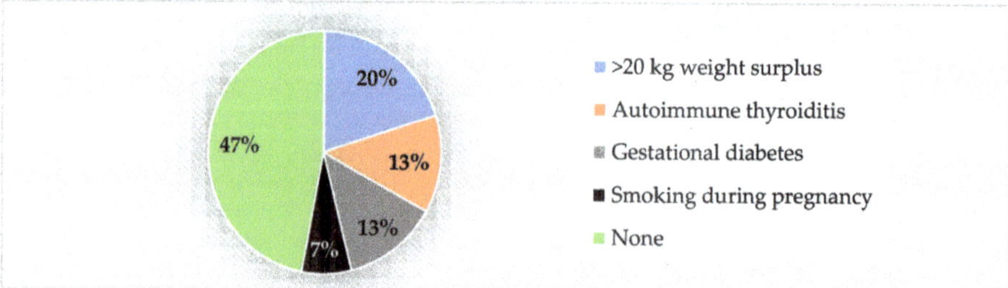

Figure 4. Mother's health during pregnancy, in overweight children.

- Family history. A total of 60% of overweight kids had an at-risk medical family history (Figure 5). Statistically significantly higher CIMT values were detected in the group with such risk factors within the immediate family, compared to children with negative family history (p = 0.049).

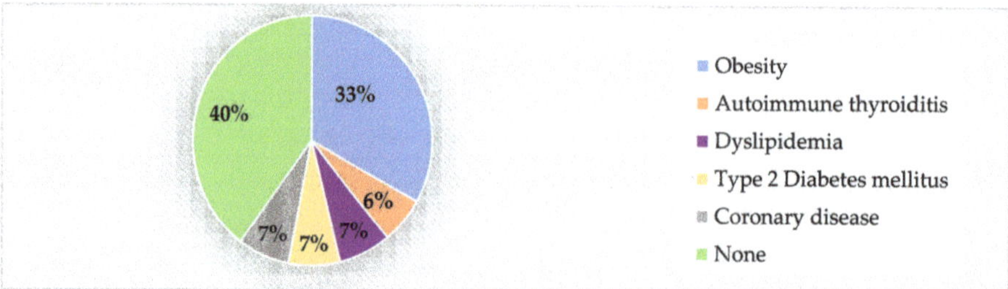

Figure 5. Family history of risk factors, in overweight children.

- High blood pressure. Even fewer children presented high blood pressure in the overweight group (13%, compared to 20% in the obese group). However, the CIMT values were significantly higher in hypertensive children than in non-hypertensive ones (p = 0.012).
- Smoking. CIMT mean was higher in smoking overweight children, but the differences were not statistically significant due to the small case sample (p = 0.06). Even so, smoking remains a solid risk factor for increased CIMT, in overweight children as well.

- Lifestyle. A total of 40% of overweight children admitted to leading a sedentary lifestyle, with less than 1 h of physical activity per day. However, the higher values of CIMT in the risk group were not statistically significant ($p = 0.324$).

3.3.3. Normal-Weight Patients

CIMT overall mean values were lower in all subgroups compared to the ones in the obese and overweight groups. Although we detected higher CIMT in all risk categories, the differences were not statistically significant, due to small samples (Table 5).

Table 5. Mean CIMT in the presence of different risk factors in normal-weight children.

Risk Factor		% of Total Normal-Weight Children	Mean CIMT (mm)	p-Value
Postnatal food	Formula Breastmilk	36% 64%	0.41 0.38	0.25
Birth weight	<2500 g >3500 g Normal	8% 40% 52%	0.39 0.40 0.38	0.5
Biological mother's health	Abnormal Normal	32% 68%	0.43 0.37	0.09
Family history	Abnormal Normal	28% 72%	0.42 0.38	0.2
Smoking	Smoker Non-smoker	4% 96%	0.56 0.38	N/A
Hypertension	Yes No	4% 96%	0.47 0.39	N/A
Lifestyle	Sedentary Normal	32% 68%	0.42 0.38	0.24
CIMT median for entire normal-weight group (mm)				0.37

- Postnatal nutrition. Only 36% of normal-weight children received formula as nurslings, and their CIMT mean values scored higher than in the breastfed group ($\overline{X} = 0.41$ mm vs. $\overline{X} = 0.37$ mm).
- Birthweight. Birth weight had no influence on the outcome of CIMT; 52% of children had normal BW and a $\overline{X} = 0.38$ mm CIMT, whereas 48% had abnormal BW (8% <2500 g and $\overline{X} = 0.39$ mm CIMT, and 40% >3500 g and $\overline{X} = 0.4$ mm CIMT, respectively).
- Biological mother's health during pregnancy. Unhealthy pregnancy seemed to be a valid risk factor for higher CIMT ($\overline{X} = 0.43$ mm vs. $\overline{X} = 0.37$ mm) even in normal-weight children; however, only 32% of normal-weight children were born from unhealthy pregnancies, compared to 70% in the case of obese children and 53% in the case of overweight children, respectively. A total of 12% of unhealthy pregnancies were due to >20 kg weight gain and 8% due to autoimmune thyroiditis, whereas gestational diabetes, gestational hypertension, and smoking during pregnancy added up to 4% each. A total of 68% of pregnancies that resulted in normal-weight children were declared completely physiological.
- Family history. Up to 72% of normal-weight children had no family history of cardiometabolic diseases. CIMT mean values scored higher in children with at-risk family history, but the differences were not statistically significant ($\overline{X} = 0.42$ mm vs. $\overline{X} = 0.3$ mm, $p = 0.2$). Out of the 28% of children with a history of cardiometabolic diseases in their immediate family, 16% declared obesity, whereas autoimmune thyroiditis, high blood pressure, and type 2 diabetes each added up 4%.
- High blood pressure. Out of the 25 overall subjects with high blood pressure, only one patient belonged to the normal-weight group.

- Smoking. The same situation was detected for the risk factor of smoking: One patient out of the 25 smoking patients belonged to the normal-weight group.
- Lifestyle. Up to 68% of normal-weight controls led healthy lifestyles with regard to physical activity (>1 h of exercise/day and/or practicing an organized sport). Those patients presented lower values for CIMT than children with sedentary lifestyles, but the differences were not statistically significant ($p = 0.24$).

3.4. Assessment of Waist Circumference and CIMT

3.4.1. Waist Circumference in Obese Children

Mean waist circumference was $\overline{X} = 100.5$ cm, with no difference between girls and boys ($p = 0.96$).

We detected a strong positive correlation between values of children's waist circumference and their CIMT values (Spearman's correlation coefficient $\rho = 0.69$, $p = 0.0006$). See Figure 3.

In addition, the higher the values of their waist circumference, the higher their blood pressure (r = 0.61, $p = 0.004$); see Table 6.

Table 6. Correlations between waist circumference and other parameters in obese children.

	CIMT Median (mm)		BMI (kg/m^2)	LDL-c (mg/dL)	Total Cholesterol (mg/dL)	Triglycerides (mg/dL)	Blood Pressure (mmHg)
Spearman's ρ	0.69 **	Pearson's r	0.88 *	0.43	0.34	0.42	0.61 *
p-value	0.0006	p-value	<0.001	0.06	0.14	0.06	0.004

* Correlation is significant at the 0.05 level (two-tailed). ** Correlation is significant at the 0.01 level (two-tailed).

3.4.2. Waist Circumference in Overweight Children

Mean waist circumference was $\overline{X} = 87.3$ cm, with no difference between girls and boys ($p = 0.6$).

A strong positive correlation was found between waist circumference and CIMT (Pearson's r = 0.64) in overweight children, but we did not detect any other correlations for other parameters.

3.4.3. Waist Circumference in Normal-Weight Children

- Mean waist circumference was $\overline{X} = 63.9$ cm, with no difference between girls and boys ($p = 0.96$).
- We detected a strong positive correlation between values of children's waist circumference and their CIMT values (Spearman's $\rho = 0.77$); see Table 7 and Figure 5.

Table 7. Correlations between waist circumference and other parameters in normal-weight children.

	CIMT Median (mm)	BMI (kg/m^2)	Triglycerides (mg/dL)		LDL-c (mg/dL)	Total Cholesterol (mg/dL)	Blood Pressure (mmHg)
Spearman's ρ	0.77 **	0.88 **	0.23	Pearson's r	0.375	0.38	0.405 *
p-value	<0.001	<0.001	0.26	p-value	0.065	0.06	0.044

* Correlation is significant at the 0.05 level (two-tailed). ** correlation is significant at the 0.01 level (two-tailed).

Waist circumference, an acknowledged marker of visceral obesity and insulin resistance [40], is positively correlated to CIMT values (Figure 6) and to other complications of obesity, especially to higher blood pressure values.

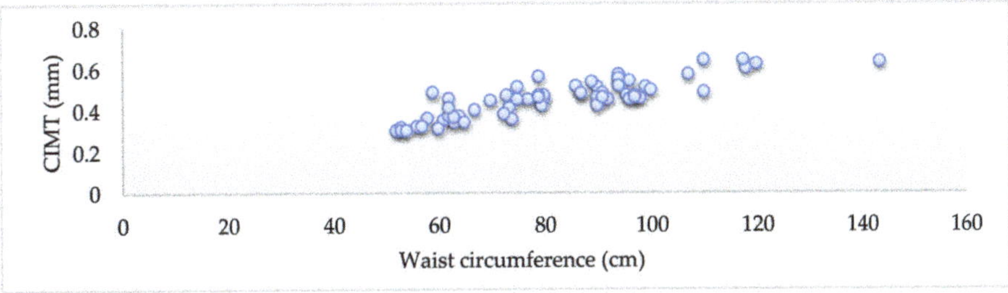

Figure 6. The correlation between waist circumference and CIMT values.

3.5. Assessment of Blood Parameters and CIMT

The focus of this part was the correlation analysis between CIMT values and the below-mentioned blood parameters.

3.5.1. HDL Cholesterol and CIMT

Obese children showed a moderate negative correlation between values of CIMT and HDL cholesterol (HDL-c), $\rho = -0.53$ (Figure 7), but there was a weaker negative correlation in normal-weight children. Mean values of HDL-c were slightly above 40 mg/dL, and a surprising result is that the lowest mean values were detected in the normal-weight group ($\overline{X}$ = 42.3 mg/dL).

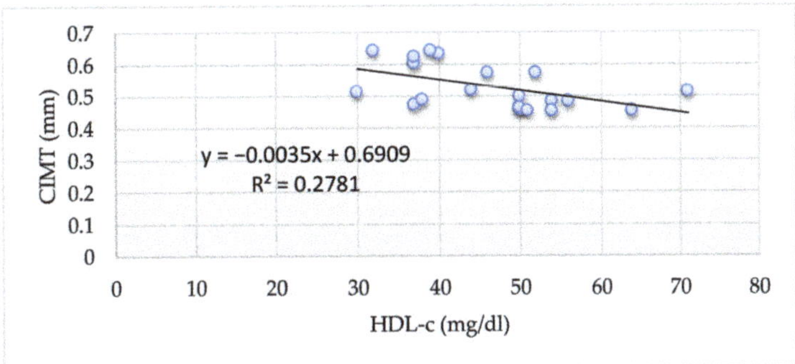

Figure 7. Correlation between CIMT and HDL-c in obese children.

3.5.2. LDL-Cholesterol and CIMT

We detected a positive correlation between CIMT and the values of LDL cholesterol (LDL-c) in the obese group, $\rho = 0.31$ (Figure 8). Neither the controls nor the overweight group showed such a correlation ($\rho = 0.04$ and r = 0.07, respectively); see Table 8. Mean LDL-c values were similar, with no statistical differences between groups. CIMT mean values were highest for obese ($\overline{X}$ = 0.53 mm) and overweight ($\overline{X}$ = 0.5 mm) children, and only $\overline{X}$ = 0.4 mm in the normal-weight group.

Table 8. Correlations between CIMT and blood parameters (Spearman's correlation for the obese and normal-weight group and Pearson correlation for the overweight group).

	Obese Group	Overweight Group	Normal-Weight Group
HDL-c—CIMT	−0.53	−0.031	−0.25
LDL-c—CIMT	0.31	0.04	0.07
Total Cholesterol—CIMT	0.23	0.151	0.07
Triglycerides—CIMT	0.45	0.6	0.08
Non-HDL-c—CIMT	0.14	0.19	0.34
TC:HDL-c ratio—CIMT	0.29	0.27	0.54
TG:HDL-c ratio—CIMT	0.16	0.54	0.56
Fasting glucose—CIMT	−0.3	0.21	−0.1

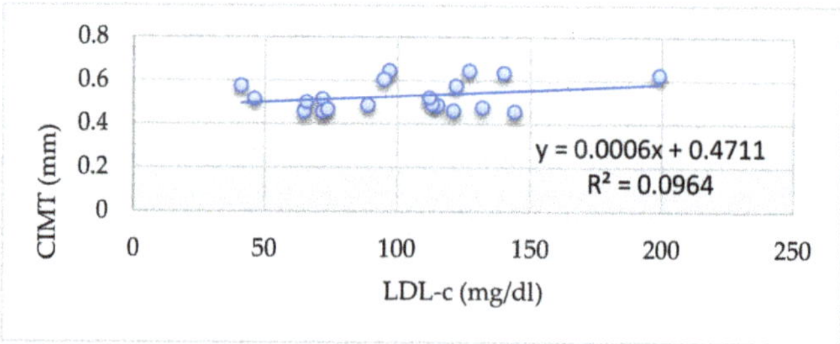

Figure 8. Correlation between CIMT and LDL-c in obese children.

3.5.3. Total Cholesterol and CIMT

We detected weak positive correlations between the values of total cholesterol (TC) and CIMT in the weight-excess groups of study (ρ = 0.23 in obese children, r = 0.15 in overweight children, and ρ = 0.07 in normal weight ones). No statistical significance was detected. We observed that the more severe the weight excess was, the stronger the correlation between the TC and CIMT (Figure 9).

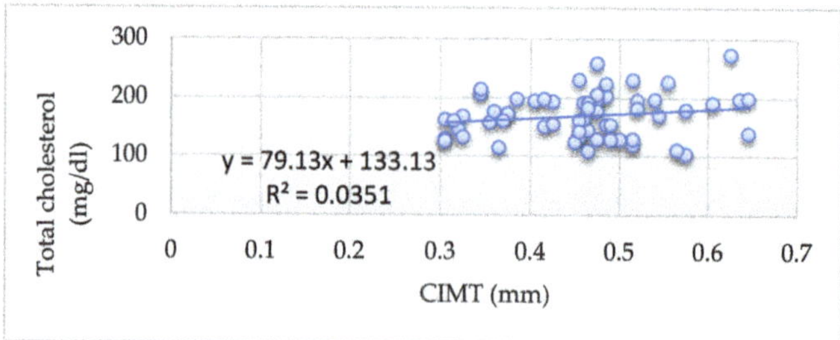

Figure 9. Correlation between CIMT and total cholesterol in all children.

3.5.4. Triglycerides and CIMT

Moderate and strong correlations between CIMT values and the levels of triglycerides (TG) were detected in the obese (Figure 10a) and overweight groups (ρ = 0.45 and r = 0.6, respectively, Figure 10b), whereas no correlation was detected in the normal-weight group (ρ = 0.08); see Table 8. Mean levels of TG were significantly higher in the overweight group, as was the strength of the correlation. TG levels are correlated to BMI values (r = 0.2) and mean values of TG in sedentary children are higher, but our findings were not statistically relevant (p = 0.47).

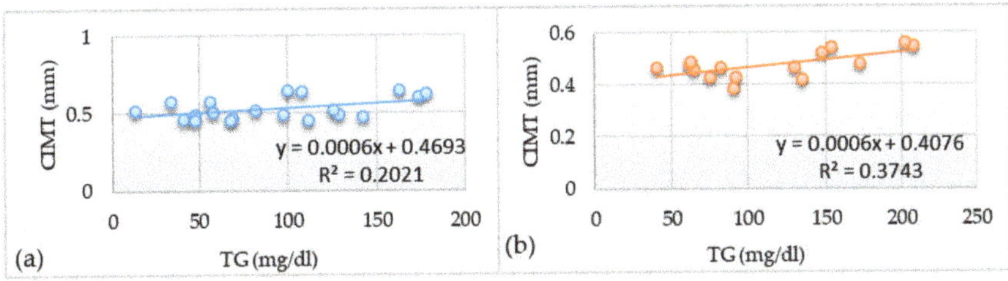

Figure 10. (**a**). Correlation between CIMT and TG in obese children. (**b**) Correlation between CIMT and TG in overweight children.

3.5.5. Non-HDL Cholesterol, Total Cholesterol/HDL-c Ratio, Triglyceride/HDL-c Ratio, and CIMT

Mean values for non-HDL-c, TC/HDL-c ratio, and TG/HDL-c ratio were less ideal as weight severity increased (Table 9). Correlation strengths between CIMT values and the mentioned parameters grew in tandem with weight excess, as well (Table 8, Figure 11). We did not detect any statistically significant difference between CIMT values of different range categories when comparing these parameters.

Table 9. Means of non-HDL-c, TC/HDL-c ratio and TG/HDL-c ratio across the three groups.

	Normal Weight	Overweight	Obese
$\overline{X}$ **Values**			
Non-HDL-c (mg/dL)	119	134	122.7
TC/HDL-c ratio	4	3.8	3.8
TG/HDL-c ratio	2	2.5	2.2

Up to 40% of obese and overweight patients presented at risk values of non-HDL-c (≥145 mg/dL). Half the patients with normal weights presented healthy values for non-HDL-c (<120 mg/dL) and only 20% presented values higher than 145 mg/dL [41].

Up to 60% of all three groups presented borderline TC/HDL-c ratio values (3.5–5) and 19% of obese children presented TC/HDL-c ratio values ≥5 [41].

Over 70% of patients in all three groups presented TG/HDL ratios ≥1.12, which, according to de Georgis et al. and Iwani et al. are values indicating higher risk [42–44]. We mention that cut-off values for TG/HDL-c in children are not fully established and differ among ethnicities [44].

3.5.6. Fasting Glucose and CIMT

The results concerning the relationship between fasting glucose (FG) and CIMT were somewhat contradictory in the sense that we detected negative correlations between FG and CIMT in the control and obese group ($r = -0.1$ and $\rho = -0.3$, respectively), and a positive correlation in the overweight group ($r = 0.21$); see Table 8. These results are either due to small group sampling, or to the fact that fasting glucose in children with ongoing pathologies like insulin resistance have fluctuations in their fasting glucose levels [45]. Mean values of FG were similar in all groups ($\overline{X}$ = 82 mg/dL).

3.6. Multivariate Regression Linear Model

We applied a multivariate linear regression model to identify independent predictors for CIMT. Independent variables included waist circumference, LDL-c, HDL-c, TC, TG, non-HDL-c, TC/HDL-c ratio, and TG/HDL-c ratio. CIMT was considered a dependent variable. Our analysis showed that waist circumference and TG/HDL-c ratio are statistically significant predictors of the dependent variable, CIMT (Tables 10 and 11).

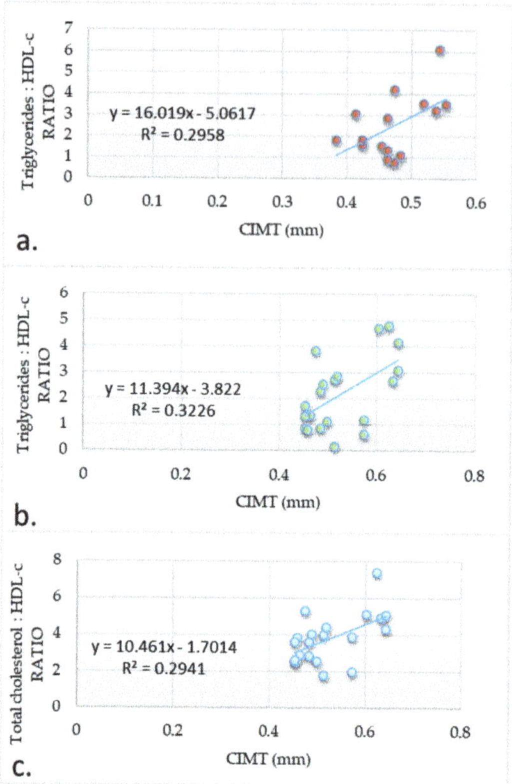

Figure 11. Correlations between CIMT and (**a**) TG:HDL-c ratio in overweight children, (**b**) TG:HDL-c ratio in obese children, and (**c**) TC:HDL-c ratio in obese children.

Table 10. WC and TG/HDL-c as significant predictors of CIMT.

Model	R	R-Square	Adjusted R-Square	Std. Error of the Estimate	Change Statistics				
					R-Square Change	F-Change	df1	df2	Sig. F-Change
1	0.855	0.732	0.727	0.04679	0.732	158.266	1	58	0.000
2	0.284	0.081	0.065	0.08663	0.081	5.097	1	58	0.028

Predictors: (constant), waist circumference. Dependent variable: CIMT.

Table 11. Independent variables excluded as predictors of CIMT.

	Beta In	t	Sig.	Partial Correlation	Collinearity Statistics		
					Tolerance	VIF	Minimum Tolerance
LDL-c	0.005	0.069	0.945	0.009	0.981	1.020	0.981
HDL-c	−0.078	−1.151	0.255	−0.151	1.000	1.000	1.000
TG	0.068	0.964	0.339	0.127	0.941	1.063	0.941
TC	0.010	0.148	0.883	0.020	0.957	1.045	0.957
Non-HDL-c	0.081	0.537	0.594	0.071	0.701	1.427	0.701
TC/HDL-c ratio	0.059	0.312	0.756	0.041	0.447	2.235	0.447

Predictors: (constant), TG/HDL-c ratio. Dependent variable: CIMT.

4. Discussion

The main purpose of this study was to show to what degree CIMT can be considered a reliable evaluation parameter along with the usual ones in obese children. To achieve this goal, we analyzed whether the presence of certain genetic and epigenetic factors influence the values of CIMT and observed how CIMT correlated to certain usual clinical and blood parameters.

We chose ultrasonography-measured CIMT as the focus of our analysis, which, although not the most accurate way of analyzing arterial walls, is the most appropriate for children as a screening tool. Studies have proven that MRI wall-thickness measurements may have better clinical utility because they have better prognostic value for cardiovascular risk than ultrasonography. In addition to intima-media measurement, MRI also images adventicia thickening and its vasa vasorum proliferation, the entire carotid artery, and the carotid bulb, where plaque forms early. However, its limitations make it more of a research tool than a day-to-day clinical one. It is over 10 times more expensive than ultrasonography, less accessible due to construction costs, requires longer scan times, and, because it is so sensible to motion, its common use in children is unrealistic [46,47]. CT angiography may detect vasa vasorum neovascularization and thus be a good predictor of vascular accidents, but its limitations with regard to radiation and poor contrast between lipid and fibrotic structures makes it unfit for a screening tool, especially in children [47]. Fluorodeoxyglucose-PET can be useful in assessing plaque vulnerability, because it targets areas of inflammation that are prone to rupture. When coupled with CT or MRI, its value increases. However, it is a method that is highly unlikely to be useful in screening children at risk [48]. That said, it is obvious that for our everyday clinical risk assessment in minor patients, CIMT ultrasonography is the only imagistic tool that is useful in assessing subclinical atherosclerosis.

Our patients were divided into three study groups—obese, overweight, and normal-weight—and subgroups depending on the object of study. CIMT values were correlated to BMI levels and there were significant differences between the three groups with regard to CIMT. In our study, excess adipose tissue was estimated by BMI and waist circumference measurements, but we acknowledge that the utility of peripheral adipose tissue ultrasound measurement and bioelectrical impedance analysis as a more accurate detection of subcutaneous fat tissue, which has been shown to have important roles in cytokine and growth-factor production and thus, in cardiometabolic risk [49].

Atherosclerosis is a process that starts in early childhood and has a lifelong progression, its pathological speed depending on genetic predisposition and environmental factors [7]. As a consequence, CIMT is a parameter that increases over time. In order to analyze how obesity affects children of different ages, we divided each study group into age categories: pre-pubertal (<12 years old), pubertal (12–15 years old), and post-pubertal (≥16 years old). In the overweight group, significant differences concerning CIMT values were detected between pre-pubertal and post-pubertal patients ($p = 0.024$, Bonferroni corrected); see Table 1. Extremely significant differences were detected in the normal-weight group (Table 1), where CIMT values increased very clearly with age, so much so that post-pubertal normal-weight children had CIMT mean values very similar to children of the same age with weight excess. Moreover, when analyzing the <12-years-old category, we detected extremely significant differences between CIMT mean values in the obese and overweight groups ($p = 0.015$), and the obese and normal-weight groups ($p << 0.001$). Therefore, in the presence of considerable excess weight, age has a smaller impact on the endothelial wall than the pathological processes caused by excess adipose tissue. Inflammation and an early onset of subclinical atherosclerosis makes the arterial walls of pre-pubertal obese children have similar characteristics to the arterial walls of normal-weight or slightly overweight post-pubertal children and even young adults.

In adults, CIMT reaches higher values in men [8]; however, in children, differences on behalf of sex are insignificant. In our analysis, in the obese group, the mean values for CIMT in boys and girls were identical, whereas in the overweight group, boys scored higher values for CIMT ($p = 0.043$); see Table 2.

An essential part of our study was the analysis of potential factors that increase risk for higher CIMT from an early age. Most of these factors overlap their influence on CIMT with that of weight excess predisposition. We analyzed each risk factor within the respective group.

Although discussions on postnatal nutrition's role in cardiometabolic risk is still being researched, most studies are in agreement that breastfeeding has a protective cardiovascular role only when it is not prolonged, being inversely associated with mortality by coronary disease [10]. A 65-year longitudinal study showed that breastfeeding reduces the progression of atherosclerosis, which translates into a lower CIMT [13]. On the other hand, artificial feeding is known to be associated not only with obesity [12], but also with high blood pressure [14] and insulin resistance [15]. Our analysis showed that although in all three groups, children who were bottle-fed presented higher CIMT values, the results were not statistically significant, except for the overweight group ($p = 0.023$). Although the results concerning CIMT were inconclusive, the proportions of formula-fed/breast-fed children in each group clearly suggest that artificial nutrition represents a risk factor for weight excess in childhood: 65% of the obese group, 47% of the overweight group, and only 37% of the normal-weight group were bottle-fed.

Data on birth weight and its influence on cardiometabolic risk over time are more conclusive. Children born large for gestational age have a predisposition for obesity and an increased CIMT in young adulthood, although they seem to have an otherwise healthy cardiovascular risk [17]. Children with low birth weight have increased CIMT values in young adulthood only if they have experienced severe intrauterine growth retardation followed by exaggerated postnatal growth [50]. A recent meta-analysis showed that in children born small for gestational age, CIMT has increased values in infants rather than in older children, which may prove that fetal growth restriction is an important factor that may increase the child's risk trajectory [51]. Our findings are in accord with previous studies, although our sample cases were smaller. In all three study groups, approximately half of the children had normal birth weights (2500–3500 g) and the other half had either low birth weight (<2500 g) or high birth weight (>3500 g) (Tables 3–5). Although we did not detect significant differences, CIMT had higher values for children with high birth weight in both overweight children and controls. However, in the obese group, the mean values for low birth weight were higher than for high birth weight (Table 3). Low birth weight might have more of impact on endothelial suffering, as it is associated with in utero cardiac remodeling and reduced arterial compliance [52]. Previous studies have shown that the more severe fetal growth impairment is, the more significant the data [51].

Because a mother's health and health-related habits during pregnancy are directly connected to the intrauterine development of the child [53], we analyzed how some maternal risk factors for child obesity and cardiovascular risk influence our subjects.

A total of 70% of our obese patients were born to unhealthy mothers. A total of 30% of them had mothers who gained >20 kg during their pregnancy, 25% had mothers with autoimmune thyroiditis, 10% had mothers with gestational diabetes mellitus, and 5% had mothers with gestational hypertension (Figure 2). These patients presented significantly increased values in CIMT compared to obese children born to healthy mothers ($p = 0.047$, Table 3). In overweight children, the proportion of cases that came from abnormal pregnancies dropped to 53%, and we also detected increased CIMT mean values compared to the overweight children with healthy mothers ($p = 0.049$, Table 4). For normal-weight children, the proportion of unhealthy pregnancies was even smaller, at 32% (Table 5). Overall, the most predominant risk factor was by far maternal obesity (>20 kg weight gain). The mechanism behind its effects on the fetus is based on placental insufficiency, which promotes intrauterine growth restriction [54]. This can lead to structural and functional alterations on the fetus' cardiovascular system that may persist throughout childhood [55] and can remain permanent risks or predispositions for high blood pressure, increased CIMT, and arterial stiffness [56].

Family medical history should play an important role in shaping an individual's cardiometabolic risk profile. Family history alone is not enough to place blame, but it surely helps to evaluate the genetic risk of the individual in order to estimate prognostics on the onset and severity of the disease's evolution [57]. In our study, 75% of obese children had a family history positive for risk, 40% had obese first-grade relatives, followed by a long distance by autoimmune thyroiditis, type 2 diabetes mellitus, and cardiovascular diseases (Figure 3). Although CIMT values were higher for children with a family history of risk in all three groups of study, significance was found only in the overweight group (p = 0.049).

Smoking is a well-known risk factor for vascular dysfunction, affecting the integrity of endothelial cells and promoting lipid and monocyte invasion of endothelial walls [58], and thus accelerating the stiffness of the arterial wall. We addressed the problem of adolescent smoking, but cannot rely on our findings due to the small amount of smoking teenagers in our group. Obese smokers had the most increased CIMT mean: $\overline{X}$ = 0.61 mm, p = 0.015 (Table 3). We note the only case of a 16-year-old male normal-weight smoker, who presented neither of the evaluated risk factors and who had a CIMT value of 0.56, higher than anyone in his group (group CIMT $\overline{X}$ = 0.39 mm); see Table 5. Because both age and the presence of metabolic syndrome are closely linked to inflammation and oxidative stress, and ultimately the damage of the vasculature structure, which is easily detectable starting in young adulthood [59], in young adults with such conditions, smoking will strengthen the adverse effects of age and all the other risk factors by directly damaging endothelial integrity.

High blood pressure in obese children and young adults is a consequence of the mixture of insulin resistance, sympathetic nervous system overactivity, and damage to the endothelial wall [60]. With regard to CIMT values, once installed, hypertension is an aggravating factor [61]. Moreover, hypertensive children present increased CIMT values, regardless of their BMI [62]. Our study confirmed the higher CIMT values among hypertensive children compared to normotensive ones (p = 0.012) with similar BMIs. Hypertensive subjects in the obese group reached 20%.

Our approach to lifestyle as a risk was focused on the level of sedentary behavior of each subject. We considered a subject sedentary when they practiced no sport and/or had less than 1 h of physical activity/day. Lack of physical activity is the factor that completes the big picture of obesity [63]. Although we did not find statistically meaningful differences, our study showed a tendency toward higher CIMT based on sedentary habits in all weight categories. However, this factor is a rather indirect one with concern to CIMT, because it affects vascular integrity by increasing the subject's susceptibility to obesity.

Abdominal visceral excess adipose tissue, clinically estimated by measurement of waist circumference, is considered a marker of insulin resistance [64] and a key component of metabolic syndrome when it reaches values >90th percentile for age and sex in children [65,66]. In the general population, CIMT values are significantly higher in men and women with a waist circumference ≥79 cm compared to values <79 cm [67]. Our study also showed a high correlation between the values of waist circumference and CIMT. In the obese group, significant correlations were detected between waist circumference and parameters like BMI, blood pressure, total cholesterol, LDL cholesterol, and triglycerides, as well (Table 6). In all three groups, the higher the mean values of waist circumference were, the stronger the correlations to CIMT and other evaluation parameters.

As for blood parameters, we analyzed the correlations between CIMT values and parameters usually used for evaluating weight excess in children: lipid panel and fasting glucose levels.

HDL cholesterol is considered the protective component of the lipid panel, with cytoprotective, anti-inflammatory, antithrombotic, and antioxidant functions and playing a large role in clearing away excess cholesterol from vessels and transporting it back to the liver [68]. Recent studies show that HDL-c's functionality is more important than its circulating quantity per se, and factors that affect its functionality should be avoided as much as possible: oxidative environment (acute phase response), metabolic syndrome, obesity, and consumption of saturated fats [69]. HDL-c in children is primarily affected by lack of physical activity [70]. Dietary habits that include enough polyunsaturated fats, in detriment to saturated fats, improve the anti-inflammatory function of HDL-c [71]. Children with increased CIMT and sedentary behavior are the ones with the lowest HDL-c, highest triglycerides levels, and highest risk of developing metabolic syndrome [67]. In our study, HDL cholesterol was significantly and negatively correlated to CIMT mean values (Table 8). It was also one of the most prevalent abnormal parameters detected, with 40% of the obese and normal-weight children and 7% of the overweight children presenting an HDL level of $\leq$40 mg/dL.

LDL cholesterol plays a major role in atherosclerotic plaque progression due to its migration under the sub-endothelial space, along with excess chylomicrons, where they are ingurgitated by macrophages and monocytes [72]. High LDL-c is usually associated with low HDL-c, depending on individual physical fitness. The diagnostic of dyslipidemia in an obese child should be taken into consideration as an aggravating factor for arterial long-term health [73]. Adolescents with high lipid levels present a greater risk of developing high CIMT as adults, and, should they have weight excess, the risk increases substantially compared to adults who did not have either risk during their childhood [74]. In obese children, LDL-c correlated to CIMT values (Table 8), a finding that supports international data. Furthermore, all three analysis groups showed correlations between CIMT and total cholesterol values (Figure 9), although mean values of total cholesterol were not suggestive for dyslipidemia. This may be because overall LDL-c values were not extremely high, and HDL-c levels were low (47% of cases presented <40 mg/dL values). We did note, however, that as weight excess severity increased, the lipid panel showed pathological, at-risk values. Individuals with familial hypercholesterolemia present high values for lipoproteins even in the context of less extreme weight excess, and affected children present higher CIMT than their unaffected siblings. This is further proof that once installed, dyslipidemia increases the progress of atherosclerosis, which can be detected early enough to evaluate the cardiovascular risk and aid the medical effort to avoid cardiovascular complications [74].

Triglyceride levels are influenced primarily by dietary habits and in part by physical activity [75]. As obese children have problems with both aspects, it is not a surprise that obese children present higher levels of TG [76]. Fasting triglycerides levels >150 mg/dL are known to represent a biomarker for cardiovascular risk [77]. Nevertheless, specific roles of TG are still controversial because there is significant personal variability involved in serum levels of TG compared to other more stable parameters like HDL-c [77]. Many studies have shown that TG levels are not directly involved in the increase of CIMT [78,79]. However, other studies suggest that serum TG's importance in estimating atherosclerotic risk is undervalued and that the lack of consistency in showing a connection could be derived from the fact that TG levels have great variability from one day to another [80]. Some studies have shown discreet correlations of TG levels to CIMT, and, even more specifically, through remnant triglyceride-rich lipoproteins associated with fatty meals [81]. As essential as fats are for human health, the modern diet of both children and adults disregards the ideal proportions of polyunsaturated fatty acids (PUFAs). This imbalance is an aggravating factor for vascular damage. Moreover, genetic variations in children with obesity may influence the metabolism of PUFAs, and the intake requirements for a healthy cardiovascular profile may therefore vary [82,83]. Our findings support these data: We detected significant correlations between TG levels and CIMT in both obese and overweight children (Table 8, Figure 10a,b). Although mean values were not surprisingly high, we did detect an association of higher TG with higher BMI and sedentary behavior. More data should be assessed in larger groups of obese children and a comparison between fasting TG and postprandial TG levels would be highly recommended.

Moreover, non-HDL cholesterol has proven to be an important tool in assessing atherosclerotic risk, as well [84]. Furthermore, a recent longitudinal study has shown that pathological levels of non-HDL-c in youths over 15 years old correlate to high CIMT in adulthood. Its predictive value for high adult CIMT was similar to LDL-c's. It has also been shown that through preventive actions, the effects of youth dyslipidemia can be attenuated if serum lipids are normalized by adulthood [85]. Total cholesterol/HDL-c ratio is one of the better predictors of ischemic coronary events. Its ability to do so is given by the cluster of metabolic disturbances present in individuals with abdominal obesity, insulin resistance, and dyslipidemia; therefore, its prediction value is even more powerful in individuals who have them all [86]. Triglyceride/HDL-c ratio could be a better predictor for cardiometabolic risk and insulin resistance in children than metabolic syndrome status [42,44,87]. Our findings showed correlations between CIMT values and non-HDL-c, TC/HDL-C ratio, and TG/HDL-C ratio, and their correlation strengths grew as weight severity based on BMI grew (Table 8, Figure 11). The mean values for these three parameters were influenced by weight excess (Table 9). Over half of the subjects in all three groups presented values higher than ideal for all three parameters, especially in the case of TG/HDL-c ratio (Table 9).

Childhood obesity disturbs glucose–insulin metabolism as well, and one of the earliest signs is the impairment of fasting glucose, which highly predicts type 2 diabetes mellitus [88,89]. However, the problem with impaired fasting glucose (IGF) as a factor for risk evaluation is that it varies substantially between different regions and ethnicities. For instance, IGF in Swedish and American obese children is considerably higher than in central European obese children [90]. Both IGF and impaired glucose tolerance increase CIMT by affecting arterial vessels [40]. As weight excess becomes more severe, insulin resistance and high glucose levels support the pathological processes of endothelial thickening. However, in this pre-diabetic situation, glucose variability is possible, with highs and lows all throughout the day [45]. This may be a reason why studies on IFG should be conducted on larger sample sizes to increase statistical significance. On this note, our research found a contradicting negative correlation between CIMT and fasting glucose levels in our obese group. However, in the overweight group, we detected a positive correlation, which suggests that the need for larger samples is indeed pertinent.

The multivariate linear regression analysis of CIMT as a dependent variable and waist circumference, LDL-c, HDL-c, TC, TG, non-HDL-c, TC/HDL-c ratio, and TG/HDL-c ratio as independent variables showed that waist circumference and TG/HDL-c ratio are statistically significant predictors of CIMT (Tables 10 and 11). This is an important finding, with concrete clinical use.

The main limitation of this study was the small sample size, but even so, our analysis was statistically significant and in line with further publications. A further longitudinal analysis would be interesting regarding CIMT and other correlated parameters, once the same subjects start to lose weight and change their lifestyle.

5. Conclusions

Weight excess in children is associated with increased values of CIMT, and the severity of the excess increases the expected values of CIMT. Abdominal adiposity of obese children, a clinical marker of metabolic distress, is very reliably positively correlated to CIMT values. Waist circumference and TG/HDL-c are significant predictors of CIMT.

Risk factors like weight gain of over 20 kg during pregnancy and overall metabolic disturbances of the mother, family history of cardiovascular risk, high blood pressure, and smoking are linked to increased CIMT. Our study supports international data on artificial postnatal nutrition, high/low birth weight, smoking, and sedentary lifestyle being linked to increased CIMT, but our analysis's statistical significance was not definitive.

All evaluated blood parameters showed correlations to CIMT, except for fasting glucose.

Author Contributions: Conceptualization, D.S. and M.-S.M.; methodology, D.S., M.-S.M. and I.M.; software, D.S. and A.C.; validation, D.S., C.P. and I.P.V.; formal analysis, M.-S.M. and I.M.; investigation, M.-S.M. and C.P.; resources, C.P., I.P.V. and D.S.; data curation, F.D.; writing—original draft preparation, M.-S.M.; writing—review and editing, D.S. and C.P.; visualization, A.C. and F.D.; supervision, D.S.; project administration, I.M. and D.S. All authors have read and agreed to the published version of the manuscript.

Funding: This research received no external funding.

Institutional Review Board Statement: The study was conducted according to the guidelines of the Declaration of Helsinki, and approved by the Ethics Committee of Scientific Research (CECS) of University of Medicine and Pharmacy Victor Babes Timisoara (Nr. 03/19.01.2021).

Informed Consent Statement: Informed consent was obtained from all subjects' legal guardians involved in the study, as well as verbal consent from all the underage subjects. Written informed consent was obtained from the patients to publish this paper.

Acknowledgments: We are grateful for the collaboration of all the patients enrolled in this study and extend our wishes of health to all of them. We acknowledge the productive contributions of all authors and the support given by our centers of affiliation to facilitate our study.

Conflicts of Interest: The authors declare no conflict of interest.

References

1. Browne, N.T.; Snethen, J.A.; Greenberg, C.S.; Frenn, M.; Kilanowski, J.F.; Gance-Cleveland, B.; Burke, P.; Lewandowski, L. When Pandemics Collide: The Impact of COVID-19 on Childhood Obesity. *J. Pediatr. Nurs.* **2021**, *56*, 90–98. [CrossRef] [PubMed]
2. Stavridou, A.; Kapsali, E.; Panagouli, E.; Thirios, A.; Polychronis, K.; Bacopoulou, F.; Psaltopoulou, T.; Tsolia, M.; Sergentanis, T.N.; Tsitsika, A. Obesity in Children and Adolescents during COVID-19 Pandemic. *Children* **2021**, *8*, 135. [CrossRef] [PubMed]
3. Thompson, M.; Mansfield, B.; Stringer, M.; Stewart, B.; Potter, J.; Fernengel, K. An evidence-based resource for the management of comorbidities associated with childhood overweight and obesity. *J. Am. Assoc. Nurse Pract.* **2016**, *28*, 559–570. [CrossRef] [PubMed]
4. Downing, R., 2nd; Michael, T.; Place, R.; Hoffman, E.; Visich, P. The Influence of Metabolic Syndrome Risk Factors on Carotid Intima Media Thickness in Children. *Glob. Pediatr. Health* **2021**, *8*, 2333794X20987453. [CrossRef] [PubMed]
5. Beauloye, V.; Zech, F.; Tran, H.T.; Clapuyt, P.; Maes, M.; Brichard, S.M. Determinants of early atherosclerosis in obese children and adolescents. *J. Clin. Endocrinol. Metab.* **2007**, *92*, 3025–3032. [CrossRef]
6. Farello, G.; Antenucci, A.; Stagi, S.; Mazzocchetti, C.; Ciocca, F.; Verrotti, A. Metabolically healthy and metabolically unhealthy obese children both have increased carotid intima-media thickness: A case control study. *BMC Cardiovasc. Disord.* **2018**, *18*, 1–6. [CrossRef]
7. Wilson, D.P. Is Atherosclerosis a Pediatric Disease? 2000. In *Endotext. South Dartmouth (MA)*; Feingold, K.R., Anawalt, B., Boyce, A., Chrousos, G., de Herder, W.W., Dhatariya, K., Dungan, K., Hershman, J.M., Hofland, J., Kalra, S., et al., Eds.; MDText.com, Inc.: South Dartmouth, MA, USA, 2000. Available online: https://www.ncbi.nlm.nih.gov/books/NBK395576/ (accessed on 3 February 2020).
8. Sinning, C.; Wild, P.S.; Echevarria, F.M.; Wilde, S.; Schnabel, R.; Lubos, E.; Herkenhoff, S.; Bickel, C.; Klimpe, S.; Gori, T.; et al. Gutenberg-Heart Study. Sex differences in early carotid atherosclerosis (from the community-based Gu-tenberg-Heart Study). *Am. J. Cardiol.* **2011**, *107*, 1841–1847. [CrossRef]
9. Van den Oord, S.C.; Sijbrands, E.J.; ten Kate, G.L.; van Klaveren, D.; van Domburg, R.T.; van der Steen, A.F.; Schinket, A.F.L. Carotid intima-media thickness for cardiovascular risk assessment: Systematic review and meta-analysis. *Atherosclerosis* **2013**, *228*, 1–11. [CrossRef]
10. Fall, C.H.D.; Barker, D.J.P.; Osmond, C.; Winter, P.D.; Clark, P.M.S.; Hales, C.N. Relation of infant feeding to adult serum cholesterol concentration and death from ischaemic heart disease. *BMJ* **1992**, *304*, 801–805. [CrossRef]
11. Lurbe, E.; Aguilar, F.; Álvarez, J.; Redon, P.; Torró, M.I.; Redon, J. Determinants of Cardiometabolic Risk Factors in the First Decade of Life: A Longitudinal Study Starting at Birth. *Hypertension* **2018**, *71*, 437–443. [CrossRef]
12. Oddy, W.H.; Mori, T.A.; Huang, R.C.; Marsh, J.A.; Pennell, C.E.; Chivers, P.T.; Hands, B.P.; Jacoby, P.; Rzehak, P.; Koletzko, B.V.; et al. Early infant feeding and adiposity risk: From infancy to adulthood. *Ann. Nutr. Metab.* **2014**, *64*, 262–270. [CrossRef]
13. Martin, R.M.; Ebrahim, S.; Griffin, M.; Davey-Smith, G.; Nicolaides, A.N.; Georgiou, N. Breastfeeding and atherosclerosis: Intima-media thickness and plaques at 65-year follow-up of the Boyd Orr cohort. *Arterioscler. Thromb. Vasc. Biol.* **2005**, *25*, 1482–1488. [CrossRef] [PubMed]
14. Martin, R.M.; McCarthy, A.; Davies, D.P.; Davey Smith, G.; Ben-Shlomo, Y. Association between infant nutrition and blood pressure in early adulthood: The Barry Caerphilly Growth cohort study. *Am. J. Clin. Nutr.* **2003**, *77*, 1489–1497. [CrossRef] [PubMed]

15. Ravelli, A.C.J.; van der Meulen, J.H.; Osmond, C.; Barker, D.J.P.; Bleker, O.P. Infant feeding and adult glucose tolerance, lipid profile, blood pressure, and obesity. *Arch. Dis. Child.* **2000**, *82*, 248–252. [CrossRef] [PubMed]
16. Belbasis, L.; Savvidou, M.D.; Kanu, C.; Evangelou, E.; Tzoulaki, I. Birth weight in relation to health and disease in later life: An umbrella review of systematic reviews and meta-analyses. *BMC Med.* **2016**, *14*, 147. [CrossRef]
17. Skilton, M.R.; Siitonen, N.; Würtz, P.; Viikari, J.S.; Juonala, M.; Seppälä, I.; Laitinen, T.; Lehtimäki, T.; Taittonen, L.; Kähönen, M.; et al. High birth weight is associated with obesity and increased carotid wall thickness in young adults: The cardiovascular risk in young Finns study. *Arterioscler. Thromb. Vasc. Biol.* **2014**, *34*, 1064–1068. [CrossRef] [PubMed]
18. Di Bernardo, S.; Mivelaz, Y.; Epure, A.M.; Vial, Y.; Simeoni, U.; Bovet, P.; Younes, S.E.; Chiolero, A.; Sekarski, N. Assessing the consequences of gestational diabetes mellitus on offspring's cardiovascular health: MySweetHeart Cohort study protocol. *BMJ Open* **2017**, *7*, e016972.
19. Benschop, L.; Schalekamp-Timmermans, S.; Roeters van Lennep, J.E.; Jaddoe, V.W.V.; Steegers, E.A.P.; Ikram, M.K. Cardiovascular Risk Factors Track from Mother to Child. *JAHA* **2018**, *7*, e009536. [CrossRef]
20. Jølving, L.R.; Nielsen, J.; Kesmodel, U.S.; Nielsen, R.G.; Nørgård, B.M.; Beck-Nielsen, S.S. Chronic diseases in the children of women with maternal thyroid dysfunction: A nationwide cohort study. *Clin. Epidemiol.* **2018**, *10*, 1381–1390. [CrossRef]
21. İşgüven, P.; Gündüz, Y.; Kılıç, M. Effects of Thyroid Autoimmunity on Early Atherosclerosis in Euthyroid Girls with Hashimoto's Thyroiditis. *J. Clin. Res. Pediatr. Endocrinol.* **2016**, *8*, 150–156. [CrossRef]
22. De Giorgis, T.; Giannini, C.; Scarinci, A.; D'Adamo, E.; Agostinelli, S.; Chiarelli, F.; Mohn, A. Family history of premature cardiovascular disease as a sole and independent risk factor for increased carotid intima-media thickness. *J. Hypertens.* **2009**, *27*, 822–828. [CrossRef]
23. Li, S.; Yun, M.; Fernandez, C.; Xu, J.; Srinivasan, S.R.; Chen, W.; Berenson, G.S. Cigarette smoking exacerbates the adverse effects of age and metabolic syndrome on subclinical atherosclerosis: The Bogalusa Heart Study. *PLoS ONE* **2014**, *9*, e96368. [CrossRef]
24. Mozos, I.; Maidana, J.P.; Stoian, D.; Stehlik, M. Gender Differences of Arterial Stiffness and Arterial Age in Smokers. *Int. J. Environ. Res. Public Health* **2017**, *14*, 565. [CrossRef] [PubMed]
25. Burton, A. Parental smoking may set up children for atherosclerosis. *Environ. Health Perspect.* **2010**, *118*, A200. [CrossRef] [PubMed]
26. Yang, B.; Li, M.; Chen, B.; Xu, Y.; Li, T.D. Deterioration of endothelial function and carotid intima-media thickness in Tibetan male adolescents exposed to second-hand smoke. *JRAAS* **2012**, *13*, 413–419. [CrossRef]
27. Vuolo, M.; Staff, J. Parent and child cigarette use: A longitudinal, multigenerational study. *Pediatrics* **2013**, *132*, e568–e577. [CrossRef] [PubMed]
28. Khalil, A.; Huffman, M.D.; Prabhakaran, D.; Osmond, C.; Fall, C.H.D.; Tandon, N.; Lakshmy, R.; Prabhakaran, P.; Biswas, S.K.; Ramji, S.; et al. New Delhi Birth Cohort. Predictors of carotid intima-media thickness and carotid plaque in young Indian adults: The New Delhi birth cohort. *Int. J. Cardiol.* **2013**, *167*, 1322–1328. [CrossRef] [PubMed]
29. Magnussen, C.G.; Venn, A.; Thomson, R.; Juonala, M.; Srinivasan, S.R.; Viikari, J.S.; Berenson, G.S.; Dwyer, T.; Raitakari, O.T. The association of pediatric low- and high-density lipoprotein cholesterol dyslipidemia classifications and change in dyslipidemia status with carotid intima-media thickness in adulthood evidence from the cardiovascular risk in Young Finns study, the Bogalusa Heart study, and the CDAH (Childhood Determinants of Adult Health) study. *J. Am. Coll. Cardiol.* **2009**, *53*, 860–869.
30. Santos, I.S.; Bittencourt, M.S.; Goulart, A.C.; Schmidt, M.I.; Diniz, M.F.H.S.; Lotufo, P.A.; Benseñor, I.M. Insulin resistance is associated with carotid intima-media thickness in non-diabetic subjects. A cross-sectional analysis of the ELSA-Brasil cohort baseline. *Atherosclerosis* **2017**, *260*, 34–40. [CrossRef]
31. Mozos, I.; Malainer, C.; Horbańczuk, J.; Gug, C.; Stoian, D.; Luca, C.T.; Atanasov, A.G. Inflammatory Markers for Arterial Stiffness in Cardiovascular Diseases. *Front. Immunol.* **2017**, *8*, 1058. [CrossRef]
32. Velea, I.P.; Albulescu, R.; Arghirescu, S.T. Obezitatea la copil. In *Pediatrie-Curs Pentru Studenții Facultății de Medicină: Obezitatea la Copil*; Velea, I., Ed.; Editura Victor Babes: Timisoara, Romania, 2016; pp. 289–297.
33. Bennett, W.E., Jr.; Hendrix, K.S.; Thompson, R.T.; Carroll, A.E.; Downs, S.M. The natural history of weight percentile changes in the first year of life. *JAMA Pediatr.* **2014**, *168*, 681–682. Available online: http://jamanetwork.com/journals/jamapediatrics/fullarticle/1867333 (accessed on 15 May 2020). [CrossRef] [PubMed]
34. Kumar, S.; Kelly, A.S. Review of Childhood Obesity: From Epidemiology, Etiology, and Comorbidities to Clinical Assessment and Treatment. In *Mayo Clinic Proceedings*; Elsevier: Amsterdam, The Netherlands, 2017; Volume 92, pp. 251–265. Available online: http://www.mayoclinicproceedings.org/article/S0025-6196(16)30595-X/fulltext (accessed on 5 June 2020).
35. Casella, I.B.; Presti, C.; Porta, R.M.; Sabbag, C.R.; Bosch, M.A.; Yamazaki, Y. A practical protocol to measure common carotid artery intima-media thickness. *Clinics* **2008**, *63*, 515–520. [CrossRef] [PubMed]
36. Gómez-Marcos, M.A.; Recio-Rodríguez, J.I.; Patino-Alonso, M.C.; Agudo-Conde, C.; Gómez-Sanchez, L.; Gómez-Sanchez, M.; Rodríguez-Sánchez, E.; García-Ortiz, L. Protocol for measuring carotid intima-media thickness that best correlates with cardiovascular risk and target organ damage. *Am. J. Hypertens.* **2012**, *25*, 955–961. [CrossRef] [PubMed]
37. El Jalbout, R. Intima-Media Thickness Measurement in Children; Techniques and Reference Values. *Am. J. Biomed. Sci. Res.* **2020**, *7*, 101–103. [CrossRef]
38. Freire, C.M.; Ribeiro, A.L.; Barbosa, F.B.; Lana, A.M.; e Silva, A.C.; Ribeiro-Oliveira, A. Comparison between automated and manual measurements of carotid intima-media thickness in clinical practice. *Vasc. Health Risk Manag.* **2009**, *5*, 811–817.

39. Vermeersch, S.; Rietzschel, E.; De Buyzere, M.; Van Bortel, L.M.; D'Asseler, Y.; Gillebert, T.C.; Verdonck, P.R.; Segers, P. Validation of a new automated IMT measurement algorithm. *J. Hum. Hypertens.* **2007**, *21*, 976–978. [CrossRef]
40. Aydin, Y.; Berker, D.; Ustun, I.; Gul, K.; Erden, G.; Kutlucan, A.; Yilmaz, L.; Guler, S. Evaluation of carotid intima media thickness in impaired fasting glucose and impaired glucose tolerance. *Minerva Endocrinol.* **2011**, *36*, 171–179. [CrossRef]
41. FOR, E.P.O.I.G.; CHILDREN, R.R.I. Expert Panel on Integrated Guidelines for Cardiovascular Health and Risk Reduction in Children and Adolescents; National Heart, Lung, and Blood Institute. Expert panel on integrated guidelines for cardiovascular health and risk reduction in children and adolescents: Summary report. *Pediatrics* **2011**, *128*, S213–S256. [CrossRef]
42. Iwani, N.A.K.Z.; Jalaludin, M.Y.; Zin, R.M.W.M.; Fuziah, M.Z.; Hong, J.Y.H.; Abqariyah, Y.; Mokhtar, A.H.; Nazaimoon, W.M.W. TG : HDL-C Ratio Is a Good Marker to Identify Children Affected by Obesity with Increased Cardiometabolic Risk and Insulin Resistance. *Int. J. Endocrinol.* **2019**, *2019*, 8586167. [CrossRef]
43. Di Giorgis, T.; Marcovecchio, M.L.; Di Giovanni, I.; Giannini, C.; Chiavaroli, V.; Chiarelli, F.; Mohn, A. Triglycerides-to-HDL ratio as a new marker of endothelial dysfunction in obese prepubertal children. *Eur. J. Endocrinol.* **2013**, *170*, 173–180. [CrossRef]
44. Iwani, N.A.K.Z.; Jalaludin, M.Y.; Zin, R.M.W.M.; Fuziah, M.Z.; Hong, J.Y.H.; Abqariyah, Y.; Mokhtar, A.H.; Nazaimoon, W.M.W. Triglyceride to HDL-C ratio is associated with insulin resistance in overweight and obese children. *Sci. Rep.* **2007**, *7*, 40055. [CrossRef]
45. Chakarova, N.; Dimova, R.; Grozeva, G.; Tankova, T. Assessment of glucose variability in subjects with prediabetes. *Diabetes Res. Clin. Pract.* **2019**, *151*, 56–64. [CrossRef]
46. Skilton, M.R.; Boussel, L.; Bonnet, F.; Bernard, S.; Douek, P.C.; Moulin, P.; Serusclat, A. Carotid intima–media and adventitial thickening: Comparison of new and established ultrasound and magnetic resonance imaging techniques. *Atherosclerosis* **2011**, *215*, 405–410. [CrossRef]
47. Zhang, Y.; Guallar, E.; Qiao, Y.; Wasserman, B.A. Arterioscler. Is Carotid Intima-Media Thickness as Predictive as Other Noninvasive Techniques for the Detection of Coronary Artery Disease? *Arterioscler. Thromb. Vasc. Biol.* **2014**, *34*, 1341–1345. [CrossRef]
48. Rudd, J.H.; Narula, J.; Strauss, H.W.; Virmani, R.; Machac, J.; Klimas, M.; Tahara, N.; Fuster, V.; Warburton, E.A.; Fayad, Z.A.; et al. Imaging atherosclerotic plaque inflammation by fluorodeoxyglucose with positron emission tomography: Ready for prime time? *J. Am. Coll. Cardiol.* **2010**, *55*, 2527–2535. [CrossRef]
49. Chirita-Emandi, A.; Papa, M.C.; Abrudan, L.; Dobrescu, M.A.; Puiu, M.; Velea, I.P.; Paul, C. A novel method for measuring subcutaneous adipose tissue using ultrasound in children-interobserver consistency. *Rom J. Morphol. Embryol.* **2017**, *58*, 115–123.
50. Oren, A.; Vos, L.E.; Uiterwaal, C.S.; Gorissen, W.H.; Grobbee, D.E.; Bots, M.L. Birth weight and carotid intima-media thickness: New perspectives from the atherosclerosis risk in young adults (ARYA) study. *Ann. Epidemiol.* **2004**, *14*, 8–16. [CrossRef]
51. Epure, A.M.; Rios-Leyvraz, M.; Anker, D.; Di Bernardo, S.; Da Costa, B.R.; Chiolero, A.; Sekarski, N. Risk factors during first 1,000 days of life for carotid intima-media thickness in infants, children, and adolescents: A systematic review with meta-analyses. *PLoS Med.* **2020**, *17*, e1003414. [CrossRef]
52. Verburg, B.O.; Jaddoe, V.W.; Wladimiroff, J.W.; Hofman, A.; Witteman, J.C.; Steegers, E.A. Fetal hemodynamic adaptive changes related to intrauteringrowth: The Generation R Study. *Circulation* **2008**, *117*, 649–659. [CrossRef]
53. Golab, B.; Santos, S.; Voerman, E.; Lawlor, D.; Jaddoe, V.; Gaillard, R. Common pregnancy complications and risk of childhood obesity-influence of maternal obesity: An individual participant data. *Lancet Child Adolesc. Health* **2019**, *2*, 812. [CrossRef]
54. Grieger, J.A.; Clifton, V.L. A review of the impact of dietary intakes in human pregnancy on infant birthweight. *Nutrients* **2015**, *7*, 153–178. [CrossRef]
55. Crispi, F.; Bijnens, B.; Figueras, F.; Bartrons, J.; Eixarch, E.; Le Noble, F.; Ahmed, A.; Gratacós, E. Fetal growth restriction results in remodeled and less efficient hearts in children. *Circulation* **2010**, *121*, 2427–2436. [CrossRef] [PubMed]
56. Crispi, F.; Miranda, J.; Gratacós, E. Long-term cardiovascular consequences of fetal growth restriction: Biology, clinical implications, and opportunities for prevention of adult disease. *Am. J. Obstet. Gynecol.* **2018**, *218*, S869–S879. [CrossRef] [PubMed]
57. Corica, D.; Aversa, T.; Valenzise, M.; Messina, M.F.; Alibrandi, A.; De Luca, F.; Wasniewska, M. Does family history of obesity, cardiovascular, and metabolic diseases influence onset and severity of childhood obesity? *Front. Endocrinol.* **2018**, *9*, 187. [CrossRef]
58. Ambrose, J.A.; Barua, R.S. The pathophysiology of cigarette smoking and cardiovascular disease: An update. *J. Am. Coll. Cardiol.* **2004**, *43*, 1731–1737. [CrossRef]
59. Costopoulos, C.; Liew, T.V.; Bennett, M. Ageing and atherosclerosis: Mechanisms and therapeutic options. *Biochem. Pharmacol.* **2008**, *75*, 1251–1261. [CrossRef] [PubMed]
60. Sorof, J.; Danies, S. Obesity Hypertension in Children-A Problem of Epidemic Proportions. *Hypertension* **2002**, *40*, 441–447. [CrossRef]
61. National High Blood Pressure Education Program Working Group on High Blood Pressure in Children and Adolescents. The fourth report on the diagnosis, evaluation, and treatment of high blood pressure in children and adolescents. *Pediatrics* **2004**, *114*, 1–22.
62. Lande, M.B.; Carson, N.L.; Roy, J.; Meagher, C.C. Effects of Childhood Primary Hypertension on Carotid Intima Media Thickness-A Matched Controlled Study. *Hypertension* **2006**, *48*, 40–44. [CrossRef]
63. Keane, E.; Li, X.; Harrington, J.M.; Fitzgerald, A.P.; Perry, I.J.; Kearney, P.M. Physical Activity, Sedentary Behavior and the Risk of Overweight and Obesity in School-Aged Children. *Pediatr. Exerc. Sci.* **2017**, *29*, 408–418. [CrossRef]

64. Bassali, R.; Waller, J.L.; Gower, B.; Allison, J.; Davis, C.L. Utility of waist circumference percentile for risk evaluation in obese children. *Int. J. Pediatr. Obes.* **2010**, *5*, 97–101. [CrossRef]
65. Freedman, D.S.; Kahn, H.S.; Mei, Z.; Grummer-Strawn, L.M.; Dietz, W.H.; Srinivasan, S.R.; Berenson, G.S. Relation of body mass index and waist-to-height ratio to cardiovascular disease risk factors in children and adolescents: The Bogalusa Heart Study. *Am. J. Clin. Nutr.* **2007**, *86*, 33–40. [CrossRef] [PubMed]
66. Stefan, N.; Kantartzis, K.; Machannm, J.; Schick, F.; Thamer, C.; Rittig, K.; Balletshofer, B.; Machicao, F.; Fritsche, A.; Häring, H.U. Identification and characterization of metabolically benign obesity in humans. *Arch. Intern. Med.* **2008**, *168*, 1609–1616. [CrossRef] [PubMed]
67. Kamon, T.; Kaneko, H.; Itoh, H.; Kiriyama, H.; Mizuno, Y.; Morita, H.; Yamamichi, N.; Komuro, I. Association Between Waist Circumference and Carotid Intima-Media Thickness in the General Population. *Int. Heart J.* **2020**, *61*, 103–108. [CrossRef]
68. Toth, P.P. Reverse cholesterol transport: High-density lipoprotein's magnificent mile. *Curr. Atheroscler. Rep.* **2003**, *5*, 386–393. [CrossRef]
69. Kosmas, C.E.; Martinez, I.; Sourlas, A.; Bouza, K.V.; Campos, F.N.; Torres, V.; Montan, P.D.; Guzman, E. High-density lipoprotein (HDL) functionality and its relevance to atherosclerotic cardiovascular disease. *Drugs Context* **2018**, *7*, 212525. [CrossRef]
70. O'Donovan, G.; Stensel, D.; Hamer, M.; Stamatakis, E. The association between leisure-time physical activity, low HDL-cholesterol and mortality in a pooled analysis of nine population-based cohorts. *Eur. J. Epidemiol.* **2017**, *32*, 559–566. [CrossRef]
71. Nicholls, S.J.; Lundman, P.; Harmer, J.A.; Cutri, B.; Griffiths, K.A.; Rye, K.A.; Barter, P.J.; Celermajer, D.S. Consumption of saturated fat impairs the anti-inflammatory properties of high-density lipoproteins and endothelial function. *J. Am. Coll. Cardiol.* **2006**, *48*, 715–720. [CrossRef]
72. Peterson, J.; Bihain, B.E.; Bengtsson-Olivecrona, G.; Deckelbaum, R.J.; Carpentier, Y.A.; Olivecrona, T. Fatty acid control of lipoprotein lipase: A link between energy metabolism and lipid transport. *Proc. Natl. Acad. Sci. USA* **1990**, *87*, 909–913. [CrossRef]
73. Hatami, M.; Tohidi, M.; Mohebi, R.; Khalili, D.; Azizi, F.; Hadaegh, F. Adolescent lipoprotein classifications according to National Health and Nutrition Examination Survey (NHANES) vs. National Cholesterol Education Program (NCEP) for predicting abnormal lipid levels in adulthood in a Middle East population. *Lipids Health Dis.* **2012**, *11*, 107. [CrossRef]
74. Kusters, D.M.; Wiegman, A.; Kastelein, J.J.P.; Hutten, B.A. Carotid Intima-Media Thickness in Children With Familial Hypercholesterolemia. *Circ. Res.* **2014**, *114*, 307–310. [CrossRef]
75. Gortmaker, S.L.; Swinburn, B.A.; Levy, D.; Carter, R.; Mabry, P.L.; Finegood, D.T.; Huang, T.; Marsh, T.; Moodie, M.L. Changing the future of obesity: Science, policy, and action. *Lancet* **2011**, *378*, 838–847. [CrossRef]
76. May, A.L.; Kuklina, E.V.; Yoon, P.W.; Centers for Disease Control and Prevention (CDC). Prevalence of abnormal lipid levels among youths–United States, 1999–2006. *Morb. Mortal. Wkly. Rep.* **2010**, *59*, 29–33.
77. Boullart, A.C.; de Graaf, J.; Stalenhoef, A.F. Serum triglycerides and risk of cardiovascular disease. *Biochim. Et Biophys. Acta (BBA)-Mol. Cell Biol. Lipids* **2012**, *1821*, 867–875. [CrossRef] [PubMed]
78. Gardener, H.; Della Morte, D.; Elkins, M.S.; Sacco, R.L.; Rundek, T. Lipids and carotid plaque in the Northern Manhattan Study. *BMC Cardiovasc. Disord.* **2009**, *9*, 55. [CrossRef]
79. Touboul, P.J.; Labreuche, J.; Bruckert, E.; Schargrodsky, H.; Prati, P.; Tosetto, A.; Hernandez-Hernandez, R.; Woo, K.S.; Silva, H.; Vicaut, E.; et al. HDL-C, triglycerides and carotid IMT: A meta-analysis of 21,000 patients with automated edge detection IMT measurement. *Atherosclerosis* **2014**, *232*, 65–71. [CrossRef]
80. Durrington, P. Triglycerides are more important in atherosclerosis than epidemiology suggested. *Atherosclerosis* **1998**, *141*, S57–S62. [CrossRef]
81. Karpe, F.; Boquist, S.; Tang, R.; Bond, G.M.; de Faire, U.; Hamsten, A. Remnant lipoproteins are related to intima-media thickness of the carotid artery independently of LDL cholesterol and plasma triglycerides. *J. Lipid Res.* **2001**, *42*, 17–21. [CrossRef]
82. Ander, B.P.; Dupasquier, C.M.; Prociuk, M.A.; Pierce, G.N. Polyunsaturated fatty acids and their effects on cardiovascular disease. *Exp. Clin. Cardiol.* **2003**, *8*, 164–172.
83. Serafim, V.; Chirita-Emandi, A.; Andreescu, N.; Tiugan, D.A.; Tutac, P.; Paul, C.; Velea, I.; Mihailescu, A.; Șerban, C.L.; Zimbru, C.G.; et al. Single Nucleotide Polymorphisms in PEMT and MTHFR Genes are Associated with Omega 3 and 6 Fatty Acid Levels in the Red Blood Cells of Children with Obesity. *Nutrients* **2019**, *11*, 2600. [CrossRef]
84. Frontini, M.G.; Srinivasan, S.R.; Xu, J.; Tang, R.; Bond, M.G.; Berenson, G.S. Usefulness of childhood non-high density lipoprotein cholesterol levels versus other lipoprotein measures in predicting adult subclinical atherosclerosis: The Bogalusa Heart Study. *Pediatrics* **2008**, *121*, 924–929. [CrossRef] [PubMed]
85. Juonala, M.; Wu, F.; Sinaiko, A.; Woo, J.G.; Urbina, E.M.; Jacobs, D.; Steinberger, J.; Prineas, R.; Koskinen, J.; Sabin, M.A.; et al. Non-HDL Cholesterol Levels in Childhood and Carotid Intima-Media Thickness in Adulthood. *Pediatrics* **2020**, *145*, e20192114. [CrossRef]
86. Lemieux, I.; Lamarche, B.; Couillard, C.; Pascot, A.; Cantin, B.; Bergeron, J.; Dagenais, G.R.; Després, J.P. Total Cholesterol/HDL Cholesterol Ratio vs LDL Cholesterol/HDL Cholesterol Ratio as Indices of Ischemic Heart Disease Risk in Men: The Quebec Cardiovascular Study. *Arch. Intern. Med.* **2001**, *161*, 2685–2692. [CrossRef]
87. Krawczyk, M.; Rumińska, M.; Witkowska-Sędek, E.; Majcher, A.; Pyrżak, B. Usefulness of the triglycerides to high-density lipoprotein cholesterol ratio (TG/HDL-C) in prediction of metabolic syndrome in polish obese children and adolescents. *Acta Biochim. Pol.* **2018**, *65*, 605–611. [CrossRef] [PubMed]

88. Hagman, E.; Ighani Arani, P.; Fischer, M.; Danielsson, P.; Marcinkiewicz, K.; Petriczko, E.; Marcus, C. Blood sugar levels are higher in young obese children in Sweden than in Poland. *Acta Paediatr.* **2014**, *103*, 1174–1178. [CrossRef] [PubMed]
89. Amer Diabet, A. Diagnosis and classification of diabetes mellitus American Diabetes Association. *Diabetes Care* **2011**, *34*, S62–S69.
90. Hagman, E.; Reinehr, T.; Kowalski, J.; Ekbom, A.; Marcus, C.; Holl, R.W. Impaired fasting glucose prevalence in two nationwide cohorts of obese children and adolescents. *Int. J. Obes.* **2014**, *38*, 40–45. [CrossRef]

applied sciences

MDPI

Technical Note

MR Imaging and Electrophysiological Features of Doxorubicin-Induced Fibrosis: Protocol Development in a Small Preclinical Pig Study with Histological Validation †

Peter Lin [1], Terenz Escartin [1,2], Melissa Larsen [2], Matthew Ng [2], Mengyuan Li [3], Jennifer Barry [2], Idan Roifman [4] and Mihaela Pop [2,*]

1 Department of Medical Biophysics, University of Toronto, 101 College Street, Toronto, ON M5G 1L7, Canada
2 Sunnybrook Research Institute, 2075 Bayview Avenue, Toronto, ON M4N 3M5, Canada
3 Institute of Biomedical Engineering, University of Toronto, 164 College Street, Toronto, ON M5S 3G9, Canada
4 Sunnybrook Health Sciences Centre, Department of Cardiology, 2075 Bayview Avenue, Toronto, ON M4N 3M5, Canada
* Correspondence: mihaela.pop@utoronto.ca

† This paper is (in part) an extended version of a conference proceeding paper published by Statistical Atlases and Computational Models of the Heart (STACOM) workshop, held in Strasbourg, France, 2021.

Abstract: A critical chemotherapeutic complication is cardiotoxicity, often leading, in time, to heart failure. In this work, we developed a novel animal protocol using magnetic resonance (MR) imaging and electrophysiology (EP) tests, designed to detect subtle structural and functional changes associated with myocardial damage in sub-chronic phases post-chemotherapy. A weekly dose of doxorubicin (DOX) was injected in four juvenile swine throughout a four-week plan, using an intravenous approach that mimics the treatment in cancer patients. We performed cardiac MR imaging as follows: in all four pigs pre-DOX; at 1 and 5 weeks post-DOX in a group of two pigs; and, at 1 and 9 weeks post-DOX in the other two pigs, using Cine imaging to assess ejection fraction (EF) and late gadolinium enhancement to quantify collagen density in the left ventricle. Additionally, X-ray-guided voltage mapping and arrhythmia tests were conducted in the group at 9 weeks post-DOX and in a healthy pig. Tissue samples were collected for histology. The results showed that EF decreased from ~46% pre-DOX to ~34% within the first 9 weeks post-DOX. This decline in LV function was explained by a gradual increase in collagen density, especially noticeable at week 9 post-DOX as derived from MRI analysis. Furthermore, ventricular fibrillation was induced via rapid pacing at 9 weeks post-DOX, most likely caused by fibrotic patches identified in voltage maps, as confirmed by MRI and collagen-sensitive histological stains. Overall, our novel preclinical protocol was able to reveal key signs of potentially-irreversible tissue changes, along with electrical remodeling and arrhythmia risk in the early months following DOX therapy. Future work will include more datasets to statistically power the study, and will use the protocol to test cardioprotective strategies.

Keywords: cardiotoxicity; MRI; fibrosis; chemotherapy; doxorubicin; voltage mapping; arrhythmia

Citation: Lin, P.; Escartin, T.; Larsen, M.; Ng, M.; Li, M.; Barry, J.; Roifman, I.; Pop, M. MR Imaging and Electrophysiological Features of Doxorubicin-Induced Fibrosis: Protocol Development in a Small Preclinical Pig Study with Histological Validation. *Appl. Sci.* **2022**, *12*, 11620. https://doi.org/10.3390/app122211620

Academic Editor: Qi-Huang Zheng

Received: 3 October 2022
Accepted: 13 November 2022
Published: 16 November 2022

1. Introduction

Clinical motivation: Cardio-oncology is an emerging multi-disciplinary field that investigates the onset and progression of myocardial injury induced in cancer patients by chemo-therapeutic or molecular agents, as well as radiation therapy [1]. Among the FDA-approved anthracyclines, doxorubicin (DOX) hydrochloride is one of the most effective chemotherapeutic drugs prescribed in the treatment of breast cancer [2]. DOX has been also used in the treatment of bladder cancer, lymphoma (Hodgkin and non-Hodgkin), Kaposi's sarcoma, and acute lymphocytic leukemia, as well as metastatic cases (e.g., gastric cancer, ovarian cancer, neuroblastoma). However, despite the therapeutic benefit offered by DOX, preclinical and clinical studies have increasingly reported progressive heart dysfunction as a critical adverse consequence of DOX-based treatments [3,4]. Furthermore, pharmacological studies showed that while DOX therapy induces cell death and cancer regression, it also triggers a problematic mechanism associated with cardiotoxicity based on a cascade of events that includes reactive oxidative stress [5] and modulation of mitochondrial function [6].

While the mechanism of cardiotoxicity remains somewhat controversial, it has been demonstrated that the cardiotoxic effects in myocytes and endothelial cells are primarily dose-dependent, leading to long-term irreversible damage of remodeled tissue, abnormal excitation–contraction coupling, and inefficient blood pumping. These major ventricular function impairments have been attributed to a gradual deposition of reactive collagenous fibrosis, causing cardiovascular complications such as cardiomyopathy or congestive heart failure and leading to collateral mortality in cancer survivors [1,2]. For decades, clinical studies have documented such cardiotoxic effects induced by anthracyclines only in late chronic stages, that is, after cardiomyopathy, heart failure and lethal events had occurred. Some recent investigations have focused on studying biological and physiological consequences in acute and subacute phases following chemotherapy, suggesting that both apoptosis and edema (i.e., fluid accumulation) occur within a time window of days and weeks post-DOX. In contrast, the irreversible myocardial damage, which is primarily represented by a gradual deposition of reactive fibrosis due to increased deposition of extracellular matrix (ECM), is believed to occur in sub-chronic and later phases following chemotherapy [1–4]. Moreover, only a limited number of clinical studies have related arrhythmia to cardiotoxicity in cancer survivors as a suspected causation of abnormal heart rhythms post-chemotherapy, despite the fact that these aberrant rhythms can degenerate into potentially lethal events such ventricular fibrillation [7].

1.1. Image-Based Methods to Evaluate Chemotherapy-Induced Cardiotoxic Effects

Clinical protocols and diagnostic methods routinely use echocardiographic imaging and have reasonable sensitivity and specificity in comprehensively monitoring and detecting DOX-mediated functional changes [8], while biopsy sampling is invasive and sparse, often missing structurally damaged tissue. Thus, the irreversible injury evolves undetected, worsening in time and leading, within months and years, to cardiomyopathy, and eventually to heart failure. Therefore, there has been a critical need to develop more accurate methods to detect earlier the cardiotoxic effects in sub-chronic phases following DOX-based therapies. With this respect, MR imaging offers excellent tissue contrast. For example, one group recently utilized MRI to assess cardiotoxicity in breast cancer survivors, showing that most patients had significantly different values of ejection fraction and global strain post-DOX compared to the baseline, and some presented evidence of diffuse fibrosis [9]. While 2D Cine and Late Gadolinium Enhancement methods are able to characterize post-DOX functional and structural changes [9], these scans typically use protocols with large slice thickness (~8–10 mm), similarly to those in acute or chronic infarct scars [10]. Current T1 and T2 mapping methods can differentiate, overall, the injured myocardium post-chemotherapy [11]; however, these low-resolution images (voxel size $2 \times 2 \times 10$ mm^3) might miss subtle structural alterations. Thus, advanced methods using high-resolution 3D imaging should be adapted for imaging protocols of post-DOX evaluation.

1.2. Role of Pre-Clinical Animal Models of Cardiotoxicity

Animal models of cardiotoxicity represent a reasonable alternative to clinical investigations, allowing us to perform detailed and controlled studies that can address unanswered clinical questions. For example, image-based longitudinal studies in animals can reveal new biomarkers of adverse myocardial remodeling at different time points following DOX therapy, improving our mechanistic understanding regarding reversible vs. irreversible effects. Such models can be used to test more effective cardioprotective strategies prior to implementation in clinical trials. Furthermore, compared to sparse in vivo biopsy sampling, animal models hold the great advantage of allowing us to study the entire heart after explantation for the purpose of histopathological assessment.

With respect to this, a large body of literature has reported the development of preclinical models of chemotherapy-induced cardiotoxicity. Many chronic models using small animals (e.g., mice [12], rabbits [13], mini-pigs [14]), have clearly demonstrated that the severity of structural myocardial remodeling is dependent on the cumulative dose of the chemotherapeutic drug, leading to mechanical dysfunction with poor ejection fraction and progressive heart failure. Unfortunately, small animal models are less relevant to the clinical translation of diagnostic imaging methods. Thus, more recently, one research group developed a swine model of cardiotoxicity and used MR imaging to demonstrate that the prolongation of T2 parameter (indicating edema) is reversible in early stages post-DOX, and that the LV function declines gradually within the first few months post-DOX [15]. However, in this study, DOX was administered intracoronary through a catheter, an invasive procedure which does not mimic the chemotherapeutic plan in patients. Therefore, alternative delivery methods, such as the intravenous injection of doxorubicin (as done in patients), might be more realistic to study cardiotoxicity in large animal models.

The specific aims of our work here are: (1) to develop a translational swine model to study DOX-induced cardiotoxicity, precisely mimicking the intravenous delivery of doxorubicin in cancer patients; and (2) to establish subtle quantitative MR imaging features and electrophysiological characteristics of DOX-induced cardiotoxic effects in the subchronic phases post-DOX (i.e., weeks and a couple of months after the DOX treatment was ceased). Specifically, in this preclinical work, we propose an MR imaging protocol that allows us to monitor, post-DOX, the temporal evolution of the left ventricular function (i.e., via Cinematic methods) and of the structural alterations (i.e., using high-resolution 3D contrast-enhanced MR to detect deposition of collagenous fibrosis), along with histological validation. In addition, our novel experimental protocol includes a complex X-ray-guided electrophysiological study to evaluate endocardial bipolar voltage maps and arrhythmia inducibility following the completion of DOX treatment.

2. Methods

All preclinical animal experiments were approved by the research ethics board of the Animal Care Committee at our Sunnybrook Research Institute (Toronto), and the DOX-related procedures were performed while ensuring all biohazard safety requirements.

The design of the experimental animal study to evaluate DOX-induced cardiotoxicity included the following: DOX treatment delivered weekly via intravenous injection; functional and structural MRI imaging (pre-DOX and post-DOX delivery) and associated image analysis; X-ray guided electrophysiology studies (voltage mapping and arrhythmia inducibility); and, lastly, histological staining of select myocardial tissue samples. The associated pipeline of our research protocol is illustrated in Figure 1, where each component is described in more detail below.

2.1. Development of a Preclinical Large Animal Model to Study DOX-Induced Cardiotoxicity

In this work, we used $n = 4$ juvenile healthy Yorkshire swine, weighing 20–25 kg prior to the commencement of DOX injections. Note that an additional healthy swine was used as control for the electrophysiology studies and histological validation. The dosage of doxorubicin (i.e., 1 mg/kg) was given based on the pig's weight, as per the typical chemotherapeutic doses. For each delivery, DOX was diluted into a 100 mL bag of saline and administered intravenously (i.v.) either into the ear or using a vascular access port (VAP) designed for large animals. The DOX solution was slowly injected over a 20–30 min period. The four pigs receiving DOX treatment were split into 2 groups: (i) Group 1 (pig #1 and pig #2, respectively), sacrificed at 5 weeks post-DOX; and (ii) Group 2 (pig #3 and pig #4, respectively), sacrificed at 9 weeks post-DOX. This grouping was carried out to enable the validation of longitudinal observations at two time points vs. histology.

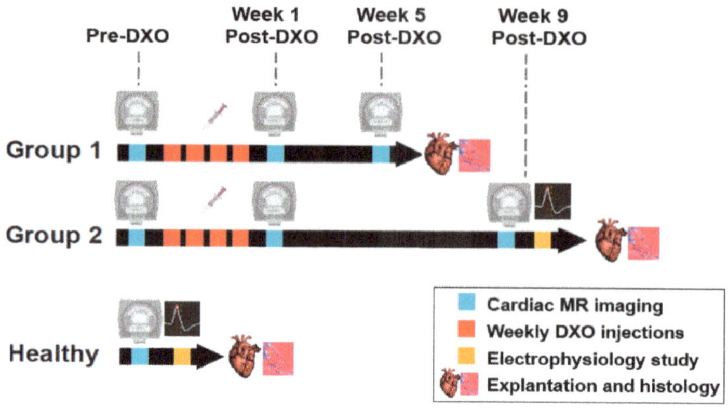

Figure 1. Pipeline of the experimental study to evaluate DOX cardiotoxicity pre- and post-treatment, as indicated by the arrows (the arrowheads point to the end of longitudinal studies).

2.2. In Vivo MR Imaging Protocol and Associated Analysis

MR imaging studies were conducted on a 3T whole body scanner (MR 750, General Electric Healthcare, Waukesha, WI, USA). Prior to imaging (Figure 2a), each animal was sedated using an anesthetic mixture of atropine (0.05 mg/kg) and ketamine (30 mg/kg), and was supported through mechanical ventilation. Anesthesia was maintained with isoflurane/O_2 (1–5%). For image acquisition, an 8-channel cardiac anterior array coil was placed on each pig. The heart rate and associated physiological signals were continuously monitored. The proposed MR imaging protocol presented in this paper included: a short-axis 2D CINE sequence for heart function evaluation, as well as a high-resolution 3D late gadolinium enhancement (LGE) for the identification and quantification of fibrosis. Amiodarone was injected in order to avoid arrhythmic events during the MR imaging study, which kept the heart rate stable and below 100 bpm. The heart rate stabilization improved image acquisition process and, consequently, the quality of reconstructed images.

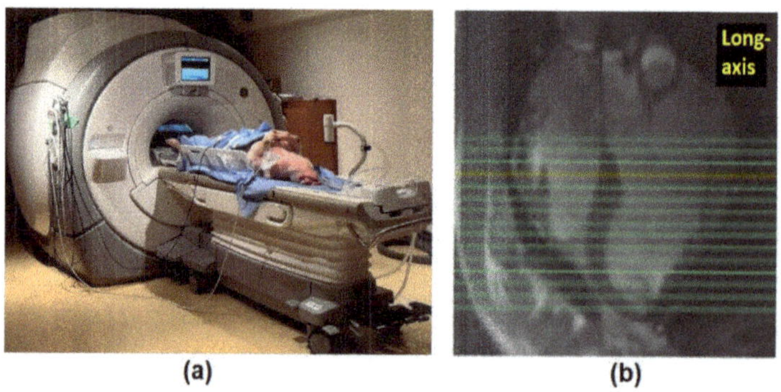

Figure 2. Example of a preclinical MR imaging study: (**a**) experimental set-up with the pig on the table of the 3T MR scanner (prior to placing the animal inside the bore); and (**b**) long-axis image serving for the prescription of the short-axis Cine images.

As per the diagram previously presented in Figure 1, the MR images were acquired at the following time points: (a) at baseline healthy state (pre-DOX injections) in both groups; (b) at one week after the completion of DOX injections in all four animals; (c) at five weeks post-DOX injections in Group 1 (i.e., in pig #1 and pig #2, respectively), and (d) at nine weeks post-DOX injections in Group 2 (i.e., in pig #3 and pig #4, respectively).

For the assessment of cardiac function, we utilized a steady-state free precession (SSFP) sequence in Cine mode, with the following MR parameters: 16–20 short-axis slices to cover the entirety of the heart (prescribed on longitudinal-axis images, as in Figure 2b), 8 views/segment, 20 cardiac phases/slice, repetition time TR= 4.2 ms, echo time TE = 1.8 ms, flip angle = 45°, matrix size = 224 × 160, in-plane resolution of 1 mm × 1 mm, and slice thickness = 5 mm (with no gap between slices). All Cinematic images were acquired using ECG- gating and breath holds.

Contrast-enhanced imaging was performed by employing a free-breathing 3D late gadolinium enhancement (LGE) method, approximately 5–6 min after injecting a bolus of gadolinium-based contrast agent Gd-DTPA (0.2 mmol/kg, Magnevist, Bayer Healthcare Pharmaceuticals, Berlin, Germany). The 3D LGE method with isotropic voxels size was based on a 3D inversion recovery fast gradient echo (IR-FGRE) sequence with fat suppression and respiratory navigation (initial inversion time TI = 300 ms, repetition and echo times TR/TE = 3.5/1.5 ms, bandwidth BW = 100 kHz, flip angle = 15°, and an isotropic spatial resolution of 1.4 mm × 1.4 mm × 1.4 mm), similarly to the method we previously used in preclinical MR imaging studies to evaluate chronic infarct scars [16].

2.3. MR Image Analysis

First, the Cine images were analyzed with the CVI42 software (Circle Cardiovascular Imaging, Calgary) [17], to assess left ventricular function. The endocardial and epicardial contours were semi-automatically delineated, and then corrected by a clinical expert (cardiologist I.R.). Using these contours, we derived the end-systolic and end-diastolic volumes (ESV and EDV, respectively). The ejection fraction (EF) functional parameter was then calculated with the well-known formula: EF(%) = (EDV − ESV)/EDV, for each dataset (i.e., each time-point), allowing us to observe the longitudinal changes over the weeks following the completion post-DOX delivery.

Second, the LGE images were analyzed using in-house custom scripts written in Matlab (Mathworks, Torrance, CA, USA), according to the image analysis pipeline illustrated in Figure 3. Briefly, for each raw MR image, we first performed manual endocardial and epicardial contouring of the LV. A region of interest (ROI) was selected from the remote myocardium on the posterior side. Subsequently, pixel-wise maps of signal intensity within the segmented LV were used as input to an algorithm able to differentiate healthy myocardium vs. fibrotic pixels based on a simple signal intensity thresholding method.

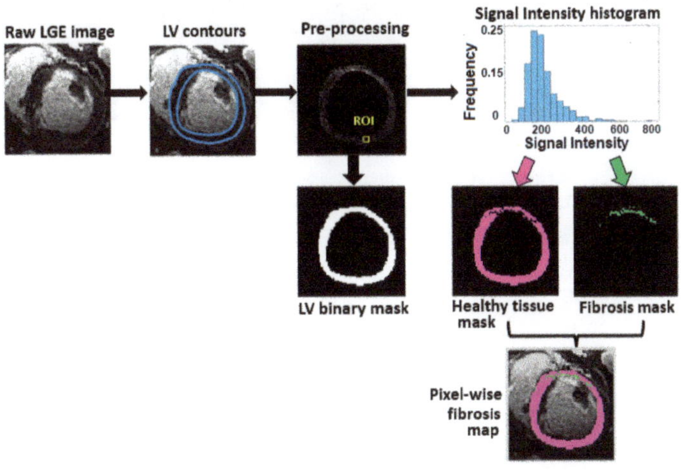

Figure 3. Diagrammatic illustration of the analysis pipeline used to generate pixel-wise fibrosis maps from high resolution LGE images (see text for more details). Note: fibrosis pixels are represented in green and healthy myocardium pixels in magenta.

Specifically, in order to cluster the pixels into the two distinct regions (i.e., fibrosis and healthy tissue), we used a 5 standard deviation (SD) threshold for signal intensity, which is clinically accepted for fibrosis assessment [18]. Moreover, the LV binary masks were used to calculate the LV volume for each heart. At each time-point, for each heart, the density of fibrosis (%) was calculated as the ratio between the fibrosis volume (derived from the total number of segmented fibrosis voxels in each LV) and the LV volume.

2.4. Electrophysiology Studies (Mapping and Arrhythmia Test)

The day following the imaging studies at 9 weeks post-DOX, catheter-based electro-anatomic mapping (EAM) and an arrhythmia inducibility test were performed in pigs #3 and #4, respectively. An additional EAM study and arrhythmia test was performed in a healthy (control) pig. All animals were intubated and sedated, and anesthesia was maintained throughout these EAM procedures using the same combination of medications as those administered prior to the MR imaging scans. However, amiodarone was not administered, since one objectives of the EP study was to evaluate arrhythmia inducibility.

All three interventional electrophysiology EP studies were carried out under X-ray guidance, using a C-arm Toshiba INFINIX VF-I/SP-S (Figure 4a). Catheter-based EAM of the left ventricular (LV) endocardial surface was performed in these animals by employing a conventional CARTO3 electrophysiology system (Biosense Webster Inc, Diamond Bar, CA, USA). For the endocardial mapping procedures, we used a PentaRay® catheter (Biosense Webster Inc., Irvine, CA, USA) inserted into the LV cavity, via femoral access.

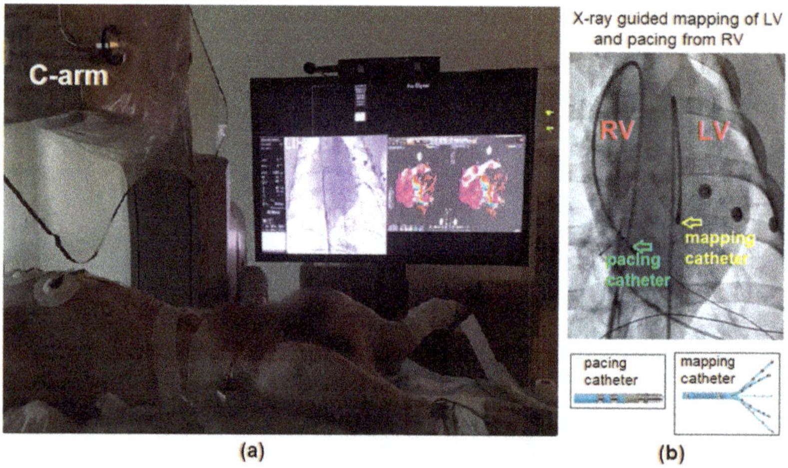

Figure 4. (**a**) Experimental set-up of the EP study (electro-anatomic mapping and arrhythmia study) shown during the intervention in pig #3, at week 9 following the completion of DOX injections; and (**b**) X-ray image of the mapping and pacing catheters inside the heart during the EP study.

The EAMs were primarily acquired for the purpose of constructing detailed bipolar voltage maps (i.e., more than one thousand points per map, in sinus rhythm and under pacing conditions). The bipolar voltages characterized by low amplitude values (0.1–1.5 mV) were attributed to patches of fibrotic tissue, whereas the areas with an amplitude voltage > 1.5 mV were considered normal, in accordance with the clinical threshold typically used to define scarred tissue [19]. The low bipolar voltage areas were qualitatively compared to those defined by LGE using ADAS 3D software, version 2.11.1-beta.2 (www.adas3D.com, accessed on 31 October 2022). Several representative points that had low voltage values (denoting fibrosis) were selected for a qualitative evaluation of the QRS and QRS-T intervals.

For the arrhythmia inducibility test, we inserted an SF Thermocool catheter (Biosense Webster Inc., Irvine, CA, USA) into the right ventricle (RV) and performed rapid pacing of the heart. Figure 4b shows an exemplary X-ray image of the heart during the EP study (in pig #3), with the pacing catheter inside the RV and the mapping catheter (with 5 prongs and multiple sensors on each prong) inside the LV. The external ECGs placed onto the animal torso for reference, are also visible.

2.5. Histology

Except for pigs #3 and #4 (in which death was caused by VF), the other three animals were euthanized as per the approved protocol. All five hearts were carefully explanted, then fixed and preserved in formaldehyde solution for at least 1 week in order to ensure a uniform preservation of tissue. Tissue samples were collected from each heart and select cross-sections (i.e., slices cut at 4 mm thickness) corresponding to short-axis MR images (as guided by anatomical landmarks) were cut at 4 μm thickness. These were mounted on large glass slides and stained with collagen-specific Mason Trichrome stain to visualize the deposition of collagenous/reactive fibrosis. Following staining, the slides were digitally scanned using a special TissueScope (Huron Technologies, St. Jacobs, ON, Canada) that accommodates large pathology slides. These digital images were visualized using Aperio ImageScope [20], an open-source software specifically designed for pathological evaluation of digital images.

2.6. Statistical Analysis

Since this is a methodological protocol paper with the animal model being tested only in a small pilot study, the study was not powered statistically. Therefore, we focused on qualitative analyses and highlighting the longitudinal evolution of individual parameters (EF and fibrosis density) per animal.

3. Results

3.1. Longitudinal Assessment of LV Function from Cine Images

Figure 5a shows examples of LV contours at end diastole (ED) and end systolic (ED) phases in a short-axis cine slice, at different time-points (i.e., pre-DOX, week 1 post-DOX, and week 9 post DOX), taken from pig #3. For qualitative comparison, the slices were selected to be at the approximate same level using anatomical markers (e.g., papillary muscle as well as geometrical shape/features of the left ventricular endocardial wall), taken into consideration that the hearts of these juvenile swine had slightly grown in size between the time points of MRI scans (starting from the first pre-DOX scan and onward).

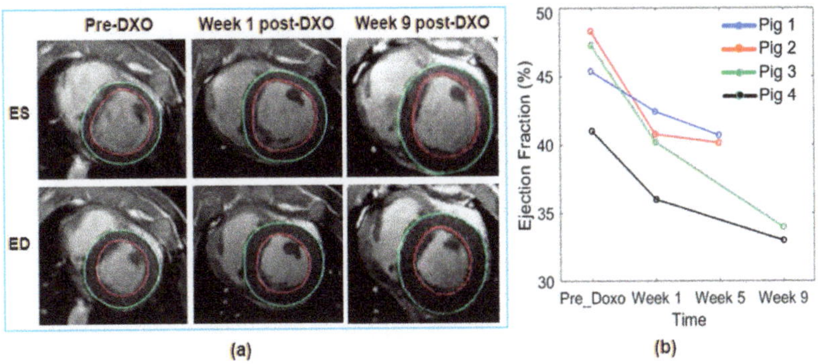

Figure 5. Characterization of the LV function through analysis of Cine MR images: (**a**) example of short-axis slice at mid-heart level, representing the end-systole and end-diastole phases during the cardiac cycle at different time points pre-DOX and post-DOX in pig #3 (ED = end-diastole, ES = end-systole); and (**b**) temporal evolution of the individual EF (%) parameter for each animal.

All animals had normal cardiac function prior to DOX injections, with EF values ranging from 41% to 48%. Overall, the EF values gradually decreased over time in each pig post-DOX. We observed that initially, all EF values clearly decreased by week 1 post-DOX. At week 5 post-DOX, the EF values for both pig #1 and pig #2 decreased, but not below

40%, whereas at week 9 post-DOX, the EF values decreased to less than 35% for both pig #3 and pig #4. The temporal evolution of the functional EF parameter in each animal is plotted in Figure 5b, where the steady DOX-induced decline in cardiac function is obvious on an individual heart basis.

3.2. Longitudinal Assessment of MRI-Defined Fibrosis Density

Figure 6 illustrates exemplary results from the segmentation of LGE images in pig #1 and pig #3 using the 5SD threshold to detect fibrosis, at different post-DOX time-points.

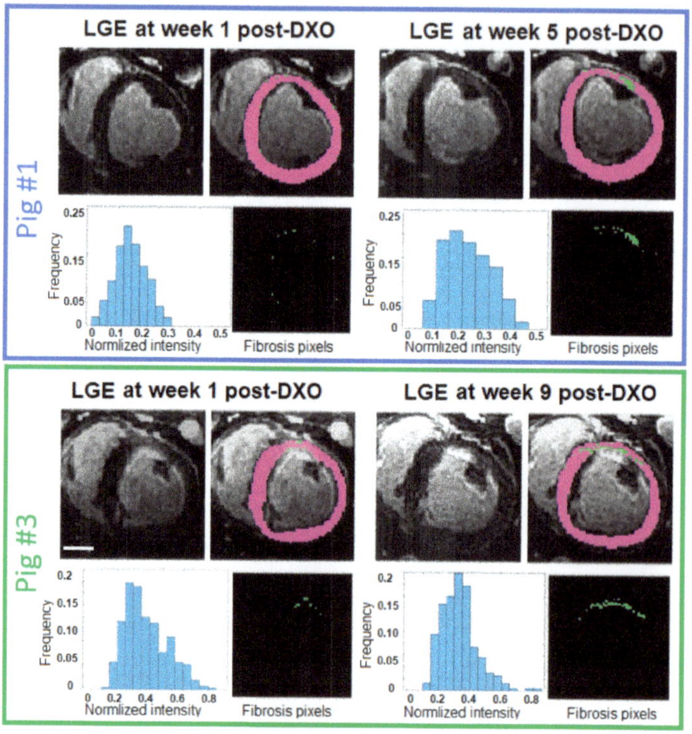

Figure 6. Representative results obtained by segmenting LGE images to distinguish fibrotic pixels (in green) from normal myocardium pixels (in magenta). For the green masks segmented in the MR slices presented, for pig #1, there are 11 fibrosis pixels out of 1003 total pixels of the LV at 1 week post-DOX, and 43 pixels out of 1092 pixels of LV at 5 weeks post-DOX. For pig #3, there are 11 pixels out of 905 LV pixels at 1-week post-DOX, and 65 pixels out of 939 LV pixels at week 9 post-DOX. White scalebar is 2 cm.

The LGE-defined pixels of fibrosis are depicted in green, whereas the healthy myocardium is depicted in magenta. The overall higher values of signal intensity in LV (as observed in the histograms) suggest an overall increase in fibrosis. We also observed that most pixels classified as 'fibrosis' (based on a higher signal intensity compared to the rest of myocardium in the LV), were scattered within the anterior territory and had a typical appearance of diffuse fibrosis in contrast-enhanced MR images.

The results from the estimation of fibrosis density from segmented LGE images are presented in Figure 7, along with representative histological images of collagen-sensitive Masson Trichrome stain. The qualitative increase in the deposition of reactive fibrosis post-DOX observed in the histopathological images was found to be in good agreement with the gradual increase in LGE-defined fibrosis. Figure 7a includes the temporal evolution of LGE-defined fibrosis density (%) post-DOX treatment, plotted individually for each animal.

Figure 7b demonstrates a gradual deposition of reactive interstitial fibrosis (caused by an increased extracellular matrix deposition) using samples selected from the healthy (control) animal, pig #1, and from pig #3. Typically, collagenous tissue stains blue-green, whereas healthy tissue stains dark red. The nuclei stain black, appearing pycnotic (i.e., smaller and condensed) or completely disappearing in the cells within more extensive fibrotic patches. In addition, we also noticed that by week 5 and week 9 post-DOX, respectively, there were no visible areas with edematous tissue or accumulated fluid.

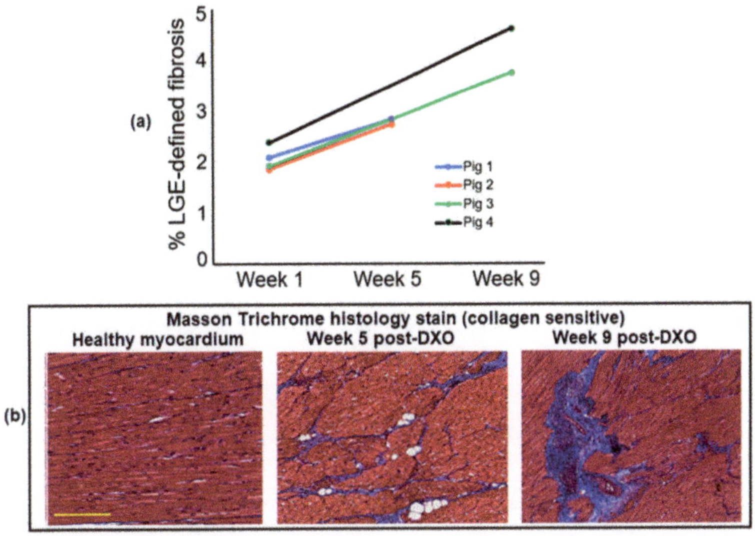

Figure 7. Results from the LGE-based evaluation of fibrosis density (calculated relative to the LV volume) for each animal, along with selected histopathological images denoting increased collagen deposition: (**a**) temporal evolution of LGE-defined fibrosis density from LGE images for each pig; (**b**) examples of Masson Trichrome stained samples from the control animal as well as from samples collected at 5 weeks and 9 weeks post-DOX, showing an increasing deposition of interstitial collagen and reactive fibrosis (in agreement with results obtained from MRI). The yellow scale-bar is 200 μm.

3.3. EP Studies (EAM Mapping and Arrhythmia Inducibility Tests)

Figure 8a–c illustrates examples of endocardial bipolar voltage maps, along with representative intracardiac electrograms (iEGMs), recorded following a retrograde aortic approach: (a) in the healthy control pig; (b) in pig #3; and (c) in pig #4, respectively. The healthy tissue (depicted in magenta) was defined by bipolar voltages with amplitude >1.5 mV, whereas small patches of fibrosis (depicted in dark red and green) had bipolar voltages <1.5 mV, as is typical for the clinical threshold that defines unexcitable fibrotic scars.

Figure 8d,e illustrates the 3D segmented LGE images for pig #3 and pig #4 after registration with endocardial bipolar voltage maps from CARTO3. The LGE-CARTO3 registration was performed in ADAS 3D, using a rigid landmark-based approach with 8 fiducial markers and the endocardial surface (i.e., taking a 5% layer closest to the surface) of the myocardium defined from the 3D LGE images.

Overall, both methods (LGE and CARTO3) revealed small patches of fibrosis scattered on the endocardial surface, especially on the anterior side of these two hearts. However, a precise geometrical correspondence between the locations of patches in the LGE shell vs. those in bipolar voltages cannot be found. This is due to the different spatial resolutions used, with the voxels in LGE images being of small size (1.4 mm × 1.4 mm × 1.4 mm) with the recorded bipolar voltage points several mm apart. The resulting slightly larger fibrotic patches defined by CARTO3 (compared to those seen in the LGE endocardial shells), were likely due to the surface interpolation of the recording points. Furthermore, it is possible that there were a few very small fibrotic points that were missed by the contact catheter, but these likely appeared in the LGE-defined endocardial shell.

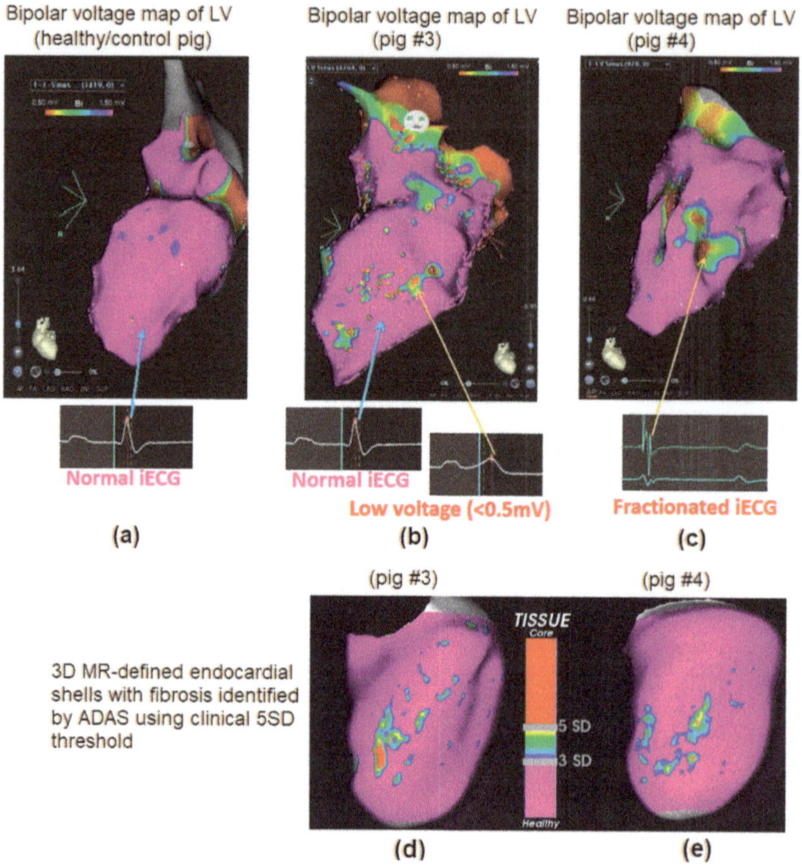

Figure 8. Examples of endocardial bipolar voltage maps, along with representative iEGM waves are presented for the healthy pig (**a**), pig #3 (**b**) and pig #4 (**c**). The location of low bipolar voltages in (**b**,**c**) agreed well with the with the corresponding location of LGE-defined fibrosis using the ADAS 3D software, as illustrated for pig #3 (**d**) and pig #4 (**e**). In both endocardial maps (i.e., from CARTO3 and segmented LGE shells), the fibrotic areas are indicated by red-green, while the healthy tissue is in magenta.

Examples of wave analysis for QRS and QRS-T intervals for a few selected fibrosis points are included in Table 1. Note that the recorded iEGM waves were evaluated for reproducible morphological appearance within three consecutive QRS heart beats prior to the acquisition window, using the 12-lead ECGs as reference. The QRS-T intervals were clearly prolonged for the points of low voltage compared to those in the heathy pig. In addition, the QRS intervals for these points were slightly longer in pig #4, but fluctuated in pig #3.

Table 1. QRS and QRS-T intervals and exemplary iEGM waves for points of low bipolar voltage (<0.5 mV) selected from pig #3 and pig #4, along with reference intervals from healthy pigs.

Animal ID	Intracardiac EGM Waves	Bipolar Voltage (mV)	QRS Interval (ms)	Q-T Interval (ms)
Healthy Ref.	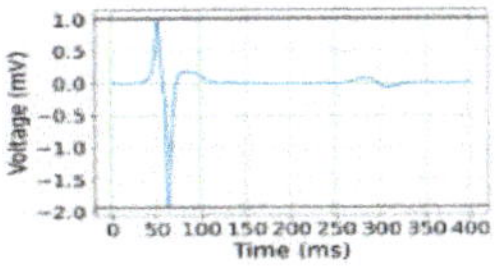	3.00	56	290
Pig 4	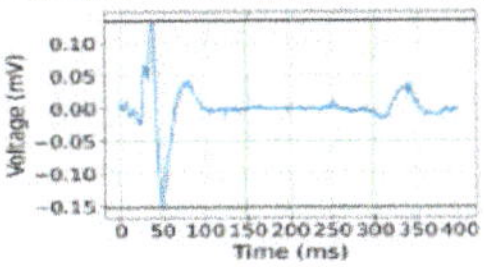	0.29	65	338
Pig 4	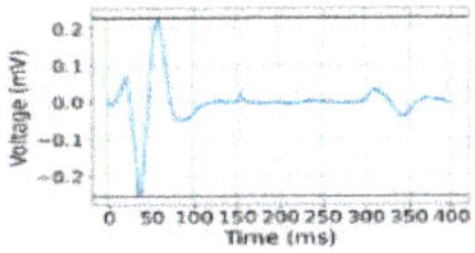	0.48	59	338
Pig 4	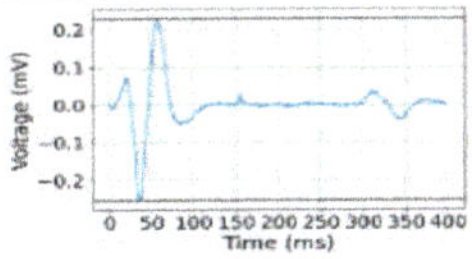	0.33	63	351
Pig 4	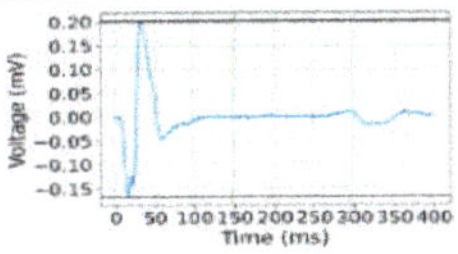	0.37	61	357
Pig 3	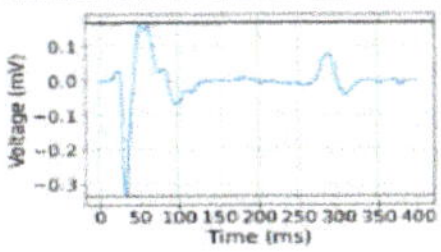	0.5	47	293
Pig 3	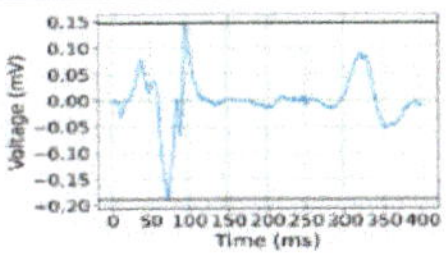	0.34	47	306
Pig 3	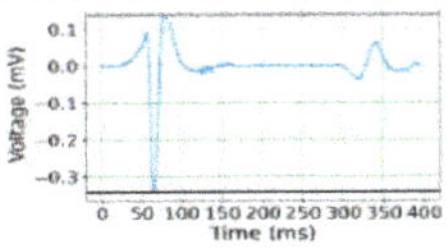	0.48	56	307

Note that CARTO typically saves 2.5 s of recording prior to the moment a bipolar voltage is acquired to be projected on the anatomical shell. In Table 1, we presented the signal captured during the acquisition window, which was set before the QRS and T waves, within the time window of 400 ms during sinus rhythm. In evaluating iEGMs for reproducible morphological appearance in the regions of low voltage identified from registered segmented LGE and bipolar maps, we identified consistent morphological appearance of low voltage intracardiac recordings at the fibrotic regions within 3 QRS beats of the projected bipolar voltage signal, suggesting acceptable reproducibility.

Figure 9 shows examples of recorded iEGMs and reference 12-lead ECGs in pig #3, during: sinus rhythm (650 ms R-R interval), pacing at 500 ms and induced ventricular fibrillation, VF. In both animals in Group 2 (i.e., pig #3 and pig #4) VF was induced via rapid pacing (<300 ms); however, it could not be induced in the healthy control animal, regardless of the pacing rate. These observations, corroborated by the recorded bipolar voltage maps and the MRI-derived fibrosis maps suggested that the small patches of diffuse fibrosis identified in histopathology images acted as small anatomical obstacles and perturbed the normal propagation of electrical waves through the heart. These scattered patches likely broke the propagating waves into smaller spiraling waves, and in conjunction with their corresponding prolonged QRS-T intervals, favored a chaotic electrical propagation specific to fast and lethal VF. Note that these two pigs were not defibrillated, as such a maneuver was out of the scope of this study.

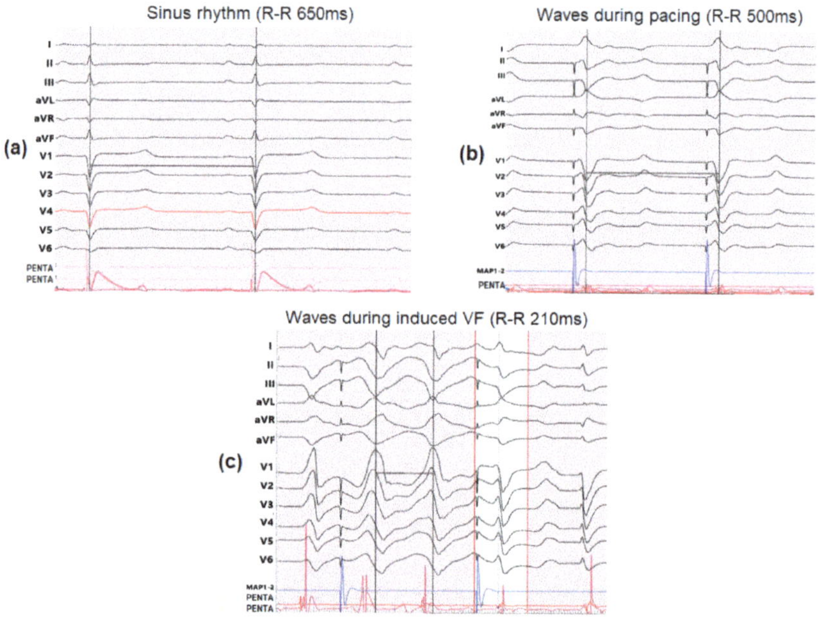

Figure 9. Examples of iEGMs and reference ECG waves in pig #3 at 9 weeks post-DOX, during: (**a**) sinus rhythm; (**b**) pacing from RV at 2 Hz resulting in a cardiac cycle length of 500 ms; and (**c**) induced ventricular fibrillation (VF) that resulted in a fast cycle length of 210 ms.

4. Discussion

Accurate identification of characteristics specific to reversible and irreversible myocardial remodeling post-chemotherapy could help clinicians to assess the long-term effects of DOX therapy and to predict the risk of sudden cardiac death associated with heart failure [21]. A major role in the identification of early cardiotoxic signs (weeks and months post-DOX) can be played by cardiac imaging. Among the various imaging methods nowadays available to characterize cardiac function [22] and structure [9,11], MR imaging may soon be the preferred technique for evaluating myocardial injury post-chemotherapy [23], owing to its robust imaging sequences and excellent tissue contrast.

In this work, we demonstrated the feasibility of a large animal (swine) model to explore functional and subtle structural cardiotoxic changes induced by DOX. As a first novel aspect of our protocol, in distinction to the catheter-based intracoronary delivery employed by other researchers in pig models [14,15], in our pilot study, we successfully used non-invasive intravenous (i.v.) injections of DOX, similarly to the clinical chemotherapeutic approaches. Our histologically validated results clearly demonstrated that reactive fibrosis occurred as a side-effect of the DOX treatment within the first couple of months, following the completion of one 4-week cycle of treatment. The deposition of fibrosis appeared to lead to an irreversible ventricular mechanical dysfunction, although we acknowledge that a monitoring period longer than 9 weeks could better support this observation. With this respect, we suggest that future studies could replace the Yorkshire swine with Yucatan pigs, since the latter animals do not gain weight (in time) and therefore they could better fit in the relatively narrow bore of 3T MR scanners.

A second novel aspect is related to the identification of scattered pixels of fibrosis in the high-resolution 3D LGE images. Compared to clinical 2D MR scans that typically use 8–10 mm slice thickness, our 3D LGE method is superior in overcoming issues associated with partial volume effects, which are problematic particularly in the setting of diffuse fibrosis. Furthermore, our Cine and LGE MR imaging methods are both superior, in terms of spatial resolution, to those used in other preclinical DOX studies. For example, for functional imaging, our voxel size was 1 mm $\times$ 1 mm $\times$ 5 mm= 5 mm^3 (which is four times smaller than the voxel size of 1.8 mm $\times$ 1.8 mm $\times$ 6 mm = 19.6 mm^3 used by Galán-Arriola et al. in [15]). For the LGE method, we used a voxel size of 1.4 mm $\times$ 1.4 mm $\times$ 1.4 mm = 2.74 mm^3, which is smaller compared to the ~3.75 mm^3 voxels used in [15]. The higher the spatial resolution is, the less significant the partial volume effects are; thus, we suggest that our MR imaging methods may provide more accurate functional parameters and structural information. Moreover, the authors of ref. [15] only presented raw LGE images, without specifically identifying the fibrotic pixels using clinically accepted algorithms as in our study.

Our free-breathing, high resolution 3D LGE method appears more adequate for pixel-wise quantitative analysis of collagen density than the 2D T1 mapping images acquired at 5 mm slice thickness [24]. However, LGE imaging is known to be sensitive to contrast injections, and the image analysis based on 5SD threshold is user-dependent (i.e., with respect to the selected ROIs), while T1 mapping techniques are more robust [25]. Given the important role of diffuse fibrosis in post-chemotherapy [26], our future work will focus on implementing a high-resolution 3D T1* mapping method. This method has the capability to distinguish dense collagenous fibrotic patches, as demonstrated in a previous study performed ex vivo in chronically infarcted porcine hearts [27], and has already been translated to preclinical in vivo free-breathing 3D imaging of infarct scars, using a 1.4 mm isotropic resolution [16]. Furthermore, in the imaging protocol, we plan to include T2-based methods to evaluate edema resorption, a reversible side-effect of DOX which we recently observed in [24] (in three animals). Lastly, with respect to cardiac functional assessment, the gradual decline in ejection fraction observed in the current work within the first 9 weeks post-DOX indicated an early sub-acute and sub-chronic occurrence of biomechanical dysfunction, in agreement with the swine study of cardiotoxicity that used intracoronary DOX injections [15]. We suggest that the collagen deposition has substantially contributed to the overall EF decline, and will likely have a critical role in the further evolvement towards heart failure.

Regarding the third novel aspect, to the best of our knowledge, this is the first pre-clinical study to report X-ray guided electro-anatomical mapping of endocardial bipolar voltages in a large animal model post-chemotherapy, as most studies have focused only on longitudinal assessment of cardiac mechanical function. The recorded intracardiac iEGM signals, acquired with a catheter-based clinical systems, revealed small patches of fibrotic tissue with reduced bipolar voltage amplitude, which is typically seen clinically in scarred myocardium post-infarction. Our findings also suggest that early electrical remodeling takes place within the first couple of months post-DOX, with prolonged QRS-T intervals accompanying the low voltages in fibrotic areas. The small patches of dense collagenous fibrosis located within areas of abnormal iEGM waves' morphology likely created unexcitable obstacles in the electrical wave pathway, generating reentrant waves, spiral wave break, and chaotic electrical propagation. This eventually led to lethal ventricular fibrillation during the rapid pacing procedure, similarly to the scar-related ventricular arrhythmias induced in post-infarction. Future work will include the monitoring of spontaneous arrhythmia episodes via MR-compatible implantable cardioverter defibrillators, in order to study the risk of lethal arrhythmia development post-chemotherapy.

We acknowledge that one study limitation is the small number of animals. However, in this work, we aimed to describe in detail the experimental protocol and qualitative results, using a minimum number of animals for tests and respecting the '3Rs principle' in animal research (i.e., replacement, reduction, and refinement). Nonetheless, we plan to expand the cohort in the future by including more animals, which will enable us to give statistical power to the study.

We envision that preclinical large animal models of cardiotoxicity will substantially help researchers better understand the mechanistic effects of cardiotoxicity. Such models can also provide a robust translational platform for testing new individual cardio-protective strategies or a synergistic combination of those [28,29]. These could slow the irreversible myocardial injury and associated dysfunction post-chemotherapy, and restrict further progression towards the heart failure stage. Our future work will also focus on testing 3D virtual models in order to predict the electro-mechanical function post-cardiotoxicity, integrating MRI-defined fibrosis areas and electrical remodeling information. Such in silico computer models can be exploited to virtually predict risk of arrhythmia and impaired mechanical contraction (including the EF index) [30] either for screening of drugs' cardiotoxicity [31] or to design more efficient therapeutic strategies, in conjunction with information provided by early imaging biomarkers and knowledge gained from animal models of cardiotoxicity [32].

5. Conclusions

Overall, this feasibility pilot study suggests that swine models of cardiotoxicity (based on i.v. DOX delivery) can be used for revealing key signs of functional and structural changes via non-invasive MRI, along with the risk of lethal arrhythmia in the sub-chronic phases following DOX injections. Our protocol can be further employed to test various cardioprotective strategies in order to minimize the myocardial injury post-DOX. Notably, MR imaging biomarkers may be able to help predict late cardiotoxic effects of therapy from the early reversible and irreversible alterations (i.e., fibrosis) of myocardial tissue. Thus, such image-based translational large animal models and protocols could play an important role in finding effective solutions to improve the delivery of cancer therapies (e.g., synergistic combinations of anti-cancer drugs with statins, mitigating their optimal combined dose and therapeutic plan), while considerably reducing the progression to heart failure, as well as the mortality and morbidity associated with it.

Author Contributions: Study design: P.L. and M.P.; preclinical animal model development: M.L. (Melissa Larsen) and J.B.; imaging and electrophysiology studies: T.E., M.L. (Melissa Larsen), J.B., P.L. and M.P.; data analysis and figures: P.L., T.E., M.N., M.L. (Mengyuan Li), I.R. and M.P.; draft writing, M.P.; final manuscript, P.L., T.E., M.N., M.L. (Melissa Larsen), J.B., M.L. (Mengyuan Li), I.R. and M.P.; funding secured by M.P. All authors have read and agreed to the published version of the manuscript.

Funding: This research was funded by Canadian Institutes of Health Research (CIHR), Project grant: PJT 153212 awarded to Principal Investigator Dr Mihaela Pop.

Institutional Review Board Statement: This study did not use sensitive data, patient information, or any other identity data. This animal study received ethics approvals.

Informed Consent Statement: This study did not require consent from patients or otherwise.

Data Availability Statement: This work in progress is focused on the development of a research protocol, with the associated methodology being described in detail for study reproducibility and the preclinical model being tested only in a small cohort of animals. The collection of more data (for statistical power) has not been completed; thus, the authors cannot make the current data available until after the full completion of the study.

Acknowledgments: The authors would like to thank Adebayo Adeeko for processing the explanted hearts and for staining the histology slides, Xiuling Qi for assistance with initial imaging scans (both from Sunnybrook Research Institute, Toronto, ON, Canada), and to Susan Camilleri (The Center for Phenogenomics, Toronto, ON, Canada) for help with histopathological interpretation of Masson Trichrome slides.

Conflicts of Interest: The authors declare no conflict of interest.

References

1. Kostakou, P.M.; Kouris, N.T.; Kostopoulos, V.S.; Damaskos, D.S.; Olympios, C.D. Cardio-oncology: A new and developing sector of research and therapy in the field of cardiology. *Heart Fail. Rev.* **2018**, *24*, 91–100. [CrossRef] [PubMed]
2. Von Hoff, D.D.; Rozencweig, M.; Piccart, M. The cardiotoxicity of anticancer agents. *Semin. Oncol.* **1982**, *9*, 23–33. [PubMed]
3. Christiansen, S.; Autschbach, R. Doxorubicin in experimental and clinical heart failure. *European. J. Cardiothorac. Surg.* **2006**, *30*, 611–616. [CrossRef] [PubMed]
4. Podyacheva, E.Y.; Kushnareva, E.A.; Karpov, A.A.; Toropova, Y.G. Analysis of models of doxorubicin-induced cardiomyopathy in rats and mice: A modern view from the perspective of the pathophysiologist and the clinician. *Front. Pharmacol.* **2021**, *12*, 670479. [CrossRef] [PubMed]
5. Angsutararux, P.; Luanpitpong, S.; Issaragrisil, S. Chemotherapy-induced cardiotoxicity: Overview of the roles of oxidative stress. *Oxid. Med. Cell Longev.* **2015**, 795602. [CrossRef] [PubMed]
6. Varga, Z.V.; Ferdinandy, P.; Liaudet, L.; Pacher, P. Drug-induced mitochondrial dysfunction and cardiotoxicity. *Am. J. Physiol. Heart Circ. Physiol.* **2015**, *309*, H1453–H1467. [CrossRef] [PubMed]
7. Herrmann, J. Adverse cardiac effects of cancer therapies: Cardiotoxicity and arrhythmia. *Nat. Rev. Cardiol.* **2020**, *17*, 474–502. [CrossRef]
8. Kang, Y.; Scherrer-Crosbie, M. Echocardiography Imaging of Cardiotoxicity. *Cardiol. Clin.* **2019**, *37*, 419–427. [CrossRef]
9. Safaei, A.M.; Kamangar, T.M.; Asadian, S.; Rezaeian, N.; Esmati, E.; Kolahdouzan, K.; Hosseini, L.; Lashkari, M.; Jafari, F.; Hashemi, F.A. Detection of the early Cardiotoxic effects of doxorubicin-containing chemotherapy regimens in patients with breast cancer through novel cardiac magnetic resonance imaging: A short-term follow-up. *J. Clin. Imaging Sci.* **2021**, *11*, 33. [CrossRef]
10. Roifman, I.; Ghugre, N.; Zia, M.; Zavodni, A.; Pop, M.; Connelly, K.; Wright, G. Assessment of the longitudinal changes in infarct heterogeneity. *BMC-Cardiovasc. Disord.* **2016**, *16*, 198. [CrossRef]
11. Tahir, E.; Azar, M.; Shihada, S.; Seiffert, K.; Goy, Y.; Beitzen-Heineke, A.; Molwitz, I.; Muellerleile, K.; Stehning, C.; Schön, G.; et al. Myocardial injury detected by T1 and T2 mapping on CMR predicts subsequent cancer therapy-related cardiac dysfunction in patients with breast cancer treated by epirubicin-based chemotherapy or left-sided RT. *Eur. Radiol.* **2022**, *32*, 1853–1865. [CrossRef] [PubMed]
12. Zeiss, C.J.; Gatti, D.M.; Toro-Salazar, O.; Davis, C.; Lutz, C.M.; Spinale, F.; Stearns, T.; Furtado, M.B.; Churchill, G.A. Doxorubicin-induced cardiotoxicity in collaborative cross (CC) mice recapitulates individual cardiotoxicity in humans. *G3 Genes | Genomes | Genet.* **2019**, *9*, 2637–2646. [CrossRef] [PubMed]
13. Hong, Y.J.; Park, H.S.; Park, J.K.; Han, K.; Park, C.H.; Kim, T.K.; Yoo, S.J.; Lee, J.Y.; Kim, P.K.; Hur, J.; et al. Early detection and serial monitoring of anthracycline-induced cardiotoxicity using T1-mapping cardiac magnetic resonance imaging: Animal study. *Sci. Rep.* **2017**, *7*, 2663. [CrossRef]
14. Balosetti, C.; Curnier, D.; Heon, H.; Dahdah, N.; Cheriet, F.; Friedrich, M.; Perie, D. Early detection of doxorubicin induced cardiotoxicity in the swine by cardiac MRI. *Can. J. Cardiol.* **2016**, *32*, S300. [CrossRef]
15. Galán-Arriola, C.; Lobo, M.; Vílchez-Tschischke, J.P.; López, G.J.; de Molina-Iracheta, A.; Pérez-Martínez, C.; Agüero, J.; Fernández-Jiménez, R.; Martín-García, A.; Oliver, E.; et al. Serial magnetic resonance imaging to identify early stages of anthracycline-induced cardiotoxicity. *J. Am. Coll. Cardiol.* **2019**, *73*, 779–791. [CrossRef] [PubMed]
16. Zhang, L.; Lai, P.; Pop, M.; Wright, G.A. Accelerated multicontrast volumetric imaging with isotropic resolution for improved peri-infarct characterization using parallel imaging, low-rank and spatially varying edge-preserving sparse modeling. *Magn. Reson.* **2018**, *79*, 3018–3031. [CrossRef]
17. Available online: https://www.circlecvi.com/ (accessed on 1 April 2021).
18. Rashed, K.; Pranav, B.; Piet, C.; Housden, R.J.; Zhong, C.; Zahr, K.; Hyon-Mok, S.; Laura, R.; Sergio, V.; Xènia, V.; et al. Evaluation of state-of-the-art segmentation algorithms for left ventricle infarct from late Gadolinium enhancement MR images. *Med. Image Anal.* **2016**, *30*, 95–107.
19. Oduneye, S.O.; Pop, M.; Biswas, L.; Ghate, S.; Flor, R.; Ramanan, V.; Barry, J.; Celik, H.; Crystal, E.; Wright, G.A. Postinfarction ventricular tachycardia substrate characterization: A comparison between late enhancement magnetic resonance imaging and voltage mapping using an MR-guided electrophysiology system. *IEEE Trans. Biomed. Eng.* **2013**, *60*, 2442–2449. [CrossRef]
20. Available online: https://aperio-imagescope.software.informer.com/ (accessed on 1 March 2021).
21. Steinherz, L.J.; Steinherz, P.G.; Tan, C.T.; Heller, G.; Murphy, M.L. Cardiac toxicity 4 to 20 years after completing anthracycline therapy. *JAMA* **1991**, *266*, 1672–1677. [CrossRef]
22. Panis, V.; Donal, E. Imaging techniques for cardiac function. *Appl. Sci.* **2021**, *11*, 10549. [CrossRef]

23. Wei, X.; Lin, L.; Zhang, G.; Zhou, X. Cardiovascular magnetic resonance imaging in the early detection of cardiotoxicity induced by cancer therapies. *Diagnostics* **2022**, *12*, 1846. [CrossRef] [PubMed]
24. Lin, P.; Escartin, T.; Ng, M.; Li, M.; Larsen, M.; Barry, J.; Roifman, I.; Pop, M. Novel imaging biomarkers to evaluate heart dysfunction post-chemotherapy: A preclinical MRI feasibility study. In *Lecture Notes in Computer Science*; Springer: Berlin/Heidelberg, Germany, 2022; Volume 13131, pp. 29–37.
25. Jordan, J.H.; Vasu, S.; Morgan, T.M.; D'Agostino, R.B., Jr.; Meléndez, G.C.; Hamilton, C.A.; Arai, A.E.; Liu, S.; Liu, C.Y.; Lima, J.A.; et al. Anthracycline-associated T1 mapping characteristics are elevated independent of the presence of cardiovascular comorbidities in cancer survivors. *Circ. Cardiovasc. Imag* **2016**, *9*, e004325. [CrossRef] [PubMed]
26. Meléndez, G.C.; Gregory Hundley, W. Is myocardial fibrosis a new frontier for discovery in cardiotoxicity related to the administration of anthracyclines? *Circul. Cardiovas. Imag.* **2016**, *9*, 12. [CrossRef] [PubMed]
27. Pop, M.; Ramanan, V.; Yang, F.; Zhang, L.; Newbigging, S.; Ghugre, N.R.; Wright, G.A. High-resolution 3-D T1*-mapping and quantitative image analysis of gray zone in chronic fibrosis. *IEEE Trans. Biomed. Eng.* **2014**, *61*, 2930–2938. [CrossRef] [PubMed]
28. Octavia, Y.; Tocchetti, C.G.; Gabrielson, K.L.; Janssens, S.; Crijns, H.J.; Moens, A.L. Doxorubicin-induced cardiomyopathy: From molecular mechanisms to therapeutic strategies. *J. Mol. Cell. Cardiol.* **2012**, *52*, 1213–1225. [CrossRef] [PubMed]
29. Dragojevic, S.; Ryu, J.S.; Hall, M.E.; Raucher, D. Targeted drug delivery biopolymers effectively inhibit breast tumor growth and prevent doxorubicin-induced cardiotoxicity. *Molecules* **2022**, *27*, 3371. [CrossRef]
30. Marchesseau, S.; Delingette, H.; Sermesant, M.; Ayache, N. Fast parameter calibration of a cardiac electromechanical model from medical images based on the unscented transform. *Biomech. Model Mechanobiol.* **2013**, *12*, 815–831. [CrossRef]
31. Yuan, Y.; Bai, X.; Luo, C.; Wang, K.; Zhang, H. The virtual heart as a platform for screening drug cardiotoxicity. *Br. J. Pharmacol.* **2015**, *172*, 5531–5547. [CrossRef]
32. Park, C.J.; Brank, M.E.; Vasu, S.; Melendez, G.C. The Role of Cardiac MRI in Animal Models of Cardiotoxicity: Hopes and Challenges. *J. Cardiovasc. Transl. Res.* **2020**, *13*, 367–376. [CrossRef]

applied sciences

MDPI

Article

Multi-View 3D Transesophageal Echocardiography Registration and Volume Compounding for Mitral Valve Procedure Planning

Patrick Carnahan [1,2,*], John Moore [1], Daniel Bainbridge [3], Elvis C. S. Chen [1,2,4,5] and Terry M. Peters [1,2,4]

1 Imaging Laboratories, Robarts Research Institute, London, ON N6A 5B7, Canada; jmoore@robarts.ca (J.M.); chene@robarts.ca (E.C.S.C.); tpeters@robarts.ca (T.M.P.)
2 School of Biomedical Engineering, Western University, London, ON N6A 3K7, Canada
3 Department of Anesthesiology, London Health Sciences Centre, Western University, London, ON N6A 3K7, Canada; daniel.bainbridge@lhsc.on.ca
4 Department of Medical Biophysics, Western University, London, ON N6A 3K7, Canada
5 Lawson Health Research Institute, London, ON N6A 3K7, Canada
* Correspondence: pcarnah@uwo.ca

Abstract: Three-dimensional ultrasound mosaicing can increase image quality and expand the field of view. However, limited work has been done applying these compounded approaches for cardiac procedures focused on the mitral valve. For procedures targeting the mitral valve, transesophageal echocardiography (TEE) is the primary imaging modality used as it provides clear 3D images of the valve and surrounding tissues. However, TEE suffers from image artefacts and signal dropout, particularly for structures lying below the valve, including chordae tendineae, making it necessary to acquire alternative echo views to visualize these structures. Due to the limited field of view obtainable, the entire ventricle cannot be directly visualized in sufficient detail from a single image acquisition in 3D. We propose applying an image compounding technique to TEE volumes acquired from a mid-esophageal position and several transgastric positions in order to reconstruct a high-detail volume of the mitral valve and sub-valvular structures. This compounding technique utilizes both fully and semi-simultaneous group-wise registration to align the multiple 3D volumes, followed by a weighted intensity compounding step based on the monogenic signal. This compounding technique is validated using images acquired from two excised porcine mitral valve units and three patient data sets. We demonstrate that this compounding technique accurately captures the physical structures present, including the mitral valve, chordae tendineae and papillary muscles. The chordae length measurement error between the compounded ultrasound and ground-truth CT for two porcine valves is reported as 0.7 ± 0.6 mm and 0.6 ± 0.6 mm.

Keywords: compounded echocardiography; volume stitching; 3D registration; mosaicing; 3D TEE; mitral valve; monogenic signal

Citation: Carnahan, P.; Moore, J.; Bainbridge, D.; Chen, E.C.S.; Peters, T.M. Multi-View 3D Transesophageal Echocardiography Registration and Volume Compounding for Mitral Valve Procedure Planning. *Appl. Sci.* **2022**, *12*, 4562. https://doi.org/10.3390/app12094562

Academic Editors: Mihaela Pop and Cristian A. Linte

Received: 14 April 2022
Accepted: 29 April 2022
Published: 30 April 2022

1. Introduction

Three-dimensional (3D) ultrasound imaging is used extensively as a diagnostic and guidance tool for cardiac procedures. Three-dimensional echocardiography allows for the acquisition of volumetric data of the heart, which can be analysed in any plane. The current standard of care for mitral valve procedures includes diagnostic imaging with a 3D transesophageal echocardiography (TEE) probe [1,2]. This method of imaging provides a clear view of the mitral valve, and including color Doppler allows the cardiologist/cardiac surgeon to identify the mitral valve pathology. While echocardiography is a powerful imaging technique, it nevertheless has some major limitations. The field of view is limited when using 3D transducers, which can limit the range of anatomy that can be easily viewed, and structures further away from the image probe may suffer from poor spatial resolution, while thin structures parallel to the ultrasound beam suffer from signal dropout artefacts.

Ultrasound compounding, or mosaicing, has been proposed by several groups to address limitations of 3D ultrasound and improve imaging capability. By registering and blending together adjacent acquisitions from different poses, we can expand the field of view and address the issue of signal dropout, producing higher quality images with greater information for the clinician. Several image compounding techniques have been proposed to register a set of ultrasound volumes, all of which demonstrate improved image quality [3] and provide an avenue for combining common cardiac ultrasound views into a single volume with reduced noise, reduced speckle, and fewer signal dropout artefacts. Researchers have demonstrated 3D ultrasound compounding techniques with applications in cardiac, fetal and breast imaging [4–6]. Common across all compounding methods are two critical steps: global registration of all volumes and blending the overlapping regions of the registered volumes to generate the resulting image [3]. Evaluation of registration frameworks has identified three main approaches, consisting of sequential alignment, semi-simultaneous and fully simultaneous registration [7]. Using a sequential alignment approach, each acquisition is registered to the next; however, this technique suffers from drift and error accumulation. The semi-simultaneous approach uses each volume as the moving volume in turn and every other volume as fixed for multiple cycles until convergence is met. This approach balances computational complexity, as only the parameters of a single transform need to be considered in optimization; furthermore, it has global alignment because every volume is considered at every step. The final approach, fully simultaneous group-wise registration, optimizes the transformation parameters of all volumes simultaneously, applying a loss function as the sum of pairwise losses. This approach is optimal for registration quality; however, it is limited by computational complexity due to the number of parameters that need to be optimized.

For mitral valve imaging using standard en-face views, the limitations of 3D ultrasound result in the structures beyond the valve, including the chordae tendineae, papillary muscles, and left-ventricular outflow tract being difficult to identify. While imaging from a different position (e.g., transgastric view) can capture these structures, the field of view limits the utility of these images as at these positions the entire mitral valve apparatus cannot be captured in a single volume. Currently, choosing the treatment plan that provides the greatest benefit to the patient is one of the biggest clinical challenges for cardiologists and cardiac surgeons [8]. Determining optimal neochord length is one of the main issues that cardiac surgeons must address in MV repair procedures [9], and it is particularly challenging to define this length due to a general lack of accurate anatomical information from standard diagnostic imaging, which includes only the en-face view of the mitral valve. Chordae tendineae length measurements are required for mitral valve procedures involving the implantation of artificial chordae [10]. It is nevertheless challenging to determine their optimal length because direct observation is limited to physical measurements made inside the flaccid heart during surgery, along with 2D transgastric long-axis images which require that the image plane be aligned with the entire chord to achieve accurate results [11]. While 3D transgastric TEE ultrasound is safe and easily acquired during routine TEE imaging (adding approximately 3–5 min to the procedure), it is rarely employed due to the limited visibility of the leaflets and field-of-view limitations preventing the entire length of the chordae from being captured. Without the ability to see the entire subvalvular apparatus in the same image data as the standard en-face view of the leaflets, transgastric image information is of very limited clinical value.

In our prior work on volume compounding using TEE volumes, we explored a workflow using only the semi-simultaneous registration strategy and a distance-based weighted average blending [12]. Our prior approach was able to successfully produce compounded volumes; however, the registration component of the workflow was not robust, could fail depending on the order in which the volumes were processed, and the blending approach induced imaging artefacts. In this study, we propose an improved method to register and compound transgastric and en-face volumes utilizing a combination of semi and fully simultaneous registration with a novel weighting function for blending overlapping regions to reduce compounding artefacts. This provides an avenue for combining common cardiac ultrasound views into a single volume with reduced noise and fewer dropout artefacts. Many image compounding techniques involve the use of a tracked probe and were targeted at combining multiple transthoracic views [6]. Our method differs from these previous methods as it does not require any external tracking of the ultrasound probe, and it has been tailored for use with TEE probes to combine 3D mid-esophageal and transgastric volumes that can be acquired as part of a standard diagnostic imaging session. In the en-face volumes, the mitral valve is clearly visible, and in the transgastric views, the chordae are very clear, as these views are nearly perpendicular to each other. By combining both the en-face and transgastric views, we can maintain optimal imaging for both structures in a single compounded volume. Integrated leaflet and chordae geometry in a single volume will greatly improve the cardiac surgeons' ability to accurately measure the length of individual chordae (a crucial factor in neochordae repair techniques [9]) and plan their repair strategy.

2. Materials and Methods

2.1. Image Registration

Following local REB approval, we adapted standard diagnostic TEE acquisition protocols to include multiple transgastric views in addition to the standard mid-esophageal view. Volumes were acquired using ECG gating to match the cardiac phase. Our imaging protocol requires a minimum of one mid-esophageal acquisition and four transgastric acquisitions with approximately 80% spatial overlap or more between adjacent volumes for successful registration of the acquisitions, an example of which can be seen in Figure 1. The transgastric acquisitions should begin at the mitral valve and proceed along the ventricle to the papillary muscles. Compounding is then accomplished by aligning over-sampled data through automated image registration, re-sampling the aligned volumes into a consistent output space, and generating the output image through the voxel-wise blending of the overlapping volumes. The compounding workflow is shown in Figure 2.

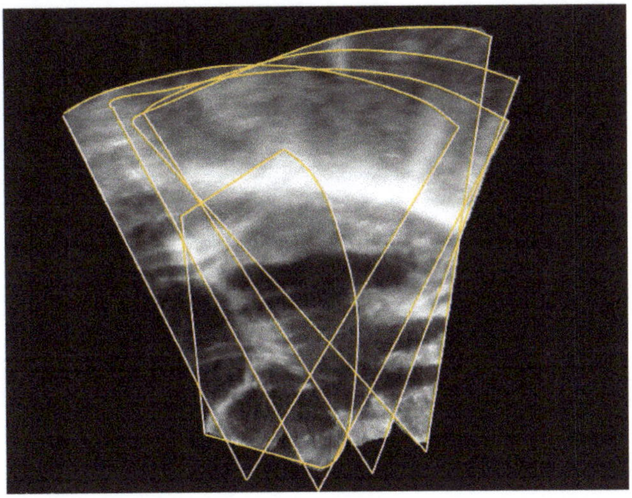

Figure 1. Compounded TEE volume of mitral valve with individual acquisitions outlined.

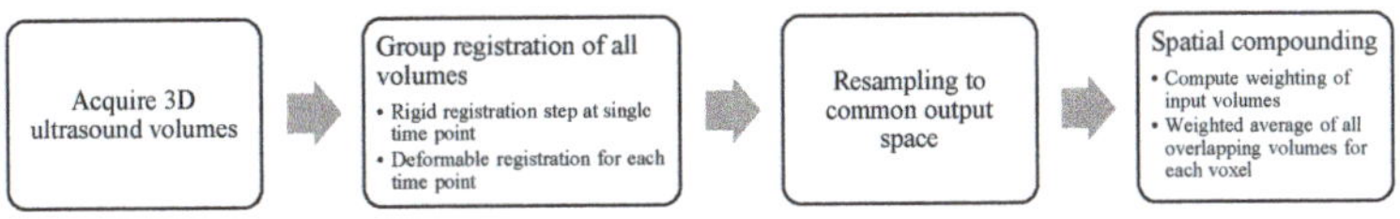

Figure 2. Workflow of TEE compounding for the mitral valve.

Performing image registration of multiple volumes can be achieved using pairwise, fully simultaneous, or semi-simultaneous approaches [13]. We implemented both the semi-simultaneous and fully simultaneous approaches described by Wachinger et al. [13]. This is done as an extension of our initial work on this method, in which only the semi-simultaneous approach was used [12]. We first perform rigid registration at the end-systole between all volumes using fully simultaneous group-wise registration with the sum of pairwise normalized cross-correlations as the loss function. This gives us a rough global alignment of the volumes and is not dependent on input order. Then, two cycles of semi-simultaneous registration are performed, which we found achieved better agreement between volumes than the fully simultaneous registration alone. Finally, we utilize non-rigid registration in the semi-simultaneous framework at each frame in the acquisitions to account for the slight deviations at different points of the cardiac cycle due to imperfect synchronization. For both semi-simultaneous steps, the loss function used was the sum of normalized cross-correlation between the moving volume and each fixed volume. At each step, optimization was performed using adaptive stochastic gradient descent in a multi-resolution registration framework with four resolution levels, each smoothing the image by a factor of 2 over the previous one. We implemented this approach using the Elastix toolkit (http://elastix.isi.uu.nl/, accessed on10 October 2021) on the 3D Slicer platform (https://www.slicer.org/, accessed on 10 October 2021). This open-source implementation of our work is available at https://github.com/pcarnah/CardiacVolumeStitching, accessed on 10 October 2021.

2.2. Image Blending

After the volumes are registered, re-sampling and compounding are performed at each cardiac phase to construct a 3D + time compounded volume with a wide field of view. Before compounding, the volumes are re-sampled to a common grid using cubic b-spline interpolation to ensure that there is complete voxel overlap between volumes so that the blending step (a weighted average of all overlapping volumes at each voxel location) can be performed. The output grid is determined by the extent of all overlapping input images, using isotropic spacing equal to the minimum spacing in any dimension in any input image.

We evaluated multiple weighting strategies including the voxel-wise maximum, average and weighted average using two different weighting schemes. The voxel weighting methods we compared were the scaled distance from the image probe and a combination of distance from the probe and a feature detector based on the monogenic signal [14]. The local phase measure from the monogenic signal was previously demonstrated as part of an application-specific loss function for ultrasound compounding [15]. For our purposes, we used the oriented symmetry measure, which returns values from -1 to $+1$, with positive values being associated with features of interest and negative values being associated with background noise, as visualized in Figure 3. We implemented the 3D extension of the monogenic signal in Python. The distance function assigns higher weights to voxels closer to the image probe, with values ranging from 25 near the probe down to 0 with an inverse-square law dropoff, which we used to approximate the reduction in resolution further away from a phased-array ultrasound probe. For a single voxel position p in a source volume i, the expression for the distance from the image probe weighting d_i is

$$d_i(p) = 25\frac{\left(10 - \frac{10\times\|p-o_i\|}{\max\limits_{q\in V_i}(\|q-o_i\|)}\right)^2}{10^2}, \tag{1}$$

where o_i is the position of the probe origin in volume i. The oriented symmetry weighting S_i is given as

$$S_i(p) = \begin{cases} 10 \times s_i(p) & \text{if } s_i(p) < 0 \\ 25 \times s_i(p) & \text{if } s_i(p) \geq 0 \end{cases}, \tag{2}$$

where s_i is the value of the oriented symmetry measure from the monogenic signal at position p in volume i. The combined weighting function W_i of distance and oriented symmetry is

$$W_i(p) = \max(0.5, S_i(p) + d_i(p)), \tag{3}$$

giving the sum of the two weights with a minimum value of 0.5. Both the distance weight d_i and symmetry weight S_i have a maximum value of 25, contributing equally to the overall voxel weighting. Finally, the expression for the final weighted average output intensity $F(p)$ is

$$F(p) = \frac{\sum_{V_i|V_i(p)>0} (V_i(p) \times w_i(p))}{\sum_{V_i|V_i(p)>0} (w_i(p))}, \tag{4}$$

where $V_i(p)$ is the intensity value in volume i at position p, and w_i is either the distance weighting alone or the combined distance and oriented symmetry weighting. A visual comparison of the results of applying the different blending approaches can be seen in Figure 4. The voxel-wise maximum approach produces a volume with very sharp features but passes through any imaging artefacts and highlights registration errors. Simple averaging produces a smoother image but lacks the definition of smaller features and boundary edges appear blurred. Distance weighted averaging further improves image quality, taking advantage of the increased line density nearer to the probe and the corresponding increase in spatial resolution, but small features and edge boundaries are still blurred and lack contrast to the background. The incorporation of the monogenic signal-based feature detector into the weighting function helps to reduce this blurring and makes the structures of interest more distinct without amplifying imaging and registration artefacts.

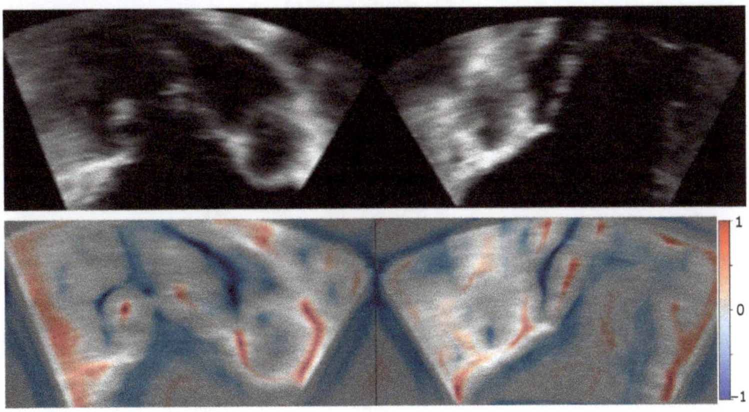

Figure 3. Original image (**top**), with oriented symmetry measure from monogenic signal (**bottom**).

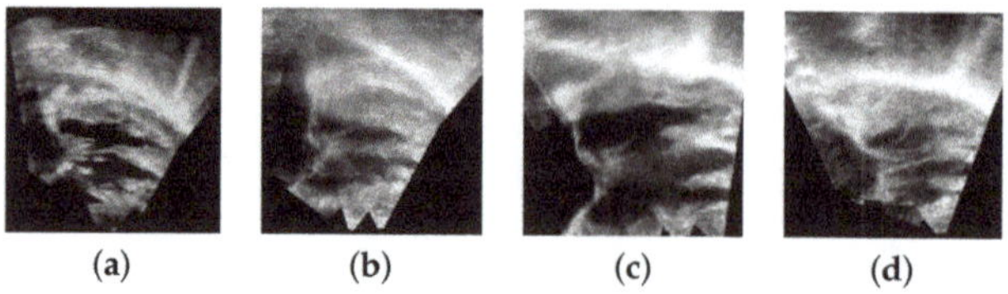

Figure 4. Results of blending functions max (a), average (b), distance weighted (c), and oriented symmetry plus distance weighted (d).

2.3. Data Acquisition

Three patients were imaged using our acquisition protocol under REB approval, using the Philips Epiq TEE system. These image sets were then registered and combined with each of the four blending approaches. Visual inspection of the resulting volumes was performed by an echo-cardiography specialist, to verify apparent anatomical correctness, image quality and clinical value. We validated the geometrical accuracy of this volume compounding approach on two excised porcine mitral valve units, shown in Figure 5. These valves were imaged using a Philips Epiq system with an X8-2T TEE probe, with volumes being captured sequentially from a mid-esophageal point, along a 3D printed path simulating the esophagus to a transgastric position. The valve was also stained with iodine and imaged with a cone-beam CT scanner (Medtronic O-Arm, Medtronic Canada, Brampton, ON, Canada) to provide ground truth data. As shown in Figure 6, the ultrasound volumes were compounded using our described registration approach with the monogenic signal-based blending method. Linear measurements were then made of the visible chordae structures in both volumes.

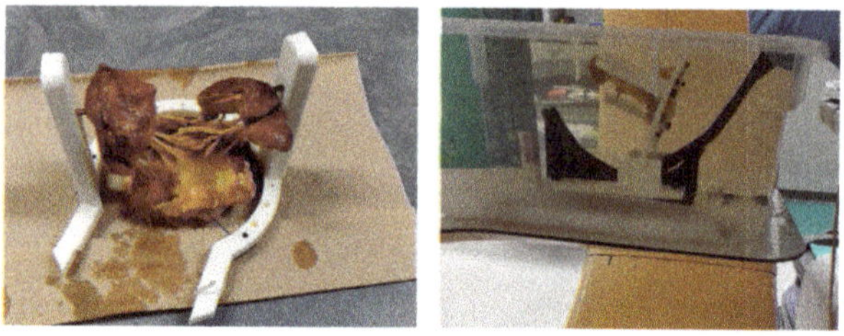

Figure 5. Excised porcine valve stained in iodine. Pictured on right is the valve being imaged using a TEE probe from a transgastric position.

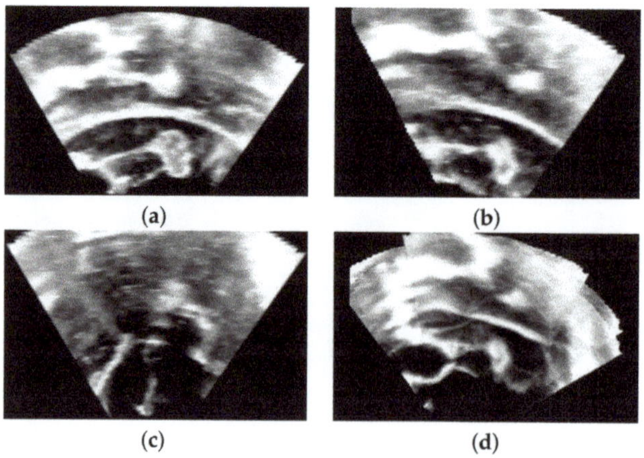

Figure 6. Original volumes simulating transgastric (a,b) and mid-esophageal (c) views of porcine valve unit. Resulting volume from compounding (d).

3. Results

3.1. Porcine Model

The compounded volumes visually replicated the anatomical structures visible in the ground truth CT scan. As shown in Figure 7, the mitral valve leaflets, papillary muscles and individual chordae are clearly visible in the compounded volume. The compounded volume and CT are compared in Table 1 for each valve. For both volumes, four chordae that were easily visible in the compounded echo and CT were measured from the papillary muscle tip to the leaflet insertion point, and the average absolute difference between US and CT length measurements was computed. The measured chordae lengths ranged from 22.0 mm to 36.0 mm.

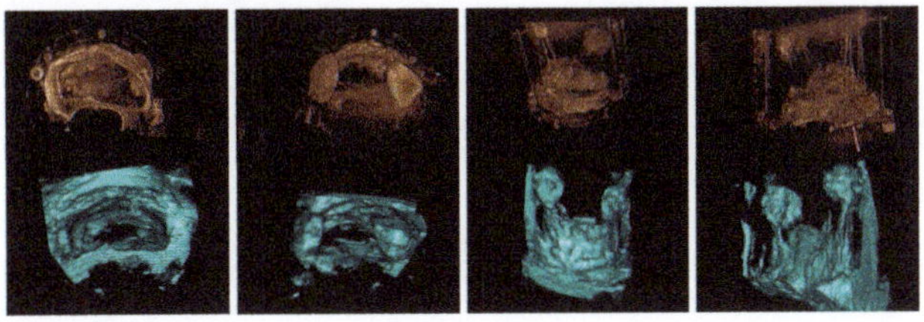

Figure 7. Side-by-side volume rendered comparisons from multiple view points of the CT data (**top**) and compounded echo (**bottom**).

Table 1. Volume comparison metrics between compounded echo and CT.

Excised Valve	Chordae Measurement Absolute Difference (mm)
Valve 1	0.7 ± 0.6
Valve 2	0.6 ± 0.6

3.2. Patient Images

We processed image volumes acquired from two patients to create compounded volumes that were visually inspected by a cardiac anaesthesiologist specializing in echocardiography. The general consensus was that both volumes maintain acceptable clinical quality for the mitral valve leaflets and that the chordae tendineae were very clearly visible in the volumes for both patients. The compounding process enabled the contrast between background noise and tissue to be more evident. Overall, compounded volumes exhibit an improvement in image quality and include a wider field of view with little signal dropout, shown in Figure 8. The overall conclusion was that these volumes represented an improvement over existing techniques, both in image quality and range of structures visible.

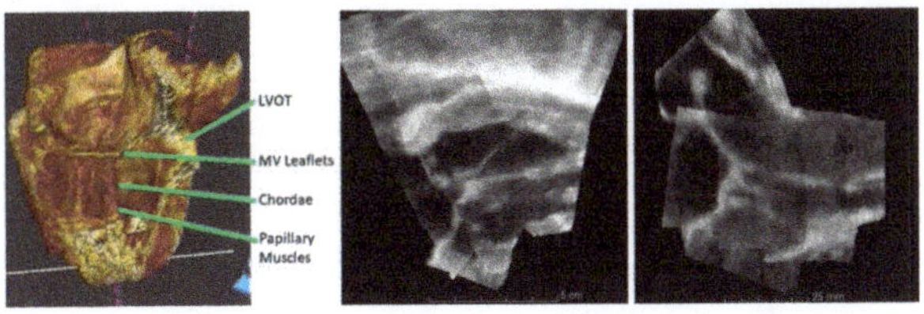

Figure 8. Visualizations of compounded TEE data from five different TEE volumes. (**Left**), a volume rendered view. (**Middle**), a commissure-commissure slice. (**Right**), an AP slice.

4. Discussion

Spatial compounding has been demonstrated for many applications to improve field of view and image quality. Incorporating image information into the weighting function shows clear improvement over prior blending approaches. The combined distance and oriented symmetry weighting improves image quality and helps eliminate blending artefacts where the separate image acquisitions did not entirely agree. Compared to our previous results for this application, the registration strategy incorporating simultaneous group registration as the initial step improved robustness and eliminated the effect of initialization order, helping to prevent registration failure where the volumes do not reach alignment.

For the application of mitral valve procedure planning, we show that spatially compounded 3D echocardiography volumes are able to capture the complex structures in the left ventricle. Utilizing spatial compounding reduces image noise and provides a single volume containing the mitral valve, chordae tendineae, and papillary muscles, enabling clinicians to work from a single volume, instead of reconciling multiple separate volumes together. We demonstrate on porcine models that our spatial compounding method using a 3D TEE probe can reproduce the structures captured by a CT scan with high geometrical accuracy. Although the chordae appear thicker in the compounded volume, the separate individual chordae can still be identified from leaflet to papillary muscle. We found that the length of the chordae can be accurately measured from the compounded volume, as the thickening artefact does not affect the measurement of the length of the chordae.

The workflow described here can be integrated into the clinical standard of care, requiring only 4–5 acquisitions with approximately 80% overlap. Standard practice currently includes diagnostic 3D TEE for patients undergoing mitral valve procedures and transgastric images are already acquired as part of this process in the form of 2D long-axis views in an attempt to capture the chordae in their entirety. This compounding technique enables the transgastric images to be acquired as a series of 3D volumes instead of the traditional 2D views, while still maintaining visibility of the entire chordae structures. Our workflow makes it possible for clinicians to map almost the entire chordal structure in 3D from the leaflet to the anchoring point in the left ventricle, greatly improving the surgeon's ability to optimize lengths for the introduced neochordae. Another instance where detailed, compounded 3D echocardiography has potential is the early diagnosis of endocarditis, where individual 3D image volume analysis can often remain ambiguous [16]. The volumes produced by applying 3D spatial compounding to TEE imaging capture the entire valve complex. Currently, many procedures require additional imaging in the form of cardiac CT/MRI to accurately perform diagnoses or plan for procedures [17]. Spatially compounded multi-view echo has the potential to add to the clinical imaging workflow by providing similar levels of information to cardiac CT at high frame rates, low cost and no radiation exposure to patients. Further validation of the clinical utility of this method will aim to demonstrate the effectiveness of the compounded echo, in particular related to cases in which cardiac CT is currently necessary.

Future work on this compounding method may include further improvements to registration speed and accuracy, as well as further exploration of blending approaches. Currently, the compounding process is performed offline due to computational requirements, with the process taking roughly 1 h for a five volume data set of two beat acquisitions. However, further optimizations may enable a real-time compounding approach where the final volume is created as the volumes are acquired, which may also allow for guidance to be provided to the operator to ensure the volumes are collected with sufficient overlap. Incorporating image-based real-time tracking algorithms and incorporating GPU acceleration may allow for active guidance of the volume acquisition relative to the initial position. Currently, this is a major drawback of the approach using offline processing, as in cases where the acquisitions have insufficient clarity or overlap, compounding will fail and the imaging session will have already concluded. With improvements to the compounding process, extensions to the tricuspid valve could also be possible, extending the range of clinical applications for this work. Additional validation of the compounding approach for mitral chordae on a larger cohort of patients will be performed to evaluate how accurately neochord sizes can be predicted from these extended volumes. Further work needs to be done to evaluate the effects of multi-view compounding on image quality using patient image data. Quantitative evaluation of image quality on a larger image set will be carried out by looking at the improvements in the contrast to noise ratio and image sharpness, as described in prior volume compounding work [5,6].

5. Conclusions

We describe a workflow for capturing a series of volumes using a TEE probe during standard diagnostic imaging that can then be registered and compounded together. Furthermore, we demonstrate improvements to the compounding process in registration robustness and final image quality. These compounded volumes capture the sub-valvular structures of interest for cardiac procedure planning. Capturing the necessary additional volumes can be performed while only adding an additional ten minutes to the time for the current standard-of-care diagnostic imaging protocol. We validate the geometrical accuracy of the compounding approach on two excised porcine valves, finding the measurement error between compounded ultrasound and ground-truth CT to be 0.7 ± 0.6 mm and 0.6 ± 0.6 mm, respectively. This method is able to provide clinicians with a single volume that captures the mitral valve and the sub-valvular structures, which will enhance the procedure planning process.

Author Contributions: Conceptualization, P.C., J.M. and D.B.; methodology, P.C.; software, P.C.; validation, P.C., J.M. and D.B.; formal analysis, P.C.; investigation, P.C. and D.B.; resources, T.M.P.; data curation, P.C.; writing—original draft preparation, P.C.; writing—review and editing, J.M., E.C.S.C. and T.M.P.; visualization, P.C.; supervision, E.C.S.C. and T.M.P.; project administration, T.M.P.; funding acquisition, T.M.P. All authors have read and agreed to the published version of the manuscript.

Funding: This research was funded by: Canadian Foundation for Innovation CFI-20994, 36199; Natural Science and Engineering Research Council of Canada NSERC RPGIN 2014-04504; Canadian Institutes for Health Research CIHR FDN 143232; and the Academic Medical Organization of Southwestern Ontario.

Institutional Review Board Statement: The study was conducted according to the guidelines of the Declaration of Helsinki and approved by the Institutional Review Board of Western University (111462, 1 October 2019).

Informed Consent Statement: Informed consent was obtained from all subjects involved in the study.

Data Availability Statement: The data presented in this study are available on request from the corresponding author. The data are not publicly available due to data confidentiality.

Conflicts of Interest: The authors declare no conflict of interest. The funders had no role in the design of the study; in the collection, analyses, or interpretation of data; in the writing of the manuscript, or in the decision to publish the results.

References

1. Shah, P.M. Current concepts in mitral valve prolapse—Diagnosis and management. *J. Cardiol.* **2010**, *56*, 125–133. [CrossRef]
2. Linden, A.; Seeburger, J.; Noack, T.; Falk, V.; Walther, T. Imaging in Cardiac Surgery: Visualizing the Heart. *Thorac. Cardiovasc. Surg.* **2017**, *65*, S213–S216. [CrossRef] [PubMed]
3. Perperidis, A. Postprocessing Approaches for the Improvement of Cardiac Ultrasound B-Mode Images: A Review. *IEEE Trans. Ultrason. Ferroelectr. Freq. Control* **2016**, *63*, 470–485. [CrossRef] [PubMed]
4. Soler, P.; Delso, G.; Villain, N.; Angelini, E.; Bloch, I. Superresolution spatial compounding techniques with application to 3D breast ultrasound imaging. In *Proceedings of the SPIE Proceedings*; Emelianov, S., Walker, W.F., Eds.; SPIE: Philadelphia, PA, USA, 2006. [CrossRef]
5. Wright, R.; Toussaint, N.; Gomez, A.; Zimmer, V.; Khanal, B.; Matthew, J.; Skelton, E.; Kainz, B.; Rueckert, D.; Hajnal, J.V.; et al. Complete Fetal Head Compounding from Multi-view 3D Ultrasound. In *Lecture Notes in Computer Science*; Springer International Publishing: Cham, Switzerland, 2019; pp. 384–392. [CrossRef]
6. Szmigielski, C.; Rajpoot, K.; Grau, V.; Myerson, S.G.; Holloway, C.; Noble, J.A.; Kerber, R.; Becher, H. Real-Time 3D Fusion Echocardiography. *JACC Cardiovasc. Imaging* **2010**, *3*, 682–690. [CrossRef] [PubMed]
7. Wachinger, C.; Wein, W.; Navab, N. Registration Strategies and Similarity Measures for Three-dimensional Ultrasound Mosaicing. *Acad. Radiol.* **2008**, *15*, 1404–1415. [CrossRef] [PubMed]

8. Taramasso, M.; Candreva, A.; Pozzoli, A.; Guidotti, A.; Gaemperli, O.; Nietlispach, F.; Barthelmes, J.; Emmert, M.Y.; Weber, A.; Benussi, S.; et al. Current challenges in interventional mitral valve treatment. *J. Thorac. Dis.* **2015**, *7*, 1536. [PubMed]
9. Grinberg, D.; Le, M.Q.; Kwon, Y.J.; Fernandez, M.A.; Audigier, D.; Ganet, F.; Capsal, J.F.; Obadia, J.F.; Cottinet, P.J. Mitral valve repair based on intraoperative objective measurement. *Sci. Rep.* **2019**, *9*, 1–7. [CrossRef] [PubMed]
10. Gillinov, A.M.; Banbury, M.K. Pre-Measured Artificial Chordae for Mitral Valve Repair. *Ann. Thorac. Surg.* **2007**, *84*, 2127–2129. [CrossRef] [PubMed]
11. Mahmood, F.; Matyal, R. A Quantitative Approach to the Intraoperative Echocardiographic Assessment of the Mitral Valve for Repair. *Anesth. Analg.* **2015**, *121*, 34–58. [CrossRef] [PubMed]
12. Carnahan, P.; Moore, J.T.; Bainbridge, D.; Chen, E.C.S.; Peters, T.M. Multi-view 3D echocardiography volume compounding for mitral valve procedure planning. In Proceedings of the Medical Imaging 2020: Image-Guided Procedures, Robotic Interventions, and Modeling, Houston, TX, USA, 15–20 February 2020; Fei, B., Linte, C.A., Eds.; SPIE: Philadelphia, PA, USA, 2020. [CrossRef]
13. Wachinger, C.; Wein, W.; Navab, N. Three-Dimensional Ultrasound Mosaicing. In Proceedings of the Medical Image Computing and Computer-Assisted Intervention—MICCAI 2007, Brisbane, Australia, 29 October– 2 November 2007; Ayache, N., Ourselin, S., Maeder, A., Eds.; Springer Berlin Heidelberg: Berlin/Heidelberg, Germany, 2007; pp. 327–335.
14. Felsberg, M.; Sommer, G. The monogenic signal. *IEEE Trans. Signal Process.* **2001**, *49*, 3136–3144. [CrossRef]
15. Grau, V.; Noble, J.A. Adaptive Multiscale Ultrasound Compounding Using Phase Information. In *Lecture Notes in Computer Science*; Springer: Berlin/Heidelberg, Germany, 2005; pp. 589–596. [CrossRef]
16. Hohmann, C.; Michels, G.; Schmidt, M.; Pfister, R.; Mader, N.; Ohler, M.; Blanke, L.; Jazmati, N.; Lehmann, C.; Rybniker, J.; et al. Diagnostic challenges in infective endocarditis: Is PET/CT the solution? *Infection* **2019**, *47*, 579–587. [CrossRef] [PubMed]
17. Roberts, W.T.; Bax, J.J.; Davies, L.C. Cardiac CT and CT coronary angiography: Technology and application. *Heart* **2008**, *94*, 781–792. [CrossRef] [PubMed]

MDPI
St. Alban-Anlage 66
4052 Basel
Switzerland
Tel. +41 61 683 77 34
Fax +41 61 302 89 18
www.mdpi.com

Applied Sciences Editorial Office
E-mail: applsci@mdpi.com
www.mdpi.com/journal/applsci

www.ingramcontent.com/pod-product-compliance
Lightning Source LLC
Chambersburg PA
CBHW061806210926
51850CB00006B/626
* 9 7 8 3 0 3 6 5 7 1 0 0 3 *